Diseases in Legume Crops

Uday Chand Jha • Harsh Nayyar •
Kamal Dev Sharma • Eric J Bishop von
Wettberg • Prashant Singh •
Kadambot H. M. Siddique
Editors

Diseases in Legume Crops

Next Generation Breeding Approaches for Resistant Legume Crops

Springer

Editors
Uday Chand Jha
Crop Improvement Division
Indian Institute of Pulses Research
Kanpur, Uttar Pradesh, India

Harsh Nayyar
Department of Botany
Panjab University
Chandigarh, India

Kamal Dev Sharma
CSKHPKV Palampur
Himachal Pradesh, India

Eric J Bishop von Wettberg
University of Vermont
Burlington, VT, USA

Prashant Singh
Banaras Hindu University
Varanasi, India

Kadambot H. M. Siddique
Institute of Agriculture
University of Western Australia
Crawley, WA, Australia

ISBN 978-981-99-3357-0 ISBN 978-981-99-3358-7 (eBook)
https://doi.org/10.1007/978-981-99-3358-7

This Springer imprint is published by the registered company Springer Nature Singapore Pte Ltd.
The registered company address is: 152 Beach Road, #21-01/04 Gateway East, Singapore 189721, Singapore

Preface

Legume crops, including grain legumes, are important for global food security due to their versatility and rich nutritional content. However, grain legumes face significant challenges from major and minor diseases that limit grain yield and production. Various approaches have been developed to combat these issues, including plant breeding and emerging next-generation breeding tools that are economically viable and eco-friendly. In addition, advances in genomics and functional genomics have helped uncover genomic regions/QTLs and candidate genes responsible for disease resistance using biparental and genome-wide association studies. Moreover, recent advances in high-throughput phenotyping technologies could facilitate disease screening in various legumes.

This book explores how plant breeding, genomics-assisted breeding, functional genomics, phenomics, and other cutting-edge techniques, like speed breeding, genomic selection, haplotype-based breeding, and genome editing, could be used to develop future disease-resistant grain legume cultivars, helping to achieve global food security and the United Nations' "zero hunger" sustainable development goal.

Various researchers working to develop grain legumes with improved resistance against various diseases have contributed 14 chapters to this book.

This book is intended for graduate and postgraduate students, researchers, and policymakers interested in understanding the disease impact on yield losses in legume crops and how plant breeding, "omics approaches," and emerging novel breeding tools can be used to design disease-resistant legumes. It will also interest those in related fields such as plant breeding, plant pathology, and plant molecular biology. We express our gratitude to the Indian Council of Agricultural Research (ICAR), New Delhi, for their support and encouragement of our scientific pursuit in the form of this book, **Diseases in Legume Crops—Next Generation Breeding**

Approaches for Resistant Legumes Crops. We also thank Dr. Himanshu Pathak, Director General, ICAR, Secretary, DARE, Ministry of Agriculture and Farmers' Welfare, Government of India, and Dr. T.R. Sharma, Deputy Director General (Crop Science), ICAR, for their immense support.

Kanpur, Uttar Pradesh, India — Uday Chand Jha
Chandigarh, India — Harsh Nayyar
Himachal Pradesh, India — Kamal Dev Sharma
Burlington, VT, USA — Eric von Wettberg
Varanasi, India — Prashant Singh
Crawley, WA, Australia — Kadambot H. M. Siddique

Contents

Editors and Contributors

About the Editors

Uday Chand Jha graduated from Visva Bharati University, Santiniketan, India. He has been working in the areas of grain legume breeding, genetics, and genomics for both biotic and abiotic stress tolerance since 2010 at the Indian Institute of Pulses Research, Kanpur, ICAR, India. He is associated with contributing eight chickpea varieties and has registered three chickpea germplasms (with specific trait) at NBPGR, New Delhi. He was awarded with the ISPRD fellow in 2016 by the ISPRD and the Young Scientist Award by the IJBSM society in 2017 at Jaipur, India.

Harsh Nayyar has M.Sc. and Ph.D. in Botany, with a specialization in Stress Physiology from Punjab Agricultural University, India. He has worked in the Faculty of the Indian Council of Agricultural Research, Himachal Pradesh Agricultural University, India, and is currently a Professor in Panjab University, India. Dr. Nayyar has been working on responses of various food legumes (chickpea, lentil, beans) to drought, cold, heat, salt, and metals, for the past 15 years, in collaboration with many national (Punjab Agricultural University, India; Himachal Pradesh Agricultural University, India) and international (World Vegetable Center, ICRISAT, ICARDA, University of Western Australia) organizations.

Kamal Dev Sharma holds a Ph.D. in Plant Pathology. He is currently Professor and Head of the Department of Agricultural Biotechnology, CSKHPKV, Palampur. His areas of expertise include plant genomics and abiotic and biotic stresses of plants.

Eric J. B. von Wettberg is an Associate Professor at the University of Vermont. He holds a Ph.D. in Ecology from Brown University, received in 2007. He has been a visiting scholar at Addis Ababa University, Ethiopia; Peter the Great St. Petersburg

Polytechnic University, St. Petersburg, Russia; and École Nationale Superieure Agronomique de Toulouse, France. He has vast experience in conducting research in the areas of population genomics and domestication of legumes, symbiosis, conservation genetics, landscape genetics, and symbiont and microbial mediation of plant traits.

Prashant Singh is currently working as an Assistant Professor in the Department of Botany at Banaras Hindu University (BHU), Varanasi. Dr. Singh has a strong background in plant physiology, molecular biology, seed dormancy, and plant molecular pathology.

Kadambot H. M. Siddique is the Hackett Professor of Agriculture Chair and Director of the UWA Institute of Agriculture, the University of Western Australia. He is also a Distinguished Professor, Shenyang Agricultural University, and Adjunct Professor, Northwest Agriculture & Forestry University and Lanzhou University. He is also a faculty member of Guangzhou University and Huazhong Agricultural University, China, and Fellow of the Australian Academy of Technological Sciences and Engineering, African Academy of Sciences, Australian Agricultural Institute, and Indian National Academy of Agricultural Sciences. Professor Siddique was designated by the UN FAO as Special Ambassador for the International Year of Pulses 2016. He has over 30 years' experience in agricultural research, teaching, and management in both Australia and overseas.

Contributors

Aasiya Nabi Sher-e-Kashmir University of Agricultural Sciences and Technology of Kashmir, Jammu, India

Seltene Abady School of Plant Sciences, Haramaya University, Dire Dawa, Ethiopia

Kousik Atta Department of Plant Pathology, Bidhan Chandra Krishi Viswavidyalaya, Mohanpur, Nadia, West Bengal, India
Mineral Nutrition Lab, Division of Plant Physiology, ICAR-Indian Agricultural Research Institute, Pusa, New Delhi, India

Rachana Bagudam International Crops Research Institute for the Semi-Arid Tropics (ICRISAT), Hyderabad, India

Melike Bakır Department of Agricultural Biotechnology, Faculty of Agriculture, Erciyes University, Kayseri, Turkey

Kiranmayee Bangaru International Crops Research Institute for the Semi-Arid Tropics (ICRISAT), Hyderabad, India

Surendra Barpete International Center for Agricultural Research in the Dry Areas (ICARDA)-Food Legumes Research Platform, Amlaha, Madhya Pradesh, India

Adfar Bashir Sher-e-Kashmir University of Agricultural Sciences and Technology of Kashmir, Jammu, India

R. Beena Department of Plant Physiology, College of Agriculture, KAU, Thiruvananthapuram, Kerala, India

Chellapilla Bharadwaj Division of Genetics, ICAR-Indian Agricultural Research Institute, New Delhi, India

Sabhyata Bhatia National Institute of Plant Genome Research, New Delhi, India

Shayla Bindra Department of Plant Breeding and Genetics, Punjab Agricultural University, Ludhiana, Punjab, India

Emmanuel Brefo Department of Plant and Soil Science, University of Vermont, Burlington, VT, USA

Zahoor A. Dar Sher-e-Kashmir University of Agricultural Sciences and Technology of Kashmir, Jammu, India

Arpita Das Bidhan Chandra Krishi Viswavidyalaya, Mohanpur, Nadia, West Bengal, India

Dnyaneshwar Bandu Deshmukh International Crops Research Institute for the Semi-Arid Tropics (ICRISAT), Hyderabad, India

Sanjeev K. Deshpande Department of Genetics and Plant Breeding, University of Agricultural Sciences, Dharwad, Karnataka, India

Sandeep Kaur Dhaliwal Department of Plant Breeding and Genetics, Punjab Agricultural University, Ludhiana, India

Girish Prasad Dixit Indian Institute of Pulses Research, Kanpur, Uttar Pradesh, India

Tabia Fayaz Sher-e-Kashmir University of Agricultural Sciences and Technology of Kashmir, Jammu, India

Gazala Gulzar Sher-e-Kashmir University of Agricultural Sciences and Technology of Kashmir, Jammu, India

Jasmine Hart Department of Plant and Soil Science, University of Vermont, Burlington, VT, USA
The Gund Institute for Environment, University of Vermont, Burlington, VT, USA

Rintu Jha Indian Agricultural Research Institute, New Delhi, India
Department of Botany, Institute of Science, Banaras Hindu University, Varanasi, Uttar Pradesh, India
Division of Genetics, ICAR-Indian Agricultural Research Institute, New Delhi, India

Uday Chand Jha Indian Institute of Pulses Research, Kanpur, India
Crop Improvement Division, ICAR-Indian Institute of Pulses Research, Kanpur, Uttar Pradesh, India

Mahsa Pourali Kahriz Department of Field Crops, Faculty of Agriculture, Ankara University, Ankara, Turkey

Parisa Pourali Kahriz Department of Field Crops, Faculty of Agriculture, Ankara University, Ankara, Turkey

Satinder Kaur School of Agricultural Biotechnology, Punjab Agricultural University, Ludhiana, India

N. M. Kavyashree Department of Genetics and Plant Breeding, University of Agricultural Sciences, Dharwad, Karnataka, India

Jasmeen Khandakar Department of Plant Pathology, Bidhan Chandra Krishi Viswavidyalaya, Mohanpur, Nadia, West Bengal, India

Bailey Kretzler Department of Plant and Soil Science, University of Vermont, Burlington, VT, USA
The Gund Institute for Environment, University of Vermont, Burlington, VT, USA

Anshu Kumar Department of Plant Pathology, Bidhan Chandra Krishi Viswavidyalaya, Mohanpur, Nadia, West Bengal, India

Shiv Kumar ICARDA-Food Legumes Research Platform, Amlaha, Madhya Pradesh, India

Tapan Kumar International Center for Agricultural Research in the Dry Areas (ICARDA), Amlaha, Madhya Pradesh, India

Krishnendu Kundu Department of Plant Pathology, Bidhan Chandra Krishi Viswavidyalaya, Mohanpur, Nadia, West Bengal, India

Chayanika Lahkar Department of Plant Breeding and Genetics, Punjab Agricultural University, Ludhiana, India

Irtifa Lateef Sher-e-Kashmir University of Agricultural Sciences and Technology of Kashmir, Jammu, India

Ajaz A. Lone Dryland Agriculture Research Station, SKUAST, Kashmir, India
Sher-e-Kashmir University of Agricultural Sciences and Technology of Kashmir, Jammu, India

Fouad Maalouf International Center for Agricultural Research in the Dry Areas (ICARDA), Rabat, Morocco

Anurag Mathew International Crops Research Institute for the Semi-Arid Tropics (ICRISAT), Hyderabad, India

Rishabh Mirchandani National Institute of Plant Genome Research, New Delhi, India

Saptarshi Mondal University of Georgia, Griffin, GA, USA

Soumik Mukherjee Department of Plant Pathology, Bidhan Chandra Krishi Viswavidyalaya, Mohanpur, Nadia, West Bengal, India

Sabyasachi Mukhopadhyay Department of Plant Pathology, Bidhan Chandra Krishi Viswavidyalaya, Mohanpur, Nadia, West Bengal, India

Harsh Nayyar Department of Botany, Panjab University, Chandigarh, India

Naziya Nabi Sher-e-Kashmir University of Agricultural Sciences and Technology of Kashmir, Jammu, India

Qadrul Nisa Sher-e-Kashmir University of Agricultural Sciences and Technology of Kashmir, Jammu, India

Susmita Oraon Department of Plant Pathology, Bidhan Chandra Krishi Viswavidyalaya, Mohanpur, Nadia, West Bengal, India

Bilal A. Padder Sher-e-Kashmir University of Agricultural Sciences and Technology of Kashmir, Jammu, India

Asish Kumar Padhy National Institute of Plant Genome Research, New Delhi, India

Abhishek Pandey Department of Plant Breeding and Genetics, Punjab Agricultural University, Ludhiana, India

A. K. Parihar ICAR-Indian Institute of Pulses Research, Kanpur, India

Janila Pasupuleti International Crops Research Institute for the Semi-Arid Tropics (ICRISAT), Hyderabad, India

Pronob J. Paul International Rice Research Institute (IRRI), South-Asia Hub, Hyderabad, India

P. V. Vara Prasad Kansas State University, Kansas, KS, USA

Ashutosh Purohit International Crops Research Institute for the Semi-Arid Tropics (ICRISAT), Hyderabad, India

Xu Qunale College of Life Sciences, Northwest A&F University, Yangling, Shaanxi, China

K. R. Ramya New Delhi, India

Upasana Rani Department of Plant Breeding and Genetics, Punjab Agricultural University, Ludhiana, Punjab, India

Shubhashish Ranjan National Institute of Plant Genome Research, New Delhi, India

Raja Junaid Rashid Sher-e-Kashmir University of Agricultural Sciences and Technology of Kashmir, Jammu, India

Zainab Rashid Sher-e-Kashmir University of Agricultural Sciences and Technology of Kashmir, Jammu, India

Gianna Sassi Department of Plant and Soil Science, University of Vermont, Burlington, VT, USA

D. R. Saxena Rafi Ahmed Kidwai College of Agriculture, Sehore, Madhya Pradesh, India

Muthappa Senthil-Kumar National Institute of Plant Genome Research, New Delhi, India

Arpana Sharma Department of Plant Pathology, Punjab Agricultural University, Ludhiana, Punjab, India

Bandana Sharma Department of Botany, Institute of Science, Banaras Hindu University, Varanasi, Uttar Pradesh, India

Kamal Dev Sharma Department of Agricultural Biotechnology, CSK Himachal Pradesh Agricultural University, Palampur, Himachal Pradesh, India

Namisha Sharma Institute of Life Sciences (ILS), Odisha, India

Shubham Sharma Department of Plant Breeding and Genetics, Punjab Agricultural University, Ludhiana, Punjab, India

Sheetal Department of Plant Breeding and Genetics, Punjab Agricultural University, Ludhiana, India

Kadambot H. M. Siddique The UWA Institute of Agriculture, University of Western Australia, Perth, WA, Australia

Baljinder Singh National Institute of Plant Genome Research, New Delhi, India

Inderjit Singh Department of Plant Breeding and Genetics, Punjab Agricultural University, Ludhiana, Punjab, India

Kartar Singh ICAR-National Bureau of Plant Genetic Resources, Jodhpur, India

Prashant Singh Department of Botany, Institute of Science, Banaras Hindu University, Varanasi, Uttar Pradesh, India

Sarvjeet Singh Department of Plant Breeding and Genetics, Punjab Agricultural University, Ludhiana, Punjab, India

Mahender Thudi Department of Agricultural Biotechnology and Molecular Biology, Dr. Rajendra Prasad Central Agricultural University, Samastipur, India

Menka Tiwari Department of Botany, Institute of Science, Banaras Hindu University, Varanasi, Uttar Pradesh, India

Neha Tiwari International Center for Agricultural Research in the Dry Areas (ICARDA), Amlaha, Madhya Pradesh, India

Kuldeep Tripathi ICAR-National Bureau of Plant Genetic Resources, New Delhi, India
Indian Council of Agricultural Research-National Bureau of Plant Genetic Resources, New Delhi, India

Shailesh Tripathi Indian Agricultural Research Institute, New Delhi, India
Division of Genetics, ICAR-Indian Agricultural Research Institute, New Delhi, India

Usma Shafi Sher-e-Kashmir University of Agricultural Sciences and Technology of Kashmir, Jammu, India

Sumiah Wani Sher-e-Kashmir University of Agricultural Sciences and Technology of Kashmir, Jammu, India

Ankush Purushottam Wankhade International Crops Research Institute for the Semi-Arid Tropics (ICRISAT), Hyderabad, India

Eric J. B. von Wettberg Department of Plant and Soil Science, University of Vermont, Burlington, VT, USA
The Gund Institute for Environment, University of Vermont, Burlington, VT, USA

Esnart Nyirenda Yohane Department of Agricultural Research Services, Lilongwe, Malawi

Kuldeep Tripathi ICAR-National Bureau of Plant Genetic Resources, New Delhi, India

1 Chickpea Diseases: Breeding and "Omics" Approaches for Designing Next-Generation Disease-Resistant Chickpea Cultivar

Uday Chand Jha, Harsh Nayyar, Kamal Dev Sharma, Rintu Jha, Mahender Thudi, Melike Bakır, Ajaz A. Lone, Shailesh Tripathi, R. Beena, Pronob J. Paul, Girish Prasad Dixit, P. V. Vara Prasad, and Kadambot H. M. Siddique

U. C. Jha (✉) · G. P. Dixit
Indian Institute of Pulses Research, Kanpur, Uttar Pradesh, India
e-mail: u9811981@gmail.com

H. Nayyar
Department of Botany, Panjab University, Chandigarh, India

K. D. Sharma
Department of Agricultural Biotechnology, CSK Himachal Pradesh Agricultural University, Palampur, India

R. Jha · S. Tripathi
Indian Agricultural Research Institute, New Delhi, India

M. Thudi
Department of Agricultural Biotechnology and Molecular Biology, Dr. Rajendra Prasad Central Agricultural University, Samastipur, India

M. Bakır
Department of Agricultural Biotechnology, Faculty of Agriculture, Erciyes University, Kayseri, Turkey

A. A. Lone
Dryland Agriculture Research Station, SKUAST, Kashmir, India

R. Beena
Department of Plant Physiology, College of Agriculture, KAU, Thiruvananthapuram, Kerala, India

P. J. Paul
International Rice Research Institute (IRRI), South-Asia Hub, Hyderabad, India

P. V. V. Prasad
Kansas State University, Kansas, KS, USA

K. H. M. Siddique
The UWA Institute of Agriculture, The University of Western Australia, Perth, WA, Australia
e-mail: kadambot.siddique@uwa.edu.au

U. C. Jha et al. (eds.), *Diseases in Legume Crops*,
https://doi.org/10.1007/978-981-99-3358-7_1

Abstract

Chickpea is an important nutritionally rich grain legume with a significant role in global food security. However, chickpea suffers significant yield losses due to various major and minor diseases, threatening global food security. Plant breeding is one of the cheapest and most environmentally friendly approaches for minimizing yield losses caused by various diseases. The chickpea gene pool contains substantial genetic variability for disease resistance, which could be harnessed to develop disease-resistant chickpea cultivars. Classical genetics-based approaches have increased our understanding of disease resistance in chickpea. In the past decade, the development of genomics resources in chickpea has helped dissect the genetic basis of fusarium wilt, ascochyta blight, dry root rot, botrytis gray mold, and other minor disease resistance and identify the underlying causal quantitative trait loci controlling disease resistance. Similarly, the availability of high-throughput SNP molecular markers has enabled genome-wide association mapping to uncover the genomic regions/haplotypes governing disease resistance across the whole genome. Furthermore, the complete chickpea genome sequence, whole genome resequencing, and pangenome sequences have provided novel insights into structural variations such as presence/absence variations and copy number variations controlling disease resistance. Likewise, advances in RNA-seq-based transcriptomics have facilitated the identification of several candidate genes related to disease resistance in chickpea and their putative functions mediating disease resistance. Moreover, emerging approaches such as genomic selection, speed breeding, and genome editing technologies could be harnessed to develop next-generation disease-resistant chickpea cultivars for sustaining chickpea yield and achieving global food security.

Keywords

Chickpea · Disease resistance · Pathogen · Genomics · Molecular markers

1.1 Introduction

Chickpea is a major grain legume with high nutritive value, including essential amino acids, fatty acids, vitamins, and essential micronutrients (Jukanti et al. 2012; Jha et al. 2022). However, global chickpea yield is severely challenged by various major diseases (ascochyta blight, fusarium wilt, botrytis gray mold, and dry root rots) and minor diseases (sclerotium rot, chlorotic stunt virus, and rusts), causing great concern for global food security. Furthermore, global climate change has predisposed minor diseases as major diseases. Several plant protection approaches, including chemical-based fungicides, are used to reduce the losses caused by these diseases. However, the high cost and lethal effects of these applied fungicides predispose the environment and ecosystems to pollution. Plant breeding approaches are cheaper and more environmentally friendly than chemical-based approaches for controlling disease infection and developing disease-resistant chickpea cultivars. Exploring the genetic variability for disease resistance in various gene pools can harness disease-resistant gene(s)/quantitative trait loci (QTLs), with pre-breeding approaches important for transferring resistant gene(s)/QTLs controlling disease resistance (Mohanty et al. 2022). Chickpea has evolved several complex molecular mechanisms for limiting pathogen infection and protecting itself against various diseases. The two most important mechanisms adopted by host plants to mediate disease resistance are PAMP-triggered immunity (PTI) and effector-triggered immunity (ETI). Classical genetics-based approaches have determined the genetic control of major gene(s) mediating disease resistance. However, in the past decade, advances in chickpea genomics, including the development of various molecular markers, have helped identify the underlying causal QTLs/genomic regions controlling resistance against various diseases. Further, the availability of high-throughput molecular markers (e.g., single nucleotide polymorphisms) has facilitated the unfolding of genomic regions/haplotypes controlling disease resistance across the whole genome in genome-wide association studies (GWAS). Likewise, chickpea reference genome maps, whole genome resequencing of global germplasm, and chickpea pangenome sequences have offered insights into the novel structural variations, including copy number variations, mediating disease resistance (Varshney et al. 2019, 2021). Thus, genomics-assisted breeding can be used to pyramid multiple pathogen-specific resistance gene(s) to develop disease-resistant chickpea cultivars. Unprecedented advances in RNA-seq-based transcriptomics have provided insights into the various underlying candidate gene(s) governing disease resistance and their precise function mediating disease resistance (Garg et al. 2019). Likewise, emerging approaches such as phenomics, speed breeding, genomic selection, and genome editing tools could expedite the development of disease-resistant chickpea cultivars for global food security.

1.2 Causal Organism of Ascochyta Blight in Chickpea, Symptoms, and Losses

Ascochyta blight (AB), caused by *Ascochyta rabiei* (Pass.) Labr. [teleomorph: *Didymella rabiei* (Kovacheski)], is the most destructive foliar fungal disease in chickpea. AB is a necrotrophic fungus belonging to class *Dothideomycetes*, order *Pleosporales*, and family *Didymellaceae* (Akamatsu et al. 2012; Jha et al. 2022). It infects aerial plant parts and under favorable environments can cause up to 100% yield losses in chickpea (Nene and Reddy 1987; Shahid et al. 2008), with AU$4.8 million of losses reported annually in Australia (Murray and Brennan 2012).

Infected chickpea seed serves as the main source of primary inoculum for disease infection (Tivoli and Banniza 2007). AB generally appears on young leaves as "water-soaked pale spots" that become enlarged and causes leaf and bud blighting symptoms. The AB pathogen has teliomorphic (sexual) and anamorphic (asexual) stages. Completion of the sexual cycle occurs on infected chickpea debris during winter under high moisture and low-temperature conditions (Trapero-Casas et al.

1996). The sexual fruiting bodies, "pseudothecia," initially remain in infected host tissue (Trapero-Casas and Kaiser 1992, 2007). Subsequently, under favorable environments, the pseudothecia discharge ascospores into the air. During the asexual cycle, infected host plant tissue, "pycnidia," produces asexual spore conidia that spread the disease (see Singh et al. 2022).

1.3 Fusarium Wilt: Symptoms and Losses

Fusarium wilt (FW), caused by *Fusarium oxysporum* f. sp. *ciceris* (*Foc*), is a major chickpea root disease causing significant yield losses globally (Jha et al. 2020), with up to 100% yield losses in favorable environments (Haware et al. 1978, 1986). FW causes two types of wilting symptoms: (1) "early wilt," dull green leaf color appears within 25 days of planting and renders up to 90% yield loss, and (2) "late wilting," leaf yellowing appears at the podding stage and accounts for up to 70% yield loss (Jiménez-Díaz et al. 2015). *Foc* races are classified into two groups based on various disease symptoms appearing on chickpea host plants during FW infection: six races (1A, 2, 3, 4, 5, and 6) are prevalent (Haware and Nene 1982; Jiménez-Díaz et al. 1993; Kelly et al. 1994; Sharma et al. 2004), and two others (0 and 1B/C) display yellowing symptoms (Jiménez-Díaz et al. 1993; Kelly et al. 1994).

1.4 Causal Organism of Chickpea Dry Root Rot, Symptoms, and Losses

Dry root rot (DRR) disease caused by *Rhizoctonia bataticola* [(Taub.) Butler] is becoming a major disease of chickpea, causing substantial chickpea yield losses in rainfed and arid regions (Karadi et al. 2021; Sharath Chandran et al. 2021). This disease was first reported in India (Mitra 1931) and subsequently in Iran, the USA, and various countries in Asia and Africa (Nene et al. 1996; Ghosh et al. 2013). DRR is a soilborne necrotic fungi prevalent mostly in dry soil and warm regions (Sharma and Pande 2013; Sharath Chandran et al. 2021). Microsclerotia of this fungus acts as the primary inoculum for disease infection (Rai et al. 2022). DRR infection and severity commence with increasing air temperatures and low/dry soil conditions during flowering and podding (Gurha et al. 2003; Sharma and Pande 2013; Sinha et al. 2021; Rai et al. 2022), causing yield losses of up to 30–40% (Sharma et al. 2016), with average annual yield losses of 20% in chickpea (Vishwadhar 2001; Gupta et al. 2012). The most common symptoms of DRR infection are minute clerotial bodies on tap roots, rotten root brittles, necrotic spots on roots, and straw-colored foliage (Maryam and Dubey 2009; Rai et al. 2022; see Table 1.1).

Table 1.1 List of various diseases in chickpea, their symptoms, and economic impacts

Disease	Causal organism	Race/pathotypes	Symptoms and economic impacts
Ascochyta blight	*Ascochyta rabiei* (anamorph); *Didymella rabiei* (teleomorph)	Three pathotypes (Udupa et al. 1998; Jamil et al. 2000); five pathotypes (Nene and Reddy 1987)	Concentric necrotic lesions on leaf, pod, and other plant parts (Pande et al. 2005); complete yield loss under congenial environment (Shahid et al. 2008)
Fusarium wilt	*Fusarium oxysporum* f.sp. *ciceri* (*Foc*)	Eight races (0, 1B/C, 1A, 2, 3, 4, 5, and 6) (Haware and Nene 1982; Jiménez-Díaz et al. 1993; Kelly et al. 1994; Sharma and Muehlbauer 2007)	Infects vascular bundle, inhibiting plant water relations, causing petiole droop, with yield losses up to 100% under congenial environment; Races 0 and 1B/C cause yellowing syndrome (Haware and Nene 1982; Jiménez-Díaz et al. 1993)
Dry root rot	*Rhizoctonia bataticola*	–	Destroys lateral roots, causes root rot, forms clerotial bodies on outer surface of tap root (Rashid et al. 2014; Khaliq et al. 2020); leaves and stems become straw-colored and lower leaves turn brown, tap root become black and devoid of lateral roots (Karadi et al. 2021); yield losses up to 30–40% under rainfed conditions (Sharma et al. 2016)
Grey mold	*Botrytis cinerea*	–	Drooping of infected terminal branches (Haware and McDonald 1992); poor or no pod set; infected seed becomes shriveled and covered with grey fungal mat (Knights and Siddique 2002); complete yield loss under favorable conditions
Rust	*Uromyces ciceris-arietini*	–	–
Collar rot	*Sclerotium rolfsii*	–	Under favorable conditions, 55–95% seedling mortality (Sharma and Ghosh 2017; Tarafdar et al. 2018); stem whitening, wilting, and stem breakage (Chen et al. 2006)

(continued)

Table 1.1 (continued)

Disease	Causal organism	Race/pathotypes	Symptoms and economic impacts
Chlorotic stunt virus	Chickpea chlorotic stunt virus	–	Small leaves, chlorosis, bushy stunted growth, and yellowing, and reddening in *desi*-type chickpea (Abraham et al. 2006; Kanakala and Kuria 2018)

1.5 Causal Organism of Chickpea Botrytis Gray Mold, Symptoms, and Losses

Botrytis gray mold (BGM), caused by *Botrytis cinerea* Pers. ex. Fr., is a foliar disease of chickpea, resulting in substantial yield losses (Pande et al. 2006b; Anuradha et al. 2011). It is prevalent in major chickpea-growing areas in India, Australia, Canada, and Spain (Haware 1998; Pande et al. 2002, 2006b; Davidson et al. 2004). Severe BGM infection can result in 100% yield loss in chickpea in favorable warm and humid environments (Pande et al. 2002, 2006b). Notable incidences of BGM epidemics have occurred in Australia (MacLeod and Sweetingham 2000), Nepal (Bakr et al. 2002), Argentina (Carranza 1965), and northern India (Grewal and Laha 1983). BGM is a seed-borne disease that can infect any plant growth stage (Cother 1977; Burgess et al. 1997). Common symptoms include drooping of terminal branches, infected flowers leading to poor pod and seed set, and gray fungal matter in infected seed (Haware and McDonald 1992; Grewal et al. 1992; Knights and Siddique 2002).

1.6 Causal Organism of Chickpea Collar Rot, Symptoms, and Losses

Collar rot, caused by *Sclerotium rolfsii*, is an emerging soilborne disease causing 55–95% mortality in chickpea seedlings under warm and sufficient soil moisture conditions (Chen et al. 2006; Sharma and Ghosh 2017). Being saprophytic in nature, the causal organism survives in plant debris as mycelium or by forming a sclerotial structure (Sharma and Ghosh 2017; Tarafdar et al. 2018). Consequently, infected host plants have abundant brown-colored sclerotia and symptoms of stem whitening, wilting, and breakage (Chen et al. 2006) (see Table 1.1).

1.7 Causal Organism of Chickpea Rust, Symptoms, and Losses

Chickpea rust, caused by *Uromyces ciceris-arietini* Jacz,. is a minor foliar disease, first reported in the USA (Venette and Stack 1987). Rust-infected leaves contain abundant uredospores and teliospores (Venette and Stack 1987). The pathogen

U. ciceris-arietini over summers on *Medicago polyceratia* in India (Stuteville et al. 2010).

1.8 Causal Organism of Chickpea Chlorotic Stunt Virus, Symptoms, and Negative Effects

Chickpea chlorotic stunt virus disease, caused by the chickpea chlorotic stunt virus belonging to the genus *Polerovirus* and family *Solemoviridae*, results in substantial yield losses of up to 95% in chickpea (Abraham et al. 2006; Kanakala and Kuria 2018; Abraham and Vetten 2022). It was first reported in Ethiopia in 2006 (Abraham et al. 2006) and is mostly prevalent in Africa, Asia, Australia, and the Middle East (Kanakala and Kuria 2018). Epidemics of chickpea chlorotic stunt virus have been reported in Ethiopia, Syria, and Tunisia (Abraham and Vetten 2022). Infected plants have small chlorotic leaves, bushy stunted growth, and yellowing (*kabuli* type) and reddening (*desi* type) symptoms (see Table 1.1) (Abraham et al. 2006; Kanakala and Kuria 2018). *Aphis craccivora* and *Acyrthosiphon pisum* serve as the vector for disease transmission (Abraham and Vetten 2022).

1.9 Brief Note on Disease Resistance Mechanism Orchestrated by the Host Plant

In general, the host plant orchestrates two lines of defense: (1) PTI basal defense mechanism (Zipfel and Robatzek 2010; Zhang and Zhou 2010) and (2) ETI (Cui et al. 2015; Yuan et al. 2021) in response to the attacking pathogen.

The host plant initiates PTI by recognizing microbial elicitors/PAMPs (Boller and He 2009) using pattern recognition receptors embedded in the host cell membrane, inducing the host plant defense mechanism (Boller and He 2009). Subsequently, the PAMP signal is transduced through mitogen-activated protein kinase signal cascades (Nakagami et al. 2005; Chinchilla et al. 2007), with the host plant activating various TF genes and downstream target genes encoding pathogenesis-related proteins and other defense-related proteins mediating disease resistance (Garg et al. 2019).

In ETI, the second tier of defense mechanism starts with the induction of the host disease resistance (R) gene(s) following recognizing pathogen-secreted effector molecules through the host plant's nucleotide-binding domain/leucine-rich repeat receptors (Cui et al. 2015; Parker et al. 2022). Eventually, it activates mitogen-activated protein kinase signaling, G-proteins, Ca^{2+}, and resistance (R) genes evoking a hypersensitive response that causes cell death and thus inhibits pathogen infection (Zhang et al. 2012; Meng and Zhang 2013; Parker et al. 2022). However, the molecular mechanisms of host plant–pathogen interactions, pathogenesis, and host resistance against various diseases in chickpea remain unclear.

1.10 Chickpea Germplasm Repertoire for Harnessing Disease Resistance

Host plant resistance is the cheapest among the various approaches adopted for controlling disease (Siddique et al. 2013). Cultivated chickpea genotypes are important sources of AB resistance. Plant-breeding-based screening approaches have identified several important sources of AB resistance under field condition. Relying on slow blighting and partial resistance, ILC482 was released as an AB-resistant chickpea cultivar (Singh and Reddy 1993). Multilocation testing of chickpea genotypes against AB infection has identified numerous AB-resistant chickpea genotypes: ICC7052, ICC4463, ICC4363, and ICC2884 in Kenya (Kimurto et al. 2013) and IC275447, IC117744, EC267301, IC248147, and EC220109 in India (Gayacharan et al. 2020). An evaluation of 75 advanced chickpea lines in the fusarium wilt sick plot for 2 years identified IPC2007–28, IPC2010 78, IPC2009–66, IPC2016–36, and IPC2016–69 (see Table 1.2) as promising FW-resistant lines (Jha et al. 2021a). Apart from cultivated chickpea, crop wild relatives (CWRs) such as *C. echinospermum* (Newman et al. 2021; Sudheesh et al. 2021), *C. reticulatum* (Collard et al. 2001), and *C. judaicum* and *C. pinnatifidum* (Singh and Reddy 1993) are important sources of AB resistance.

Significant genetic variability for FW resistance has been reported in cultivated chickpea, including ICC-2862, −9023, −9032, ILC-5411, FLIP 85-20C, BG-212 ICC17109, and WR315 (see Table 1.2) (Haware et al. 1990; van Rheenen et al. 1992; Sharma et al. 2005; Gaur et al. 2006). Sharma et al. (2019) reported ICCV 93706, ICCV 07118, ICCV 08124, and ICCV 08113 as AB-resistant genotypes based on multilocation testing. Likewise, CWRs such as *C. reticulatum*, *C. echinospermum*, *C. bijugum*, *C. judaicum*, *C. pinnatifidum*, and *C. cuneatum* (Nene and Haware 1980; Kaiser et al. 1994; Singh et al. 1998) harbor gene/alleles conferring FW resistance, which could be used to develop FW-resistant chickpea cultivars.

Potential sources of DRR resistance have been reported (Iftikhar and Ilyas 2000; Pande et al. 2004, 2006a). Iftikhar and Ilyas (2000) screened 108 chickpea lines and identified ICCV 97112 as DRR-resistant. Similarly, Pande et al. (2006a) evaluated 211 mini-core chickpea lines and identified ICC1710 and ICC 2242 as DRR-resistant genotypes. Subsequently, Jayalaxmi et al. (Jayalakshmi et al. 2008) identified four chickpea genotypes, GCP-101, GBM-2, GBM-6, and ICCV-10, that tolerated DRR (see Table 1.2).

1.11 Genetics of Disease Resistance in Chickpea

Before the advent of molecular marker technology, classical genetic approaches provided insights into the major gene(s) controlling AB, FW, and BGM resistance in chickpea (Tiwari et al. 1985; Rewal and Grewal 1989; Dey and Singh 1993; Kumar 2006). Vir et al. (1975) reported monogenic and single-dominant gene inheritance of AB resistance. Likewise, Singh and Reddy (1983) found a single dominant gene

Table 1.2 List of resistance sources for various disease resistance in chickpea

Disease	Source of resistance	References
Ascochyta blight	*C. judaicum* and *C. pinnatifidum* (wild species)	Singh and Reddy (1993)
	C. echinospermum and *C. reticulatum* (wild species)	Collard et al. (2001)
	HOO-108 and GL92024	Dubey and Singh (2003)
	PI 559361, PI 559363, and W6 22589	Chen et al. (2004)
	RIL58-ILC72/Cr5	Rubio et al. (2006)
	Almaz, ICC 3996, and ILWC 118	Danehloueipour et al. (2007)
	FLIP 98-133C and FLIP 98-136C	Chandirasekaran et al. (2009)
	FLIP 97-121C	Kaur et al. (2012)
	FLIP 4107, FLIP 1025, and FLIP 10511	Benzohra et al. (2013)
	EC 516934, ICCV 04537, ICCV 98818, EC 516850, and EC 516971	Pande et al. (2013)
	ICC7052, ICC4463, ICC4363, ICC2884, ICC7150, ICC15294, and ICC11627	Kimurto et al. (2013)
	10A and 28B	Duzdemir et al. (2014)
	ILC72, ILC182, ILC187, ILC200, and ILC202	Benzohra et al. (2015)
	C. echinospermum accessions S2Drd_ 061 (wild species)	Newman et al. (2021)
	Deste_064, *C. reticulatum* accession Bari1_062 (wild species)	
	C. echinospermum accession Karab_063 (wild species)	
	Cicer echinospernum (wild species)	Sudheesh et al. (2021)
	IC275447, IC117744, EC267301, IC248147, and EC220109	Gayacharan et al. (2020)
Fusarium wilt	ICC 11322 (WR 315)	Singh et al. (1974)
	C. bijugum, *C. judaicum*, *C. pinnatifidum*, and *C. reticulatum* (wild species)	Nene and Haware (1980)
	C. echinospermum and *C. cuneatum* (wild species)	Singh et al. (1998)
	ICCV2, ICCV3, ICCV4, and ICCV5 (against race 1)	Kumar et al. (1985)
	ICC 11322, 14424, and 14433 (against to race 1)	Nene et al. (1989)
	ICC-2862, 9023, 9032, 10803, 11550, and 11551	Haware et al. (1990)
	FLIP 84-43C (against race 0), ILC-5411, FLIP 85-20C (against race 5), and FLIP 85-29C	van Rhenen et al. (1992)
	FLIP 85-30C, ILC-127 (against race 0), ILC-219 (against race 0), ILC-237, ILC-267, and ILC-513 (against race 0)	Jiménez-Díaz, Singh (1991)
	C. canariense (PI553457 resistant against race 0), *C. chorassanicum* (PI458553 resistant against race 0), *C. cuneatum*, *C. judaicum*, and C. *pinnatifidum* (PI 458555, PI458556 resistant to race 0) (wild species)	Kaiser et al. (1994)

(continued)

Table 1.2 (continued)

Disease	Source of resistance	References
	ICCV 2 and UC 15, FLIP 85-20C, FLIP 85-29C, and FLIP 85-30C	Ali et al. (2002)
	CA-334.20.4, CA-336.14.3.0, and ICC- 14216 K (race 5)	Castillo et al. (2003)
		Navas-Cortes et al. (1998)
	Andoum 1 and Ayala (race 0)	Halila and Harrabi (1990)
		Landa et al. (2006)
	Surutato-77, Sonora-80, Tubutama, UC-15 and UC-27, Gavilan	Sharma et al. (2005)
		Buddenhagen and Workneh (1988)
		Helms et al. (1992); Morales (1986)
	BG-212	Sharma et al. (2005)
	ICC-7520	Sharma et al. (2005)
	Annigeri	Sharma et al. (2005)
	ICC 7537 resistant to all races (except race 4)	Sharma et al. (2005)
	ICC14194, ICC17109, WR315	Gaur et al. (2006)
	CM418-1/01, CM446-1/01, CM499/01, CM499-1/01, CM499-2/01, CM554-1/01, CM554-2/01, CM557-2/01, CM557-5/01, CM557-6/01, CM557-7/01, CM5578/01, and CM499-5/01	Shah et al. (2009)
	ICCV 09118, ICCV 09113, ICCV 09115, ICCV 09308, ICCV 09314	Sharma et al. (2010)
	ICCV 05527, ICCV 05528, ICCV 96818	Sharma et al. (2012)
	Three lines derived from MABC-based C 214 × WR 315 cross	Varshney et al. (2014)
	ICCVs 98505, 07105, 07111, 07305, 08113, and 93706 (highly resistant), ICCVs 08123, 08125, 96858, 07118, 08124, 04514, 08323, and 08117 (moderately resistant)	Sharma et al. (2019)
	Digvijay	Upasani et al. (2016)
	SCGP-WR 28, H 10-05, GL 10023, IPC 2006-77, and CSJK 72	Dubey et al. (2017)
	Super annigeri and improved JG74 (resistant against foc4)	Mannur et al. (2019)
	IPC2007-28, IPC2010 78, IPC2009-66, IPC2016-36, IPC2016-69	Jha et al. (2021a)
Collar rot	ICCV 05530	Tarafdar et al. (2018)
Dry root rot	ICC 1710 and ICC 2242	Pande et al. (2006a)
Dry root rot	GCP- 101, GBM-2, GBM-6, and ICCV-10	Jayalakshmi et al. (2008)
Dry root rot	ICCV 97112	Iftikhar and Ilyas (2000)

(continued)

Table 1.2 (continued)

Disease	Source of resistance	References
Dry root rot	PG06102, BG2094, and IC552137	Talekar et al. (2017)
Dry root rot	ICCV 08305	Karadi et al. (2021)

governing AB resistance in ILC 72, ILC 183, ILC 200, and ILC 4935 and a single recessive gene controlling AB resistance in ILC191. Further, Singh and Reddy (1989) noted a single dominant gene controlling AB (race 3) resistance. However, two dominant complementary genes controlling AB resistance have been reported (Dey and Singh 1993). Danehloueipour et al. (2007) reported quantitative inheritance of AB resistance (both additive and dominance gene action).

For FW resistance, three independent loci, namely h_1, h_2, and H_3, have been reported to support the monogenic inheritance of *Foc1* resistance (Upadhyaya et al. 1983a, b; Singh et al. 1987) and *Foc3* and *Foc*5 resistance (Tekeoglu et al. 2000; Sharma et al. 2004). Resistance for each race 1A, 2, 3, 4, and 5 are controlled by a single gene (Sharma et al. 2005). Monogenic and recessive resistance against *Foc4* in WR315 genotype was reported (Tullu et al. 1998, 1999). However, Tullu et al. (1999) and Rubio et al. (2003) reported digenic inheritance of *Foc0* and *Foc1* resistance, while Tullu et al. (1998, 1999) reported digenic and recessive inheritance against *Foc4* in Surutato-77 genotype.

For BGM resistance, Tiwari et al. (1985) reported a single dominant gene "*Bor1*," Rewal and Grewal (1989) reported two genes with epistasis interaction (13:3 ratio), and Chaturvedi et al. (1995) reported two duplicate dominant genes with epistasis interaction (15:1 ratio) controlling BGM resistance. However, the complex nature of this disease and the impact of the environmental effect cause complexity in working out its genetic resistance.

DRR is governed by monogenic inheritance with resistance dominant over susceptibility based on F_1 and F_2 populations developed from crossing DRR-resistant and susceptible parents (Rao 1987).

1.12 Genomic Resources: QTL Mapping for Disease Resistance in Chickpea

Advances in chickpea genomic resource development and the advent of molecular marker technology have helped identify QTLs controlling disease resistance and dissect disease resistance genetics (Sabbavarapu et al. 2013; Tar'an et al. 2003; Patil et al. 2014; Varshney et al. 2014; Talekar et al. 2017; for details, see Jha 2018; Jha et al. 2020, 2022). Two QTLs (*QTL-1* and *QTL-2*) contributing to AB resistance were tagged using RAPD markers (Santra et al. 2000). Subsequently, AB resistance QTLs were discovered using simple sequence repeat (SSR) markers: *ar1*, *ar2a*, and *ar2b* QTLs (Udupa and Baum 2003), *Ar19* (Cho et al. 2004), QTL_{AR3} (Iruela et al.

2007), QTL(1–5) (Anbessa et al. 2009), *Abr QTL 3, QTL* (*AR1*) (Madrid et al. 2012), *Abr QTL 4* (Tar'an et al. 2013), and *AB-Q-SR-4-1*, *AB-Q-SR-4-2*, *AB-Q-APR-6-1*, *AB-Q-APR-6-2*, *AB-Q-APR-4-1*, and *AB-Q-APR-5B* (Sabbavarapu et al. 2013) (see Table 1.3). Likewise, the availability of high-throughput SNP markers enabled the identification of nine AB resistance QTLs explaining 9–19% phenotypic variation (Daba et al. 2016). Similarly, *qABR4.1*, *qABR4.2*, and *qABR4.3* QTLs reported by Kumar et al. (2018), *AB_echino_2014* and *AB_echino_2015* QTLs reported by Sudheesh et al. (2021), and *qab-4.1*, *qab-4.2*, and *qab-7.1* QTLs reported by Kuswah et al. (2021) attributing AB resistance could be used in genomics-assisted breeding for AB resistant chickpea cultivars.

For FW, random amplified polymorphic DNA(RAPD) and SSR markers linked to various gene(s)/genomic regions conferring resistance to various FW races were developed, including a RAPD marker linked to *Foc0* (Mayer et al. 1997; Rubio et al. 2003; Cobos et al. 2005) and SSR markers linked to *Foc0* (Tekeoglu et al. 2000; Halila et al. 2009; Jendoubi et al. 2016), *Foc1* (Winter et al. 2000; Gowda et al. 2009; Sabbavarapu et al. 2013; Patil et al. 2014; Varshney et al. 2014), *Foc2* (Gowda et al. 2009; Halila et al. 2009), *Foc3* (Sharma et al. 2004; Gowda et al. 2009), *Foc4* (Winter et al. 2000), and *Foc5* (Tekeoglu et al. 2000; Castro et al. 2010). Likewise, several major QTLs controlling FW resistance have been identified from mapping populations. Sabbavarapu et al. (2013) discovered two major QTLs, *FW-Q-APR-6-1* and *FW-Q-APR-6-2*, controlling FW(*Foc1*) resistance that explained 10.4–18.8% of the phenotypic variation by assessing recombinant lines derived from a C214 × WR315 mapping population subjected to FW in the field (see Table 1.3). Likewise, Garg et al. (2018) discovered three QTLs, *FW-Q-APR-2-1*, *FW-Q-APR-4-1*, and *FW-Q-APR-6-1*, for FW (*Foc1*) resistance on LG2, LG4, and LG6, and two QTLs, *FW-Q-APR-2-1* and *FW-Q-APR-4-1*, for FW(*Foc3*) resistance on LG2 and LG4. Using marker-assisted backcrossing, Mannur et al. (2019) recently transferred the FW (*Foc4*) resistance genomic region to Annigeri and JG74 and developed a super Annigeri elite chickpea cultivar with improved FW resistance.

Limited genomics resources are available for BGM resistance. Three QTLs contributing to BGM resistance were elucidated from an ICCV2 × JG62 mapping population (Anuradha et al. 2011). *QTL1* located on LG6A explained 12.8% of the phenotypic variance and *QTL2* and *QTL3* located on LG3 explained 9.5 and 48%, respectively (Anuradha et al. 2011). Similarly, a phenotypic assessment of recombinant inbred lines developed from GPF 2 × *C. reticulatum* acc. ILWC 292 and genotyping of this mapping population identified three consistent BGM resistance QTLs over 2 years, *qbgm-4.1*, *qbgm-4.2*, and *qbgm-5.1*, explaining 7.2–12% of the phenotypic variation (Kuswah et al. 2021).

Talekar et al. (2017) reported monogenic inheritance of DRR resistance in a bulk segregation analysis of individuals from an $F_{2:3}$ population developed from an L550 × PG06102 cross. The authors also identified ICCM0299, TR29, CaM111, and ICCM0120b markers for distinguishing resistant and susceptible bulks of DRR. Recently, Karadi et al. (2021) genotyped a mapping population BG 212 × ICCV08305 using an Affymetrix Axiom CicerSNP array, uncovering a minor DDR resistance *qDRR-8* QTL flanked by *Ca8_3970986* and *Ca8_3904895*

Table 1.3 List of various disease-resistant QTLs reported in chickpea

Name of gene/QTL(s)	Type of marker	Linkage group	References
Fusarium wilt			
H1 locus of *Foc 1*	RAPD	–	Mayer et al. (1997)
Single recessive gene (race 1 and race 4)	RAPD	–	Tullu et al. (1998)
Races 4 and 5	STMS and SCAR	–	Winter et al. (2000)
foc-0, *foc-4*, and *foc-5*	STS	–	Tekeoglu et al. (2000)
Foc 01/foc 01 and *Foc 02/foc 02*	RAPD	–	Rubio et al. (2003)
One gene for fusarium wilt race 0 (*Foc0*)	RAPD and STMS	LG3	Cobos et al. (2005)
Three loci (race 2)	–		Kumar (2006)
foc-3, *foc-1* (syn. H(1)), and *foc-4*	STMS, STS, and STMS	–	Sharma et al. (2004); Winter et al. (2000); Gowda et al. (2009)
Single gene (race 0)	RAPD		Rubio et al. (2003)
Foc5	SSR	–	Cobos et al. (2009)
Foc1	STMS		Gowda et al. (2009)
Foc2	STMS		
Foc3	STMS		
$Foc0_2$ /$foc0_2$	STMS	LG2	Halila et al. (2009)
foc-5	STMS	LG2	Castro et al. (2010)
FW-Q-APR-6-1 (Foc1) and *FW-Q-APR-6-2 (Foc1)*	STMS	LG6	Sabbavarapu et al. (2013)
Wilt 1 (race 1), *wilt 2* (race 1)	STMS	LG2	Patil et al. (2014)
Genomic region resistance for *foc1* and *foc3*	TR19, TA194, TAA60, GA16, TA110, and TS82	LG2	Varshney et al. (2014)
Five QTLs	GSSR 11-EST SSR 3, TR 24-EST SSR 21, EST SSR 21-EST SSR 65, and GSSR 18-TC14801	LG1	Jingade and Ravikumar (2015)
Foc01/foc01	H2I20 and TS43 (STMS)	LG5	Jendoubi et al. (2016)
LOC101514038 and *LOC101499491*	CaGM20820, CaGM20889		
Candidate genes			
Foc 2	TA 37 and TA110	–	Pratap et al. (2017)

(continued)

Table 1.3 (continued)

Name of gene/QTL(s)	Type of marker	Linkage group	References
Three QTLs (race 1), *FW-Q-APR-2-1*	TR19 and H2B061	CaLG02	Garg et al. (2018)
FW-Q-APR-4-1, *FW-Q-APR-6-1*	TA132 and TA46	CaLG04 and CaLG06	
2QTLs (race3)	TA80 and CaM0594	CaLG02 and CaLG04	
FW-Q-APR-2-1 and *FW-Q-APR-4-1*	CKAM1256 and TS72		
LOC101511605 (*Foc5*)	TA59, CaGM07922, SNPs	LG2	Caballo et al. (2019a)
Genomic region conferring resistance against foc4	TA59, TA96, TR19, and TA27,	LG2	Mannur et al. (2019)
	GA16 and TA96		
Foc2	CESSR433, NCPGR21, and ICCM0284		Jha et al. (2021a)
Ascochyta blight			
QTL-1 and QTL-2	RAPD, ISSR	–	Santra et al. (2000)
Two QTL	STMS	LG4	Collard et al. (2003)
ar1, *ar2a*, and *ar2b*	SSR	LG 2 and 4	Udupa and Baum (2003)
Three QTL + Ar19 (or Ar21d) gene	SSR	(LG)4A and LG2 + 6	Cho et al. (2004)
One QTL	RAPD, ISSR, STMS, isozyme	LG2	Cobos et al. (2006)
QTL_{AR3}	STMS	LG2	Irulea et al. (2007)
Three QTLs	SSR	LG3, 4, and 6	Tar'an et al. (2007)
Five QTLs (QTL1–5)	SSR	LG 2, 3, 4, 6, and 8	Anbessa et al. (2009)
Three QTLs	SSR	LG 3 and4	Kottapalli et al. (2009)
Three QTLs	SSR	LG 3 and 4	Aryamanesh et al. (2010)
QTL(AR1), EIN4-like sequence	SSR	LG4	Madrid et al. (2012)
Abr QTL 3 and *Abr QTL 4*	SSR	LG 4 and 8	Tar'an et al. (2013)
AB-Q-SR-4-1, *AB-Q-SR-4-2*, *AB-Q-APR-6-1*, *AB-Q-APR-6-2*, *AB-Q-APR-4-1*, and *AB-Q-APR-5B*	SSR	LG4, 5, and 6	Sabbavarapu et al. (2013)

(continued)

Table 1.3 (continued)

Name of gene/QTL(s)	Type of marker	Linkage group	References
42 candidate genes Ein3, Avr9/Cf9 and Argonaute 4	SNP	Ca2	Madrid et al. (2014)
ab_QTL1, *ab_QTL2*	EST, SNP	–	Stephens et al. (2013)
qtlAb-1.1, *qtlAb-2.1*, *qtlAb-3.1*	SNP	LG 1, 2, 3, 4, 6, 7, and 8	Daba et al. (2016)
qtlAb-4.1, *qtlAb-6.1*, *qtlAb-7.1*			
qtlAb-8.1, *qtlAb-8.2*, *qtlAb-8.3*	SNP		
AB4.1 QTL along with 12 candidate genes	SNP	LG 4	Li et al. (2017)
One QTL for seedling resistance; minor QTLs for SR and adult plant resistance	SSR, SNP	–	Garg et al. (2018)
qABR4.1 and *qABR4.2*	SNP	LG 4	Kumar et al. (2018)
qABR4.3 QTLs and *CaAHL18*			
Candidate gene			
WRKY TF *(Cr_02657.1)*, *(Cr_09847.1)* encodes a TF of ARF family	SNP	LG 3, 4, 6	Newman et al. (2021)
CPR01-qAB1.1, *CPR01-qAB1.2*, *CPR01-qAB1.3*, *CPR01-qAB1.4*, *CPR01-qAB4.1*, *CPR01-qAB4.2*, *CPR01-qAB4.3*, *CPR01-qAB4.4*, *CPR01-qAB4.5*, *CPR01-qAB6.1*, *CPR01-qAB6.2*, and *CPR01-qAB7.1*	SNP	LG 1, 2, 3, 4, 6, 7, and 8	Deokar et al. (2019a)
Eight QTLs	SNP	LG 2, 3, 4, 5 and 6	Deokar et al. (2019b)
AB_echino_2014 and *AB_echino_2015*	SNP	LG4	Sudheesh et al. (2021)
Dry root rot			
Mono gene	SSR	–	Talekar et al. (2017)
qDRR-8	SNP	CaLG08	Karadi et al. (2021)
Botrytis gray mold			
QTL1, QTL2 and QTL3	SSR	LG 3 and 6A	Anuradha et al. (2011)
qbgm-4.1, *qbgm-4.2*, and *qbgm-5.1*	SNP	LG 4 and 5	Khuswah et al. (2021)

RAPD random amplified polymorphic DNA, *SSR* simple sequence repeat, *STM* sequence tagged microsatellite, *SCAR* sequence cleaved amplified region, *ISSR* inter simple sequence repeat, *SNP* single nucleotide polymorphism markers on CaLG8 explaining 6.7% of the phenotypic variation (see Table 1.3). The increasing repertoire of chickpea genomics has helped develop FW- and AB-resistant chickpea genotypes using marker-assisted selection (Varshney et al. 2014; Mannur et al. 2019).

Genome-wide association mapping offers a great opportunity to discover disease-resistant QTLs in large sets of genotypes across the whole genome, including several QTLs controlling FW and AB resistance in chickpea (Jha et al. 2020, 2021c, 2022). Association mapping of 75 chickpea genotypes using SSR markers identified 9 (2016) and 8 (2017) significant marker-trait associations for *Foc2* resistance. Combined analysis of whole genome resequencing and GWAS of 132 chickpea genotypes unraveled AB-resistant QTL *AB4.1* with 12 underlying candidate genes on LG4 (Li et al. 2017), notably *Ca_05515*, *Ca_05511*, *Ca_05517*, *Ca_05521*, and *Ca_05522*. Likewise, GWAS analysis of a set of 146 (*C. reticulatum*) and 44 (*C. echinospermum*) chickpea genotypes uncovered four significant candidate genes—*Cr_ 02657.1*, *Cr_09847.1*, *Cr_16402.1*, and *Cr_08467.1*—contributing to AB resistance (Newman et al. 2021). Thus, these genomics resources could facilitate genomics-assisted breeding for developing disease-resistant chickpea cultivars.

1.13 Advances in Functional Chickpea Genomics Contributing to Disease Resistance

In the last decade, rapid advances in functional chickpea genomics have enabled the functional characterization of various gene(s)/genomic regions conferring disease resistance (Coram and Pang 2005a, b; Kumar et al. 2018). Microarrays and deepSuperSAGE-based analyses have been performed in chickpea to gain insights into the underlying candidate genes and their function contributing to disease resistance and host–pathogen interactions (Coram and Pang 2005a, 2006; Gupta et al. 2009; Gurjar et al. 2012; Xue et al. 2015). A microarray study elucidated 20 defense-related ESTs from ICC3996 (AB resistant) and Lasseter (AB susceptible) genotypes in response to AB reaction (Coram and Pang 2005a). In another microarray analysis, the same authors found 97 differentially expressed genes mostly related to encoding pathogenesis-related proteins, leucine zipper protein, and Ca-binding protein in response to AB reaction (Coram and Pang 2006).

Subsequent advances in RNA-seq-based transcriptomics have facilitated the identification of novel candidate genes conferring disease resistance. Sagi et al. (2017) performed a functional genomic analysis to investigate the participatory role of NBS-LRR genes mediating AB resistance, identifying some important candidate genes: *LOC101509145* and *LOC101498915* upregulated in CDC Corinne chickpea and *LOC101512894*, *LOC101513745*, and *LOC101497042* upregulated in ICCV 96029 and CDC Luna in response to AB infection. Similarly, in an RNA-seq analysis, Kumar et al. (2018) reported high expression of the *CaAHL18* gene underlying AB-resistant *qABR4.1* QTL in the AB-resistant chickpea genotype. Apart from these gene(s)/QTLs, numerous noncoding RNAs mediate disease resistance in various crops, including legumes (Jha et al. 2021b). To explore the role of noncoding RNAs involved in mediating disease resistance in chickpea, Garg et al. (2019) conducted a transcriptome analysis of AB-resistant and AB-susceptible lines. The authors obtained a plethora of upregulated and downregulated noncoding miRNAs (e.g., nov_miR126 and nov_miR131a, nov_miR9, miR319l, and miR167a) and their corresponding target genes contributing to AB resistance.

A functional analysis of contrasting chickpea genotypes in response to FW revealed the involvement of various gene(s) ranging from cellular transporters, sugar metabolism, lignification, and hormonal homeostasis involved in inducing defense signaling and reactive oxygen species (ROS) (Gupta et al. 2013; Upasani et al. 2017). Caballo et al. (2019b) discovered several candidate genes, including *LOC101499873* and *LOC101490851*, and three novel candidate genes, *LOC101509359*, *LOC101495941*, and *LOC101510206*, expressing 24 h postinoculation (hpi) that contributed to FW resistance. Elucidating the involvement of microRNAs (miRNAs) in mediating FW resistance, Kohli et al. (2014) performed an RNA-seq-based transcriptome analysis of FW-infected ICC 4958 genotype and discovered several known and novel miRNAs involved in chickpea FW resistance. The authors also noted that miR530 exhibited 17-fold enhanced expression and slightly higher expression of miR156_1 and miR156_10 than those of the other identified miRNAs in response to FW infection.

1.14 Proteomics and Metabolomics Insights into Disease Resistance in Chickpea

A proteomics approach can identify the proteins involved in host–pathogen interactions during the pathogenesis process, secreted by virulent pathogens during host invasion, and the host proteins mediating resistance against the virulent pathogen (Kumar et al. 2016). The involvement of various proteins, ranging from chitinases, xylem proteinases, and pathogenesis-related (PR) proteins to cellulose synthases, is notable during pathogen attachment to the host and host-mediated resistance against the invading pathogen (Kumar et al. 2016; Silvia Sebastiani et al. 2017; Chen et al. 2019). In addition, the host plant secretes various enzymes, such as phosphoglucomutase, transaldolase, enolase, pyruvate dehydrogenase, citrate synthase, aconitase, and phenylalanine ammonia lyase, that help resist the attaching pathogen (Kumar et al. 2016; Garcia-Limones et al. 2009; Chen et al. 2019). In a proteomics analysis, Kumar et al. (2016) noted a higher abundance of several ROS-activating enzymes (glutathione peroxidase, ascorbate peroxidase, peroxiredoxin) in response to FW reaction in resistant host Digvijay than FW-sensitive JG62. Similarly, host Digvijay had a higher abundance of endo b-1,3-glucanase, chitinase, and profilin cytoskeleton proteins than FW-sensitive JG62 during FW invasion (Kumar et al. 2016). However, few proteomic studies have investigated host–pathogen interactions in chickpea; dedicated efforts in this area will offer greater insights into the role of various proteins mediating disease resistance in chickpea.

Metabolomics can assist in identifying various metabolites, including sugars, proteins, and organic acids, hormonal crosstalk, and various signaling molecules in response to host–pathogen interactions during pathogen invasion and host-mediated resistance against pathogens (Kumar et al. 2016). Several metabolites are produced in chickpea exposed to FW: Kumar et al. (2016) reported higher upregulation of various enzymes (sucrose synthase, phosphoglucomutase, transaldolase) and higher abundance of defense-related metabolites (chitinases, caffeic acid O-methyltransferase, antifungal clotrimazole, phytosterol) in Digvijay (FW resistant) than JG 62 (FW susceptible), which could mediate FW resistance in Digvijay.

1.15 Role of High-Throughput Phenotyping Approaches for Detecting Disease

Current advances in high-throughput phenotyping (HTP) platforms offer the opportunity to monitor host plant and disease interactions with high precision and examine disease-resistant host plant genotypes on a large scale under artificial and field conditions. Among the various HTP technologies available, thermal cameras, hyperspectral image sensing, remote sensing, unmanned aerial vehicles, and drone-based phenotyping are primarily used for detecting disease onset and disease infection and measuring disease severity (Jarolmasjed et al. 2019; Rousseau et al. 2013; Danilevicz et al. 2021). Zhang et al. (2019) used an unmanned aircraft system in association with various multispectral cameras to assess the severity of AB infection in chickpea.

1.16 Novel Breeding Approaches for Developing Disease-Resistant Genotypes

Marker-assisted backcross breeding (MABB) was implemented to improve varieties for disease resistance (Varshney et al. 2014; Pratap et al. 2017; Mannur et al. 2019). Improved varieties with fusarium wilt resistance [Super Annigeri 1, Pusa Chickpea 20211, and IPCMB 19-3 (Samriddhi)] were released for commercial cultivation across various agro-ecologies in India. In 2020, the fusarium wilt-resistant variety Pusa Manav (Pusa Chickpea 20,211), developed by ICAR-IARI, New Delhi, India, was released for commercial cultivation in the Central Zone of India. Pusa Manav (Pusa Chickpea 20,211), with about 2.4 q/ha yield and resistance to fusarium, recorded a 28% yield advantage over its recurrent parent (Pusa 391).

Next-generation sequencing-based efforts of genome sequencing and whole genome resequencing of chickpea genotypes, facilitated in developing huge large number of SNP markers. These genome-wide SNP markers could be potentially harnessed for "genomic selection" to predict genomic-estimated breeding values of untested individuals based on using various prediction models (Meuwissen et al. 2001). This approach could be beneficial for selecting individuals with disease resistance from early generations and has been used to select disease (AB)-resistant lines in pea but not chickpea. Likewise, speed breeding protocols (Ghosh et al. 2018) have been developed to reduce crop generation time and thus expedite the screening of disease-resistant lines in chickpea breeding programs. The emerging CRISPR/Cas9-based genome editing approach (Altpeter et al. 2016) could be used to manipulate target host plant disease resistance gene(s)/virulence genes of the invading pathogen with high precision. In chickpea, this technology has been used to edit *4-coumarate ligase* (4CL) and *Reveille 7* (RVE7) genes (Badhan et al. 2021) related to drought tolerance but is yet to be applied to mediate disease resistance.

1.17 Conclusion and Future Perspective

The global human population is projected to be 9–10 billion by 2050 (O'Neill et al. 2010), with an estimated 70–110% increase in food production needed to meet the growing food demand (Ray et al. 2013). Biotic stress caused by various diseases significantly reduces annual chickpea yields, and global climate change has increased the incidence of various diseases. Large amounts of chemical-based pesticides are used to control these diseases, resulting in environmental pollution. Plant breeding approaches could significantly reduce this environmental pollution and minimize disease attack. By harnessing chickpea's genetic diversity, several disease-resistant chickpea genotypes have been developed; however, few studies have exploited chickpea CRWs for developing disease-resistant cultivars. Thus, breeding programs should incorporate CWRs to develop pre-breeding lines for use in the main breeding program to sustain chickpea yields under the increasing events of various major and minor diseases (Mohanty et al. 2022). The increase in chickpea genomic resources has offered opportunities to embrace genomics-assisted breeding for pyramiding multiple race-specific resistant gene(s)/alleles into a single elite chickpea cultivar for broad-spectrum disease resistance. Likewise, GWAS, WGRS, and pangenome sequence approaches could help unfold disease-resistant haplotypes/genomic regions across the whole genome in a large set of global chickpea germplasm (Varshney et al. 2019, 2021; Thudi et al. 2016). In parallel, the complete pathogen genome sequence and pangenome sequence information could be used to investigate novel effector encoding gene(s) or virulence gene (s) of pathogens (Badet and Croll 2020). Simultaneously, advances in functional genomics could uncover various resistance candidate gene(s) with putative functions, improving our understanding of the complex molecular mechanisms of host plant and pathogen interactions during pathogen infection and elucidating host plant immunity/resistance mechanisms. Emerging approaches such as drone-based HTP may facilitate the screening of infected plants sown across large areas, enabling us to take preventive measures to reduce disease spread. Likewise, genomic selection based on various prediction models could be used to select disease-resistant lines from early segregating plant progenies. Speed breeding protocols could be used to fast-forward the mapping population for mapping disease resistance QTLs. Moreover, genome editing tools could be used to precisely manipulate the host S gene(s) and pathogen-encoding effector gene(s) to improve the host plant resistance of elite yet disease-susceptible cultivars. Thus, leveraging these breeding and "omics"-based technologies could expedite the development of disease-resistant chickpea cultivars for global food security.

References

Abraham AD, Menzel W, Lesemann DE, Varrelmann M, Vetten HJ (2006) Chickpea chlorotic stunt virus: a new polerovirus infecting cool-season food legumes in Ethiopia. Phytopathology 96:437–446

Abraham A, Vetten HJ (2022) Chickpea chlorotic stunt virus: a threat to cool-season food legumes. Arch Virol 167:21–30

Akamatsu HO, Chilvers MI, Kaiser WJ, Peever TL (2012) Karyotype polymorphism and chromosomal rearrangement in populations of the phytopathogenic fungus, *Ascochyta rabiei*. Fungal Biol 116:1119–1133

Ali MEK, Inanga S, Sugimoto Y (2002) Source of resistance to *Fusarium* wilt of chickpea in Sudan. Phytopathol Mediterr 41:163–169

Altpeter F, Springer NM, Bartley LE, Blechl AE, Brutnell TP, Citovsky V et al (2016) Advancing crop transformation in the era of genome editing. Plant Cell 28:1510–1520

Anbessa Y, Taran B, Warkentin T, Tullu A, Vandenberg A (2009) Genetic analyses and conservation of QTL for Ascochyta blight resistance in chickpea (*Cicer arietinum* L.). Theor Appl Genet 119:757–765

Anuradha C, Gaur PM, Pande S, Gali KK, Ganesh M, Kumar J, Varshney RK (2011) Mapping QTL for resistance to botrytis grey mould in chickpea. Euphytica 182:1–9

Aryamanesh N, Nelson MN, Yan G, Clarke HJ, Siddique KH (2010) Mapping a major gene for growth habit and QTLs for ascochyta blight resistance and flowering time in a population between chickpea and Cicer reticulatum. Euphytica 173:307–319

Badet T, Croll D (2020) The rise and fall of genes: origins and functions of plant pathogen pangenomes. Curr Opin Plant Biol 56:65–73. https://doi.org/10.1016/j.pbi.2020.04.009

Badhan S, Ball AS, Mantri N (2021) First report of CRISPR/Cas9 mediated DNA-free editing of 4CL and RVE7 genes in chickpea protoplasts. Int J Mol Sci 22:396

Bakr MA, Hussain SA, Afzal MA, Rahman MA (2002) Chickpea status and production constraints in Bangladesh. In: Bakr MA, Siddique KHM, Johansen C (eds) Integrated management of botrytis grey mould of chickpea in Bangladesh and Australia, Summary Proceedings of a Project Inception Workshop. Bangladesh Agricultural Research Institute, Joydebpur, Gazipur, Bangladesh, pp 19–32

Benzohra IE, Bendahmane BS, Benkada MY, Labdi M (2015) Screening of 15 chickpea germplasm accessions for resistance to Ascochyta rabiei in north West of Algeria. Am-Eurasian J Agric Environ 15:109–114

Benzohra IE, Bendahmane BS, Labdi M, Benkada MY (2013) Sources of resistance in chickpea germplasm to three pathotypes of Ascochyta rabiei (pass.) Labr. In Algeria. World Appl Sci J 21:873–878

Boller T, He SY (2009) Innate immunity in plants: an arms race between pattern recognition receptors in plants and effectors in microbial pathogens. Science 324(5928):742–744

Buddenhagen IW, Workneh F (1988) Bosque-Perez NA (1988) chickpea improvement and chickpea diseases in California. Int Chickpea Newsl 19:9–10

Burgess DR, Bretag T, Keane PJ (1997) Seed-to-seedling transmission of *Botrytis cinerea* in chickpea and disinfestations of seed with moist heat. Aust J Exp Agric 37:223–229

Caballo C, Castro P, Gil J, Millan T, Rubio J, Die JV (2019a) Candidate genes expression profiling during wilting in chickpea caused by Fusarium oxysporum f. sp. ciceris race 5. PLoS One 14(10):e0224212

Caballo C, Castro P, Gil J, Millan T, Rubio J, Die JV (2019b) Candidate genes expression profiling during wilting in chickpea caused by *Fusarium oxysporum* f. sp. *ciceris* race 5. PLoS One 14: e0224212

Carranza JM (1965) Wilt of chickpea (*Cicer arietinum* L.) caused by *B. cinerea* (in Spanish). In: Revista de la Facultad de Agronomia, vol 41. Universidad Nacional de la Plata, Provincia de Buenos Aires, pp 135–138

Castillo P, Navas-Cortés JA, Gomar-Tinoco D, Di Vito M, Jiménez-Díaz RM (2003) Interactions between Meloidogyne artiellia, the cereal and legume root-knot nematode, and Fusarium oxysporum f. sp. ciceris race 5 in chickpea. Phytopathology 93:1513–1523

Castro P, Pistón F, Madrid E, Millan T, Gil J, Rubio J (2010) Development of chickpea near-isogenic lines for *Fusarium* wilt. Theor Appl Genet 121:1519–1526

Chandirasekaran R, Warkentin TD, Gan Y, Shirtliffe S, Gossen BD, Tar'an B, Banniza S (2009) Improved sources of resistance to ascochyta blight in chickpea. Can J Plant Sci 89:107–118

Chaturvedi R, Singh IS, Gupta AK (1995) Inheritance of resistance to botrytis grey mould in chickpea (*Cicer arietinum* L.). Legum Res 18:1–4

Chen W, Coyne CJ, Peever TL, Muehlbauer FJ (2004) Characterization of chickpea differentials for pathogenicity assay of ascochyta blight and identification of chickpea accessions resistant to Didymella rabiei. Plant Pathol 53:759–769

Chen W, Schatz B, Henson B, McPhee KE, Muehlbauer FJ (2006) First report of Sclerotinia stem rot of chickpea caused by *Sclerotinia sclerotiorum* in North Dakota and Washington. Plant Dis 90:114

Chen L, Wu Q, He W, He T, Wu Q, Miao Y (2019) Combined *de novo* transcriptome and metabolome analysis of common bean response to *Fusarium oxysporum* f. sp. *phaseoli* infection. Int J Mol Sci 20:6278. https://doi.org/10.3390/ijms20246278

Chinchilla D, Zipfel C, Robatzek S, Kemmerling B, Nürnberger T, JDG J et al (2007) A flagellin-induced complex of the receptor FLS2 and BAK1 initiates plant defence. Nature 448:497–500

Cho S, Chen W, Muehlbauer FJ (2004) Pathotype-specific genetic factors in chickpea (Cicer arietinum L.) for quantitative resistance to ascochyta blight. Theor Appl Genet 109:733–739

Cobos M, Fernandez M, Rubio J, Kharrat M, Moreno M, Gil J et al (2005) A linkage map of chickpea (*Cicer arietinum* L.) based on populations from Kabuli ⍰ desi crosses: location of genes for resistance to *Fusarium* wilt race 0. Theor Appl Genet 110:1347–1353

Cobos MJ, Rubio J, Strange RN, Moreno MT, Gil J, Millan T (2006) A new QTL for Ascochyta blight resistance in an RIL population derived from an interspecific cross in chickpea. Euphytica 149:105–111

Cobos MJ, Winter P, Kharrat M, Cubero JI, Gil J, Millan T, Rubio J (2009) Genetic analysis of agronomic traits in a wide cross of chickpea. Field Crop Res 111(1–2):130–136

Collard BCY, Ades P, Pang E, Brouwer J, Taylor P (2001) Prospecting for sources of resistance to Ascochyta blight in wild Cicer species. Australas Plant Pathol 30:271–276

Collard BC, Pang EC, Ades PK, Taylor PW (2003) Preliminary investigation of QTLs associated with seedling resistance to ascochyta blight from Cicer echinospermum, a wild relative of chickpea. Theor Appl Genet 107:719–729

Coram TE, Pang ECK (2005a) Isolation and analysis of candidate ascochyta blight defence genes in chickpea. Part I. generation and analysis of an expressed sequence tag (EST) library. Physiol Mol Plant Pathol 66:192–200

Coram TE, Pang ECK (2005b) Isolation and analysis of candidate ascochyta blight defence genes in chickpea. Part II. Microarray expression analysis of putative defence-related ESTs. Physiol Mol Plant Pathol 66:201–210

Coram TE, Pang EC (2006) Expression profiling of chickpea genes differentially regulated during a resistance response to *Ascochyta rabiei*. Plant Biotechnol J 4:647–666

Cother EJ (1977) Isolation of important pathogenic fungi from seeds of *Cicer arietinum*. Seed Sci Technol 5:593–597

Cui H, Tsuda K, Parker JE (2015) Effector-triggered immunity: from pathogen perception to robust defense. Annu Rev Plant Biol 66:487–511

Daba K, Deokar A, Banniza S, Warkentin TD, Tar'an B (2016) QTL mapping of early flowering and resistance to Ascochyta blight in chickpea. Genome 59:413–425

Danehloueipour N, Yan G, Clarke H, Siddique K (2007) Diallel analyses reveal the genetic control of resistance to Ascochyta blight in diverse chickpea and wild Cicer species. Euphytica 154: 195–205

Danilevicz MF, Bayer PE, Nestor BJ, Bennamoun M, Edwards D (2021) Resources for image-based high-throughput phenotyping in crops and data sharing challenges. Plant Physiol 187: 699–715

Davidson JA, Pande S, Bretag TW, Lindbeck KD, Kishore GK (2004) Biology and management of *botrytis* spp. legume crops. In: Elad Y, Williamson B, Tudzynski P, Delen N (eds) Botrytis: biology, pathology and control. Kluwer Academic Publishers, Hingham, pp 295–318

Deokar A, Sagi M, Daba K, Tar'an B (2019a) QTL sequencing strategy to map genomic regions associated with resistance to ascochyta blight in chickpea. Plant Biotechnol J 17:275–288

Deokar A, Sagi M, Tar'an B (2019b) Genome-wide SNP discovery for development of high-density genetic map and QTL mapping of ascochyta blight resistance in chickpea (Cicer arietinum L). Theor Appl Genet 132(6):1861–1872

Dey SK, Singh G (1993) Resistance to Ascochyta blight in chickpea-genetic basis. Euphytica 68: 147–153

Dubey SC, Birendra S, Srinivasa N (2017) Evaluation of chickpea genotypes against Fusarium wilt for resistant sources. Indian Phytopathol 70:254–255

Dubey SC, Singh B (2003) Evaluation of chickpea genotypes against Ascochytablight. Indian Phytopathol 56:505

Duzdemir O, Selvi B, Yanar Y, Yıldırımı A (2014) Sources of resistance in chickpea (Cicer arietinum L.) land races against Ascochyta rabiei causal agent of Ascochyta blight disease. Pak J Bot 46(4):1479–1483

Garcia-Limones C, Dorado G, Navas-Cortes JA, Jimenez-Diaz RM, Tena M (2009) Changes in the redox status of chickpea roots in response to infection by *Fusarium oxysporum* f. sp. *ciceri*: apoplastic antioxidant enzyme activities and expression of oxidative stress-related genes. Plant Biol 11:194–203

Garg V, Khan AW, Kudapa H, Kale SM, Chitikineni A, Qiwei S, Sharma M, Li C, Zhang B et al (2019) Integrated transcriptome, small RNA and degradome sequencing approaches provide insights into Ascochyta blight resistance in chickpea. Plant Biotechnol J 17:914–931

Garg T, Mallikarjuna BP, Thudi M, Samineni S, Singh S, Sandhu JS et al (2018) Identification of QTLs for resistance to *Fusarium* wilt and Ascochyta blight in a recombinant inbred population of chickpea (*Cicer arietinum* L.). Euphytica 214:45

Gaur PM, Pande S, Upadhyaya HD, Rao BV (2006) Extra-large kabuli chickpea with high resistance to fusarium wilt. J SAT Agric Res 2(1):1–2

Gayacharan RU, Singh S, Basandrai AK, Rathee VK, Tripathi K et al (2020) Identification of novel resistant sources for ascochyta blight (*Ascochyta rabiei*) in chickpea. PLoS One 15:e0240589

Ghosh R, Sharma M, Telangre R, Pande S (2013) Occurrence and distribution of chickpea diseases in central and southern parts of India. Am J Plant Sci 4:940–944

Ghosh S, Watson A, Gonzalez-Navarro OE, Ramirez-Gonzalez RH, Yanes L, Mendoza-Suárez M (2018) Speed breeding in growth chambers and glasshouses for crop breeding and model plant research. Nat Protoc 13:2944–2963

Gowda SJM, Radhika P, Kadoo N, Mhase L, Gupta V (2009) Molecular mapping of wilt resistance genes in chickpea. Mol Breed 24:177–184

Grewal JS, Laha SK (1983) Chemical control of botrytis blight in chickpea. Indian Phytopathol 36: 516–520

Grewal JS, Pal M, Rewal N (1992) Botrytis gray mold of chickpea in India. In: Haware MP, Faris DG, Gowda CLL (eds) Botrytis gray mold of chickpea. ICRISAT, Patan-cheru, pp 6–8

Gupta S, Bhar A, Chatterjee M, Das S (2013) *Fusarium oxysporum* f.sp. *ciceri* race 1 induced redox state alterations are coupled to downstream defense signaling in root tissues of chickpea (*Cicer arietinum* L.). PLoS One 8:e73163

Gupta S, Chakraborti D, Rangi RK, Basu D, Das S (2009) A molecular insight into the early events of chickpea (*Cicer arietinum*) and *Fusarium oxysporum* f. sp. *ciceri* (race 1) interaction through cDNA-AFLP analysis. Phytopathology 99:1245–1257

Gupta O, Rathi M, Mishra M (2012) Screening for resistance against Rhizoctonia bataticola causing dry root rot in chickpea. J Food Legumes 25:139–141

Gurha SN, Singh G, Sharma YR (2003) Diseases of chickpea and their management. In: Ali M, Kumar S, Singh NB (eds) Chickpea research in India. Army Printing Press, Lucknow, pp 195–227

Gurjar G, Giri AP, Gupta VS (2012) Gene expression profiling during wilting in chickpea caused by *Fusarium oxysporum* f. sp. ciceri. Am J Plant Sci 3:190–201. https://doi.org/10.4236/ajps.2012.32023

Halila I, Cobos MJ, Rubio J, Millaìn T, Kharrat M, Marrakchi M et al (2009) Tagging and mapping a second resistance gene for fusarium wilt race 0 in chickpea. Eur J Plant Pathol 124:87–92

Halila MH, Harrabi MM (1990) Breeding for dual resistance to Ascochyta and wilt diseases in chickpea. Options Mediterr Ser Semin 9:163–170

Haware MP (1998) Diseases of chickpea. In: Allen DJ, Lenne JM (eds) The pathology of food and pasture legumes. CAB International, Wallingford, pp 473–516

Haware MP, Jimenez-Diaz RM, Amin KS, Phillips J, Halila H (1990) Integrated management of wilt and root rots of chickpea. In: Chickpea in the nineties: proceedings of the second international work shop on chickpea improvement, Patancheru, pp 129–137

Haware MP, McDonald D (1992) Integrated management of botrytis gray mold of chickpea. In: Haware MP, Faris DG, Gowda CLL (eds) Botrytis gray mold of chickpea. ICRISAT, Patancheru, pp 3–6

Haware MP, Nene YL (1982) Races of *Fusarium oxysporum* f. sp. ciceris. Plant Dis 66:809–810. https://doi.org/10.1094/PD-66-80

Haware MP, Nene YL, Natarajan M (1986) Survival of Fusarium oxysporum f. sp. cieer; in soil in absence of chickpea. Page I in Management of soilborne diseases of crop plants: Proceeding of National Seminar. Tamil Nadu Agricultural University, Coimbatore

Haware MP, Nene YL, Rajeshwari R (1978) Eradication of *Fusarium oxysporum* f. sp. ciceri transmitted in chickpea seed. Phytopathology 68:1364–1367. https://doi.org/10.1094/phyto-68-1364

Helms D, Panella L, Buddenhagen IW, Workneh F, Tucker CL, Foster KW (1992) Registration of UC15'chickpea. Crop Sci 32(2):500

Iftikhar K, Ilyas MB (2000) Screening of chickpea germplasm against dry root rot disease (*Macrophomina phaseolina*) in pots/glass house. Pak J Phytopathol 12:66–70

Iruela M et al (2007) Validation of a QTL for resistance to ascochyta blight linked to resistance to fusarium wilt race 5 in chickpea (*Cicer arietinum* L.). In: Tivoli B, Baranger A, Muehlbauer FJ, Cooke BM (eds) Ascochyta blights of grain legumes. Springer, Dordrecht

Jamil FF, Sarwar N, Sarwar M, Khan JA, Geistlinger J, Kahl G (2000) Genetic and pathogenic diversity within Ascochyta rabiei (pass.) lab. Populations in Pakistan causing blight of chickpea (Cicer arietinum L.). Physiol Mol. Plant Pathol 57(6):243–254

Jarolmasjed S, Sankaran S, Marzougui A, Kostick S, Si Y, Quirós Vargas JJ, Evans K (2019) High-throughput phenotyping of fire blight disease symptoms using sensing techniques in apple. Front Plant Sci 10:576

Jayalakshmi SK, Usharani S, Benagi VI, Mannur DM (2008) Sources of resistance to dry root rot of chickpea caused by *Rhizoctonia bataticola*. Agri Sci Digest 28:147–148

Jendoubi W, Bouhadida M, Millan T, Kharrat M, Gil J, Rubio J (2016) Identification of the target region including the Foc0 1/foc0 1 gene and development of near isogenic lines for resistance to *Fusarium* wilt race 0 in chickpea. Euphytica 210:119–133

Jha UC (2018) Current advances in chickpea genomics: applications and future perspectives. Plant Cell Rep 37:947–965

Jha UC, Bohra A, Pandey S, Parida S (2020) Breeding, genetics and genomics approaches for improving Fusarium wilt resistance in major grain legumes. Front Genet 11:1001. https://doi.org/10.3389/fgene.2020.01001

Jha UC, Jha R, Bohra A, Manjunatha L, Saabale PR, Parida SK, Chaturvedi SK, Thakro V, Singh NP (2021c) Association mapping of genomic loci linked with Fusarium wilt resistance (Foc2) in chickpea. Plant Gent Resour 19:195–202. https://doi.org/10.1017/S1479262121000228

Jha UC, Nayyar H, Mantri N, Siddique KHM (2021b) Non-coding RNAs in legumes: their emerging roles in regulating biotic/abiotic stress responses and plant growth and development. Cell 10:1674

Jha UC, Saabale PR, Chatuvedi SK, Singh NP (2021a) Advanced chickpea lines resistant against fusarium wilt (*Fusarium oxysporum*). Indian J Agric Sci 91(1)

Jha UC, Sharma KD, Nayyar H, Parida SK, Siddique KHM (2022) Breeding and genomics interventions for developing ascochyta blight resistant grain legumes. Int J Mol Sci 23:2217

Jiménez-Díaz RM, Alcala-Jimenez A, Hervás A, Trapero-Casas JL (1993) Pathogenic variability and host resistance in the Fusarium oxysporum f. sp. ciceris-Cicer arietinum pathosystem. Hodowla Roślin, Aklimatyzacja i Nasiennictwo 37(3)

Jiménez-Díaz RM, Castillo P, del Mar J-GM, Landa BB, Navas-Cortés JA (2015) Fusarium wilt of chickpeas: biology, ecology and management. Crop Prot 73:16–12

Jiménez-Díaz RM, Singh KB, Trapero-Casas A, Trapero-Casas JL (1991) Resistance in kabuli chickpeas to *Fusarium* wilt. Plant Dis 75:914–918. https://doi.org/10.1094/pd-75-0914

Jingade P, Ravikumar RL (2015) Development of molecular map and identification of QTLs linked to *Fusarium* wilt resistance in chickpea. J Genet 94:723–729. https://doi.org/10.1007/s12041-015-0589-7

Jukanti AK, Gaur PM, Gowda CL, Chibbar RN (2012) Nutritional quality and health benefits of chickpea (*Cicer arietinum* L.): a review. British J Nutri 108:S11–S26

Kaiser WJ, Alcalá-Jiménez AR, Hervás-Vargas A, Trapero-Casas JL, Jiménez-Díaz RM (1994) Screening wild species for resistance to races 0 and 5 of *Fusarium oxysporum* f. sp. *ciceris*. Plant Dis 78:962–967

Kanakala S, Kuria P (2018) Chickpea chlorotic dwarf virus: an emerging monopartite dicot infecting mastrevirus. Viruses 11:5

Karadi A, Samineni S, Sajja S, Sharma M, Thudi M, Mallikarjuna BP, Viswanatha KP, Varshney RK, Gaur PM (2021) Molecular mapping of dry root rot resistance genes in chickpea (*Cicer arietinum* L.). Euphytica 217:1–3

Kaur L, Sandhu JS, Malhotra RS, Imtiaz M, Sirari A (2012) Sources of stable resistance to Ascochyta blight in exotic kabuli chickpea. J Food Legumes 25(1):79–80

Kelly A, Alcalá-Jiménez AR, Bainbridge BW, Heale JB, Perez-Artes E, Jimenez-Diaz RM (1994) Use of genetic fingerprinting and random amplified polymorphic DNA to characterize pathotypes of Fusarium oxysporum f. sp. ciceris infecting chickpea. Phytopathology 84: 1293–1298

Khaliq A, Alam S, Khan IU, Khan D, Naz S, Zhang Y, Shah AA (2020) Integrated control of dry root rot of chickpea caused by *Rhizoctonia bataticola* under the natural field condition. Biotech Reps 25:e00423

Kimurto PK, Towett BK, Mulwa RS, Njogu N, Jeptanui LJ, Rao GNVPR, Silim S, Kaloki P, Korir P, Macharia JK (2013) Evaluation of chickpea genotypes for resistance to Ascochyta blight (*Ascochyta rabiei*) disease in the dry highlands of Kenya. Phytopathol Mediterr 52:212–221

Knights EJ, Siddique KHM (2002) Manifestation of *Botrytis cinerea* on chickpeas in Australia. In: Workshop proceedings integrated management of botrytis grey mould of chickpea in Bangladesh and Australia. Bangladesh Agricultural Research Institute, Joydebpur, Gazipur, pp 70–77

Kohli D, Joshi G, Deokar AA, Bhardwaj AR, Agarwal M, Katiyar-Agarwal S, Srinivasan R, Jain PK (2014) Identification and characterization of wilt and salt stress-responsive microRNAs in chickpea through high-throughput sequencing. PLoS One 9:e108851

Kottapalli P, Gaur PM, Katiyar SK, Crouch JH, Buhariwalla HK, Pande S, Gali KK (2009) Mapping and validation of QTLs for resistance to an Indian isolate of Ascochyta blight pathogen in chickpea. Euphytica 165:79–88

Kumar S (2006) Inheritance of resistance to *Fusarium* wilt (race 2) in chickpea. Plant Breed 117: 139–142. https://doi.org/10.1111/j.1439-0523.1998.tb01467.x

Kumar J, Haware MP, Smithson JB (1985) Registration of four short duration *Fusarium* wilt-resistant kabuli (garbanzo) chickpea germplasms. Crop Sci 25:576–577. https://doi.org/10.2135/cropsci1985.0011183x002500030047x

Kumar K, Purayannur S, Kaladhar VC, Parida SK, Verma PK (2018) mQTL-seq and classical mapping implicates the role of an AT-HOOK MOTIF CONTAINING NUCLEAR LOCALIZED (AHL) family gene in Ascochyta blight resistance of chickpea. Plant Cell Environ 41:2128–2140

Kumar Y, Zhang L, Panigrahi P, Dholakia BB, Dewangan V, Chavan SG, Kunjir SM, Wu X, Li N, Rajmohanan PR (2016) *Fusarium oxysporum* mediates systems metabolic reprogramming of chickpea roots as revealed by a combination of proteomics and metabolomics. Plant Biotechnol J 14:1589–1603

Kuswah A, Bhatia D, Rani U, Yadav IS, Singh I, Bharadwaj C, Singh S (2021) Molecular mapping of quantitative trait loci for ascochyta blight and botrytis grey mould resistance in an inter-specific cross in chickpea (*Cicer arietinum* L.) using genotyping by sequencing. Breed Sci 71: 229–239

Landa BB, Navas-Cortés JA, del Mar J-GM, Katan J, Retig B, Jiménez-Díaz RM (2006) Temperature response of chickpea cultivars to races of Fusarium oxysporum f. sp. ciceris, causal agent of Fusarium wilt. Plant Dis 90(3):365–374

Li Y, Ruperao P, Batley J, Edwards D, Davidson J, Hobson K, Sutton T (2017) Genome analysis identified novel candidate genes for Ascochyta blight resistance in chickpea using whole genome re-sequencing data. Front Plant Sci 8:359

MacLeod W, Sweetingham M (2000) Botrytis grey mould of chickpea. In: Regan K, White P, Siddique KHM (eds) Proceedings of pulse research and industry development in Western Australia. Agriculture Western Australia, Perth, pp 87–88

Madrid E, Rajesh PN, Rubio J, Gil J, Millán T, Chen W (2012) Characterization and genetic analysis of an EIN4-like sequence (CaETR-1) located in QTLAR1 implicated in ascochyta blight resistance in chickpea. Plant Cell Rep 31(6):1033–1042

Madrid E, Seoane P, Claros MG, Barro F, Rubio J, Gil J, Millan T (2014) Genetic and physical mapping of the QTLAR3 controlling blight resistance in chickpea (*Cicer arietinum* L). Euphytica 198:69–78

Mannur DM, Babbar A, Thudi M, Sabbavarapu MM, Roorkiwal M, Sharanabasappa BY et al (2019) Super Annigeri 1 and improved JG 74: two *Fusarium* wilt-resistant introgression lines developed using marker-assisted backcrossing approach in chickpea (*Cicer arietinum* L.). Mol Breed 39:2

Maryam A, Dubey SC (2009) Morphological and pathogenic variation among isolates of *Rhizoctonia bataticola* causing dry root rot of chickpea. Indian Phytopathol 62:183–189

Mayer M, Tullu A, Simon C, Kumar J, Kaiser W, Kraft J et al (1997) Development of a DNA marker for *Fusarium* wilt resistance in chickpea. Crop Sci 37:1625–1637

Meng X, Zhang S (2013) MAPK cascades in plant disease resistance signaling. Annu Rev Phytopathol 51:245–266. https://doi.org/10.1146/annurev-phyto-082712-102314

Meuwissen THE, Hayes BJ, Goddard ME (2001) Prediction of total genetic value using genome-wide dense marker maps. Genetics 157:1819–1829

Mitra M (1931) Report of the imperial mycologist. Sci Rep Agric Res Inst 1929–1930:58–71

Mohanty JK, Jha UC, Dixit GP, Parida SK (2022) Harnessing the hidden allelic diversity of wild Cicer to accelerate genomics-assisted chickpea crop improvement. Mol Biol Rep 16:1–9

Morales GJA (1986) Chickpea breeding programme in Sonora, Mexico. Int Chickpea Newsl 15:11–12

Murray G, Brennan J (2012) The current and potential costs from diseases of pulse crops in Australia; GRDC Research Code: CER00002. Grains Research and Development Corporation, Barton

Nakagami H, Pitzschke A, Hirt H (2005) Emerging MAP kinase pathways in plant stress signalling. Trends Plant Sci 10:339–346

Navas-Cortés JA, Hau B, Jiménez-Díaz RM (1998) Effect of sowing date, host cultivar, and race of Fusarium oxysporum f. sp. ciceris on development of Fusarium wilt of chickpea. Phytopathology 88:1338–1346

Nene YL, Haware MP (1980) Screening chickpea for resistance to wilt. Plant Dis 64:379–380. https://doi.org/10.1094/pd-64-379

Nene YL, Haware MP, Reddy MV, Philps JP, Castro EL, Kotasthane SR, Gupta OM, Singh G, Shukia P, Sah RP (1989) Identification of broad based and stable resistance to wilt and root-rots in chickpea. Indian Phytopathol 42(4):499–505

Nene YL, Reddy MV (1987) Chickpea diseases and their control. In: Saxena MC, Singh KB (eds) Chickpea diseases and their control, pp 233–270

Nene YL, Shelia VK, Sharma SB (1996) A world list of chickpea and pigeonpea pathogens, 5th edn. International Crops Research Institute for Semi-Arid Tropics, Hyderabad, p 27

Newman TE, Jacques S, Grime C, Kamphuis FL, Lee RC, Berger J, Kamphuis LG (2021) Identification of novel sources of resistance to ascochyta blight in a collection of wild Cicer accessions. Phytopathology 111:369–379

O'Neill BC, Dalton M, Fuchs R, Jiang L, Pachauri S, Zigova K (2010) Global demographic trends and future carbon emissions. Proc Natl Acad Sci 107:17521–17526

Pande S, Galloway JJ, Gaur PM, Siddique KHM, Tripathi HS, Taylor P, MacLeod MWJ, Basandrai AK, Bakr A, Joshi S, Krishna Kishore G, Isenegger DA, Narayana Rao J, Sharma M (2006b) Botrytis grey mould of chickpea: a review of biology, epidemiology, and disease management. Aust J Agric Res 57:1137–1150

Pande S, Kishore GK, Rao JN (2004) Evaluation of chickpea lines for resistance to dry root rot caused by *Rhizoctonia bataticola*. Int Chickpea Pigeonpea Newsl 11:37–38

Pande S, Kishore GK, Upadhyaya HD, Rao JN (2006a) Identification of sources of multiple disease resistance in mini-core collection of chickpea. Plant Dis 90:1214–1218

Pande S, Sharma M, Gaur PM, Basandrai AK, Kaur L, Hooda KS, Basandrai D, Kiran Babu T, Jain SK, Rathore A (2013) Biplot analysis of genotype × environment interactions and identification of stable sources of resistance to Ascochyta blight in chickpea (Cicer arietinum L.). Australas Plant Pathol 42:561–571

Pande SK, Siddique KH, Kishore GK, Bayaa B, Gaur PM, Gowda CL, Bretag TW, Crouch JH (2005) Ascochyta blight of chickpea (Cicer arietinum L.): a review of biology, pathogenicity, and disease management. Austr J Agric Res 56:317–332

Pande S, Singh G, Rao JN, Bakr MA, Chaurasia PCP, Joshi S, Johansen C, Singh SD, Kumar J, Rahman MM, Gowda CLL (2002) Integrated management of botrytis gray mold of chickpea. Information bulletin no. 61. ICRISAT, Andhra Pradesh

Parker JE, Hessler G, Cui H (2022) A new biochemistry connecting pathogen detection to induced defense in plants. New Phytol 234:819–826

Patil BS, Ravikumar RL, Bhat JS, Soregaon D (2014) Molecular mapping of QTLs for resistance to early and late *Fusarium* wilt in chickpea. Czech J Genet Plant Breed 50:171–176

Pratap A, Chaturvedi SK, Tomar R, Rajan N, Malviya N, Thudi M et al (2017) Marker-assisted introgression of resistance to *Fusarium* wilt race 2 in Pusa 256, an elite cultivar of desi chickpea. Mol Gen Genomics 293:1237–1245

Rai A, Irulappan V, Senthil-Kumar M (2022) Dry root rot of chickpea: a disease favored by drought. Plant Dis 106:346–356

Rao PA, Haware MP (1987) Inheritance of dry root rot (*Rhizoctonia bataticola*) resistance in chickpea (*Cicer arietinum* L.). Plant Breed 98:349–352

Rashid MH, Hossain MA, Kashem MA, Kumar S, Rafii MY, Latif MA (2014) Efficacy of combined formulations of fungicides with different modes of action in controlling botrytis gray mold disease in chickpea. Sci World J 2014:639246

Ray DK, Mueller ND, West PC, Foley JA (2013) Yield trends are insufficient to double global crop production by 2050. PLoS One 8:e66428

Rewal S, Grewal JS (1989) Inheritance of resistance to *Botrytis cinerea* Pers in *Cicer arietinum* L. Euphytica 44:61–63

Rousseau C, Belin E, Bove E, Rousseau D, Fabre F, Berruyer R, Guillaumès J, Manceau C, Jacques MA, Boureau T (2013) High throughput quantitative phenotyping of plant resistance using chlorophyll fluorescence image analysis. Plant Methods 9:1–3

Rubio J, Hajj-Moussa E, Kharrat M, Moreno MT, Millan T, Gil J (2003) Two genes and linked RAPD markers involved in resistance to *Fusarium oxysporum* f. sp. *ciceris* race 0 in chickpea. Plant Breed 122:188–191

Rubio J, Moreno MT, Moral A, Rubiales D, Gil J (2006) Registration of RIL58-ILC72/Cr5, a chickpea germplasm line with rust and Ascochyta blight resistance. Crop Sci 46(5):2331

Sabbavarapu MM, Sharma M, Chamarthi SK, Swapna N, Rathore A, Thudi M, Gaur PM, Pande S, Singh S, Kaur L (2013) Molecular mapping of QTLs for resistance to Fusarium wilt (race 1) and Ascochyta blight in chickpea (*Cicer arietinum* L.). Euphytica 193:121–133

Sagi MS, Deokar AA, Tar'an B (2017) Genetic analysis of NBS-LRR gene family in chickpea and their expression profiles in response to Ascochyta blight infection. Front Plant Sci 8:838

Santra DK, Tekeoglu M, Ratnaparkhe M, Kaiser WJ, Muehlbauer FJ (2000) Identification and mapping of QTLs conferring resistance to ascochyta blight in chickpea. Crop Sci 40(6): 1606–1612

Shah TM, Atta BM, Mirza JI, Haq MA (2009) Screening of chickpea (Cicer arietinum) induced mutants against Fusarium wilt. Pak J Bot 41:1945–1955

Shahid AA, Husnain T, Riazuddin S (2008) Ascochyta blight of chickpea: production of phytotoxins and disease management. Biotechnol Adv 26(6):511–515

Sharath Chandran US, Tarafdar A, Mahesha HS, Sharma M (2021) Temperature and soil moisture stress modulate the host defense response in chickpea during dry root rot incidence. Front Plant Sci 12:932

Sharma M, Babu TK, Gaur PM, Ghosh R, Rameshwar T, Chaudhary RG, Upadhyay JP, Gupta O, Saxena DR, Kaur L, Dubey SC (2012) Identification and multi-environment validation of resistance to Fusarium oxysporum f. sp. ciceris in chickpea. Field Crop Res 135:82–88

Sharma KD, Chen W, Muehlbauer FJ (2005) Genetics of chickpea resistance to five races of *Fusarium* wilt and a concise set of race differentials for *Fusarium oxysporum* f. sp. *ciceris*. Plant Dis 89:385–395. https://doi.org/10.1094/pd-89-0385

Sharma M, Ghosh R (2017) Heat and soil moisture stress differentially impact chickpea plant infection with fungal pathogens. In: Senthil-Kumar M (ed) Plant tolerance to individual and concurrent stresses. Springer, New Delhi, pp 47–57

Sharma M, Ghosh R, Pande S (2016) Dry root rot [*Rhizoctonia bataticola* (Taub.) Butler] an emerging disease of chickpea—where do we stand? Arch Phytopathol Plant Protect 48:1–16

Sharma M, Ghosh R, Tarafdar A, Rathore A, Kumar AV, Chobe DR et al (2019) Exploring the genetic cypher of chickpea (*Cicer arietinum* L.) through identification and multi-environment validation of resistant sources against *Fusarium* wilt (*Fusarium oxysporum* f. sp *ciceris*). Front Sust Food Syst 3:78. https://doi.org/10.3389/fsufs.2019.00078

Sharma KD, Muehlbauer FJ (2007) Fusarium wilt of chickpea: physiological specialization, genetics of resistance and resistance gene tagging. Euphytica 157(1):1–4

Sharma M, Pande S (2013) Unravelling effects of temperature and soil moisture stress response on development of dry root rot [Rhizoctonia bataticola (Taub.)] Butler in chickpea. Am J Plant Sci 4(3):584–589

Sharma M, Pande S, Mangala UN, Tripathi S, Gaur PM (2010) Genetic resistance in desi and kabuli chickpea lines to Fusarium wilt caused by Fusarium oxysporum f. sp. ciceris. Indian J Plant Prot 38(1):57–62

Sharma KD, Winter P, Kahl G, Muehlbauer FJ (2004) Molecular mapping of *Fusarium oxysporum* f. sp. *ciceris* race 3 resistance gene in chickpea. Theor Appl Genet 108:1243–1256

Siddique KH, Erskine W, Hobson K, Knights EJ, Leonforte A, Khan TN, Paull JG, Redden R, Materne M (2013) Cool-season grain legume improvement in Australia—use of genetic resources. Crop Pasture Sci 64:347–360

Silvia Sebastiani M, Bagnaresi P, Sestili S, Biselli C, Zechini A, Orrù L, Cattivelli L, Ficcadenti N (2017) Transcriptome analysis of the melon-*Fusarium oxysporum* f. sp. *melonis* race 1.2 pathosystem in susceptible and resistant plants. Front Plant Sci 8:362

Singh R, Kumar K, Purayannur S, Chen W, Verma PK (2022) *Ascochyta rabiei*: a threat to global chickpea production. Mol Plant Pathol 23:1241–1261

Singh H, Kumar J, Smithson JB, Haware MP (1987) Complementation between genes for resistance to race 1 of *Fusarium oxyspomm* f. sp. *ciceri* in chickpea. Plant Pathol 36:539–543. https://doi.org/10.1111/j.1365-3059.1987.tb02271.x

Singh DV, Lal S, Singh SN (1974) Breeding gram (Cicer arietinum L.) for resistance to wilt. Indian J genet Plant Breed 34:267–270

Singh KB, Ocampo B, Robertson LD (1998) Diversity for abiotic and biotic stress resistance in the wild annual *Cicer* species. Genet Resour Crop Evol 45:9–17

Singh KB, Reddy MV (1983) Inheritance of resistance to Ascochyta blight in chickpea. Crop Sci 23:9–10

Singh KB, Reddy MV (1989) Genetics of resistance to Ascochyta blight in four chickpea lines. Crop Sci 29:657–659

Singh K, Reddy M (1993) Sources of resistance to Ascochyta blight in wild Cicer species. Neth J Plant Pathol 99:163–167

Sinha R, Irulappan V, Patil BS, Reddy PC, Ramegowda V, Mohan-Raju B, Rangappa K, Singh HK, Bhartiya S, Senthil-Kumar M (2021) Low soil moisture predisposes field-grown chickpea plants to dry root rot disease: evidence from simulation modeling and correlation analysis. Sci Rep 11: 1–2

Stephens A, Lombardi M, Cogan NO, Forster JW, Hobson K, Materne M, Kaur S (2013) Genetic marker discovery, intraspecific linkage map construction and quantitative trait locus analysis of Ascochyta blight resistance in chickpea (Cicer arietinum L.). Mol Breed 33(2):297–313

Stuteville DL, Graves WL, Dixon LJ, Castlebury LA, Minnis AM (2010) *Uromyces ciceris-arietini*, the cause of chickpea rust: new hosts in the Trifolieae, Fabaceae. Plant Dis 94:293–297

Sudheesh S, Kahrood HV, Braich S, Dron N, Hobson K, Cogan NO, Kaur S (2021) Application of genomics approaches for the improvement in Ascochyta blight resistance in chickpea. Agronomy 11:1937

Talekar SC, Lohithaswa HC, Viswanatha KP (2017) Identification of resistant sources and DNA markers linked to genomic region conferring dry root rot resistance in chickpea (*Cicer arietinum* L.). Plant Breed 136:161–166

Tar'an B, Warkentin TD, Tullu A, Vandenberg A (2007) Genetic mapping of ascochyta blight resistance in chickpea (Cicer arietinum L.) using a simple sequence repeat linkage map. Genome 50(1):26–34

Tar'an B, Warkentin TD, Vandenberg A (2013) Fast track genetic improvement of ascochyta blight resistance and double podding in chickpea by marker-assisted backcrossing. Theor Appl Genet 126:1639–1647

Tarafdar A, Rani TS, Chandran US, Ghosh R, Chobe DR, Sharma M (2018) Exploring combined effect of abiotic (soil moisture) and biotic (*Sclerotium rolfsii* Sacc.) stress on collar rot development in chickpea. Front. Plant Sci 9:1154

Tar'an B, Warkentin T, Somers DJ, Miranda D, Vandenberg A, Blade S, Woods S, Bing D, Xue A, DeKoeyer D, Penner G (2003) Quantitative trait loci for lodging resistance, plant height and partial resistance to mycosphaerella blight in field pea (Pisum sativum L.). Theor Appl Genet 107(8):1482–1491

Tekeoglu M, Tullu A, Kaiser WJ, Muehlbauer FJ (2000) Inheritance and linkage of two genes that confer resistance to Fusarium wilt in chickpea. Crop Sci 40:1247–1256

Thudi M, Khan AW, Kumar V, Gaur PM, Katta K, Garg V et al (2016) Whole genome resequencing reveals genome-wide variations among parental lines of 16 mapping populations in chickpea (*Cicer arietinum*). BMC Plant Biol 16:10. https://doi.org/10.1186/s12870-015-0690-3

Tivoli B, Banniza S (2007) Comparison of the epidemiology of ascochyta blights on grain legumes. In: Ascochyta blights of grain legumes. Springer, Dordrecht, pp 59–76

Tiwari SK, Pandey MP, Pandya BP, Chaube HS (1985) Inheritance of resistance to botrytis grey mould in chickpea. Int Chickpea Newsl 12:11–12

Trapero-Casas A, Kaiser WJ (1992) Development of *Didymella rabiei*, the teleomorph of *Ascochyta rabiei*, on chickpea straw. Phytopathology 82:1261–1266

Trapero-Casas A, Kaiser WJ (2007) Differences between ascospores and conidia of *Didymella rabiei* in spore germination and infection of chickpea. Phytopathology 97:1600–1607

Trapero-Casas A, Navas-Cortés JA, Jimenez-Diaz RM (1996) Airborne ascospores of *Didymella rabiei* as a major primary inoculum for Ascochyta blight epidemics in chickpea crops in southern Spain. Eur J Plant Pathol 102:237–245

Tullu A, Kaiser WJ, Kraft JM, Muehlbauer FJ (1999) A second gene for resistance to race 4 of Fusarium wilt in chickpea and linkage with a RAPD marker. Euphytica 109(1):43–50

Tullu A, Muehlbauer FJ, Simon CJ, Mayer MS, Kumar J, Kaiser WJ, Kraft JM (1998) Inheritance and linkage of a gene for resistance to race 4 of fusarium wilt and RAPD markers in chickpea. Euphytica 102:227–232

Udupa SM, Baum M (2003) Genetic dissection of pathotype-specific resistance to ascochyta blight disease in chickpea (Cicer arietinum L.) using microsatellite markers. Theor Appl Genet 106(7): 1196–1202

Udupa SM, Weigand F, Saxena MC, Kahl G (1998) Genotyping with RAPD and microsatellite markers. Theoret Applied Genet 97:299–307. https://doi.org/10.1007/s001220050899

Upadhyaya H, Haware M, Kumar J, Smithson J (1983a) Resistance to wilt in chickpea. I. Inheritance of late-wilting in response to race 1. Euphytica 32:447–456

Upadhyaya H, Smithson J, Haware M, Kumar J (1983b) Resistance to wilt in chickpea. Further evidence for two genes for resistance to race 1. Euphytica 32:749–755

Upasani ML, Gurjar GS, Kadoo NY, Gupta VS (2016) Dynamics of colonization and expression of pathogenicity related genes in Fusarium oxysporum f. sp. ciceri during chickpea vascular wilt disease progression. PLoS One 11(5):e0156490

Upasani ML, Limaye BM, Gurjar GS, Kasibhatla SM, Joshi RR, Kadoo NY, Gupta VS (2017) Chickpea-*Fusarium oxysporum* interaction transcriptome reveals di erential modulation of plant defense strategies. Sci Rep 7:7746

van Rheenen HA, Reddy MV, Kumar J, Haware MP (1992) Breeding for resistance to soil-borne diseases of chickpea. In: Singh KB, Saxena MC (eds) Disease resistance breeding in chickpea. International Center for Research in dry areas (ICARDA), Aleppo, p 89e112

Varshney R, Mohan S, Gaur P, Chamarthi S, Singh V, Srinivasan S et al (2014) Marker-assisted backcrossing to introgress resistance to *Fusarium* wilt race 1 and Ascochyta blight in C 214, an elite cultivar of chickpea. Plant Genome 7:278–289

Varshney RK, Roorkiwal M, Sun S, Bajaj P, Chitikineni A, Thudi M, Singh NP, Du A et al (2021) A chickpea genetic variation map based on the sequencing of 3,366 genomes. Nature 599:622–627

Varshney RK, Thudi M, Roorkiwal M, He W, Upadhyaya HD, Yang W, Bajaj P, Cubry P, Rathore A, Jian J et al (2019) Resequencing of 429 chickpea accessions from 45 countries provides insights into genome diversity, domestication and agronomic traits. Nat Genet 51:857–864

Venette JR, Stack RW (1987) First report of rust (*Uromyces ciceris-arietini*) on garbanzo (chickpea) in the United States. Plant Dis 71:101

Vir S, Grewal JS, Gupta VP (1975) Inheritance of resistance to Ascochyta blight in chickpea. Euphytica 24:209–211

Vishwadhar CSG (2001) Disease resistance in pulse crop-cur- rent status and future approaches. In: Nagarajan S, Singh DP (eds) The role of resistance in intensive agriculture. Kalyani, New Delhi, pp 144–157

Winter P, Benko-Iseppon AM, Hüttel B, Ratnaparkhe MI, Tullu AB, Sonnante GA, Pfaff TH, Tekeoglu MU, Santra DI, Sant VJ, Rajesh PN (2000) A linkage map of the chickpea (Cicer arietinum L.) genome based on recombinant inbred lines from a C. arietinum× C. reticulatum cross: localization of resistance genes for fusarium wilt races 4 and 5. Theor Appl Genet 101: 1155–1163

Xue R, Wu J, Zhu Z, Wang L, Wang X, Wang S et al (2015) Differentially expressed genes in resistant and susceptible common bean (*Phaseolus vulgaris* L.) genotypes in response to *Fusarium oxysporum* f. sp. *phaseoli*. PLoS One 10:e0127698

Yuan MH, Ngou BPM, Ding PT, Xiu-Fan X (2021) PTI-ETI crosstalk: an integrative view of plant immunity. Curr Opin Plant Biol 62:102030

Zhang C, Chen W, Sankaran S (2019) High-throughput field phenotyping of Ascochyta blight disease severity in chickpea. Crop Prot 125:104885

Zhang H, Gao Z, Zheng X, Zhang Z (2012) The role of G-proteins in plant immunity. Plant Signal Behav 7:1284–1288. https://doi.org/10.4161/psb.21431

Zhang J, Zhou JM (2010) Plant immunity triggered by microbial molecular signatures. Mol Plant 3: 783–793

Zipfel C, Robatzek S (2010) Pathogen-associated molecular pattern triggered immunity: veni, vidi… ? Plant Physiol 154:551–554

Ascochyta Blight of Chickpea: A Menace to Be Managed by Resistance Breeding

2

Shayla Bindra, Shubham Sharma, Arpana Sharma, Upasana Rani, Inderjit Singh, Chellapilla Bharadwaj, and Sarvjeet Singh

Abstract

Chickpea (*Cicer arietinum*) is cultivated in more than 50 countries and is one of the most valued legumes due to its nutritional content. Ascochyta blight (AB) is a major disease that significantly affects crop yield leading to a large gap between demand and production. Approaches for disease management like cultural practices and chemical control either have limited effectiveness or are not ecofriendly. The only available environment-friendly approach to improve crop resistance with complete efficacy is breeding resistant genotypes. The vital prerequisite for sustainable agricultural production is the development of durable host resistance. Owing to the diversity of the pathogen population and prevalence of partial resistance in known sources of resistance, chickpea is susceptible to several races of *Ascochyta rabiei*. Hence, it is challenging to breed varieties with effective and stable resistance. Recent advances in the genetic and genomic know-hows have provided better understanding of the complex host-pathogen interactions. In addition, several AB-resistant gene(s)/QTL/genomic regions have been identified on various linkage groups. These genomic resources could be precisely utilized in genomic-assisted breeding by the plant breeders to develop and/or transfer AB-resistant genomic regions to elite cultivars.

S. Bindra (✉) · S. Sharma · U. Rani · I. Singh · S. Singh
Department of Plant Breeding and Genetics, Punjab Agricultural University, Ludhiana, Punjab, India
e-mail: shaylabindra@pau.edu; shubham-2014009@pau.edu; upasanarani@pau.edu; inderjitpb@pau.edu; sarvjeet62@pau.edu

A. Sharma
Department of Plant Pathology, Punjab Agricultural University, Ludhiana, Punjab, India
e-mail: arpana-2015010@pau.edu

C. Bharadwaj
Division of Genetics, ICAR-Indian Agricultural Research Institute, New Delhi, India

U. C. Jha et al. (eds.), *Diseases in Legume Crops*,
https://doi.org/10.1007/978-981-99-3358-7_2

Keywords

Chickpea · Disease · Ascochyta blight · Screening · Resistance · Breeding

2.1 Introduction

Chickpea (*Cicer arietinum*), aka garbanzo bean and Bengal gram, is the second most important *rabi* (winter) pulse crop after the common bean (*Phaseolus vulgaris*). It originated in the southern Caucasus and northern Persia. As revealed from Middle Eastern archaeological sites dated to the eighth millennium BC, it is one of the first legume crops domesticated and cultivated by humans (Zohary and Hopf 2006). The cultivation of chickpea spreads over a wide range of environmental conditions such as tropical, subtropical as well as temperate regions of the world. On the basis of its seed shape, size, and color, chickpea is categorized into two distinct classes, namely, *desi* and *kabuli* (Rao and Van den Maesen 1985). Chickpeas are a rich source of carbohydrates and proteins with starch content ranging from 41–50% to 12.4–31.5% crude protein (Hirdyani 2014). It is also a source of vitamins like riboflavin (B_2) and pantothenic acid (B_5) in addition to minerals chiefly including calcium, iron, zinc, copper, and potassium (Lebiedzińska and Szefer 2006). Phytosterols present in chickpea seeds are known to exhibit antifungal, anti-inflammatory, anti-bacterial, and anti-turmeric properties (Jukanti et al. 2012). Seeds of chickpea also contain polyunsaturated fats, beneficial in lowering the cholesterol level of the blood stream (Pittaway et al. 2008).

Though chickpea is cultivated worldwide, India is the largest producer, as it alone contributes about 90% to the global area with an average productivity of 1088.3 kg/ha. Overall, the global productivity of 1057.8 kg/ha level of chickpea is sufficiently low compared to food grain crops such as wheat with global productivity of 3491.9 kg/ha and national productivity of 3466.9 kg/ha (FAOSTAT 2022). The probable reason for the low production and productivity of chickpea are the various biotic and abiotic factors restricting chickpea crop growth and production. Among the biotic stresses, chickpea production is largely constrained by *Ascochyta rabiei* (Pass.) Labr. (teleomorph: *Didymella rabiei* (Kov.) V. Arx). As, ascochyta blight (AB) caused by *A. rabiei* infects all the aboveground parts during vegetative, flowering, and podding stages. The yield losses could be as high as up to 100% with poor seed quality (Sharma et al. 2013) or no seed set. The disease has been reported from Algeria, Australia, Afghanistan, Bangladesh, Bulgaria, China, Colombia, Canada, Cyprus, Egypt, Ethiopia, France, Greece, Hungary, India, Israel, Italy, Iran, Iraq, Jordan, Kenya, Libya, Mexico, Morocco, Pakistan, Portugal, Romania, Sudan, Syria, Spain, Tanzania, Tunisia, Turkey, the USA, and the former USSR (Nene and Reddy 1987; Nene et al. 1996; Khan et al. 1999). AB was identified in India for the very first time in the former Punjab province of British India in the year 1911 (Butler 1918), and since then it has spread to the entire Indian subcontinent covering all the chickpea growing areas. In India, the best documented epidemics were observed in northern states from 1981 to 1983 (Singh et al. 1984b),

2014 to 2015 (Shah et al. 2015), and 2018 to 2020 (Iqbal et al. 2018). The major reason for the outbreak of epidemic condition is the breakdown of resistance in available cultivars by *A. rabiei*.

Such loss in yield cannot be effectively managed by cultural practices and chemical control. One of the most effective, cost-efficient, and environment-friendly approache to manage AB is the development and deployment of resistant cultivars. With the advancement in biotechnological approaches, modern breeding techniques have eased the development of resistant cultivars by significantly reducing the duration of the development process. Not only this but they have even aided in the identification of quantitative trait loci (QTL) with an insight into precise location by means of biparental mapping and genome-wide association studies (GWAS). In the recent past, approaches such as next-generation sequencing, whole-genome resequencing (WGRS), pangenome assembly, and RNA-seq have enabled the construction of the genome assemblies, detection of AB-resistant candidate gene(s), and their plausible functions. The chapter describes the AB and its causal organism and management by conventional and modern breeding tools.

2.2 AB Causal Organism and its Symptoms

Ascochyta blight is also known as gram blight, chickpea blight, anthracnose, rabia, or scorch of chickpea; ascochytosis is caused by *Ascochyta rabiei* (Pass.) Labr, a necrotrophic fungus belonging to the class *Dothideomycetes*, order *Pleosporales*, and family *Didymellaceae*. Artificial inoculation of *A. rabiei* on lentil, fieldpea, vetch, normal bean, and cowpea uncovered that the pathogen is pathogenic to all these species (Zachos et al. 1963; Nene and Reddy 1987; Khan et al. 1999). It exists both as an anamorph and a teleomorph on chickpea crop. In the anamorph stage, round or pear-shaped black fruiting bodies called pycnidia are formed. A pycnidium contains several hyaline unicellular and occasionally bicellular spores termed as pycnidiospores, or conidia, developed on short conidiophores (stalks) embedded in a mucilaginous mass. These pycnidiospores are straight, oval to oblong, or slightly bent at one or both ends and measure around 6–12 by 4–6 μm (Nene 1982). The teleomorph, *Didymella rabiei* (Kovacheski) var. Arx (Syn. *Mycosphaerella rabiei* Kovacheski), is a bipolar heterothallic ascomycete characterized by the development of pseudothecia on chickpea crop residues over-wintering in the field. For successful sexual reproduction, the teleomorph requires pairing of two compatible mating types (MAT1-1 and MAT1-2), which are widely distributed in several major chickpea-growing areas of the world (Haware 1987; Kaiser 1997; Armstrong et al. 2001). Pseudothecia are dark brown to black, subglobose, 120–270 μm in diameter, erupting from the host tissue, and without a conspicuous ostiole. Binucleate asci are cylindrical to subclavate surrounded by paraphyses and contain 8 hyaline unequally bicellular ascospores. Ascospores are ellipsoid to biconic with a constriction at the septum and measure 9.5–16 by 4.5–7 μm (Crociara et al. 2022). The fungus grows readily on a variety of nutrient media, the best being chickpea meal dextrose agar. *A. rabiei* generally produces a pale cream-colored mycelium in which pale brown to black pycnidia are immersed. Cultures are variable in morphology and color, with isolates often producing a prevalence of unicellular conidia (CAB International 2000).

Talking about the symptoms, *A. rabiei* affects all aboveground parts of the host plant due to rapid cell collapse and spread of necrotic lesions, defoliation, and breakage followed by death of the plant (Singh et al. 2022a). In the field, it initially appears as small patches (foci) of blighted plants, rapidly spread over night across entire fields under favorable environmental conditions (Pande et al. 2005). Nene and Reddy (1987) reported that disease via airborne inoculum results in little, necrotic spots in the young leaves. Under ideal epiphytotic conditions, the spots rapidly enlarge and amalgamate to form round, brown spots bearing pycnidia organized in concentric rings on leaflets resulting in necrosis of young leaves (Spoel et al. 2007). On shoots and petiole, the lesions appear as elongated and irregular with several pycnidia. The lesions that develop on stems differ in size and in later stages girdle the affected plant parts. The area above the girdled portion is killed and even breaks off (Manjunatha et al. 2018). The circular brown spots with numerous pycnidia arranged in concentric rings develop on pods. The fungus penetrates the pod wall and infects the seed resulting in shriveled, brown discolored seeds (Sally 2005; Pande et al. 2010; Islam et al. 2017). In other plant parts such as petioles, leaflets, and immature branches, the symptoms of the disease initially appear as loss of turgor and epinasty, followed by water soaking and necrosis (Alam et al. 1989).

2.3 Prevalent *Ascochyta rabiei* Races

To unravel the races prevalent for a pathogen, periodic race profile analysis is required as they keep on coevolving with plant pathosystem. Hence, race profile analysis of plant pathogen population is necessary to understand co-evolving plant pathosystems. Otherwise, to breed resistant chickpea cultivars against chickpea blight is very challenging owing to the incessant evolutionary behavior of the pathogen, thereby necessitating the determination of physiological races for the development of resistant cultivars. In India, Singh (1990) identified 12 races (3072, 4080, 3844, 3492, 3522, 3968, 4064, 3968, 3560, 3744, 3904, 4088, and 1744) of *A. rabiei* by means of 12 differential chickpea lines. Approximately 8 pathotypes were identified by 16 differential lines by Basandrai et al. (2005). Baite and Dubey (2018) identified 7 races using 10 differential lines (ICC11879, ICC4991, ICC3996, ICC15978, ICC1467, ICC1903, ICC1527, H00108, GL26054, and GPF2) (Table 2.1).

Table 2.1 Indian race scenario of *A. rabiei*

Collection site	No. of isolates studied	Classification	No. of differential lines	References
India	50	2 races	5 (1-13, EC26435, C235, F-8, and V-138)	Vir and Grewal (1974)
India (North Indian states)	348	12 races	12 (L550, C235, ICC5124, C8, NEC138-C, JM595, ICC76, ICC7000, ICC7002, *Cicer pinnatifidum*, ICC2165 and ICC1467)	Singh (1990)
India	11	5 races	7 (P1343-1, P5292-1, C235, V-138, ILC1929, ILC249, and I-13)	Singh and Pal (1993)
India	348	12 races	12 (L550, C235, ICC5124, C8, NEC138-C, JM595, ICC76, ICC7000, ICC7002, *Cicer pinnatifidum*, ICC2165 and ICC1467)	Singh and Sharma (1998)
India	16	6 races	16 (ICC12, ICC607, ICC2165, ICC3918, ICC4200, ICC4475, ICC5124, ICC6306, ICC7002, ICC13754, ICC14911, ICCX810800, ICCX910028-39ABR-BP-10ABR-BP, ILC3870, FLIP 82–258, and Pb7 [ICC4991])	Basandrai et al. (2005)
India	25	7 races	10 (ICC11879, ICC4991, ICC3996, ICC15978, ICC1467, ICC1903, ICC1527, H00108, GL26054, and GPF2)	Baite and Dubey (2018)

2.4 Epidemiology of the Disease

2.4.1 Mode of Spread and Survival

Chickpea crop is disease-ridden via seed-borne, soil-borne, and air-borne infections (Chattopadhyay et al. 2001). *A. rabiei* spreads via primary and secondary sources of inoculum. Contaminated seeds and ascospores aid as a major source of primary inoculum whereas conidia act as a secondary source of inoculum (Dey and Singh 1994; Kimber et al. 2007). This pathogen survives 13 years on seeds stored at 4 °C (Kaiser 1997) and 5 months at 10–35 °C (Singh et al. 1995). During the growing season, conidia formed in pycnidia are assumed to disband over a short distance (~10 m) in a combination of strong wind and rain splashes (Khaliq et al. 2020). Consequently, the disease blowout mainly ensues over the anthropogenic movement of seeds, dissemination of spores via wind and water, and infected chickpea rubble (Singh et al. 2022a) (Fig. 2.1).

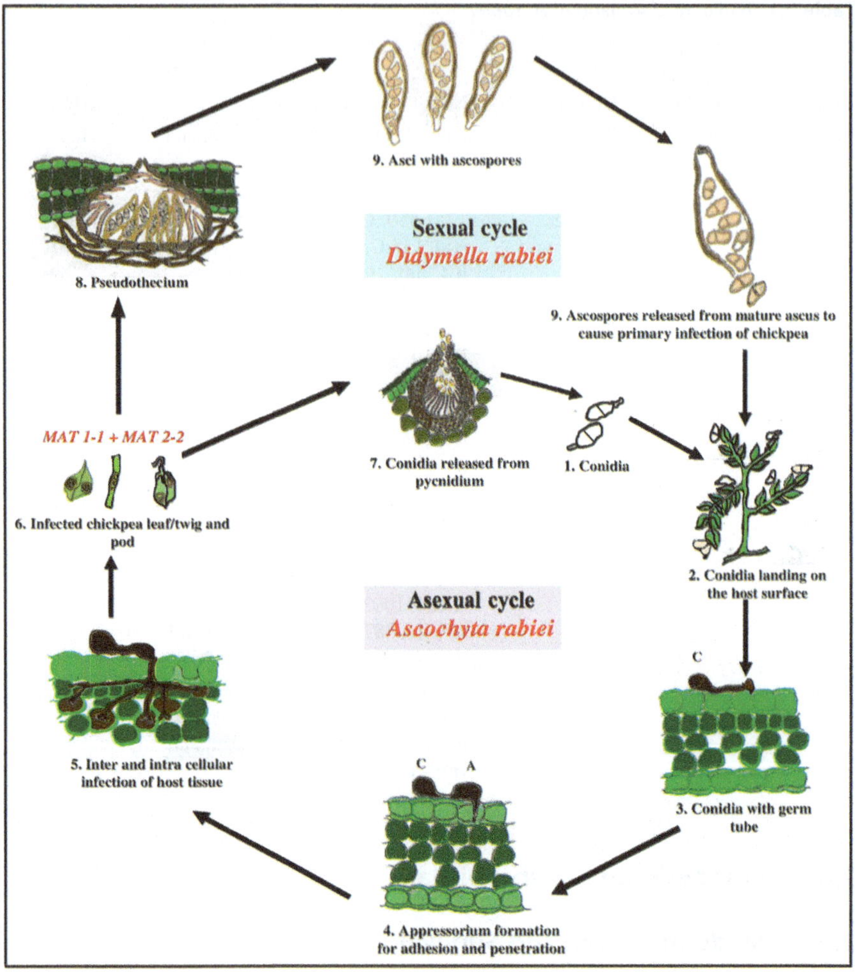

Fig. 2.1 Life cycle of *Ascochyta rabiei* (asexual) and *Didymella rabiei* (sexual) causing ascochyta blight. In asexual cycle: (1) Conidia (C) from infected chickpea plant (2) comes in contact with healthy host tissue (3) followed by germination and (4) appressorium (A) formation at the tip of hyphae/germ tube for adhesion and penetration by rupturing host epidermal layer (5) and subepidermal spread of hyphae between (inter) and within (within) host tissue for infection, and (6) mycelial aggregation in infected chickpea tissue (leaf/twig/pod) (7) leads to the formation of the asexual fruiting body (pycnidium) having asexual spores (conidia) which are being released and spread via air currents and rain splashes. In sexual cycle: (8) Specialized structure called as pseudothecium (an open ascocarp), also called ascostroma, formed in the cavities within a stroma/matrix of mycelium on the infected plant under unfavorable conditions/maturity of the crop having (9) sac like structures, asci formed from the hymenium consisting of eight ascospores, and (10) these ascospores are released from mature asci in the spring season (carried by wind to over 10 km distance) to cause primary infection of chickpea

2.4.2 Disease development

Illness and disease development of AB occurs at temperatures ranging from 5 to 30 °C, with an optimum of 20 °C, and 17 h of wetness to produce severe infection. Dry periods of 6–48 h immediately after inoculation occasionally increase disease severity; however, dry periods of >12 h after an initial dampening period of 6 h generally have an adversative effect on disease development. Jhorar et al. (1998) reported that dry periods proximately post-inoculation, followed by a wet interval, lowered disease severity leading to the decline in severity level. Following inoculation, the disease gets more severe as the duration of dark period increases. On infected leaves, the number of pycnidia and conidia increases when leaf wetness is sustained over 8 days. Towards the end of the season, AB can spread and develop rapidly in cool weather, with an incubation period as short as 6 days (Pandey et al. 1987). Different chickpea genotypes differ significantly in their sensitivity to AB infection, liable on the age of the plant wherein the podding stage is considered to be the most vulnerable (Chongo and Gossen 2001).

2.4.3 Disease Prediction Models

The disease prediction models have been developed for various agro-geographical zones and different growth stages based on the climatic factors favoring disease development to assess the disease risk. Jhorar et al. (1997) developed a model to predict chickpea blight by comparing AB incidence and weather variables over a 15-year period at two different locations (one with a regular disease frequency and the other with absenteeism of disease incidence). The model presented that maximum temperature and RH in the afternoon were the two most important factors for the disease forecast. A ratio of these two weather variables, denoted as the humid thermal ratio (HTR), stood as the best prognosticator of AB outbreaks.

2.5 Physiological Basis of Host-Pathogen Interaction

As soon as the pathogen spores land on the host's surface, they germinate to form germ tubes with appressoria and penetration pegs. The penetration pegs breach the first layer of the plant defense mechanism provided by the host cuticle and epidermal cells (Clulow et al. 1991; Carrillo et al. 2013; Fondevilla et al. 2015). Additionally, to transmit infection inside host plant cells, pathogen secrete enzymes like glycoside hydrolases, glycosyl transferases, and carbohydrate esterases to break down cellulose, hemicellulose, and chitin (Verma et al. 2016). Succeeding infiltration, the pathogen advances to have organic associations with the host cell, multiplies its hyphae, and spreads infection all over the host system (Ilarslan and Dolar 2002). In response to pathogen attack, the host produces oxidative pressure as an exertion to destroy or impair the pathogen hyphae (Jayakumar et al. 2005). *A. rabiei* harbors genes essential to overcome oxidative stress produced by the host all through pathogen invasion. A transcriptome profiling of *A. rabiei* under oxidative stress discovered the contribution of genes including *ST47_g10291*, *ST47_g9396*, *ST47_g10294*, *ST47_g4395*, and *ST47_g7191* responsible for pathogenicity and survival under oxidative stress (Maurya et al. 2020). For the successful establishment of the infection, it is essential that the pathogen must overcome the host's PTI and ETI-mediated defense mechanisms (Jha 2018).

2.6 Host-Plant Resistance

The most economical and effective AB control approach is the utilization of host-plant resistance (HPR), either alone or as a substantial part of cohesive AB management approaches. The first step to implement HPR is the development of reliable and repeatable resistance screening techniques. For AB, a variety of screening methods under both field and controlled epiphytotic conditions have been reported (Nene 1982; Weising et al. 1991; Nasir et al. 2000).

2.6.1 Screening Techniques

Several screening techniques for screening of chickpea genotypes against *A. rabiei* under natural and artificial conditions have been developed and refined over time.

2.6.1.1 Field Screening

Globally, field screening for AB is carried out at hot spots. In India, Gurdaspur and Ludhiana in Punjab, Dhaulakuan in Himachal Pradesh, and Hisar in Haryana have been recognized as hot spots with conducive environmental conditions for the development of disease. Nene et al. (1981) gave a comprehensive account of the development of screening procedures against AB for chickpea germplasm. Since then, the components of these techniques have been further refined and improvised (Singh et al. 1982; Reddy 1984; Haware et al. 1995; Pande et al. 2011). The basic steps followed for field screening are as under:

- Field testing of chickpea genotypes for AB resistance entails planting of the test material with a 40 cm row space with interplanting of susceptible cultivar as an indicator/spreader line after every 4–8 rows.
- Dispersal of infected material between rows at flowering or inoculation with a spore suspension (at 4×10^4 spores/mL) in the evening on cloudy days.
- The field is irrigated in the morning prior to inoculations, and in the evening, *A. rabiei* spore suspension is sprayed.
- Maintenance of RH above 85% starting the very next day from 10.00–16.00 h for 21 consecutive days by means of a perfo-spray system (Fig. 2.2). The diseases symptoms appear 10–12 days after inoculations.
- Using a 1–9 disease scale (Gurha et al. 2003), the observations are recorded twice, once in the middle of the growing season and then at the end of the season.

Fig. 2.2 Field screening by perfo-spray system at Punjab Agricultural University, Ludhiana

Disease scale (Gurha et al. 2003)

Score	Symptoms	Reaction
1.0	No infection	Highly resistant
1.1–2.0	Minute water-soaked lesions on leaves and stems	
2.1–3.0	Minute water-soaked lesions seen after careful examination	
3.1–4.0	Few small and few large lesions (>5 mm^2)	Resistant
4.1–5.0	Many small and large lesions;	
5.1–6.0	Many small and large lesions, lesions coalescing (50–75% plant area infected)	Moderately susceptible
6.1–7.0	Many small and large lesions, lesions coalescing, stem girdled (75–90% plant area infected)	
7.1–8.0	Many small and large lesions, lesions coalescing, girdling stem breakage (>90% plant area infected)	Highly susceptible
8.1–9.0	100% plants dead	

2.6.1.2 Controlled Environment Screening

In India, screening under controlled conditions is carried out at ICRISAT, Patancheru (India) by utilizing a controlled environment facility (CEF) with adjustable temperature, humidity, and photoperiod established to evaluate chickpea germplasm for AB resistance (Fig. 2.3). Haware et al. (1995) were the first to work out the physical arrangements, temperature, and humidity required in CEF. Later, it has been modified over the years with established resistance screening methodologies under controlled climatic conditions essential for AB penetration, infection, colonization, and development. The following are the rationalized screening procedures employing CEF:

Fig. 2.3 Screening under controlled conditions in CEF at ICRISAT, Hyderabad

1. Whole plant screening
   - The saplings of test genotypes are grown in plastic trays (35 × 25 × 8 cm) filled with a 10:1 mixture of sanitized river sand and vermiculite in a greenhouse maintained at a constant 25 ± 1 °C for 10 days. In each tray, 10 genotypes are sown, 9 being test lines (8 seedlings/line) along with one susceptible check.
   - 10 days old seedlings are then transferred to CEF, maintained at 20 ± 1 °C and a 12h photoperiod. Seedlings are then acclimatized for 24 h. After 24 h, spore suspension of *A. rabiei* (5×10^4 conidia/mL) is sprayed on test genotypes and susceptible controls until runoff.
   - To avoid dislodging of spores after inoculation, the seedlings are partially dried for 30 min after inoculation.
   - The photoperiod of 12 h (~1500 lux light intensity by fluorescent lights), 100% relative humidity (96 h for 6–8 h per day), and air temperature of 20 ± 1 °C are maintained throughout the experiment.
   - On death of check, the disease severity on a 1–9 scale is recorded.
2. Cut twig screening technique: To screen breeding lines and segregate germplasm without damaging the plants, the cut twig screening technique (CTST) was developed. Hence, it is a nondestructive method particularly to maintain the plant for seed production. The resistant plants identified using this method can then be utilized in crossing programs during the same crop season. Sharma et al. (1995) devised this technique by maintaining excised twigs from the test plant, in test tubes containing water, followed by incubation in a moist—muslin—cloth chamber. However, using the CEF at ICRISAT, the technique has been further modified. The following are the steps involved in CTST standardization.
   - Tender shoots of 10–15 cm long from test genotypes are excised from 30 to 60 days plant using a sharp-edged blade with instant submergence in water.
   - Each excised twig's bottom half is wrapped in a cotton plug and placed in a test tube (15 × 100 mm) containing fresh water.

- The test tubes with excised branches are then transferred to the CEF with a 20 ± 1 °C and 1500 lux light intensity (12 h a day).
- After 24 h, the twigs are inoculated by spraying *A. rabiei* spore suspension (5×10^4 conidia/mL).
- The disease severity is then recorded based on a 1–9 rating scale.

2.6.2 Mechanism of Host-Plant Resistance

To prevent ascochyta blight (AB) pathogen intrusion, the host plant employs two types of defenses. The first one is PTI (pathogen-associated molecular pattern [PAMP] triggered immunity), and the second one is ETI (effector-triggered immunity) (Maurya et al. 2020). To understand the molecular responses of *A. rabiei*, Coram and Pang (2006) dissected the resistance mechanism of chickpea AB based on transcriptome analysis. They identified resistant genes, *Snakin-2* and *DRRG49-C* encoding for antimicrobial peptide. Further, Singh et al. (2012) identified the solanopyrone biosynthesis gene cluster (*Sol1–Sol5*). By transcriptome analysis, Maurya et al. (2020) identified five resistant genes (*ST47_g10291*, *ST47_g9396*, *ST47_g10294*, *ST47_g4395*, and *ST47_g7191*). The disease resistance genes like chitinases (*Ca_04405*), CC-NBS-LRR (*Ca_08361*), dirigent protein (*Ca_20726*), DOF zinc finger (*Ca_19433*), TCP transcription factor (*Ca_12866*), and cellulose synthase (*Ca_08607*) were identified by Garg et al. (2019) on exploiting integrated transcriptome and degradome sequencing approaches.

2.6.3 Sources of Resistance

Considering the rate of resistance breakdown in current chickpea varieties by means of the rapid pathogen's evolution into new pathotypes, the resistant sources currently available necessitate periodic supplementation with new ones (Gayacharan et al. 2020). To strengthen the breeding programs at ICRISAT, Pande et al. (2011) suggested to exploit 29 resistant lines. These lines had a wide range of maturity (112–142 days). Recently, Gayacharan et al. (2020) and Pastor et al. (2022) screened a large number of chickpea accessions for AB and identified around 2.75% resistant genotypes. At Punjab Agricultural University (PAU), Ludhiana, screening of large germplasm sets comprising exotic collections, indigenous lines, and intra and interspecific derivative lines is regularly done to identify resistant sources. Few of them have been reported by Singh et al. 1984a, b; Singh and Kapoor 1986; Kaur et al. 2011, 2012a, 2012b, Kumar et al. 2021. An updated list of resistant cultivars to AB in chickpea is provided in Table 2.2.

Developing chickpea varieties resistant to AB has been a challenging proposition owing to the paucity of high resistance levels in the primary gene pool. The improvement of crop species is often restricted due to the lack of genetic diversity, as a result of genetic bottlenecks associated with domestication. The prime-

Table 2.2 An updated list of AB-resistant chickpea sources (2010–2021)

Resistant source	Country	References
53628, 53225, 53227, 53230, 53231, 53233, 53235, 53244, 53380,53436,53643, 54247, 53045, 53217, 53218, 53323, 53651, 53398	Pakistan	Iqbal et al. (2010)
FLIP97-121C	India	Kaur et al. (2012a, b)
K-60013, K-98008, D-97092, K-96001, K-96022, D-91055, D-90272, D-96050, D-Pb2008, D-Pu502–362	Pakistan	Ahmad et al. (2013)
FLIP4107, FLIP1025, FLIP10511	Algeria	Benzohra et al. (2013)
ICC7052, ICC4463, ICC4363, ICC2884, ICC7150, ICC15294, ICC11627	Kenya	Kimruto et al. (2013)
EC516934, ICCV 04537, ICCV98818, EC516850, EC516971	India	Pande et al. (2013)
10A, 28B	Turkey	Duzdemİr et al. (2014)
ILC72, ILC182, ILC187, ILC200, ILC202	Algeria	Benzohra et al. (2015)
K0058-09, K0062-09, K0066-09, D095-09, K07A005, BK05A015, BK04A013	Pakistan	Shah et al. (2015)
ICCV-96836, Arerti	Ethiopia	Zewdie and Tadesse (2018)
FLIP02-04C, FLIP06-65C	Iran	Farahani et al. (2019)
IC275447, EC267301, IC117744, IC248147, EC220109	India	Gayacharan et al. (2020)
D-17001, D-17005, D-17008, D-17009, D-17011, D-17023, D-17024, D-17032	Pakistan	Shah et al. (2021)
GPF2, GL29098, GL12003, GL12021, GL14001, GL14002, GL15034, GL15056, GL15100, GL16014, GL16012, GL16043, GL16059, GL16062, GL16068, GL17033, GL17020, GG1362, GG1390, ICC1915, ICCV04509, IPCK93, ICC4200, ICCV04512, ICCV04537	India	Kumar et al. (2021)

evolutionary bottlenecks pertinent to chickpeas are the limited distribution of the wild progenitor *C. reticulatum*, the founder effect, a shift from autumn to spring sowing in the Early Bronze Age, and the replacement of landraces by elite cultivars (Abbo et al. 2003).

It has been estimated that contemporary chickpea breeding programs lack about 93.5–97.5% of the genetic diversity available in the wild progenitor, *C. reticulatum* (von Wettberg et al. 2018). Due to the dearth of adequate level of genetic resistance in cultivated genotypes, different gene pools of *Cicer* species, such as *C. bijugum*, *C. echinospermum*, *C. pinnatifidum*, *C. judaicum*, and *C. montbretii*, have been subjugated to transfer AB resistance. Two *C. echinospermum* (S2Drd_061 & Deste_064) and one *C. reticulatum* (Baril_062) have displayed both leaf and stem resistance against different AB pathotypes (Newman et al. 2021), and being cross-compatible with *C. arietinum*, they could serve as valuable sources of resistance (Ahmad and Slinkard 2004). Additionally, wild accessions from the tertiary gene pool, *C. judaicum*, viz., ILWC185, ILWC95, ILWC61, and *C. pinnatifidum* accessions ILWC188, ILWC199, and ILWC212 were reported to harbor resistance to AB (Kaur et al. 2013). Interestingly, wild *Cicer* accessions have multi-stress (four or five stresses) tolerance simultaneously viz., *C. reticulatum* (ILWC81,112), *C. echinospermum* (ILWC39,181), *C. judaicum* (ILWC46), *C. bijugam* (ILWC32,62,73,79), and *C. pinnatifidum* (ILWC236) (Mohanty et al. 2022). The exploitation of this vast amount of additional genetic diversity holds the key to chickpea improvement.

Hence, pre-breeding efforts play a significant role in crop improvement programs as they provide the base on which all molecular breeding efforts can flourish. Since wild *Cicer* is indeed an excellent genetic reservoir, they have been utilized as promising donors to introgress a range of alien characteristics into cultivated chickpea. These include resistance to ascochyta blight, dry root rot, botrytis grey mold, and pod borer. In the case of chickpea, most of the wide hybridizations are performed with *C. echinospermum* and *C. reticulatum* accessions as the crossing of the cultigen with other species has remained challenging even on following embryo rescue techniques. In the past breeding efforts, hybridization of the wild species with different popular cultivated varieties (i.e., JG11, Pusa372, PBG5, ICKG96029, and BGD72) has resulted in the development of several transgressive segregants for different agronomic traits (Singh et al. 2018). Successful *Cicer arietinum* x *C. pinnatifidum* hybridization has resulted in the development of interspecific derivative lines exhibiting significant variability for yield attributes and disease resistance (Kaur et al. 2013; Salaria 2020). Recently from an interspecific cross between *C. arietinum* and *C. judaicum*, the high-yielding variety PBG8 has been developed by Punjab Agricultural University, Ludhiana, released for commercial cultivation in the Punjab state. PBG8 has a higher level of tolerance to pod borer and is moderately resistant to ascochyta blight and botrytis grey mold (Singh et al. 2022b).

2.7 Genetics of Resistance

The genetic analysis of AB resistance performed initially revealed the contribution of two dominant complementary genes controlling AB disease in chickpea (Ahmad et al. 1952). At that time, studies on the F_2 population indicated the involvement of a single dominant gene pair, imparting AB resistance (Hafiz and Ashraf 1953; Vir et al. 1975). These studies were conducted on four crosses wherein *desi* varieties served as the source of resistance.

Singh and Reddy (1983) used preliminary screening of F_1 and F_2 populations of 12 crosses between 5 resistant and 11 susceptible parents to assert that AB resistance is inherited as a single dominant gene (*rar2*) in ILC72, ILC183, ILC200, and ILC4935 and a single recessive gene (*rar1*) in ILC191. This was the first report about the identification of recessive genes conferring AB resistance. Similarly, based on allelic analysis, Tewari and Pandey (1986) revealed fairly alike findings after screening F_2, BC_1, BC_2, and F_3 generations from crosses between six resistant and four susceptible parents under field and glasshouse conditions. Three independent genes influencing AB resistance were identified by the test of allelism: one dominant gene each in P1215-1, EC26446, and PG82-1 and one recessive gene in BRG8.

However, Dey and Singh (1993) addressed three main points on the basis of generation mean analysis, viz., number, difference, and nature of gene action. Based on this the resistance to AB was said to have an additive gene effect with gene action chiefly reliant on the genotype. The earlier studies in chickpea primarily reported monogenic (Singh and Reddy 1983) and oligogenic resistance to AB (Pieters and Tahiri 1986). In addition to one dominant (*Arc5*) and one independent recessive gene from the ICC1468 genotype, two dominant complementary genes (*Arc1* and *Arc2*) were reported from GLG84038 and (*Arc3* and *Arc4*) from GL84099 genotypes. The results of two previously reported dominant complementary resistance genes (Dey and Singh 1993) were verified in a subsequent investigation by Pal et al. (1999) employing more virulent pathotypes of the pathogen. The mechanism of resistance and expression was found to be highly genotype specific. In the ensuing century, digenic recessive, monogenic recessive, digenic dominant and recessive, monogenic dominant, trigenic dominant, and recessive resistance control was detected in 15 different crosses by utilizing 6 genotypes (two susceptible, four resistant genotypes) (Bhardwaj et al. 2010). In yet another study, three resistant genotypes with one dominant gene and a minor recessive gene or genes segregating in monogenic patterns conferred resistance. One recessive gene, two complementary recessive genes, two complementary dominant genes, and two recessive genes with epistatic interaction were reported to confer blight resistance in the majority of *C. arietinum* accessions, except ILC3279, ILC3856, and ILC4421 with either three recessive or two recessive duplicate genes governing resistance (Labdi et al. 2013) against *A. rabiei* (race 4). Polygenic resistance with additive and dominance gene action has also been detected by Danehloueipour et al. (2007) in 5×5 half-diallel cross sets involving seven genotypes of chickpea (ICC3996, Almaz, Lasseter, Kaniva, 24B-Isoline, IG9337, and Kimberley Large), three accessions of *C. reticulatum* (ILWC118, ILWC139, and ILWC184), and one accession of *C. echinospermum* (ILWC181) under field conditions.

Despite progress in understanding the genetics of AB, the identification of stable resistant cultivars has gained limited success. This emerging scenario was vague and perplexing, as in reports based on AB resistance, wherein either one dominant gene, one recessive gene, or their combination, and a set of complementary dominant or complementary recessive genes were described. Due to the absence of allelic assays, the position of the claimed resistance genes was also unknown (Winter et al. 2000). It was still unclear whether the resistance in different parts of the plant of host genotypes was controlled by the same or different set of genes and whether the recognized genes were resistant to all or few races of the pathogen (Akem 1999). Other issues arose on account of the likelihood that a gene could influence and regulate the expression of other resistance genes. A further complication was posed by the possibility of the coexistence of different *A. rabiei* pathotypes and races in the same field or even within the same infected area of the host plant. According to Barve et al. (2003), this possibly results in haphazard mating between different pathotypes of dissimilar mating-type alleles. Hence, this leads to enhanced genetic diversity on the origin of new genetic recombinants conferring new and modified potential to infect the host. This fact prompted the development of new approaches, technologies, and tools to elucidate the indistinct evidence about the number and diversity of genes, genomic regions, expression profiles, mechanisms, and related factors.

2.8 Resistance Breeding

2.8.1 Conventional Breeding

One of the main objectives of many breeding programs for chickpeas in different countries, including Canada, the USA, Australia, Turkey, and Pakistan, is breeding resistance to AB. AB resistance breeding in India started as early as in the 1930s. The first AB-resistant cultivar developed was about 60 years ago (Luthra et al. 1941). In later reports from the Soviet Union (Gushkin 1946), it was stated that three AB-resistant cultivars, Skorospelka, Alpha, and Mogucii, were developed and released. On the other hand, in the Mediterranean region, introduction efforts were undertaken until 1984 for AB-resistant cultivar. The emergence of novel strains of *A. rabiei* and the dearth of readily available genetic sources harboring high levels of resistance have delayed the progress in resistance breeding for AB.

Beginning in 1978, hybridization efforts at ICARDA chiefly intended to combine high yield with AB and cold resistance. More than 3000 AB-resistant and high-yielding lines were developed between 1981 and 2002 at Terbol in the Beqa'a valley of Lebanon using off-season generation advancement facilities. Most of the previously released cultivars were hence exposed to new races or pathotypes of *A. rabiei* after their initial success. This shortened the span of resistant cultivars. The bulk-pedigree method of breeding AB-resistant chickpeas was popular at ICARDA until 1998. Research demonstrated that the efficacy of selection for AB resistance and large seed size was augmented by single seed descent (SSD) for F_2 and F_3 generations, followed by the pedigree method from the F_4 generation.

Attempts have been undertaken to combine genes conferring resistance to different races of *A. rabiei* in one genotype. Using a stepwise breeding approach, the chickpea breeders of ICARDA have successfully pyramided a few genes from several sources. On a similar line, the focus of ICRISAT has been on the development of AB-resistant lines of desi chickpea. To pyramid resistance genes from various sources, multiple crossing approach was undertaken. Many advanced breeding lines developed by means of multiple crossing have demonstrated resistance to four isolates of *A. rabiei* on evaluation under controlled environmental conditions (Gowda 2005). Hence, as sources of AB resistance, ICARDA and ICRISAT's breeding lines and germplasm have been widely utilized. The *desi*-type cultivar "Howzat," released in 2001, was the first variety in Australia with moderate levels of resistance to AB. Additionally, a number of *desi* and *kabuli* lines with moderate to high levels of AB resistance have been selected by Australian chickpea breeders. Breeding lines from ICRISAT and ICARDA, such as ICCV96836, FLIP94-508C, FLIP94-90, FLIP94-92C, S95362, and S95342, as well as selections from already-existing Australian varieties, such as Heera, Sona, and Barwon, are among the commonly utilized resistant sources (Materne et al. 2002).

In Pakistan, chickpea industry was able to survive through the development of AB-tolerant cultivars following mutation breeding. The first variety from this program was CM72 (*desi* type) released in 1983. The other mutant variants that were eventually made available to the public included the *desi* type, CM88 and CM98, as well as the *kabuli* type, CM2000. Other AB-tolerant cultivars have been developed using traditional breeding techniques, including Dashat and NIFA88 (Pande et al. 2005).

In India, significant work has been done for the development of resistant varieties by Punjab Agricultural University, Ludhiana. The first AB-resistant variety, C235, was developed in 1960. The variety became very popular among the farmers of the Punjab state as well as neighboring states due to its wide adaptability. Though, a new blight-resistant variety, G543 in 1978, was released for AB-prone areas of Punjab State. Epidemic conditions prevailed during 1980–1981 in northwestern states of India due to the breakdown of resistance in C235 (the then popular variety of chickpea). Some other resistant chickpea varieties, namely, PBG1 (Verma et al. 1992), GPF2, PBG5 (Sandhu et al. 2004), PBG7 (Singh et al. 2015), and PBG8 (Singh et al. 2022b), have been released, and the development of new ones is underway to avoid any further epidemic like conditions. This is done to avoid any resistance breakdown by any novel virulent pathotype(s) evolved under the selection pressure posed by new resistant varieties.

Consequently, it is comprehensible that in the absence of highly resistant AB sources, no single strategy in chickpea breeding is likely to succeed. There is a need to develop and employ a variety of strategies collectively. The release of several cultivars, possibly with known reactions in different races/pathotypes, would be the convenient strategy in such cases wherein the breakdown of resistance is frequent.

2.8.2 Marker-Assisted Breeding

Studies of genetic resistance have hitherto been limited to genes with significant phenotypic impact. Advanced approaches and statistical tools have resulted in the development of novel approaches to investigate the complex genetic architecture underlying quantitative traits conferring resistance. The application of these technologies has permitted the identification of quantitative trait loci (QTLs) responsible for the expression of AB resistance. Owing to recent advances, the resistance to AB in chickpea is now considered to be a quantitative trait conferred by multiple QTLs. This has in turn made it possible to map AB resistance QTLs on various linkage groups (LGs) by deploying a variety of markers with various genetic backgrounds. Owing to an upsurge in sources conferring resistance, the development of stable-resistant genotypes by pyramiding multiple genes in a short duration is the major benefit of these markers.

Collard et al. (2003) reported two QTLs contributing to AB resistance on LG4 using Lasseter × *C. echinospermum* (PI527930) mapping population. Two major QTLs on LG 2, close to the GA16 and TA37, control resistance to *A. rabiei* pathotype I and II, respectively (Cho et al. 2004). Cho et al. (2004) also identified an additional SSR marker (TA46) located on LG2 explaining 69.2% variations for resistance to Ar21d of pathotype I and 59.2% to a mixture of pathotype II isolates under controlled environments. Iruela et al. (2007) mapped one QTL, *QTLAR3* on LG2 flanked by the STMS markers TR58 and TS82. Madrid et al. (2014) further fine-mapped this *QTLAR3* onto Ca2 with a physical position of 32–33 Mb, comprising 42 candidate genes including *Ein3*, *Avr9/Cf9*, and *Argonaute* 4 genes participating in the disease resistance mechanism. Similarly, Hamwieh et al. (2013) identified 14 microsatellite markers on 5 chickpea linkage groups (LG2, LG3, LG4, LG6, and LG8) linked to 7 QTLs (*Ar2a*, *Ar2c*, *Ar3c*, *Ar4a*, *Ar4b*, *Ar6b*, and *Ar8a*) conferring resistance to *A. rabiei*.

Anbessa et al. (2009) found five resistance QTLs, *QTL1*(LG2), *QTL2*(LG3), *QTL3*(LG4), *QTL4*(LG6), and *QTL5*(LG8), by utilizing wide-range divergent resistant parental lines of *desi* and *kabuli* with varied origin and countries, demonstrating 12–38% of the phenotypic variation in four F_2 populations. *QTL1* on LG2 continued to have a major effect of about 38%, as reported by erstwhile authors (Udupa and Baum 2003; Cobos et al. 2006), while *QTL2* (LG3) was determined to be of only diffident significance and *QTL3* (LG4) as intensified in earlier findings (Tekeoglu et al. 2002; Flandez-Galvez et al. 2003; Lichtenzveig et al. 2006). Additionally, it has been demonstrated that *QTL3* consists of many linked resistance genes of variable genetic background, i.e., the interrelated genes might be positioned on LG4 or at diverse genomic positions serving as a significant source of genetic resistance if known.

Daba et al. (2016) identified 8 QTLs on all LGs except LG5 using 92 recombinant inbred lines (CP-RIL-1 population) developed from a cross between ICCV96029 (highly susceptible to AB) and CDC Frontier (resistant to AB). Progress of affordable sequencing approaches for genome-wide genetic characterization has been sparked by technological advancement. Whole genome sequencing-based BSA (bulked segregant analysis) is one of the low-cost, rapid methods to identify QTLs of desired interests and is an alternative to traditional QTL analysis. Subsequently, 11 QTLs in CPR-01 and 6 QTLs in CPR-02 populations have been mapped on Ca1, Ca2, Ca4, Ca6, and Ca7 using next-generation sequencing (NGS)-based BSA approach (Deokar et al. 2019a). Notably, the QTLs *CPR01-qAB1.1*, *CPR01-qAB4.1*, and *CPR01-qAB4.2* identified in this study displayed an intersection with the AB-resistant QTLs previously reported by Daba et al. (2016) from a CPR-01 population using conventional mapping approach. This paved preliminary evidence in favor of the effective use of NGS-based BSA as a quick and affordable method for the discovery of QTLs relevant to AB-resistance in chickpea. FST genome-scan and genome-wide association studies have been used to reveal genes conferring AB resistance related primarily to major QTL at chromosome 4 (Li et al. 2017). The

AB4.1 (100 kb) region containing 12 predicted genes on chromosome number 4 demonstrated connotation with blight resistance. A QTL interval (30 Mb) was detected in three RIL populations (Lichtenzveig et al. 2006; Sabbavarapu et al. 2013; Stephens et al. 2014) as described in earlier reports. These genes included four serine/threonine receptor-like kinases (RLK) previously reported to resist many diseases in different species, including *Arabidopsis*, one wall-associated receptor kinase gene (WAK) that was highly induced in resistant genotypes. A CRK gene, predicted to be a cysteine-rich receptor-like kinase, was not involved in AB resistance but a zinc finger protein, and most expressively a leucine-rich repeat (LRR) receptor-like kinase was upregulated in all resistant genotypes.

A biparental mapping population was developed by crossing Amit × ICCV96029 by Deokar et al. (2019b) and genotyped via genotyping-by-sequencing (GBS) as well as Illumina® GoldenGate array. This unraveled eight QTLs conferring AB resistance positioned on LG2, 3, 4, 5, and 6 explaining 7–40% of the phenotypic variations. Recently, Kushwah et al. (2021) detected two AB-resistant QTLs, *qab-4.2* on LG4 explaining 10.6% PV, and *qab-7.1* on LG7 explaining 8.2% PV in GPF2 × ILWC292 mapping population. A brief list of a few AB-resistant QTLs identified in chickpea has been presented in Table 2.3.

To pyramid traits of interest governed by several major genes/QTLs in a specific genetic background, marker-assisted backcrossing (MABC) is the most preferred approach. With complex inheritance patterns of recessive genetic resistance to AB of chickpea, the MABC technique has made it possible to unravel AB resistance QTLs. Among various chickpea cultivars, CDC Xena, CDC Leader, and FLIP98-135C have been successfully introgressed with AB resistance along with double-podding traits following MABC (Taran et al. 2013).

Varshney et al. (2014) have demonstrated the stepwise utilization of the MABC approach for the development of superior lines resistant to AB. To develop resistant lines, two QTLs (*ABQTL-I* and *ABQTL-II*) for AB were targeted for introgression into C214 (elite cultivar). For this, eight markers linked to QTL regions were used for foreground selections in different segregating generations. In addition to the foreground, background selections were performed to have a high recovery of recurrent parent genome with evenly distributed 40 SSR markers. By means of three backcrosses and three rounds of selfing, 14 MABC AB-resistant lines were generated. Further, phenotypic screening of these lines resulted in the identification of seven stable AB resistance lines.

2.8.3 Genomic-Assisted Breeding

Genomic-assisted breeding (GAB) refers to the amalgamation and application of genomic techniques in breeding programs for the development of superior lines with enhanced biotic or abiotic stress tolerance along with higher yield levels. Although conventional breeding techniques were able to increase yield, they were unable to break the yield plateau and address the problems posed by the narrow genetic base. Up until 2005, chickpea was commonly regarded as an orphan crop due to the

Table 2.3 List of identified QTLs conferring resistance to ascochyta blight of chickpea

Mapping approach	Mapping population	QTL/gene	Type of marker used	Linkage group	Phenotypic variation	Reference
Biparental	*C. arietinum* (ICC3996) × *C. reticulatum* (ILWC184)	*QTL3*, *QTL4.1*, *QTL4.2*	SSR	LG3, LG4	49	Aryamanesh et al. (2010)
GWAS	132 advanced breeding lines	*AB4.1* QTL and 12 candidate genes (*Ca_05515*, *Ca_05520*, *Ca_05511*, *Ca_05516*, *Ca_05517*, *Ca_05521*, *Ca_05522*, *Ca_05512*, *Ca_05513*, *Ca_05514*, *Ca_05518*, and *Ca_05519*)	SNP	LG4	–	Li et al. (2017)
Biparental	CRIL-3 (250 RILs) FLIP84-92C (3) × PI599072 (217RILs)	Two major QTL: *qABR4.1*, *qABR4.2*, one minor QTL: *qABR4.3* and *CaAHL18* candidate gene	SNP	LG4	42	Kumar et al. (2018)
Biparental	JG62 × ICCV05530 (188 RILs)	*AB-Q-SR-4-1* and *AB-Q-APR-4-1*	SSR	LG4	6.44–6.98	Garg et al. (2018)
Biparental	Amit × ICCV96029 (RIL)	*qAB2.1*, *qAB2.2*, *qAB2.3*, *qAB3.1*, *qAB4.1*, *qAB4.2*, *qAB5.1*, *qAB6.1*	SNP	LG2, 3, 4, 5 and 6	7–40	Deokar et al. (2019a)
Biparental	ICCV96029 × CDCFrontier (92 RILs), ICCV96029 × Amit (139 RILs)	*CPR01-qAB1.1*, *CPR01-qAB1.2*, *CPR01-qAB1.3*, *CPR01-qAB1.4*, *CPR01-qAB4.1*, *CPR01-qAB4.2*, *CPR01-qAB4.3*, *CPR01-qAB4.4*, *CPR01-qAB4.5*, *CPR01-qAB6.1*, *CPR01-qAB6.2*, *CPR01-qAB7.1*	SNP	LG1, 2, 3, 4, 6, 7, 8	–	Deokar et al. (2019b)
Biparental	GPF2 × *C. reticulatum* acc. ILWC292, 187 RIL	*qab-4.1*, *qab-4.2*, *qab-7.1*	SNP	LG4, LG7	7–11	Kushwah et al. (2021)
GWAS	146 (*C. reticulatum*) + 44 (*C. echinospermum*)	WRKY TF (*Cr_02657.1*), (*Cr_09847.1*) encodes a TF of the ARF family	SNP	LG3, 4, 6	6.7–15.2	Newman et al. (2021)
Biparental	GL769 (*C. arietinum*) × ILWC129 (*C. reticulatum*), (157 $F_{2:3}$)	arr_8	SSR	LG1	–	Lekhi et al. (2022)

(continued)

Table 2.3 (continued)

Mapping approach	Mapping population	QTL/gene	Type of marker used	Linkage group	Phenotypic variation	Reference
Biparental	ILC1929 × ILC3279 (116 RILs) AB_{482} (135 RILs)	*AB3279-LG2-QTL-1*, *AB3279-LG2-QTL-2*, *AB3279-LG4-QTL-1*, *AB3279-LG4-QTL-2*, *AB3279-LG4-QTL-3*, *AB3279-LG4-QTL-4*, *AB3279-LG4-QTL-5*, *AB3279-LG4-QTL-6*, *AB3279-LG4-QTL-7*, *AB482-LG4-QTL-1*, *AB482-LG4-QTL-2*, *AB482-LG4-QTL-3*, *AB482-LG4-QTL-4*, *AB482-LG4-QTL-5*, *AB482-LG4-QTL-6*, *AB482-LG4-QTL-7*, *AB482-LG4-QTL-8* and *AB482-LG4-QTL-9*	SNP	LG2 and LG4	11.2–39.3	Alo et al. (2022)

scarcity of genetic and genomic resources. However, the development of substantial genetic, genomic, and transcriptome based resources over the past 10 years has changed the scenario. Now, chickpea is regarded as a genomic resource-rich crop instead of an orphan crop (Varshney et al. 2009; Nayak et al. 2010; Thudi et al. 2011; Mashaki et al. 2018). It is suggested to integrate omics data from several platforms, including transcriptomics, proteomics, and metabolomics, in order to bridge the genome-to-phenome gap to eventually identify the phenotype based on their genetic contribution (Choi 2019). The information generated from these omics techniques will help complement genomic information required to manipulate several biological processes in breeding programs.

The functional participation of genes is revealed by a technique termed as transcriptome profiling. It helps to identify genes that are expressed under a biological phenomenon, such as disease pressure. Coram and Pang (2005) used microarray to examine the expression pattern of 20 defense-related ESTs in two different chickpea parents, ICC3996 (AB-resistant) and Lasseter (AB-sensitive). Compared to control samples, 10 ESTs were found to be differentially regulated in resistant genotypes. Further, these ESTs were classified into clusters based on analogous annotations wherein upregulatory action of "leucine zipper protein," "SNAKIN2 antimicrobial peptide precursor," and "elicitor-induced receptor protein" genes was revealed in ICC3996. In a subsequent large-scale gene expression study on AB-resistant, susceptible, and moderately resistant chickpea lines and wild species, Coram and Pang (2006) investigated the alterations in gene expression for AB. Microarray analysis showed that 97 (genes) out of 715 chickpea cDNAs exhibited differential expression in at least one genotype at one time-point, signifying the differential regulation of genes reacting to blight. The genes involved in conferring AB resistance were "pathogenesis-related proteins," "proline-rich protein," SNAKIN2 antimicrobial peptide, disease resistance response protein DRRG49-C, leucine zipper protein, polymorphic antigen membrane protein, and Ca-binding protein. Leo et al. (2016) identified 6 differentially expressed genes among 10 chickpea genotypes by expression profiling of 15 defense-related genes in response to infection by *A. rabiei*. Kumar et al. (2018) identified a candidate gene *CaAHL*18 (AT-HOOK MOTIF CONTAINING NUCLEAR LOCALIZED 18) with a higher level of expression within the *qAB4.1* region on AB infection at 12 hpi (hours post-infection) and 72 hpi, signifying its probable involvement in resistance.

Based on the localization of the R-genes in the plant cell and their putative protein domain organization, several types of R-genes have been discovered and categorized (van Ooijen et al. 2007). The most commonly known and well-studied family of R-genes is NBS-LRR genes. They code for proteins with a central nucleotide-binding site (NBS) and a carboxyl/C-terminal leucine-rich repeat (LRR) domain (McHale et al. 2006). The majority of plant NBS-LRR proteins are intracellular receptors able to detect the presence of pathogen effectors either directly by binding to the pathogen effector proteins or indirectly by recognizing any modification in the pathogen effector target proteins in the host. Thereby, triggering defense signal transductions frequently leading to hypersensitive responses and other biochemical changes able to restrict the pathogen growth (Meyers et al. 2003; DeYoung and

Innes 2006). Aiming to understand how AB resistance is mediated by chickpea, NBS-LRR resistance genes for the first time were identified by Sagi et al. (2017). They found 121 NBS-LRR genes spread over the whole chickpea genome and colocalized 30 (out of a total of 121) NBS-LRR genes with ascochyta blight QTLs. Even variable expression of 27 candidate genes in resistant CDC Corinne and CDC Luna genotypes and one susceptible ICC96029 genotype was observed at different time points on AB infection. Five NBS-LRR genes showed genotype-specific expression with *LOC101512894*, *LOC101513745*, and *LOC101497042* with an upregulation in ICCV96,029 and CDC Luna and downregulation or no regulation in CDC Corinne. Similarly, *LOC101509145* and *LOC101498915* were upregulated in CDC Corinne at 48 and 72 hpi and downregulated or not regulated in ICCV96029 and CDC Luna at all time points.

2.9 Way Forward

Breeders persistently strive to find novel sources of resistance across various gene pools. This is primarily pursued to transfer novel alien genes into elite cultivars to confer resistance to AB along with stabilized yield levels. Being a cost-effective, environment-friendly approach to confer sustainable disease management, it is regarded as the method of choice. Due to posing global climate change and utilization of resistant host cultivars the plant pathogens, particularly AB pathogens, keep on coevolving by developing novel virulence. This results in the breakdown of host resistance leading to heavy yield losses. The genetics of host-pathogen interactions and the mechanism of gene action in AB-resistance have both been the subject of several investigations. Genetics of AB-resistance is now recognized as polygenic as it is conferred by some major genes with several QTLs located on each of the eight linkage groups. Such involvement of several regions and genes controlling AB-resistance specifies not only intricacy but also a network of interdependent components. These regions and/or genes undoubtedly influence functioning in unison to provide the plant with an effective level of resistance. Owing to quick breakthroughs in functional genomics, particularly RNA-seq, a better knowledge of the intricate molecular processes underlying host-plant relationships, disease progression, and host-plant resistance mechanisms have been made available. Numerous genes exhibiting differential expression in response to AB stress have been analyzed, and some of them were given putative functions based on comparisons with known genes in other species. However, so far only a small number of these genes were given candidate status. Though vast, this information is still only a drop in the ocean, and the real mechanism needs to be elaborated. To develop resilient resistant cultivars, it is important to understand the mechanisms fundamental to resistance. This will pave the way to keep track of resistance breakdown when the population structure gradually changes. After marker-assisted selection, genomic selection is the emerging innovative breeding technique that could be utilized to choose superior recombinants and progeny with high breeding value and stable AB resistance. With this view, it is imperative that the ultimate objective of breeders is to have durable resistance along with the effectiveness of vertical resistance. This could be achieved by pyramiding several genomic regions and/or genes identified in chickpea genotypes on employing advanced genomic approaches.

References

Abbo S, Berger J, Turner NC (2003) Viewpoint: evolution of cultivated chickpea: four bottlenecks limit diversity and constrain adaptation. Funct Plant Biol 30:1081–1087. https://doi.org/10.1071/FP03084

Ahmad GD, Hafez A, Ashaf M (1952) Association of morphological characters with blight resistance. In: Proceedings of 4th Pakistan Science Conference. Pakistan Association for the Advancement of Science, Peshawar, pp 17–19

Ahmad S, Khan MA, Sahi ST, Ahmad R (2013) Evaluation of chickpea germplasm against *Ascochyta rabiei* (Pass) Lab. J Animal Plant Sci 23:440–443

Ahmad F, Slinkard AE (2004) The extent of embryo and endosperm growth following interspecific hybridization between *Cicer arietinum* L. and related annual wild species. Genet Resour Crop Evol 51:765–772. https://doi.org/10.1023/B:GRES.0000034580.67728.E4/METRICS

Akem C (1999) Ascochyta blight of chickpea: Present status and future priorities. Int J Pest Manag 45:131–137. https://doi.org/10.1080/096708799227923

Alam SS, Bilton JN, Slawin AMZ et al (1989) Chickpea blight: Production of the phytotoxins solanapyrones A and C by *Ascochyta rabiei*. Phytochemistry 28:2627–2630. https://doi.org/10.1016/S0031-9422(00)98054-3

Alo F, Rani AR, Baum M et al (2022) Novel genomic regions linked to ascochyta blight resistance in two differentially resistant cultivars of chickpea. Front Plant Sci 13:762002. https://doi.org/10.3389/FPLS.2022.762002/FULL

Anbessa Y, Taran B, Warkentin TD et al (2009) Genetic analyses and conservation of QTL for ascochyta blight resistance in chickpea (*Cicer arietinum* L.). Theor Appl Genet 119:757–765. https://doi.org/10.1007/S00122-009-1086-2

Armstrong CL, Chongo G, Gossen BD, Duczek LJ (2001) Mating type distribution and incidence of the teleomorph of *Ascochyta rabiei* (*Didymella rabiei*) in Canada. Can J Plant Pathol 23:110–113. https://doi.org/10.1080/07060660109506917

Aryamanesh N, Nelson MN, Yan G et al (2010) Mapping a major gene for growth habit and QTLs for ascochyta blight resistance and flowering time in a population between chickpea and *Cicer reticulatum*. Euphytica 173:307–319. https://doi.org/10.1007/S10681-009-0086-2

Baite MS, Dubey SC (2018) Pathogenic variability of *Ascochyta rabiei* causing blight of chickpea in India. Physiol Mol Plant Pathol 102:122–127. https://doi.org/10.1016/j.pmpp.2018.01.001

Barve MP, Arie T, Salimath SS et al (2003) Cloning and characterization of the mating type (MAT) locus from *Ascochyta rabiei* (teleomorph: *Didymella rabiei*) and a MAT phylogeny of legume-associated Ascochyta spp. Fungal Genet Biol 39:151–167. https://doi.org/10.1016/S1087-1845(03)00015-X

Basandrai AK, Pande S, Kishore GK et al (2005) Cultural, morphological and pathological variation in Indian isolates of *Ascochyta rabiei*, the chickpea blight pathogen. Plant Pathol J 21:207–213. https://doi.org/10.5423/PPJ.2005.21.3.207

Benzohra IE, Seddik Bendahmane B, Labdi M, Youcef Benkada M (2013) Sources of resistance in chickpea germplasm to three pathotypes of *Ascochyta rabiei* (Pass.) labr in Algeria. World Appl Sci J 21:873–878. https://doi.org/10.5829/idosi.wasj.2013.21.6.2874

Benzohra IE, Seddik Bendahmane B, Youcef Benkada M, Labdi M (2015) Screening of 15 chickpea germplasm accessions for resistance to *Ascochyta rabiei* in North West of Algeria. J Agric & Environ Sci 15:109–114. https://doi.org/10.5829/idosi.aejaes.2015.15.1.12505

Bhardwaj R, Sandhu AJS, Livinder AE et al (2010) Genetics of ascochyta blight resistance in chickpea. Euphytica 171:337–343. https://doi.org/10.1007/s10681-009-0020-7

Butler EJ (1918) Fungi and disease in plants: an introduction to the diseases of field and plantation crops, especially those of India and the east. Thacker, Spink, Calcutta and Simla, p 547

CAB International (2000) Crop protection compendium. Wallingford, UK, CAB International

Carrillo E, Rubiales D, Pérez-de-Luque A, Fondevilla S (2013) Characterization of mechanisms of resistance against *Didymella pinodes* in *Pisum* spp. Eur J Plant Pathol 135:761–769. https://doi.org/10.1007/s10658-012-0116-0

Chattopadhyay P, Glick WH, Huber GP et al (2001) Organizational actions in response to threats and opportunities. Acad Manag J 44:937–955. https://doi.org/10.5465/3069439

Cho S, Chen W, Muehlbauer FJ (2004) Pathotype-specific genetic factors in chickpea (*Cicer arietinum* L.) for quantitative resistance to ascochyta blight. Theor Appl Genet 109:733–739. https://doi.org/10.1007/S00122-004-1693-X

Choi HK (2019) Translational genomics and multi-omics integrated approaches as a useful strategy for crop breeding. Genes Genomics 41:133–146. https://doi.org/10.1007/S13258-018-0751-8/FIGURES/2

Chongo G, Gossen BD (2001) Effect of plant age on resistance to *Ascochyta rabiei* in chickpea. Can J Plant Pathol 23:358–363. https://doi.org/10.1080/07060660109506956

Clulow SA, Lewis BG, Matthews P et al (1991) A pathotype classification for *Mycosphaerella pinodes*. J Phytopathol 131:322–332. https://doi.org/10.1111/j.1439-0434.1991.tb01203.x

Cobos MJ, Rubio J, Strange RN et al (2006) A new QTL for Ascochyta blight resistance in an RIL population derived from an interspecific cross in chickpea. Euphytica 149:105–111. https://doi.org/10.1007/S10681-005-9058-3/METRICS

Collard BCY, Pang ECK, Ades PK, Taylor PWJ (2003) Preliminary investigation of QTLs associated with seedling resistance to ascochyta blight from *Cicer echinospermum*, a wild relative of chickpea. Theor Appl Genet 107:719–729. https://doi.org/10.1007/S00122-003-1297-X

Coram TE, Pang ECK (2005) Isolation and analysis of candidate ascochyta blight defence genes in chickpea. Part I. Generation and analysis of an expressed sequence tag (EST) library. Physiol Mol Plant Pathol 66:192–200. https://doi.org/10.1016/J.PMPP.2005.08.003

Coram TE, Pang ECK (2006) Expression profiling of chickpea genes differentially regulated during a resistance response to *Ascochyta rabiei*. Plant Biotechnol J 4:647–666. https://doi.org/10.1111/j.1467-7652.2006.00208.x

Crociara C, Valetti L, Bernardi Lima N et al (2022) Morphological and molecular characterization, pathogenicity and sexual reproduction of *Ascochyta rabiei* isolates of chickpea fields in Argentina. J Phytopathol 170:221–232. https://doi.org/10.1111/jph.13073

Daba K, Deokar A, Banniza S et al (2016) QTL mapping of early flowering and resistance to ascochyta blight in chickpea. Genome 59:413–425. https://doi.org/10.1139/GEN-2016-0036

Danehloueipour N, Yan G, Clarke HJ, Siddique KHM (2007) Diallel analyses reveal the genetic control of resistance to ascochyta blight in diverse chickpea and wild *Cicer* species. Euphytica 154:195–205. https://doi.org/10.1007/S10681-006-9287-0

Deokar A, Sagi M, Daba K, Tar'an B (2019a) QTL sequencing strategy to map genomic regions associated with resistance to ascochyta blight in chickpea. Plant Biotechnol J 17:275–288. https://doi.org/10.1111/PBI.12964

Deokar A, Sagi M, Tar'an B (2019b) Genome-wide SNP discovery for development of high-density genetic map and QTL mapping of ascochyta blight resistance in chickpea (*Cicer arietinum* L.). Theor Appl Genet 132:1861–1872. https://doi.org/10.1007/S00122-019-03322-3/TABLES/4

Dey SK, Singh G (1993) Resistance to ascochyta blight in chickpea—genetic basis. Euphytica 68:147–153. https://doi.org/10.1007/BF00024163/METRICS

Dey SK, Singh G (1994) Seedborne infection of *Ascochyta rabiei* in chickpea and its transmission to aerial plant parts. Phytoparasitica 22:31–37. https://doi.org/10.1007/BF03158589

DeYoung BJ, Innes RW (2006) Plant NBS-LRR proteins in pathogen sensing and host defense. Nat Immunol 7:1243–1249. https://doi.org/10.1038/NI1410

Duzdemİr O, Selvİ B, Yanar Y, Yıldırımı A (2014) Sources of resistance in chickpea (*Cicer arietinum* L.) land races against *Ascochyta rabiei* causal agent of Ascochyta blight disease. Pak J Bot 46:1479–1483

FAOSTAT (2022) Statistical Database of the United Nation Food and Agriculture Organization (FAO) Statistical Division Rome. https://www.fao.org/faostat/en/#data/QCL. Accessed December 2022

Farahani S, Talebi R, Maleki M et al (2019) Pathogenic diversity of *Ascochyta rabiei* isolates and identification of resistance sources in core collection of chickpea germplasm. Plant Pathol J 35: 321. https://doi.org/10.5423/PPJ.OA.12.2018.0299

Flandez-Galvez H, Ades PK, Ford R et al (2003) QTL analysis for ascochyta blight resistance in an intraspecific population of chickpea (*Cicer arietinum* L.). Theor Appl Genet 107:1257–1265. https://doi.org/10.1007/s00122-003-1371-4

Fondevilla S, Krezdorn N, Rotter B et al (2015) In planta identification of putative pathogenicity factors from the chickpea pathogen *Ascochyta rabiei* by De novo transcriptome sequencing using RNA-Seq and massive analysis of cDNA ends. Front Microbiol 6:1329. https://doi.org/10.3389/fmicb.2015.01329

Garg V, Khan AW, Kudapa H et al (2019) Integrated transcriptome, small and degradome sequencing approaches provide insights into Ascochyta blight resistance in chickpea. Plant Biotechnol J 17:914–931. https://doi.org/10.1111/pbi.13026

Garg T, Mallikarjuna BP, Thudi M et al (2018) Identification of QTLs for resistance to Fusarium wilt and Ascochyta blight in a recombinant inbred population of chickpea (*Cicer arietinum* L.). Euphytica 214(3):1–11. https://doi.org/10.1007/S10681-018-2125-3

Gayacharan RU, Singh S et al (2020) Identification of novel resistant sources for ascochyta blight (*Ascochyta rabiei*) in chickpea. PLoS One 15:e0240589. https://doi.org/10.1371/JOURNAL.PONE.0240589

Gowda CLL (2005) Global theme on crop improvement, management and utilization for food security and health. Archival Report 2005. Patancheru, Andhra Pradesh, India

Gurha SN, Singh G, Sharma YR (2003) Diseases of chickpea and their management. In: Ali M, Kumar S, Singh NB (eds) ChicBBkpea research in India. Indian Institute of Pulses Research, Kanpur, pp 195–227

Gushkin JV (1946) Varieties of the Krasnoukut State Selection Station (in Russian). Sots Zernovoe Khoz 4:35–40

Hafiz A, Ashraf M (1953) Studies on the inheritance of resistance to Mycosphaerella blight in gram. Phytopathology 43:580–581

Hamwieh A, Imtiaz M, Hobson K, Kemal SA (2013) Genetic diversity of microsatellite alleles located at quantitative resistance loci for Ascochyta blight resistance in a global collection of chickpea germplasm. Phytopathol Mediterr 52:191–199

Haware MP (1987) Occurrence of perfect stage of *Ascochyta rabiei* in Syria. International Chickpea Newsletter 17:29–30

Haware MP, van Rheenen HA, Prasad NS (1995) Screening for Ascochyta blight resistance in chickpea under controlled environment and field conditions. Plant Dis 79:132–135. https://doi.org/10.1094/PD-79-0132

Hirdyani H (2014) Nutritional composition of Chickpea (*Cicer arietinum*-L) and value-added products-a review. Indian J Community Health 26:102–106

Ilarslan H, Dolar FS (2002) Histological and ultrastructural changes in leaves and stems of resistant and susceptible chickpea cultivars to *Ascochyta rabiei*. J Phytopathol 150:340–348. https://doi.org/10.1046/j.1439-0434.2002.00763.x

Iqbal SM, Ali S, Ghafoor A (2010) Development of resistance in chickpea to Ascochyta blight. Mycopathologia 8:61–64

Iqbal J, Zafar SA, Ashraf A, Hassan A (2018) Pathogenic variability in chickpea genotypes against chickpea blight. J Agri Sci and Basic Sci 3:19–22

Iruela M, Castro P, Rubio J et al (2007) Validation of a QTL for resistance to ascochyta blight linked to resistance to fusarium wilt race 5 in chickpea (*Cicer arietinum* L.). Ascochyta blights

of grain legumes. Springer, New York, pp 29–37. https://doi.org/10.1007/978-1-4020-6065-6_4/COVER
Islam W, Qasim M, Noman A et al (2017) Genetic resistance in chickpea against Ascochyta blight: Historical efforts and recent accomplishments. J Anim Plant Sci 27:1941–1957
Jayakumar P, Gan YT, Gossen BD et al (2005) Ascochyta blight of chickpea: infection and host resistance mechanisms. Can J Plant Pathol 27:499–509. https://doi.org/10.1080/07060660509507251
Jha UC (2018) Current advances in chickpea genomics: applications and future perspectives. Plant Cell Rep 37:947–965. https://doi.org/10.1007/S00299-018-2305-6
Jhorar OP, Butler DR, Mathauda SS (1998) Effects of leaf wetness duration, relative humidity, light and dark on infection and sporulation by *Didymella rabiei* on chickpea. Plant Pathol 47:586–594. https://doi.org/10.1046/J.1365-3059.1998.0280A.X
Jhorar OP, Mathauda SS, Singh G et al (1997) Relationships between climatic variables and Ascochyta blight of chickpea in Punjab, India. Agric For Meteorol 87:171–177. https://doi.org/10.1016/S0168-1923(97)00014-2
Jukanti AK, Gaur PM, Gowda CLL, Chibbar RN (2012) Nutritional quality and health benefits of chickpea (*Cicer arietinum* L): a review. Br J Nutr 108:11–26
Kaiser WJ (1997) Inter- and intranational spread of ascochyta pathogens of chickpea, faba bean, and lentil. Can J Plant Pathol 19:215–224. https://doi.org/10.1080/07060669709500556
Kaur L, Pande S, Sandhu JS et al (2011) Characterization of chickpea lines for resistance to Ascochyta blight. J Food Leg 24:310–312
Kaur L, Sandhu JS, Malhotra RS et al (2012a) Sources of stable resistance to Ascochyta blight in exotic *kabuli* chickpea. J Food Leg 25:79–80
Kaur L, Singh VP, Sandhu JS (2012b) Characterisation of *Ascochyta rabiei* isolates and evaluation of genotypic stability in chickpea. Archives Phytopath Pl Prot 45:83–89. https://doi.org/10.1080/03235401003672954
Kaur L, Sirari A, Kumar D et al (2013) Combining ascochyta blight and botrytis grey mould resistance in chickpea through interspecific hybridization. Phytopathol Mediterr 52(1):157–165
Khaliq I, Fanning J, Melloy P et al (2020) The role of conidia in the dispersal of *Ascochyta rabiei*. Eur J Plant Pathol 158:911–924. https://doi.org/10.1007/s10658-020-02126-2
Khan MS, Ramsey MD, Corbiere R, Infantino A, Porta-Puglia A, Bouznad Z, Scott ES et al (1999) Ascochyta blight of chickpea in Australia: identification, pathogenicity and mating type. Plant Pathol 48:230–234. https://doi.org/10.1046/j.1365-3059.1999.00338.x
Kimber RBE, Shtienberg D, Ramsey MD, Scott ES (2007) The role of seedling infection in epiphytotics of ascochyta blight on chickpea. Eur J Plant Pathol 117:141–152. https://doi.org/10.1007/s10658-006-9080-x
Kimruto P, Towett B, Mulwa R et al (2013) Evaluation of chickpea genotypes for resistance to Ascochyta blight (*Ascochyta rabiei*) disease in the dry highlands of Kenya. Phytopathol Mediterr 52:212–221. https://doi.org/10.14601/PHYTOPATHOL_MEDITERR-11825
Kumar K, Purayannur S, Kaladhar VC et al (2018) mQTL-seq and classical mapping implicates the role of an AT-Hook motif containing nuclear localized (AHL) family gene in Ascochyta blight resistance of chickpea. Plant Cell Environ 41:2128–2140. https://doi.org/10.1111/PCE.13177
Kumar AS, Rani U, Bindra S et al (2021) Identification of novel sources for resistance against Ascochyta blight and botrytis grey mould of chickpea. Agric Res J 58:581–585. https://doi.org/10.5958/2395-146X.2021.00082.X
Kushwah A, Bhatia D, Rani U et al (2021) Molecular mapping of quantitative trait loci for ascochyta blight and botrytis grey mould resistance in an inter-specific cross in chickpea (*Cicer arietinum* L.) using genotyping by sequencing. Breed Sci 71:229–239. https://doi.org/10.1270/JSBBS.20085
Labdi M, Malhotra RS, Benzohra IE, Imtiaz M (2013) Inheritance of resistance to *Ascochyta rabiei* in 15 chickpea germplasm accessions. Plant Breed 132:197–199. https://doi.org/10.1111/PBR.12038

Lebiedzińska A, Szefer P (2006) Vitamins B in grain and cereal–grain food, soy-products and seeds. Food Chem 95:116–122. https://doi.org/10.1016/j.foodchem.2004.12.024

Lekhi D, Kaur A, Rani U et al (2022) Inheritance and mapping of QTLs for Ascochyta blight resistance in chickpea (*Cicer arietinum*). Indian J Agric Sci 92:316–319. https://doi.org/10.56093/IJAS.V92I3.122677

Leo AE, Linde CC, Ford R (2016) Defence gene expression profiling to *Ascochyta rabiei* aggressiveness in chickpea. Theor Appl Genet 129:1333–1345. https://doi.org/10.1007/S00122-016-2706-2

Li Y, Ruperao P, Batley J et al (2017) Genome analysis identified novel candidate genes for ascochyta blight resistance in chickpea using whole genome re-sequencing data. Front Plant Sci 8:359. https://doi.org/10.3389/FPLS.2017.00359/BIBTEX

Lichtenzveig J, Bonfil DJ, Bin ZH et al (2006) Mapping quantitative trait loci in chickpea associated with time to flowering and resistance to *Didymella rabiei* the causal agent of Ascochyta blight. Theor Appl Genet 113:1357–1369. https://doi.org/10.1007/S00122-006-0390-3

Luthra JC, Sattar A, Bedi KS (1941) *Mycosphaerella rabiei* Kovacevski = *Ascochyta rabiei* (Pass.) Lab. in gram types. Ind J agric Sci 11:249–264

Madrid E, Seoane P, Claros MG et al (2014) Genetic and physical mapping of the QTLAR3 controlling blight resistance in chickpea (*Cicer arietinum* L). Euphytica 198:69–78. https://doi.org/10.1007/S10681-014-1084-6

Manjunatha L, Saabale PR, Srivastava AK et al (2018) Present status on variability and management of *Ascochyta rabiei* infecting chickpea. Indian Phytopathol 71:9–24. https://doi.org/10.1007/s42360-018-0002-6

Mashaki KM, Garg V, Nasrollahnezhad Ghomi AA et al (2018) RNA-Seq analysis revealed genes associated with drought stress response in *kabuli* chickpea (*Cicer arietinum* L.). PLoS One 13:e0199774. https://doi.org/10.1371/JOURNAL.PONE.0199774

Materne M, Knights T, Meredith K, Bretag T, Brouwer JB, Hawthorne BW (2002) Update Ascochyta-resistant chickpeas. Ground Cover 39(April 2002). http://www.grdc.com.au/growers/gc/gc39/pulses.htm#update

Maurya R, Singh Y, Sinha M et al (2020) Transcript profiling reveals potential regulators for oxidative stress response of a necrotrophic chickpea pathogen *Ascochyta rabiei*. 3 Biotech, 10:117. https://doi.org/10.1007/s13205-020-2107-8

McHale L, Tan X, Koehl P, Michelmore RW (2006) Plant NBS-LRR proteins: adaptable guards. Genome Biol 7:1–11. https://doi.org/10.1186/GB-2006-7-4-212/FIGURES/5

Meyers BC, Kozik A, Griego A et al (2003) Genome-wide analysis of NBS-LRR-encoding genes in *Arabidopsis*. Plant Cell 15:809–834. https://doi.org/10.1105/TPC.009308

Mohanty JK, Jha UC, Dixit GP, Parida SK (2022) Harnessing the hidden allelic diversity of wild *Cicer* to accelerate genomics-assisted chickpea crop improvement. Mol Biol Rep 49:5697–5715. https://doi.org/10.1007/S11033-022-07613-9

Nasir M, Bretag TW, Kaiser WJ et al (2000) Screening chickpea germplasm for ascochyta blight resistance. Australas Plant Pathol 29:102–107. https://doi.org/10.1071/AP00018/METRICS

Nayak SN, Zhu H, Varghese N et al (2010) Integration of novel SSR and gene-based SNP marker loci in the chickpea genetic map and establishment of new anchor points with *Medicago truncatula* genome. Theor Appl Genet 120:1415. https://doi.org/10.1007/S00122-010-1265-1

Nene YL (1982) A review of ascochyta blight of chickpea∗. Trop Pest Manag 28:61–70. https://doi.org/10.1080/09670878209370675

Nene YL, Haware MP, Reddy MV (1981) Chickpea diseases: resistance screening techniques International crops research institute for the semi-arid tropics. Information Bulletin No. 10 Patencheru 502 324. International Crops Research Institute for the Semi-Arid Tropics 10, Andhra Pradesh, India

Nene YL, Reddy MV (1987) Chickpea diseases and their control. In: Saxena MC, Singh KB (eds) The chickpea. Wallingford (UK), Commonwealth Agricultural Bureaux International, pp 233–270

Nene YL, Sheila VK, Sharma SB, et al (1996) A world list of chickpea and pigeon pea pathogens Pp 27 ICRISAT, Patancheru, Andhra Pradesh

Newman TE, Jacques S, Grime C et al (2021) Identification of novel sources of resistance to Ascochyta blight in a collection of wild *Cicer* accessions. Phytopathology 111:369–379. https://doi.org/10.1094/PHYTO-04-20-0137-R

Pal M, Kumar SPRJ, Singh B (1999) Genetics of resistance to Ascochyta blight in chickpea. Indian Phytopathol 52(4):403–407

Pande S, Sharma M, Gaur PM, Gowda CLL (2010) Host plant resistance to Ascochyta blight of chickpea. Information bulletin No. 82. International Crops Research Institute for the Semi-Arid Tropics

Pande S, Sharma M, Gaur PM et al (2011) Development of screening techniques and identification of new sources of resistance to Ascochyta blight disease of chickpea. Australas Plant Pathol 40: 149–156. https://doi.org/10.1007/S13313-010-0024-8

Pande S, Sharma M, Gaur PM et al (2013) Biplot analysis of genotype × environment interactions and identification of stable sources of resistance to Ascochyta blight in chickpea (*Cicer arietinum* L.). Australas Plant Pathol 42:561–571. https://doi.org/10.1007/S13313-013-0219-X

Pande S, Siddique KHM, Kishore GK et al (2005) Ascochyta blight of chickpea (*Cicer arietinum* L.): a review of biology, pathogenicity, and disease management. Aust J Agric Res 56:317. https://doi.org/10.1071/AR04143

Pandey BK, Singh US, Chaube HS et al (1987) Mode of infection of Ascochyta blight of chickpea caused by *Ascochyta rabiei*. J Phytopathol 119:88–93. https://doi.org/10.1111/j.1439-0434.1987.tb04387.x

Pastor S, Crociara C, Valetti L et al (2022) Screening of chickpea germplasm for Ascochyta blight resistance under controlled environment. Euphytica 218:12. https://doi.org/10.1007/S10681-021-02963-0

Pieters R, Tahiri A (1986) Breeding chick-pea for horizontal resistance to Ascochyta blight in Morocco. FAO Plant Protection Bulletin (FAO). https://doi.org/10.3/JQUERY-UI.JS

Pittaway JK, Robertson IK, Ball MJ et al (2008) Chickpeas may influence fatty acid and fiber intake in an ad libitum diet, leading to small improvements in serum lipid profile and glycemic control. J Am Diet Assoc 108:1009–1013. https://doi.org/10.1016/j.jada.2008.03.009

Rao NK, Van den Maesen LJG (1985) Distribution of qualitative traits in the world germplasm of chickpea (*Cicer arietinum* L). Euphytica 34:697–670

Reddy MV (1984) Screening techniques for Ascochyta blight of chickpeas. In: Proceedings of a Training course, PARC/ICARDA, Islamabad, Pakistan, ICARDA, Aleppo, Syria, pp 67–76

Sabbavarapu MM, Sharma M, Chamarthi SK et al (2013) Molecular mapping of QTLs for resistance to Fusarium wilt (race 1) and Ascochyta blight in chickpea (*Cicer arietinum* L.). Euphytica 193:121–133. https://doi.org/10.1007/S10681-013-0959-2/FIGURES/4

Sagi MS, Deokar AA, Tar'an B (2017) Genetic analysis of NBS-LRR gene family in chickpea and their expression profiles in response to ascochyta blight infection. Front Plant Sci 8:838. https://doi.org/10.3389/FPLS.2017.00838/BIBTEX

Salaria S (2020) Morphological and molecular profiling of interspecific derivatives of *Cicer arietinum* L. x C. pinnatifidum. M. Sc. Thesis. Punjab Agricultural University, Ludhiana

Sally V (2005) Population studies of *Ascochyta rabiei* on chickpea in Saskatchewan. Doctoral dissertation, University of Saskatchewan

Sandhu JS, Singh G, Bains TS et al (2004) PBG 5, a new multiple resistant *desi* chickpea variety for Punjab (India). Int Chickpea Pigeonpea News 11:18–20

Shah TM, Imran M, Atta BM et al (2015) Screening of chickpea advanced lines for sources of resistance against blight and wilt two major diseases of chickpea. Pak J Bot 47:2443–2448

Shah JA, Iqbal A, Mahmood MT et al (2021) Screening of elite chickpea germplasm against Ascochyta blight under controlled conditions. Pak J Agric Sci 34:774–780. https://doi.org/10.17582/JOURNAL.PJAR/2021/34.4.774.780

Sharma YR, Singh G, Kaur L (1995) A rapid technique for Ascochyta blight resistance in chickpea. ICPN 2:34–35

Sharma S, Upadhyaya HD, Roorkiwal M, Varshney RK, Gowda CL (2013) Chickpea. In: Genetic and genomic resources of grain legume improvement. Elsevier, Amsterdam, pp 81–111

Singh G (1990) Identification and designation of physiologic races of *Ascochyta rabiei* in India. Indian Phytopathol 43:48–52

Singh S, Kapoor S (1986) Gram lines with high degree of blight resistance to pod infection. Indian J Mycol Plant Path 16:197–198

Singh G, Kapoor S, Singh K (1982) Screening chickpea for gray mould resistance. Int Chickpea Newsletter 7:13–14

Singh G, Kapoor S, Singh K, Gill AS (1984a) Screening for resistance to gram wilt. Indian Phytopathol 37:393–394

Singh R, Kumar K, Purayannur S et al (2022a) *Ascochyta rabiei*: a threat to global chickpea production. Mol Plant Pathol 23:1241–1261. https://doi.org/10.1111/mpp.13235

Singh K, Nizam S, Sinha M, Verma PK (2012) Comparative Transcriptome Analysis of the Necrotrophic Fungus *Ascochyta rabiei* during Oxidative Stress: Insight for Fungal Survival in the Host Plant. PLoS One 7:e33128. https://doi.org/10.1371/journal.pone.0033128

Singh R, Pal M (1993) Pathogenic variability in *Ascochyta rabiei* (Pass) Labr causing chickpea blight India. J Mycol Plant Pathol 23:51–57

Singh M, Rani S, Malhotra N et al (2018) Transgressive segregations for agronomic improvement using interspecific crosses between *C. arietinum* L. x *C. reticulatum* Ladiz. and *C. arietinum* L. x *C. echinospermum* Davis species. PLoS One 13:e0203082. https://doi.org/10.1371/JOURNAL.PONE.0203082

Singh KB, Reddy MV (1983) Inheritance of resistance to Ascochyta blight in chickpea. Crop Sci 23:9–10. https://doi.org/10.2135/CROPSCI1983.0011183X002300010003X

Singh G, Sharma YR (1998) Ascochyta blight of chickpea. In: Upadhyay RK, Mukherji KG, Rajak RL (eds) IPM system in agriculture. Aditya Books Pvt Ltd, New Delhi, pp 163–195

Singh S, Singh I, Kumar A et al (2022b) Variety PBG8. Indian J Genet Plant Breed 82:875

Singh S, Singh I, Sandhu JS et al (2015) PBG7: A new high yielding variety of *desi* gram (*cicer arietinum* L.). Agri Res J 52:212–213

Singh G, Verma MM, Gill AS et al (1984b) Screening of gram varieties against Ascochyta blight. Crop Improv 11:153–154

Singh G, Kaur P, Kumar A, Verma MM, Kaur L, Sharma YR (1995) Primary and secondary spread of Ascochyta blight of gram. In: Gupta VK, Sharma RC, Johansen C (eds) Integrated disease management and plant health. Scientific Publishers, Jodhpur, India, pp 65–69

Spoel SH, Johnson JS, Dong X et al (2007) Regulation of tradeoffs between plant defenses against pathogens with different lifestyles. Proc Natl Acad Sci India Sect B Biol Sci 104:18842–18847. https://doi.org/10.1073/pnas.0708139104

Stephens A, Lombardi M, Cogan NOI et al (2014) Genetic marker discovery, intraspecific linkage map construction and quantitative trait locus analysis of ascochyta blight resistance in chickpea (*Cicer arietinum* L.). Mol Breed 33:297–313. https://doi.org/10.1007/S11032-013-9950-9

Taran B, Warkentin TD, Vandenberg A (2013) Fast track genetic improvement of ascochyta blight resistance and double podding in chickpea by marker-assisted backcrossing. Theor Appl Genet 126:1639–1647. https://doi.org/10.1007/S00122-013-2080-2

Tekeoglu M, Rajesh PN, Muehlbauer FJ (2002) Integration of sequence tagged microsatellite sites to the chickpea genetic map. Theor Appl Genet 105:847–854. https://doi.org/10.1007/S00122-002-0993-2

Tewari SK, Pandey MP (1986) Genetics of resistance to ascochyta blight in chickpea (*Cicer arietinum* L.). Euphytica 35:211–215. https://doi.org/10.1007/BF00028559/METRICS

Thudi M, Bohra A, Nayak SN et al (2011) Novel SSR markers from BAC-end sequences, DArT arrays and a comprehensive genetic map with 1,291 Marker Loci for Chickpea (*Cicer arietinum* L.). PLoS One 6:e27275. https://doi.org/10.1371/journal.pone.0027275

Udupa SM, Baum M (2003) Genetic dissection of pathotype-specific resistance to ascochyta blight disease in chickpea (*Cicer arietinum* L.) using microsatellite markers. Theor Appl Genet 106: 1196–1202. https://doi.org/10.1007/S00122-002-1168-X

van Ooijen G, van den Burg HA, Cornelissen BJC, Takken FLW (2007) Structure and function of resistance proteins in solanaceous plants. Annu Rev Phytopathol 45:43–72. 101146/annurev.phyto45062806094430

Varshney RK, Hiremath PJ, Lekha P et al (2009) A comprehensive resource of drought- and salinity-responsive ESTs for gene discovery and marker development in chickpea (*Cicer arietinum* L.). BMC Genomics 10:1–18. https://doi.org/10.1186/1471-2164-10-523/TABLES/4

Varshney RK, Mohan SM, Gaur PM et al (2014) Marker-assisted backcrossing to introgress resistance to fusarium wilt race 1 and Ascochyta blight in C 214, an elite cultivar of chickpea. Plant Genome 7(1):10. https://doi.org/10.3835/PLANTGENOME2013.10.0035

Verma MM, Brar HS, Sandhu JS, Singh G (1992) PBG 1–A new promising chickpea cultivars that beats blight. Food Legumes Coarse Grain News 19:2–5

Verma S, Gazara RK, Nizam S et al (2016) Draft genome sequencing and secretome analysis of fungal phytopathogen *Ascochyta rabiei* provides insight into the necrotrophic effector repertoire. Sci Rep 6:24638. https://doi.org/10.1038/srep24638

Vir S, Grewal JS (1974) Physiologic specialization in *Ascochyta rabiei* the causal organism of gram blight. Indian Phytopathol 27:355–360

Vir S, Grewal JS, Gupta VP (1975) Inheritance of resistance to Ascochyta blight in chickpea. Euphytica 24:209–211. https://doi.org/10.1007/BF00147187/METRICS

von Wettberg EJB, Chang PL, Başdemir F et al (2018) Ecology and genomics of an important crop wild relative as a prelude to agricultural innovation. Nat Commun 9:649–649. https://doi.org/10.1038/S41467-018-02867-Z

Weising K, Kaemmer D, Epplen JT et al (1991) DNA fingerprinting of Ascochyta rabiei with synthetic oligodeoxynucleotides. Curr Genet 19:483–489. https://doi.org/10.1007/BF00312740/METRICS

Winter P, Benko-Iseppon A-M, Hüttel B et al (2000) A linkage map of the chickpea (Cicer arietinum L.) genome based on recombinant inbred lines from a *C. arietinum* × *C. reticulatum* cross: localization of resistance genes for fusarium wilt races 4 and 5. Theor Appl Genet 101:1155–1163

Zachos DG, Panagopolos CG, Markis SA et al (1963) Research on the biology, epidemiology and control of anthracnose of chickpea (in French). Annales de l'Institut Phytopathologique Benaki 5:67–192

Zewdie A, Tadesse N (2018) Screening of chickpea for resistance to Ascochyta blight (*Didymella rabies*) under Field Conditions. Ethiop J Agric Sci 6:39–49

Zohary D, Hopf M (2006) Pulses-in domestication of plants in the old world: the origin and spread of cultivated plants in West Asia, Europe, and the Nile Valley, 3rd edn. Oxford University Press, New York, p 108

Fusarium Wilt of Chickpea: Breeding and Genomic Approaches for Designing Wilt-Resistant Chickpea

3

Kamal Dev Sharma

Abstract

Chickpea (*Cicer arietinum* L.), an important legume crop in the world, is sensitive to several biotic and abiotic stresses. The wilt caused by fungal pathogen *Fusarium oxysporum* f. sp. *ciceris* (*Foc*) is the most important disease of chickpea and is prevalent in all major chickpea growing areas of the world. It can cause up to 99.7% grain yield losses in sensitive genotypes of chickpea. Eight races of the pathogen have been reported worldwide, however, sporadic reports of occurrence of more races exist. Chickpea resistance to *Foc* is complete and is governed by one to three vertical resistance genes. Apart from complete resistance, two other types of host resistances (both partial), i.e., late wilting and slow wilting against *Foc* also exist. The resistance to wilt has been introgressed into susceptible but high yielding varieties using traditional breeding methods. Following development of several types of DNA-based markers in chickpea, e.g., SSRs, Intron targeted primers, diversity arrays technology, Insertion–deletion markers, single nucleotide polymorphisms, markers linked closely to fusarium wilt resistance genes were identified and used for marker assisted breeding. The markers were also used to pyramid genes in single germplasm line and chickpea varieties bred through marker-assisted selection/marker-assisted breeding have been released for cultivation, e.g., Pusa Manav/Pusa Chickpea 20211 in India. Efforts are also being made to exploit genomic selection and genomics assisted breeding to develop designer chickpea possessing wilt resistance.

K. D. Sharma (✉)
Department of Agricultural Biotechnology, CSK Himachal Pradesh Agricultural University, Palampur, Himachal Pradesh, India

U. C. Jha et al. (eds.), *Diseases in Legume Crops*,
https://doi.org/10.1007/978-981-99-3358-7_3

3.1 Introduction

Chickpea (*Cicer arietinum* L.), belonging to the family Fabaceae and subfamily Papilionaceae, is one of the most important grain legumes in the word. It ranks third in terms of total global production (14.78 MT) and area under cultivation (14.56 Mha) after soybean (352.64 MT from 123.55 Mha) and dry bean (31.40 MT from 36.46 Mha) (FAOSTAT 2017). India is the topmost producer of chickpea. It contributes about 70% (116.2 lakh tonnes) to the total world production of chickpea and for over 45% of total pulse production within India. Being leguminous in nature, its seeds contain high-quality proteins for human as well as animal consumption. Chickpea seeds serve as a rich source of not only proteins but also essential minerals and dietary fiber. Chickpea is consumed as *dal* (soup with seeds) primarily in the Indian subcontinent whereas it is consumed as sprouts or various preparations such as hummus (paste of garbanzo beans with tahini and olive oil that is used as a dip for veggies or crackers or spread on sandwiches) in Arabic world, roasted chickpeas, soup and stews, as a thickener in soups, splashed in sandwiches (after boiling or softening in water), as salad, meatless protein in pasta and rice dishes, and as Aquafaba (liquid in cans of chickpea is whipped to froth and used in place of egg white). Being super rich in protein and other nutrients like vitamins A, B6, and C, folate, manganese, zinc, iron, and magnesium, chickpeas offer several health benefits such as promotion of skin health; prevention of cancer, diabetes, and hair loss; and regulation of blood sugar levels, to lose weight, boost digestion, improve heart health, and eliminate wrinkles.

Chickpea also enriches soil with nitrogen. Chickpea roots form nodules through association with bacteria of species *Rhizobium* wherein the nodules fix atmospheric nitrogen. In fact, root nodules contain bacteria surrounded by plant cells. These bacteria in nodules capture NO_2 from the atmosphere and convert it to ammonia (NH_4). The nitrogen is then assimilated into the plant system. The plants in return provide nutrients primarily as carbohydrates for growth and development of bacteria. Chickpea, thus, is not only a good source of proteinaceous food for human beings but also a plant that enriches soil with nitrogen. Due to nodule formation, chickpea also requires less nitrogenous fertilization for its growth. Chickpea tolerates drought and can be grown successfully on marginal soils unsuitable for crops that require high inputs as well as high water volumes. It is a common practice with farmers in several parts of the world including India, the largest producer of chickpea, to grow chickpea on marginal soils and in areas with no sources of irrigation.

The yield per unit area of chickpea (0.85 tonnes/ha) is far less than the potential yield (40.0 tonne/ha) of chickpea. Narrow genetic base and susceptibility to biotic and abiotic stresses are the major factors in the realization of the full yield potential of chickpea. Chickpea originated from a wild progenitor—*Cicer reticulatum*—that is limited in its distribution to warmer areas of the Mediterranean and has very low genetic variability. Consequently, chickpea accessions also lack adequate diversity in its populations. Among the biotic stresses, insect pest *Helicoverpa armigera* and fungal diseases, namely, fusarium wilt caused by *Fusarium oxysporum* f. sp. *ciceris*

(*Foc*) and ascochyta blight caused by *Ascochyta rabiei*, cause considerable yield losses in chickpea. Among abiotic stresses, cold, heat, and drought are the major yield-limiting factors. Sources of resistance to *H. armigera* are not available in chickpea germplasm whereas resistance/tolerance to ascochyta blight and abiotic stresses is partial, multigenic, and governed by several QTLs, major and minor. Resistance to fusarium wilt exists in chickpea. It is vertical (qualitative or complete), race-specific in nature, and governed by monogenes to oligogenes. The main emphasis of breeding in chickpea is to increase yield either by broadening the genetic base or incorporating resistance/tolerance to biotic and abiotic stresses, thereby increasing the yield by reducing losses by these stresses. Among the biotic and abiotic stresses of chickpea, fusarium wilt is a fairly better-studied stress with major emphasis on identification of race flora, identification of sources of resistance, and deployment of resistant genes.

3.2 Chickpea Wilt: Causal Organism and Losses

Wilt of chickpea is caused by a fungal pathogen *Fusarium oxysporum (Schlechtend.: Fr.) f.* sp. *ciceris (Padwick) Matuo & K. Sato* that belongs to family *Nectriaceae*, order *Hypocreales*, division *Ascomycota*, and kingdom fungi. This fungus reproduces by asexual means, and no sexually reproductive stage or teleomorph is reported so far. It is known to produce three types of asexual spores, i.e., macroconidia, microconidia, and chlamydospores. The chickpea wilt was first reported by Butler in 1918 from India, and the correct etiology of the disease was determined after several years, i.e., in 1940 by Padwick (Cunnington et al. 2007). The causal agent of the disease was named as *Fusarium orthoceras* Appel & Wollenw. var. *ciceri* by Padwick. The pathogen was renamed as *F. oxysporum* Schl. f. sp. *ciceri* (Padwick) Snyder & Hansen by Chattopadhyay and Sen Gupta (Jiménez-Díaz et al. 2015). The pathogen name was again revised as *Fusarium oxysporum* f.sp. *ciceris* by Holliday in 1980 (Jalali and Chand 1992; Nene and Reddy 1987). Most of the *formae speciales* of *Fusarium oxysporum* are polyphyletic in origin; however, *Fusarium oxysporum* f.sp. *ciceris* (*Foc*) is one of the few *F. oxysporum* with monophyletic origin (Jiménez-Gasco et al. 2002; Jiménez-Gasco and Jiménez-Díaz 2003; Demers et al. 2014) as this *formae speciales* infects only chickpea. The origin of the pathogen has also been debated. It either originated in India or in the Mediterranean (Jiménez-Gasco et al. 2004). The Mediterranean is also the center of origin of chickpea, and hence, it is suspected that mild races such as race 0 of *Foc* have originated in the Mediterranean and subsequently spread to other parts of the world along with migration of chickpea (Jiménez-Gasco et al. 2004).

Chickpea wilt is a widespread disease. It has been reported from all the major chickpea-growing regions of the world, i.e., 32 countries spread over all 6 continents (Singh et al. 2014). The disease is highly devastating and can cause yield losses of up to 100% under severe epiphytotic conditions (Haware and Nene 1980, 1982; Halila and Strange 1997; Jiménez-Díaz et al. 2015; Trapero-Casas and Jiménez-Díaz 1985). Worldwide annual crop losses from fusarium wilt in chickpea are estimated to be 10–15% (Jiménez-Díaz et al. 2015). In India, the country that occupies 70% of chickpea area in the world, wilt incidence varies from 14.1 to 32.0% (Dubey et al. 2010) with annual yield losses to the tune of 10% (Singh and Dahiya 1973; Jalali and Chand 1992). In Mexico, the third largest producer of chickpea, the disease caused 60% losses in the Sonora region and 20% in the Sinaloa region (Manjarrez-Sandoval et al. 2004) whereas in Ethiopia yield losses are 30% (Shehabu et al. 2008). In the year 2000, an elaborate study was conducted to estimate the losses caused by fusarium wilt in chickpea (Navas-Cortés et al. 2000). A total of 108 epidemics of fusarium wilt of chickpea were created in the field using microplots, 2 germplasm lines of chickpea (most susceptible cv. P-2245 and least susceptible cv. PV-61), and 2 races (race 0, race 5) of the pathogen. Race 5 resulted in higher yield losses [highest losses were 99.7% in the most susceptible cv. P-2245 and lowest (81.9%) in the least susceptible cv. PV-61] as compared to those in race 0 (65.6 and 30.6%, respectively, for cvs. P-2245 and PV-61) (Navas-Cortés et al. 2000). Evidently, the disease is highly devastating under severe epiphytotic conditions if a virulent race of the pathogen is present in the soil. It also points to possible epidemics of disease if chickpea cultivars possessing resistance genes are not employed in chickpea-growing areas of the world. Early wilting causes more losses (77–94%) as compared to those caused by late wilting (24–65%) (Haware and Nene 1980) suggesting that late wilting cultivars can be used for cultivation if cultivars with complete resistance are not available.

3.3 Physiological Specialization in *F. oxysporum* f. sp. *ciceris*

Worldwide populations of *Foc* are comprised of eight races, 0, 1A, 1B/C, 2, 3, 4, 5, and 6 (Haware and Nene 1979; Jimenez-Diaz et al. 1993; Sharma et al. 2004a, 2005; Sharma and Muehlbauer 2007). Races 0, 1B/C, 5, and 6 have limited geographical distribution as these are restricted to Spain, the Mediterranean region, and the USA. In contrast to this, races 1A, 2, 3, and 4 are widespread. Races 2, 3, and 4 are primarily restricted to India and race 1A to India and the USA. Of late, race 6 is also identified from India (Dubey and Singh 2008) suggesting its spread from the Mediterranean to other parts of the world. Thus in India, five races of the pathogen (1A, 2, 3, 4, and 6) exist (Haware and Nene 1979; Dubey and Singh 2008) whereas in Iraq four races, 0, 1B/C, 4, and 5, have been identified (Al-Taae et al. 2013), in Syria four races (0, 1B/C, 5, and 6) (Alloosh et al. 2019), and in Spain two races (0 and 1B/C) (Jimenez-Diaz et al. 1993). The information on race flora of *Foc* in the world is, however, far from clear as it has not been studied adequately in some parts of the world, especially Pakistan, Iran, etc.

In India, the largest producer of chickpea having a maximum area under chickpea cultivation, the complexity of races is also far from resolved. Dubey et al. (2012) used a set of differentials with 10 lines and evaluated 70 isolates of *Foc* from 13 states of India and 4 crop cultivation zones for physiological specialization. The study identified eight races (race 1 to race 8) of *Foc*; however, it is not clear if these races match the eight races already identified in chickpea or not. In other studies, existence of at least three additional races than already identified has been reported from India (Dubey et al. 2010; Honnareddy and Dubey 2006). It was also suggested to add additional differentials to the existing set to correctly identify all races of the pathogen. Similar to these studies, some new races apart from the already reported races (1–4, 6, and) were identified from India by Dubey and Singh (2008). The same was true for Syria where "four races (0, 1B/C, 5 and 6) were identified and 12 isolates were not designated to any of the known races" (Alloosh et al. 2019). These studies clearly suggest that efforts must be made to develop an international set of differentials for physiological specialization of *Foc* and classify all the available isolates into well-characterized races. Even within India, different workers used different differential sets, and no comprehensive efforts have been made so far to develop a common set of differentials for identification of all races of *Foc*. At the international level, there is a need to develop a common set of differentials for identification of races of *Foc* prevalent throughout the world. To achieve it, a worldwide consortium may be created with the task to analyze isolates of *Foc* across the world for pathogenicity on several lines and develop a comprehensive differential set that can differentiate all races prevalent in the world. Such set would resolve the complexity of the worldwide race flora of *Foc* and would allow pathologists to have accurate nomenclature of the *Foc* races. Since chickpea wilt is managed primarily by deployment of host-resistant genes, correct nomenclature of races will also allow scientists to identify and deploy race-specific sources of resistance across geographic and country boundaries.

Based on symptoms produced in the host chickpea, *Foc* has been divided into two pathotypes named as "yellowing syndrome" and "wilting syndrome." In yellowing syndrome, the leaves of susceptible hosts show gradual yellowing, and plants die after a considerable period of time following infection. In wilting syndrome, leaves show chlorosis, and plants wilt much early as compared to yellowing syndrome pathotypes. Of the eight races of the pathogen reported so far, only two (race 0, race 1B/C) cause yellowing syndrome whereas the rest of the races cause wilting syndrome. Apparently the majority of the pathogen isolates are associated with wilting syndrome. The yellowing syndrome pathotypes not only produce mild symptoms but are also less devastating as compared to the wilting syndrome pathotypes as yield losses by the yellowing syndrome race (race 0) were lower than that by the wilting syndrome race (race 5) (Navas-Cortés et al. 2000).

As described above, *Foc* has a monophyletic origin, and all races of *Foc* have originated from a common ancestor (Jiménez-Gasco et al. 2002; Jiménez-Gasco and Jiménez-Díaz 2003; Demers et al. 2014) probably from race 0 in Spain (race with mild symptoms) or race 1A in the Indian subcontinent (race with severe wilt symptoms). Stepwise evolution led to differentiation of the common ancestor into different races (Jiménez-Gasco and Jiménez-Díaz 2003). Once the complexity of the race flora of *Foc* is resolved, renewed studies might be conducted to confirm the monophyletic origin of *Foc*.

Efforts have also been made to characterize different isolates and races of the pathogen using DNA-based markers or using vegetative compatibility groups with the objective to elucidate diversity in pathogen populations and develop race-specific markers for molecular diagnostics. As early as 1994, random amplified polymorphic DNA (RAPD) markers were used for differentiation of yellowing syndrome and wilting syndrome pathotypes (Kelly et al. 1994). Subsequently, RAPD, simple-sequence repeats (SSR), direct amplified minisatellite DNA (DAMD), amplified fragment length polymorphism (AFLP), BOX-and rep-PCR markers were used for estimation of genetic diversity and population structure of *Foc* (Jiménez-Gasco et al. 2001; Bayraktar et al. 2008; Dubey and Singh 2008; Kashyap et al. 2016; Alloosh et al. 2019). RAPD markers were also used for identification of races 0, 1B/C, 5, and 6 of *Foc* (Jiménez-Gasco et al. 2001). DNA fingerprinting using different genes of the fungus, i.e., *translation elongation factor 1α (EF1α)*, *β-tubulin, histone 3, actin*, and *calmodulin*, was also carried out to demonstrate the monophyletic origin of the pathogen (Jiménez-Gasco et al. 2002; Jiménez-Gasco and Jiménez-Díaz 2003; Demers et al. 2014).

3.4 Inheritance for Wilt Resistance, Late Wilting, Slow Wilting, and Wilt Differentials

Initially, inheritance of resistance to wilt was studied for race 1, and resistance was found to be governed by a recessive gene (Ayyar and Iyer 1936; Kumar and Haware 1982; Sindhu et al. 1983). Soon after it, a new type of chickpea resistance to wilt termed as "late wilting" was discovered following inoculation of plants with race 1 (Upadhyaya et al. 1983b). Late wilting refers to the delayed appearance of symptoms following inoculation and is characterized by a long latent period as compared to susceptible lines. Following the discovery of late wilting, complexity in the genetics of late wilting and complete resistance was explored. Late wilting is inherited by single gene(s) whereas complete resistance was inherited by a combination of two genes (Upadhyaya et al. 1983a; Singh et al. 1987a, b). Three genes independently govern late wilting and were named as *h1*, *h2*, and *H3*. Combination of any of these two genes, i.e., *h1h2*, *h1H3*, and *h2H3*, confers complete resistance to race 1 (renamed as race 1A to differentiate it from race 1B/C, see Sharma et al. 2005) of *Foc*. Genetics of resistance to race 2 is also complex like race 1 and is conferred by three genes, two recessive and one dominant. Initially, the resistance to race 2 was found to be monogenic recessive (Pathak et al. 1975). Subsequently, it was found to be digenic (Gumber et al. 1995) and trigenic (Kumar 1998). The phenomenon of late wilting against race 2 is also reported and is governed by three single genes—*a*, *b*, and *C*—independently wherein *a* and *b* in recessive form and *C* in dominant form govern late wilting, i.e., A.bb .., aaB ...., and A. B.C. (Kumar 1998). The combination of *aabb* confers complete resistance whereas the third gene must be in homozygous recessive form for susceptibility. Some studies doubted the involvement of two or three genes for race 2 resistance (Sharma and Muehlbauer 2007), primarily because the F_2 data of Gumber et al. (1995) and F_3 data of Kumar (1998) did not fit well with the two-gene and three-gene theories. Sharma et al. (2005) also studied the genetics of resistance to race 2 using recombinant inbred line and F_2 population from cross WR315 × C104 and found the resistance to be governed by a single recessive gene.

Unlike Kumar (1998) who used the sick plot, Sharma et al. (2005) used the root dip method (after trimming lower one-third of the roots of the plants). The root dip method ensures uniform spore penetration and infection of all plants at the same time whereas in sick plot host penetration by fungus and subsequent infection in different plants cannot take place at the same time pointing to higher authenticity of studies by Sharma et al. (2005). The root dip method used by Sharma et al. (2005) also ensures equal inoculum load to all plants whereas inoculum density will presumably vary in sick plots. It has been established that disease intensity and disease progress have a direct and positive correlation with inoculum density (Navas-Cortés et al. 2000). However, confusion in the inheritance of resistance to race 2 can only be put to rest by conducting further studies using standardized techniques.

Resistance to race 0 has also been studied extensively. It is governed by two genes named as *Foc01/foc01* and *Foc02/foc02* (Rubio et al. 2003; Cobos et al. 2005; Halila et al. 2010), although Tekeoglu et al. (2000) have found it to be monogenic. The gene identified by Tekeoglu et al. (2000) seems to be the one identified in JG 62 by Rubio et al. (2003). The gene *Foc01/foc01* is situated in the linkage group LG5 whereas Foc02/foc02 is located in the `linkage group LG2 (Halila et al. 2009). Resistance to race 3 is monogenic in nature with no information on dominant or recessive nature as recombinant inbred lines were used (Sharma et al. 2004a, 2005) whereas resistance to race 4 is monogenic recessive in some lines (Tullu et al. 1998; Sharma et al. 2005) and digenic recessive in a chickpea line Surutato-77 (Tullu et al. 1999). Resistance to race 5 was inherited by a single gene (Tekeoglu et al. 2000; Sharma et al. 2005). Apparently, genetics of resistance to races 3, 4, and 5 is not studied adequately whereas no study existed for races 1B/C and 6.

A third type of chickpea resistance to wilt named as "slow wilting" was also discovered (Sharma et al. 2005). Following this, slow wilting has been reported for different races of *Foc*, several crops other than chickpea and another species of *Fusarium*. Slow wilting in chickpea is reported for different races of *Foc*, i.e., race 2 and race 3 (Sharma et al. 2005; Sharma and Muehlbauer 2007), race 0 (Halila et al. 2009, 2010), and in near-isogenic lines for race 5 (Castro et al. 2010), tomato to *F. oxysporum* (Pshibytko et al. 2006), pigeonpea to *Fusarium udum* (Sinha and Biswas 2010), and pea to *F. oxysporum* f. sp. *pisi* (Sharma et al. 2010, Mc Phee et al. 2012). Slow wilting is characterized by the slower progress of wilt following inoculation as compared to susceptible lines, and no difference in latent period between slow wilting and susceptible lines was observed (Sharma et al. 2005, Sharma and Muehlbauer 2007). The differences between slow wilting and late wilting were outlined by Sharma et al. (2005) as "slow wilting differs from late wilting in three aspects: latent period, disease progress rate, and final disease severity." In late wilting, disease symptoms appear late, the rate of progression of the disease is, however, comparable to the susceptible lines thereafter and 100% of plants wilt in due course of time. In slow wilting, the disease symptoms appear at the same time as those appear in the susceptible genotype but the disease progresses slowly thereafter and disease incidence never reaches 100%. Until the discovery of the phenomena of slow wilting and late wilting, the resistance to wilt in chickpea was considered complete. Genetics of resistance studies revealed that major genes governed late wilting (Upadhyaya et al. 1983a, b). The genetics of slow wilting is still unresolved despite its discovery about 18 years ago. Slow wilting is suspected to be governed by minor genes owing to circumstantial evidence of the occurrence of slow wilting in progenies derived from a cross of resistant and susceptible genotypes (Sharma et al. 2005, Sharma and Muehlbauer 2007). It also points to the noninvolvement of vertical resistance genes in this phenomenon, though resistance in this case is race specific unlike the race nonspecific horizontal resistance.

3.5 Breeding for Wilt Resistance Using Traditional Breeding Technologies

Several studies have been conducted to identify sources of resistance to wilt in chickpea and transfer the resistance from resistant to susceptible lines. Usually screening has been carried out either in fields with a known history of wilt or in wilt sick plots and rarely using the root dip method. Wilt sick plots were initially developed at ICRISAT, Hyderabad, to screen chickpea germplasm (Nene and Haware 1980). Such plots have subsequently been developed at several places for routine screening of germplasm, and germplasm lines resistant to wilt were identified (Nene and Haware 1980; Haware et al. 1992; Sharma et al. 2005, 2019; Ayana et al. 2019; Mohamed and Mohamed 2020; Bekele et al. 2021; Khalifa et al. 2022). Overall studies indicate that the *desi* type of chickpea possesses a higher degree of resistance to wilt as compared to the *kabuli* type. The breeders also transferred wilt resistance from known sources of resistance to susceptible but agronomically superior lines using hybridization and developed advanced breeding lines with resistance to wilt. Castro et al. (2010) developed near-isogenic lines (NILs), from several crosses, for resistance to races 1A, 2, 3, 4, and 5 of *Foc*. Jha et al. (2021a) also evaluated advanced breeding lines for resistance to wilt using a sick plot at Kanpur and identified 34 fusarium wilt-resistant lines that can be used by breeders. Similarly, advanced recombinant inbred lines generated from several crosses were used to identify wilt-resistant chickpea (Bekele et al. 2021). In another study, elite chickpea breeding lines were evaluated across several locations in India with the aim to identify lines with multi-race resistance (Srivastava et al. 2021). The broader aim was to develop pre-breeding lines/varieties that can be used at the majority of the location in India. The study led to the identification of seven lines that showed resistant at all the locations tested. Mutation breeding was also attempted to develop wilt-resistant chickpea, and a variety "CM 88" was developed following mutation of susceptible chickpea variety CS 27 using gamma irradiation and was released for cultivation in Pakistan (Haq et al. 2001). The CM88 possessed resistance to both the important diseases of chickpea, i.e., wilt and blight. Race 2- and race 4-resistant mutants of three chickpea varieties were also generated (Chobe et al. 2016). Salimath et al. (2011) combined hybridization with mutation to obtain wilt-resistant chickpea breeding lines.

To increase the efficiency of the transfer of wilt resistance genes, gametophytic selection was also attempted. A cross was made between WR 315 ($h_1h_1h_2h_2$, resistant) and JG 62 ($H_1H_1H_2H_2$, susceptible), and F_1 plants were subjected to toxin stress (Ravikumar et al. 2007). In F_2 and BC_1 progenies, there was a deviation from the expected ratios, and the proportion of resistant plants was more as compared to the control (Ravikumar et al. 2007). A subsequent study also demonstrated the utility of this technology for selection of wilt-resistant progeny plants (Ravikumar et al. 2013) suggesting that gametophytic selection can be a powerful tool for wilt breeding in chickpea and may reduce the time for development of wilt-resistant varieties.

3.6 Molecular Markers in Chickpea

3.6.1 Simple-Sequence Repeat (SSR) Markers

Morphological markers that can be visualized with the naked eye were the first markers used by human beings for identification and characterization purposes. The discoveries in the field of biomolecules led mankind to use protein and isozyme markers, the first molecular markers. Even the first linkage map of chickpea was developed using isozyme markers (Gaur and Slinkard 1990; Gaur and Stinkard 1990). Isozyme markers, however, could not be exploited vigorously owing to the limited number of those. The double helical structure of the DNA was discovered in 1953 by Watson and Crick and subsequent discoveries of enzymes such as ligases, restriction enzymes, and technologies for the separation of fragments of DNA using agarose gels and polyacrylamide gels wherein DNA molecules of different sizes can be separated from each other and visualized on gel, which led to the discovery of DNA-based markers. The first DNA-based marker was the restriction fragment polymorphism (RFLP, Botstein et al. 1980). In RFLP, the DNA is restricted with restriction enzymes and electrophoresed on gels followed by detection of desirable fragments using small DNA/RNA fragments called as probes. Following the discovery of polymerase chain reaction in 1983 (PCR, Mullis et al. 1986), PCR-based markers called as random amplified polymorphic DNA (RAPD) were developed (Williams et al. 1990). The RFLP requires previous knowledge of DNA sequences whereas RAPD does not and, hence, can be used even if there is no prior knowledge of sequences of DNA. In RAPD, the DNA is amplified using 10-mer primers, and amplified fragments are separated on agarose gels. It was, however, soon realized that RAPD markers lacked specificity as the technology uses short-length primers. After several years of use, these markers were declared unsuitable for most of the scientific purposes. Following RAPD, inter simple-sequence repeat (ISSR) markers were developed that amplify the region between two closely spaced microsatellites. Amplified fragment length polymorphism (AFLP) markers are the other marker systems developed and were used considerably owing to high reproducibility. These are seldom used now owing to the complexity of procedure and availability of better markers with simple procedures.

The development of simple-sequence repeat (SSR) markers that are versatile and robust with high specificity revolutionized the use of DNA-based marker technology in plant sciences, animal sciences, as well as forensic sciences, and they are being used routinely even today. In the case of SSR markers, the primers are designed from sequences flanking the satellites, and the satellite region is amplified using PCR followed by gel electrophoresis. The sequences flanking the microsatellites are conserved not only within the species but also across species of the same genera and sometimes even across genera, and hence, SSR primers can be used within a species, within a genus, and to some extent within a family. Another marker system even more versatile than the SSR markers is single nucleotide polymorphisms (SNP). Earlier, the high costs of sequencing limited the use of SNPs. The costs of sequencing are now fairly low, and SNPs are being used widely in animal as well as plant sciences.

In chickpea, RAPD, ISSR, RFLP, and SSR were used for construction of linkage maps, to map genes and to study genetic diversity in different populations; however, SSRs are the most commonly used marker systems in chickpea. Of late, SNPs have also become the markers of choice in chickpea. Chickpea has low genetic diversity, and hence, markers that reveal low polymorphism were not proved highly useful in chickpea. Both the SSRs and SNPs are highly polymorphic and appropriate for use in chickpea. The development of SSR markers started with two studies on repetitive sequences in chickpea (Weising et al. 1992; Sharma et al. 1995) and the use of PCR to amplify SSRs using "microsatellite primed polymerase chain reaction." In a first step to develop SSR markers, Huttel et al. (Hüttel et al. 1999) developed two bacterial artificial chromosome (BAC) libraries followed by the development of 16 polymorphic markers named as "*Cicer arietinum* sequence-tagged microsatellite site (abbreviation: CaSTMS) markers." Subsequently, 120 polymorphic STMS markers were developed and used for construction of chickpea linkage map (Winter et al. 1999). These STMS markers were used for mapping chickpea genes/QTLs conferring disease resistance (Sharma et al. 2004b; Sharma and Muehlbauer 2005, 2007; Anbessa et al. 2009; Cho et al. 2002). The second set of chickpea SSR primers was developed in the year 2006 (Sethy et al. 2006). This was followed by development of other sets of SSR primers (Qadir et al. 2007; Gaur et al. 2011). The slow pace of development of SSRs can be attributed to two factors, i.e., the high costs of DNA sequencing during the end of the twentieth century and the fact that the countries with high acreage did not undertake lead in this direction. The costs of sequencing gradually lowered and India, the major producer of chickpea, initiated studies in this direction. One of the major initiatives to enrich chickpea marker repository was by Thudi et al. (2011) who contributed 1344 polymorphic SSR markers and 5397 polymorphic DArT markers and by Parida et al. (2015) who contributed 1470 polymorphic chickpea SSR markers.

Identification of expressed sequence tags (Jayashree et al. 2005; Dinari et al. 2013; Sharma and Nayyar 2014) in chickpea led to the development of EST-SSRs markers in chickpea. A total of 106 EST-SSR markers were developed from 477 ESTs; however, only 44 were polymorphic (Buhariwalla et al. 2005). Following this, 2131 ESTs were searched for SSRs leading to identification of 246 SSRs, 183 primer pairs, and 60 functional EST SSR markers (Choudhary et al. 2009). The search for more ESTs and EST-SSR markers in chickpea continued (Choudhary et al. 2012, Varshney et al. 2009) leading to development of 302 additional functional markers (Choudhary et al. 2012; Varshney et al. 2009). Subsequently, more EST-SSR markers (330) were added to the repository of already available markers (Stephens et al. 2014; Gupta et al. 2015). The search for ESTs in chickpea is continued, and possibly more EST-SSR markers will be added to the database of these markers.

In search for additional genomic resources, the transcription factor genes are also being identified and used to search SSRs and develop EST-SSR markers named as "Transcription factor gene-derived microsatellite (TFGMS)" markers. Kujur et al. (2013) identified 707 transcription factor genes and 1108 TFGMS within these genes. These TFGMS were used to develop 161 transcription factor functional domain-associated microsatellite (TFFDMS) markers (Kujur et al. 2013). In total, about 2500 transcription factor genes have been identified in chickpea (Garg et al. 2011; Kujur et al. 2013).

To facilitate the use of SSRs as markers and to help chickpea community searching for markers, a web interface named as "Chickpea Microsatellite Database" (CicArMiSatDB http://cicarmisatdb.icrisat.org) was also developed (Doddamani et al. 2014).

3.7 Markers Other Than SSRs

Following reduction in the cost of DNA-sequencing, the technology was used widely in chickpea to identify sequences of several regions of the chickpea genome. Whole genome sequences of chickpea were also identified for several germplasm of chickpea. This information led to the discovery of several other categories of markers than SSRs. Several other types of markers were derived from the genic regions, e.g., intron-targeted primers (ITPs), expressed sequence tag polymorphisms (ESTPs), or from other parts of the genome, e.g., diversity arrays technology (DArT), insertion-deletion (InDel) markers, and single nucleotide polymorphisms (SNPs). ITP markers are developed from the boundaries of the introns and exploit variation in length of introns. ITPs were identified in chickpea from seed and embryo (Choudhary et al. 2012; Gupta et al. 2015). ESTPs, on the other hand, exploit variation in the length of ESTs and have been developed in chickpea (Choudhary et al. 2012, Gupta et al. 2015). There are limited reports on DArT markers in chickpea (Thudi et al. 2011; Sharma et al. 2014); however, InDels have been reported in abundance (Das et al. 2015; Srivastava et al. 2016; Jain et al. 2019). InDels were also found to be highly polymorphic in chickpea (Srivastava et al. 2016).

The major focus in chickpea as of now is on SNPs owing to the abundance of those in the genome, biallelic nature, and codominance. The first report on SNPs in chickpea was by Hiremath et al. (2012) who used a panel of 70 genotypes to fish out 2486 SNPs. It was followed by several reports on SNPs in chickpea. Usually sequencing has been used to identify the SNPs; however, EcoTILLING and

Illumina® GoldenGate SNP array were also used to identify SNPs in chickpea (Deokar et al. 2014; Bajaj et al. 2016).

3.8 Markers Linked to Chickpea Genes Governing Resistance to *F. oxysporum* f. sp. *ciceris*

Resistance to wilt in chickpea is qualitative and is governed by major vertical resistance genes. Several of the wilt resistance genes in chickpea have been mapped and gene linked markers have been used to introgress those into susceptible backgrounds using a technique called as marker assisted selection. Despite significant developments in marker technologies, marker density on chickpea linkage maps is still low and markers tightly linked to wilt resistance genes are infrequently available. Lack of tightly linked markers in this case cannot be attributed to limited number of markers but to limited polymorphism in chickpea that results from its narrow genetic base. The first fusarium wilt resistance gene to be mapped in chickpea was *h1* (syn. *Foc*-1), the race 1 resistance gene. The gene was mapped by using two markers, CS-27_{700} (RAPD) and UBC-170_{550} (ISSR marker), that were at a distance of 7 cM from the gene (Mayer et al. 1997). The second gene to be mapped was the race 4 resistance gene (Ratnaparkhe et al. 1998b). The gene was situated on LG 6 and was mapped using the marker UBC-855_{500} (ISSR marker) at a distance of 5.2 cM. The UBC-855_{500} was at a distance of 0.6 cM to CS-27_{700} implying that both genes were in the same region. During the same year, the race 4 resistance gene was also reported to be linked to these markers (UBC-855_{500} and CS-27_{700}) at a distance of 5.2 cM (Ratnaparkhe et al. 1998a, b) and to UBC-170_{550} at a distance of 9 map units (Tullu et al. 1998). Using another set of parents for crosses, resistance to race 4 was revealed to be governed by two genes, and the loci were mapped using already used marker CS-27_{700} (Tullu et al. 1999). Genes for resistance to race 0 and race 5 were also mapped during the same time (Tekeoglu et al. 2000). Since RAPD markers lack high specificity, the RAPD marker obtained using the CS-27 primer was sequenced and converted to an allele-specific associated primer (ASAP) (CS-27R/CS-27F), a more specific and robust marker. The race 5 resistance gene was mapped at a distance of 7.2 cM from the race 4 resistance gene and 4.0 cM from the race 5 resistance gene whereas the two genes were separated by 11.2 cM (Tekeoglu et al. 2000). Gene for resistance to race 0 was not linked to these markers and might be on some other linkage groups. Tekeoglu et al. (2000) also proposed gene symbols *foc-0*, *foc-4*, and *foc-5* for genes governing resistance to races 0, 4, and 5, respectively. All these studies used either the RAPD markers or the ISSR markers as SSR markers were not developed at that time. These studies also inferred that genes *foc-1*, *foc-4*, and *foc-5* were in close proximity and linked to each other suggesting the possibility of a cluster of fusarium wilt resistance genes in chickpea.

Mapping of wilt resistance genes was fastened following the discovery of SSRs by Winter et al. (1999). The SSR markers offered four discrete advantages over RAPDs and ISSRs, i.e., higher number as compared to other markers, high polymorphism, robustness in results/reproducibility, and distribution all over the genome. To overcome the scarcity of adequate polymorphism in *C. arietinum*, mapping populations were developed from crosses of two species of *Cicer*, i.e., *C. arietinum* and *C. reticulatum*. Using RIL mapping populations from these two species and 354 markers (STMSs, DAFs, AFLPs, ISSRs, RAPDs, isozymes, cDNAs, SCARs, and loci for fusarium wilt resistance), a fairly good linkage map of chickpea comprising of 8 large and 8 small groups with size of 2077.9 cM was developed (Winter et al. 2000). STMS markers linked to race 4 and 5 were also identified. DNA amplification fingerprinting (DAF) markers were also used to map race 4 and 5 wilt resistance genes, and a DAF marker (R-2609-1) linked closely (2.0 cM) to race 4 resistance gene was identified (Benko-Iseppon et al. 2003). *Foc 5* was mapped to the same region (Benko-Iseppon et al. 2003). Using two mapping populations, resistance to race 0 of *Foc* was found to be governed by two genes, and these genes were mapped using the RAPD marker, $OPJ20_{600}$ (Rubio et al. 2003). Cobos et al. (2005) added another marker, STMS TA39, flanking the gene, *Foc01/ foc01*, already mapped by Rubio et al. (2003). The linkage map of chickpea with 10 linkage groups was also developed, and the gene was mapped to LG 3. The second gene governing resistance to race 0 (*foc-0*$_2$) was mapped to LG 2 in close proximity to *foc-5* and linked tightly to STMS marker TA59 (Palomino et al. 2009). An intraspecific mapping population derived from a cross of WR 315 (possesses resistance genes *Foc 1*, *Foc 2*, *Foc 3*, *Foc 4*, and *Foc 5*) was crossed with C 104 (susceptible) and recombinant inbred lines revealed the presence of one gene each for resistance to each of the five races (Sharma et al. 2005). These five genes were mapped using the STMS and STS markers (Sharma et al. 2004b; Sharma and Muehlbauer 2005, 2007). All five genes were clustered in a region of 8.2 cM on LG2 with two subclusters of 2.0 and 2.8 cM (Sharma and Muehlbauer 2007). Evidently, chickpea wilt resistance genes are present in two clusters, one situated on LG2 (chromosome F or G) and another (race 0 resistance genes) on LG3 (syn LG5, chromosome C or D). Additional markers to the LG 2 region of wilt resistance genes were also added, some of which were in close proximity to *foc-2* and *foc-3* (Gowda et al. 2009). The region surrounding foc-*2* was saturated with additional markers (Caballo et al. 2019) with the aim to identify closely linked markers. To narrow down the genomic region having *foc-5*, markers in a region of 25 Mbp were genotyped in several types of mapping populations, and a region of 820 Kbp was selected. The probable gene for *foc-2* resistance was selected in this region followed by development of SNPs, five of which were selected for future studies (Caballo et al. 2019). The markers used by Sharma et al. (2004b), Sharma and Muehlbauer (2005), and Sharma and Muehlbauer (2007) have been validated in several studies, and all except TA27 have been consistent across several populations (Lal et al. 2022). Clearly these markers are useful for breeding studies. Some of these markers were used to introgress fusarium wilt resistance to elite cultivars of chickpea using marker-assisted selection (Varshney et al. 2014a; Pratap et al. 2017; Mannur et al. 2019). Recently association mapping was also used to map the *foc-2* gene (Jha et al. 2021b).

At least two studies have reported QTLs for fusarium wilt resistance (Sabbavarapu et al. 2013; Garg et al. 2018). While Sabbavarapu et al. (2013) found QTL on LG 6, Garg et al. (2018) identified major QTL on CaLG02 (syn

LG 2) and two minor QTLs, one each on CaLG04 and CaLG06, for resistance to race 1 of *Foc*. Race 3 resistance was conferred by two QTLs situated on CaLG02 and CaLG04 (Garg et al. 2018). Usually, resistance governed by major genes is detected by studying the linkage of the gene to markers, and QTLs are employed only if a trait is multigenic. Since resistance to wilt is governed by a major and quantifiable number of genes, the mapping of individual genes rather than QTLs should be a preferable strategy. Marker-assisted selection for individual genes has so far been highly successful in chickpea. Many of the *Foc*-resistant genes have been transferred to the agronomically superior chickpea cultivars, and in some cases varieties possessing resistance to *Foc* have been released. These findings stipulate that mapping of individual *Foc* resistance genes is an effective strategy for marker-assisted breeding.

3.9 Bacterial Artificial Chromosome Libraries in Chickpea

The first bacterial artificial chromosome (BAC) library in chickpea was constructed in 2004 (Rajesh et al. 2004). The library was constructed from chickpea genotype FLIP 84-92C for positional cloning of important genes and physical mapping of the genome. The library had 23,780 clones with 3.8 haploid genome equivalents (Rajesh et al. 2004) and has been used frequently by several workers after its construction. Soon after this, a BAC library (14,976 clones) and a plant-transformation-competent binary BAC (BIBAC) library (23,040 clones) both covering 7.0× genome of chickpea were developed from cv. Hadas (Lichtenzveig et al. 2005). Following this two more BAC libraries (a BAC library and a BIBAC library) were constructed (Zhang et al. 2010). Three other BAC libraries were constructed from the chickpea cultivar ICC 4958, one by Thudi et al. (2011) and two by Varshney et al. (2014b). The libraries were used for mining of candidate genes including those present in the QTL hotspots as well as gene cloning and development of DNA-based markers leading to saturation of chickpea genetic maps. These libraries have facilitated to narrow down the QTL regions with the aim to identify candidate genes governing a particular trait such as ascochyta blight resistance, and drought tolerance.

3.10 Marker-Assisted Breeding for Wilt Resistance in Chickpea

Traditionally disease resistance breeding is carried out by screening the plants with pathogen races followed by hybridization of resistant plants with agronomically superior variety, advancement of generations including backcross breeding, and selection of disease-resistant superior plants for release as variety. Disease resistance screening involves the pathogen races, inoculations, and scoring for disease. The environment plays a crucial role in the final outcome of the disease. Marker technologies allow screening of plants without pathogen inoculation, and the outcome is not influenced by the environment. The use of markers in breeding has increased considerably in the recent past. The breeding carried out with the help of molecular markers is called marker-assisted breeding (MAB) whereas selection of plants with the help of markers is called marker-assisted selection. In MAB, the progenies are advanced in the same way as is done in traditional breeding; however, plants are evaluated using molecular markers instead of traditional technologies or sometimes a combination of both. Molecular markers offer several advantages over traditional breeding, e.g., elimination of the influence of the environment, screening at young plant stage for adult plant traits, evaluation for the presence of several disease resistance genes (usually traditional pathological procedures do not allow evaluation with more than one race on individual plants), pyramiding of disease resistance genes, background recovery or background selection (marker-assisted backcross breeding), identification of homozygotes and heterozygotes to speed up the identification of genetically superior plants, and shortening of breeding cycle. Marker-assisted selection (MAS) in chickpea has been used to develop near-isogenic lines, screen chickpea germplasm for the presence of wilt resistance (R) genes, and develop elite lines/varieties possessing wilt R genes and gene pyramiding. MAS was used to develop near-isogenic lines possessing resistance to different races of *Foc* (Castro et al. 2010; Jendoubi et al. 2016) and to screen chickpea germplasm for identification of resistant genes (Ahmad et al. 2014).

Markers linked to different races of the pathogen by Sharma et al. (2004b), Sharma and Muehlbauer (2007), and others were later validated by Ali et al. (2012), and it was reported that "Most of the SSR markers showed good correlation with phenotypic evaluation of genotypes to different races of *Foc* and may be used effectively in resistance breeding, except those markers for race 3." The marker in question was TA27. The validation showed that the markers linked to wilt resistance genes can be used widely and effectively in marker-assisted wilt-resistant breeding in chickpea.

In central India, race 2 of *Foc* is highly prevalent and causes considerable yield losses. With the aim to develop a race 2-resistant variety of chickpea for this region, Pratap et al. (2017) transferred the race 2 resistance gene from "Vijay" to Pusa 256, an elite chickpea cultivar of the *desi* type, using SSR markers TA37 and TA110. MAS for foreground as well as background selection coupled with backcross breeding was employed leading to the selection of race 2-resistant high-yielding lines (Pratap et al. 2017). Similarly, marker-assisted breeding was carried out to introgress wilt resistance into two varieties of chickpea, namely, Annigeri 1 and JG 74, both of which are elite varieties but with susceptibility to *Foc* (Mannur et al. 2019). The race 4 resistance gene was transferred into the background of these two varieties from the donor WR 315 that was shown to possess resistance to five races of the pathogen by Sharma et al. (2005). The markers reported by Sharma et al. (2004b), Sharma and Muehlbauer (2005), and Sharma and Muehlbauer (2007) were exploited for foreground screening of the progeny plants coupled with backcross breeding and background selection using SSR markers (Mannur et al. 2019). The superior lines are being evaluated at multilocation trials in India and *foc-4*-resistant varieties are expected to be released soon.

The marker-assisted selection was also used to pyramid several wilt resistance genes in the elite backgrounds with the purpose to enhance the durability of resistance and provide protection against several races of the pathogen. In India, Pusa 391 is a mega desi chickpea variety in the central zone of India. During the passage of time, it became susceptible to wilt. To develop a multi-race-resistant variety, the Pusa 391 was crossed with WR 315 (multi-race-resistant variety). Foreground selection of the progeny lines was carried out with 3 wilt gene-resistant markers (GA16, TA 27, and TA 96) whereas background selection was carried out by the 48 SSR markers (Bharadwaj et al. 2022). BC_3F_3 lines selected by this procedure were evaluated at multilocations, and a high-yielding variety possessing resistant to several races of the pathogen named as Pusa Manav/Pusa Chickpea 20211 was released for cultivation in Central India replacing the traditional Pusa 391 (Bharadwaj et al. 2022).

Marker-assisted backcrossing was also used to introgress resistance to two important diseases of chickpea, i.e., wilt and blight caused by *Ascochyta rabiei* into the background of an elite line C 214 of chickpea (Varshney et al. 2014a). For introgression of wilt resistance genes, markers linked to locus *foc1* were used whereas for ascochyta blight resistance, the markers linked to two resistant QTLs, i.e., ABQTL-I and ABQTL-II, were used. The donor for *foc1* locus was WR 315 whereas the donor for ascochyta blight resistance was ILC 3279. The BC_3F_4 lines possessing resistance to wilt and blight were identified (Varshney et al. 2014a).

3.11 Genomic-Assisted Breeding

Of late genomic-assisted breeding (GAB) is being debated widely among chickpea breeders and biotechnologists. In genomic-assisted breeding, genomic tools are integrated and used in breeding for biotic/abiotic stress tolerance and increased yields. Basically the relationship between genotype and phenotype is exploited in GAB, and various genomic tools such as genomics, transcriptomics, proteomics, and markers linked to traits are exploited to predict phenotype and assist in breeding. GAB works on a larger scale than that of the marker-assisted selection where only one or a few traits are targeted. In some cases such as blight resistance or cold tolerance, several QTLs govern a trait, and the transfer of such traits needs wider genomics or marker information. The approaches of the GAB are marker-assisted backcrossing (MABC), marker-assisted recurrent selection (MARS), and genomic selection (GS). Usually breeders aim to introgress some traits into the background of some superior line and intend to select plants with transferred traits and the background of the recurrent parent. MABC is employed by breeders to achieve this. Development of near-isogenic lines and gene pyramiding are examples of this and have been achieved for wilt resistance in chickpea (see preceding section, Castro et al. 2010; Pratap et al. 2017; Mannur et al. 2019; Bharadwaj et al. 2022). Availability of markers linked closely to the genes of interest as well as adequate markers to recover the background of the recurrent parent is important for MABC. The Department of Biotechnology, Government of India, has funded projects on MABC to introgress resistance to different races of *Foc* to develop varieties possessing wilt resistance but having elite backgrounds. In MARS, F_2 or F_3

populations are genotyped whereas F_2-derived F_4 or F_5 progenies are phenotyped based on the presence of marker alleles for some QTLs. Lines with different QTLs are identified and crossed to develop QTL pyramids. The lines with QTL pyramids are then tested at multilocations and released as varieties based on performance. MARS aims to capture several genomic regions and major and minor QTLs, and hence, genetic gain is higher than the MABC. In chickpea, no information is available for wilt or biotic stresses, but MARS is being exploited for combining favorable alleles for drought tolerance. In GS or genomic-wide selection (GWS), identification of superior lines is pursued based on genome-wide marker data. The GS is, thus, superior to MABC or MARS. The GS employs two types of populations called "training population" and "candidate population." The training population refers to the breeding lines that are used in a breeding program, and data for overall performance (e.g., yield and yield components) are available across the environments. The "candidate population" is the one being used currently by breeders. In the first step, the individuals in the training populations are genotyped using a large number of markers considering the linkage disequilibrium. In the second step, statistical models are generated for estimating genomic-assisted breeding values (GEBVs) based on marker and phenotype data. Similar to training populations, marker genotyping data, models, and GEBVs are generated on candidate population. In subsequent crosses, selection of the superior lines is done based on these GEBVs, and the lines with higher GEBVs are selected. These lines are again genotyped with the same set of markers. Finally the lines are tested at multilocations and are released for cultivation if found superior.

3.12 Conclusions and Future Perspective

Considerable advancements have been made in breeding for wilt resistance in chickpea. Markers are being used routinely to introgress wilt resistance genes, and wilt-resistant varieties developed through marker-assisted selection have been released for cultivation or are in the pipeline for release. A major concern is the lack of sufficient markers linked to wilt resistance genes in chickpea, specifically because markers linked closely to all wilt resistance genes are still unavailable. Despite the development of a considerable number of markers, marker density in the wilt resistance gene region on the genome is considerably low. To have an effective marker-assisted selection, marker density in the wilt resistance region on LG2 needs to be increased, and polymorphic markers for resistance genes need to be developed. LG2 also harbors QTLs for ascochyta blight resistance and, thus, is a hotspot for disease resistance. Breeders or biotechnologists may thus aim to increase marker density on the LG2 to achieve effective wilt and blight resistance introgression into agronomically superior lines. Another major concern is the unresolved race flora in several parts of the world especially the major chickpea-growing regions in Asia and Africa. Race identification will not only resolve the complexity of pathogen diversity in the world but will also allow breeders to develop varieties possessing resistance to race or races prevalent in a particular region. The objective of race identification can't be achieved until a suitable differential set that can differentiate all races in the world is developed. The germplasm can then be tested against races prevalent in that particular region and resistant line used for resistance gene transfer to elite lines/varieties. There is a need to form a worldwide consortium to develop a chickpea differential set and identify races of *Foc*. While the marker-assisted selection is being exploited for chickpea wilt, genomic-assisted selection and genomic-assisted breeding are yet to be exploited. Genomic-assisted breeding, however, requires mapping of a majority of the traits or at least the agronomically superior ones so that markers can be used to develop designer chickpea with stress resistance/tolerance coupled with agronomic superiority. Significant advances in marker technologies in chickpea have been made in the recent past, and the use of these technologies in marker-assisted breeding and genomic-assisted breeding will be usual practice in the near future.

References

Ahmad Z, Mumtaz AS, Ghafoor A, Ali A, Nisar M (2014) Marker assisted selection (MAS) for chickpea *Fusarium oxysporum* wilt resistant genotypes using PCR based molecular markers. Mol Biol Rep 41:6755–6762

Ali H, Ul Haq MA, Shah TM, Ur Rahman M, Chen W (2012) Validation of molecular markers for resistance among Pakistani chickpea germplasm to races of *fusarium oxysporum* f. sp. *ciceris*. Eur J Plant Pathol 132:237–244

Alloosh M, Hamwieh A, Ahmed S, Alkai B (2019) Genetic diversity of *fusarium oxysporum* f. sp. *ciceris* isolates affecting chickpea in Syria. Crop Prot 124:104863

Al-Taae AK, Hadwan HA, Al-Jobory SAE (2013) Physiological races of *fusarium oxysporum* f. sp. *ciceris* in Iraq. *Journal of.* Life Sci 7(10):1070

Anbessa Y, Taran B, Warkentin TD, Tullu A, Vandenberg A (2009) Genetic analyses and conservation of QTL for ascochyta blight resistance in chickpea (*Cicer arietinum* L.). Theor Appl Genet 119:757–765

Ayana A, Hailu N, Taye W (2019) Screening desi and kabuli chick pea varieties against fusarium wilt resistance in West Gojam, northwestern Ethiopia. J Plant Pathol Microbiol 10:474

Ayyar VR, Iyer RB (1936) A preliminary note on the mode of inheritance of reaction to wilt in *Cicer arietinum*. Proc Indian Acad Sci 3:438–443

Bajaj D, Srivastava R, Nath M, Tripathi S, Bharadwaj C, Upadhyaya HD, Tyagi AK, Parida SK (2016) EcoTILLING-based association mapping efficiently delineates functionally relevant natural allelic variants of candidate genes governing agronomic traits in chickpea. Front Plant Sci 7:450

Bayraktar H, Dolar FS, Maden S (2008) Use of RAPD and ISSR markers in detection of genetic variation and population structure among *fusarium oxysporum* f. sp. *ciceris* isolates on chickpea in Turkey. J Phytopathol 156(3):146–154

Bekele D, Tesfaye K, Fikre A, Cook DR (2021) The extent and association of chickpea fusarium wilt and root rot disease pressure with major biophysical factors in Ethiopia. J Plant Pathol 103(2):409–419

Benko-Iseppon AM, Winter P, Huettel B, Staginnus C, Muehlbauer FJ, Kahl G (2003) Molecular markers closely linked to fusarium resistance genes in chickpea show significant alignments to pathogenesis-related genes located on *Arabidopsis* chromosomes 1 and 5. Theor Appl Genet 107(2):379–386

Bharadwaj CH, Jorben J, Rao A, Roorkiwal M, Patil BS, Ahammed SK, Saxena DR, Yasin M, Jahagirdar JE, Sontakke PL, Pithia MS (2022) Development of high yielding fusarium wilt resistant cultivar by pyramiding of "genes" through marker-assisted backcrossing in chickpea (*Cicer arietinum* L). Front Genet 13:924287

Botstein D, White RL, Skolnick M, Davis RW (1980) Construction of a genetic linkage map in man using restriction fragment length polymorphisms. Am J Hum Genet 32(3):314

Buhariwalla HK, Jayashree B, Eshwar K, Crouch JH (2005) Development of ESTs from chickpea roots and their use in diversity analysis of the *Cicer* genus. BMC Plant Biol 5(1):16

Caballo C, Madrid E, Gil J, Chen W, Rubio J, Millan T (2019) Saturation of genomic region implicated in resistance to *fusarium oxysporum* f. sp. *ciceris* race 5 in chickpea. Mol Breed 39(2):16

Castro P, Pistón F, Madrid E, Millan T, Gil J, Rubio J (2010) Development of chickpea near-isogenic lines for fusarium wilt. Theor Appl Genet 121:1519–1526

Cho S, Kumar J, Shultz JL, Anupama K, Tefera F, Muehlbauer FJ, F.J. (2002) Mapping genes for double podding and other morphological traits in chickpea. Euphytica 128(2):285–292

Chobe DR, Om G, Maruti P (2016) Radiation induced mutation for resistance against races/ pathotypes of *fusarium oxysporum* f. sp. *ciceris* in chickpea (*Cicer arietinum* L.). Indian Phytopathol 69:699–701

Choudhary S, Sethy NK, Shokeen B, Bhatia S (2009) Development of chickpea EST-SSR markers and analysis of allelic variation across related species. Theor Appl Genet 118:591–608

Choudhary S, Gaur R, Gupta S, Bhatia S (2012) EST-derived genic molecular markers: development and utilization for generating an advanced transcript map of chickpea. Theor Appl Genet 124:1449–1462

Cobos MJ, Fernández MJ, Rubio J, Kharrat M, Moreno MT, Gil J, Millan T (2005) A linkage map of chickpea (*Cicer arietinum* L.) based on populations from Kabuli×desi crosses: location of genes for resistance to fusarium wilt race 0. Theor Appl Genet 110(7):1347–1353

Cunnington J, Lindbeck K, Jones RH (2007) National diagnostic protocol for the detection of fusarium wilt of chickpea (*fusarium Oxysporum* f. sp. *ciceris*). Plant health Australia: Camberra, Australia

Das S, Upadhyaya HD, Srivastava R, Bajaj D, Gowda CLL, Sharma S, Singh S, Tyagi AK, Parida SK (2015) Genome-wide insertion-deletion (InDel) marker discovery and genotyping for genomics-assisted breeding applications in chickpea. DNA Res 22(5):377–386

Demers JE, Garzón CD, Jiménez-Gasco MDM (2014) Striking genetic similarity between races of *fusarium oxysporum* f. sp. *ciceris* confirms a monophyletic origin and clonal evolution of the chickpea vascular wilt pathogen. Eur J Plant Pathol 139:309–324

Deokar AA, Ramsay L, Sharpe AG, Diapari M, Sindhu A, Bett K, Warkentin TD, Tar'an B (2014) Genome wide SNP identification in chickpea for use in development of a high density genetic map and improvement of chickpea reference genome assembly. BMC Genomics 15(1):708

Dinari A, Niazi A, Afsharifar AR, Ramezani A (2013) Identification of upregulated genes under cold stress in cold-tolerant chickpea using the cDNA-AFLP approach. PLoS One 8(1):e52757

Doddamani D, Katta MA, Khan AW, Agarwal G, Shah TM, Varshney RK (2014) CicArMiSatDB: the chickpea microsatellite database. BMC Bioinformatics 15(1):212

Dubey SC, Singh SR (2008) Virulence analysis and oligonucleotide fingerprinting to detect diversity among Indian isolates of *fusarium oxysporum* f. sp. *ciceris* causing chickpea wilt. Mycopathologia 165:389–406

Dubey SC, Singh SR, Singh B (2010) Morphological and pathogenic variability of Indian isolates of *fusarium oxysporum* f. sp. *ciceris* causing chickpea wilt. Arch Phytopathol Plant Protect 43(2):174–190

Dubey SC, Priyanka K, Singh V, Singh B (2012) Race profiling and molecular diversity analysis of *fusarium oxysporum* f. sp. *ciceris* causing wilt in chickpea. J Phytopathol 160(10):576–587

FAOSTAT (2017) http://www.fao.org/faostat/en/#data/QC

Garg R, Patel RK, Jhanwar S (2011) Gene discovery and tissue-specific transcriptome analysis in chickpea with massively parallel pyrosequencing and web resource development. Plant Physiol 156:1661–1678

Garg T, Mallikarjuna BP, Thudi M, Samineni S, Singh S, Sandhu JS, Kaur L, Singh I, Sirari A, Basandrai AK, Basandrai D (2018) Identification of QTLs for resistance to fusarium wilt and Ascochyta blight in a recombinant inbred population of chickpea (*Cicer arietinum* L.). Euphytica 214(3):45. https://doi.org/10.1007/s10681-018-2125-3

Gaur PM, Slinkard AE (1990) Genetic control and linkage relations of additional isozyme markers in chickpea. Theor Appl Genet 80(5):648–656

Gaur PM, Stinkard AE (1990) Inheritance and linkage of isozyme coding genes in chickpea. J Hered 81(6):455–461

Gaur R, Sethy NK, Choudhary S, Shokeen B, Gupta V, Bhatia S (2011) Advancing the STMS genomic resources for defining new locations on the intraspecific genetic linkage map of chickpea (*Cicer arietinum* L.). BMC Genomics 12:117

Gowda SJM, Radhika P, Kadoo NY, Mhase LB, Gupta VS (2009) Molecular mapping of wilt resistance genes in chickpea. Mol Breed 24(2):177–183

Gumber RK, Kumar J, Hanaware MP (1995) Inheritance of resistance to fusarium wilt in chickpea. Plant Breed 114(3):277–279

Gupta S, Kumar T, Verma S, Bharadwaj C, Bhatia S (2015) Development of gene-based markers for use in construction of the chickpea (*Cicer arietinum* L.) genetic linkage map and identification of QTLs associated with seed weight and plant height. Mol Biol Rep 42(11):1571–1580

Halila MH, Strange RN (1997) Screening of Kabuli chickpea germplasm for resistance to fusarium wilt. Euphytica 96:273–279

Halila I, Cobos MJ, Rubio J, Millán T, Kharrat M, Marrakchi M, Gil J (2009) Tagging and mapping a second resistance gene for fusarium wilt race 0 in chickpea. Eur J Plant Pathol 124:87–92

Halila I, Rubio J, Millán T, Gil J, Kharrat M, Marrakchi M (2010) Resistance in chickpea (*Cicer arietinum*) to fusarium wilt race '0'. Plant Breed 129:563–566

Haq MA, Hassan Mahmudul and Sadiq M (2001) 'CM 88'–a multiple disease resistant chickpea mutant variety (INIS-XA–427). International Atomic Energy Agency (IAEA)

Haware MP, Nene YL (1979) Physiologic races of the chickpea wilt pathogen. Int Chickpea Newsl 1:7–8

Haware MP, Nene YL (1980) Influence of wilt at different stages on the yield loss in chickpea. Trop Grain Legume Bull 19:38–40

Haware MP, Nene YL (1982) Races of *fusarium oxysporum* f.sp. *ciceri*. Plant Dis 66:809–810

Haware MP, Nene YL, Pundir RPS, Narayana Rao J (1992) Screening of world chickpea germplasm for resistance to fusarium wilt. Field Crop Res 30:147–154

Hiremath PJ, Kumar A, Penmetsa RV, Farmer A, Schlueter JA, Chamarthi SK, Whaley AM, Carrasquilla-Garcia N, Gaur PM, Upadhyaya HD, Kishor KPB (2012) Large-scale development of cost-effective SNP marker assays for diversity assessment and genetic mapping in chickpea and comparative mapping in legumes. Plant Biotechnol J 10(6):716–732

Honnareddy N, Dubey SC (2006) Pathogenic and molecular characterization of Indian isolates of *fusarium oxysporum* f. sp. *ciceris* causing chickpea wilt. Curr Sci 91:661–666

Hüttel B, Winter P, Weising K, Choumane W, Weigand F, Kahl G (1999) Sequence-tagged microsatellite site markers for chickpea (*Cicer arietinum* L.). Genome 42(2):210–217

Jain A, Roorkiwal M, Kale S, Garg V, Yadala R, Varshney RK (2019) InDel markers: an extended marker resource for molecular breeding in chickpea. PLoS One 14(3):e0213999

Jalali BL, Chand H (1992) Chickpea wilt. In: Singh US, Mukhopadhayay AN, Kumar J, Chaube HS (eds) Plant diseases of international importance. Vol. I. Diseases of cereals and pulses. Prentice Hall, Englewood Cliffs, NJ, pp 429–444

Jayashree B, Buhariwalla HK, Shinde S, Crouch JH (2005) A legume genomics resource: the chickpea root expressed sequence tag database. Electron J Biotechnol 8(2):8–13

Jendoubi W, Bouhadida M, Millan T, Kharrat M, Gil J, Rubio J, Madrid E (2016) Identification of the target region including the *Foc0 1/foc0 1* gene and development of near isogenic lines for resistance to fusarium wilt race 0 in chickpea. Euphytica 210(1):119–133

Jha UC, Saabale PR, Chaturvedi SK, Singh NP (2021a) Advanced chickpea lines resistant against fusarium wilt (*fusarium oxysporum*). Indian J Agric Sci 91(1):176–178

Jha UC, Jha R, Bohra A, Manjunatha L, Saabale PR, Parida SK, Chaturvedi SK, Thakro V, Singh NP (2021b) Association mapping of genomic loci linked with fusarium wilt resistance (*Foc2*) in chickpea. Plant Genetic Resources 19:195–202

Jimenez-Diaz RM, Alcala-Jimenez AR, Hervar A and Trapero-Casas JL (1993) Pathogenic variability and host resistance in the *fusarium oxysporum* f. sp. *ciceri/Cicer arietinum* pathosystem. In third proceedings of European seminar: fusarium-mycotoxins taxonomy, pathogenicity and host resistance. pp 57–94

Jiménez-Díaz RM, Castillo P, del Mar Jiménez-Gasco M, Landa BB, Navas-Cortés JA (2015) Fusarium wilt of chickpeas: biology, ecology and management. Crop Prot 73:16–27

Jiménez-Gasco MDM, Jiménez-Díaz RM (2003) Development of a specific polymerase chain reaction-based assay for the identification of *fusarium oxysporum* f. sp. *ciceris* and its pathogenic races 0, 1A, 5, and 6. Phytopathology 93(2):200–209

Jiménez-Gasco MM, Pérez-Artés E, Jiménez-Díaz RM (2001) Identification of pathogenic races 0, 1B/C, 5, and 6 of *fusarium oxysporum* f. sp *ciceris* with random amplified polymorphic DNA (RAPD). European J Plant Pathol 107:237–248

Jiménez-Gasco MM, Milgroom MG, Jiménez-Díaz RM (2002) Gene genealogies support fusarium oxysporum f. sp ciceris as a monophyletic group. Plant Pathol 51:72–77

Jiménez-Gasco M, Navas-Cortés JA, Jiménez-Díaz RM (2004) The *fusarium oxysporum* f. sp. *ciceris/Cicer arietinum* pathosystem: a case study of the evolution of plant-pathogenic fungi into races and pathotypes. Int Microbiol 7(2):95–104

Kashyap PL, Rai S, Kumar S, Srivastava AK (2016) Genetic diversity, mating types and phylogenetic analysis of Indian races of *fusarium oxysporum* f. sp. *ciceris* from chickpea. Arch Phytopathol Plant Protect 49:533–553

Kelly A, Alcala-Jiminez AR, Brainbridge BW, Heale JB, Perez-Artes E, Jiminez-Diaz RM (1994) Use of genetic fingerprinting and random amplified polymorphic DNA to characterize pathotypes of *fusarium oxysporum* f. sp. *ciceris* infecting chickpea. Phytopathology 84:1293–1298

Khalifa MW, Hannane A, Noureddine R (2022) Screening for chickpea germplasm resistant to fusarium wilt disease under natural conditions of infection. J Crop Protection 11(2):229–242

Kujur A, Bajaj D, Saxena MS, Tripathi S, Upadhyaya HD, Gowda CLL, Singh S, Jain M, Tyagi AK, Parida SK (2013) Functionally relevant microsatellite markers from chickpea transcription factor genes for efficient genotyping applications and trait association mapping. DNA Res 20(4):355–374

Kumar S (1998) Inheritance of resistance to fusarium wilt (race 2) in chickpea. Plant Breed 117(2): 139–142

Kumar J, Haware MP (1982) Inheritance of resistance to fusarium wilt in chickpea. Phytopathology 72:1035–1036

Lal D, Ravikumar RL, Jingade P, Subramanya S (2022) Validation of molecular markers linked to fusarium wilt resistance (Foc 1) in recombinant inbred lines of chickpea (*Cicer arietinum*). Plant Breed 141(3):429–438

Lichtenzveig J, Scheuring C, Dodge J, Abbo S, Zhang HB (2005) Construction of BAC and BIBAC libraries and their applications for generation of SSR markers for genome analysis of chickpea, *Cicer arietinum* L. Theor Appl Genet 110:492–510

Manjarrez-Sandoval P, Agramon-Rocha JR, Perez RAS, Gomez-Gomez L (2004) Reports. Technology in the production of chickpea. Fundación Produce Sinaloa A.C., Mexico. http://www.fps.org.mx/divulgacion/images/stories/que_hacemos/publicaciones/test/Jornada%20de%20Transferencia%20de%20Tecnologia%20del%20Cultivo%20del%20Garbanzo.pdf

Mannur DM, Babbar A, Thudi M, Sabbavarapu MM, Roorkiwal M, Sharanabasappa BY, Bansal VP, Jayalakshmi SK, Yadav SS, Rathore A, Chamarthi SK (2019) Super Annigeri 1 and improved JG 74: two Fusarium wilt-resistant introgression lines developed using marker-assisted backcrossing approach in chickpea (*Cicer arietinum* L.). Mol Breed 39(1):2

Mayer MS, Tullu A, Simon CJ, Kumar J, Kaiser WJ, Kraft JM, Muehlbauer FJ (1997) Development of a DNA marker for fusarium wilt resistance in chickpea. Crop Sci 37(5):1625–1629

Mohamed OE, Mohamed AE (2020) Screening of chickpea (*Cicer arietinum*) genotypes to three identified races of *fusarium oxysporum* f. sp *ciceris* in Sudan. J Plant Pathol Microbiol 11:507

Mullis K, Faloona F, Scharf S, Saiki RK, Horn GT, Erlich H (1986) Specific enzymatic amplification of DNA in vitro: the polymerase chain reaction. In: Cold spring harbor symposia on quantitative biology, vol 51. Cold Spring Harbor Laboratory Press, pp 263–273

Navas-Cortés JA, Hau B, Jiménez-Díaz RM (2000) Yield loss in chickpeas in relation to development of fusarium wilt epidemics. Phytopathology 90:1269–1278

Nene YL, Haware MP (1980) Screening chickpea for resistance to wilt. Plant Dis 64:379–380

Nene YL, Reddy MV (1987) Chickpea diseases and their control. In: Saxena MC, Singh KB (eds) The chickpea. Oxon, UK, C.A.B. Wallingford, pp 233–270

Palomino C, Fernández-Romero MD, Rubio J, Torres A, Moreno MT, Millán T (2009) Integration of new CAPS and dCAPS-RGA markers into a composite chickpea genetic map and their association with disease resistance. Theor Appl Genet 118:671–682

Parida SK, Verma M, Yadav SK, Ambawat S, Das S, Garg R, Jain M (2015) Development of genome-wide informative simple sequence repeat markers for largescale genotyping applications in chickpea and development of web resource. Front Plant Sci 6:645

Pathak MM, Singh KP, Lal SB (1975) Inheritence of resistance to wilt (*F. oxysporum* f. sp. *ciceris*) in gram. Indian J Farm Sci 3:10–11

Phee M, Kevin E, Inglis DA, Gundersen B, Coyne CJ (2012) Mapping QTL for fusarium wilt race 2 partial resistance in pea (*Pisum sativum*). Plant Breed 131:300–306

Pratap A, Chaturvedi SK, Tomar R, Rajan N, Malviya N, Thudi M, Saabale PR, Prajapati U, Varshney RK, Singh NP (2017) Marker-assisted introgression of resistance to fusarium wilt race 2 in Pusa 256, an elite cultivar of desi chickpea. Mol Gen Genomics 292(6):1237–1245

Pshibytko NL, Zenevich LA, Kabashnikova LF (2006) Changes in the photosynthetic apparatus during fusarium wilt of tomato. Russ J Plant Physiol 53:25–31

Qadir SA, Datta S, Singh NP, Kumar S (2007) Development of highly polymorphic SSR markers for chickpea (*Cicer arietinum* L.) and their use in parental polymorphism. Indian J Genet Plant Breeding 67(4):329–333

Rajesh PN, Coyne C, Meksem K, Sharma KD, Gupta V, Muehlbauer FJ (2004) Construction of a HindIII bacterial artificial chromosome library and its use in identification of clones associated with disease resistance in chickpea. Theor Appl Genet 108(4):663–669

Ratnaparkhe MB, Tekeoglu M, Muehlbauer FJ (1998a) Inter-simple-sequence-repeat (ISSR) polymorphisms are useful for finding markers associated with disease resistance gene clusters. Theor Appl Genet 97(4):515–519

Ratnaparkhe MB, Santra DK, Tullu A, Muehlbauer FJ (1998b) Inheritance of inter-simple-sequence-repeat polymorphisms and linkage with a fusarium wilt resistance gene in chickpea. Theor Appl Genet 96(3–4):348–353

Ravikumar RL, Patil BS, Soregaon CD, Hegde SG (2007) Genetic evidence for gametophytic selection of wilt resistant alleles in chickpea. Theor Appl Genet 114:619–625

Ravikumar RL, Chaitra GN, Choukimath AM, Soregaon CD (2013) Gametophytic selection for wilt resistance and its impact on the segregation of wilt resistance alleles in chickpea (*Cicer arietinum* L.). Euphytica 189:173–181

Rubio J, Hajj-Moussa E, Kharrat M, Moreno MT, Millan T, Gil J (2003) Two genes and linked RAPD markers involved in resistance to *fusarium oxysporum* f. sp. *ciceris* race 0 in chickpea. Plant Breed 122(2):188–191

Sabbavarapu MM, Sharma M, Chamarthi SK, Swapna N, Rathore A, Thudi M, Gaur PM, Pande S, Singh S, Kaur L, Varshney RK (2013) Molecular mapping of QTLs for resistance to fusarium wilt (race 1) and Ascochyta blight in chickpea (*Cicer arietinum* L.). Euphytica 193(1):121–133

Salimath PM, Dharwad N, Gaur PM, Sridevi O, Suma B, Reddy KS, Laxuman T (2011) Impact of hybridization and induced mutagenesis on variability of traits and resistance to fusarium wilt in chickpea *(Cicer arietmum* L.). Indian J Genet Plant Breeding 71(4):329–332

Sethy NK, Shokeen B, Edwards KJ, Bhatia S (2006) Development of microsatellite markers and analysis of intraspecific genetic variability in chickpea (*Cicer arietinum* L.). Theor Appl Genet 112(8):1416–1428

Sharma KD, Muehlbauer FJ (2005) Genetic mapping of fusarium oxysporum f. sp. ciceris race-specific resistance genes in chickpea (*Cicer arietinum* L.). In: Abstract of the international food legume research conference—IV. Indian Agricultural Research Institute, New Delhi, pp 18–22

Sharma KD, Muehlbauer FJ (2007) Fusarium wilt of chickpea: physiological specialization, genetics of resistance and resistance gene tagging. Euphytica 157(1–2):1–14

Sharma KD, Nayyar H (2014) Cold stress alters transcription in meiotic anthers of cold tolerant chickpea (*Cicer arietinum* L.). BMC Res Notes 7(1):717

Sharma PC, Winter P, Bünger T, Hüttel B, Weigand F, Weising K, Kahl G (1995) Abundance and polymorphism of di-, tri-and tetra-nucleotide tandem repeats in chickpea (*Cicer arietinum* L.). Theor Appl Genet 90(1):90–96

Sharma KD, Chen W, Muehlbauer FJ (2004a) A consensus set of differential lines for identifying races of *fusarium oxysporum* f. sp. *ciceris*. Int Pigeonpea Chickpea Newsl 11:34–36

Sharma KD, Winter P, Kahl G, Muehlbauer FJ (2004b) Molecular mapping of *fusarium oxysporum* f. sp. ciceris race 3 resistance gene in chickpea. Theor Appl Genet 108(7):1243–1248

Sharma KD, Chen W, Muehlbauer FJ (2005) Genetics of chickpea resistance to five races of fusarium wilt and a concise set of race differentials for *fusarium oxysporum* f. sp. *ciceris*. Plant Dis 89(4):385–390

Sharma A, Rajeev Rathour P, Plaha VK, Khalsa GS, Patial V, Yudhvir Singh NK, Pathania. (2010) Induction of fusarium wilt (*fusarium oxysporum* f. sp. *pisi*) resistance in garden pea using induced mutagenesis and in vitro selection techniques. Euphytica 173:345–356

Sharma M, Avuthu N, Mahendar T, Raju G, Suresh P, Varshney RK (2014) Development of DArT markers and assessment of diversity in *fusarium oxysporum* f. sp. *ciceris*, wilt pathogen of chickpea *Cicer arietinum* L. BMC Genomics 15:454

Sharma M, Raju G, Avijit T, Abhishek R, Devashish RC, Anil VK, Pooran MG, Srinivasan S, Om G, Singh NP, Saxena DR, Saifulla M, Pithia MS, Ghante PH, Deyanand MM, Upadhyay JB, Harer PN (2019) Exploring the genetic cipher of chickpea *Cicer arietinum* L. through identification and multi-environment validation of resistant sources against fusarium wilt (*fusarium oxysporum* f. sp. *ciceris*). Front Sustain food Sys 3:78

Shehabu M, Ahmed S, Sakhuja PK (2008) Pathogenic variability in Ethiopian isolates of *fusarium oxysporum* f. sp. *ciceris* and reaction of chickpea improved varieties to the isolates. Int J Pest Manage 54(2):143–149

Sindhu JS, Singh KP, Slinkard AE (1983) Inheritance of resistance to fusarium wilt in chickpeas. J Hered 74:68–68

Singh KB, Dahiya BS (1973) Breeding for wilt resistance in chickpea. In: Symposium on wilt problem and breeding for wilt resistance in Bengal gram. Indian Research Institute, New Delhi, pp 13–14

Singh H, Kumar J, Haware MP, Smithson JB (1987a) Genetics of resistance to fusarium wilt in chickpeas. In: Day PR, Jellis GJ (eds) Genetics and plant pathogenesis. Blackwell, Oxford, pp 339–341

Singh H, Kumar J, Smithson JB, Haware MP (1987b) Complementation between genes for resistance to race 1 of *fusarium oxyspomm* f. sp. *ciceri* in chickpea. Plant Pathol 36(4):539–543

Singh S, Singh I, Kapoor K, Gaur PM, Chaturvedi SK, Singh NP, Sandhu JS (2014) Chickpea. In: Broadening the Genetic Base of grain legumes. National Bureau of Plant Genetic Resources, New Delhi

Sinha P, Biswas SK (2010) Slow wilting components in pigeonpea (*Cajanus cajan* (L.) Millsp.). Eur J Plant Pathol 128:503–509

Srivastava R, Singh M, Bajaj D, Parida SK (2016) A high-resolution InDel (insertion–deletion) markers-anchored consensus genetic map identifies major QTLs governing pod number and seed yield in chickpea. Front Plant Sci 7:1362

Srivastava AK, Saxena DR, Saabale PR, Raghuvanshi KS, Anandani VP, Singh RK, Sharma OP, Wasinikar AR, Sahni S, Varshney RK, Singh NP (2021) Delineation of genotype-by-environment interactions for identification and validation of resistant genotypes in chickpea to fusarium wilt using GGE biplot. Crop Prot 144:105571

Stephens A, Lombardi M, Cogan NO, Forster JW, Hobson K, Materne M, Kaur S (2014) Genetic marker discovery, intraspecific linkage map construction and quantitative trait locus analysis of Ascochyta blight resistance in chickpea (*Cicer arietinum* L.). Mol Breed 33:297–313

Tekeoglu M, Tullu A, Kaiser WJ, Muehlbauer FJ (2000) Inheritance and linkage of two genes that confer resistance to fusarium wilt in chickpea. Crop Sci 40(5):1247–1251

Thudi M, Bohra A, Nayak SN, Varghese N, Shah TM, Penmetsa RV, Thirunavukkarasu N, Gudipati S, Gaur PM, Kulwal PL, Upadhyaya HD (2011) Novel SSR markers from BAC-end sequences, DArT arrays and a comprehensive genetic map with 1,291 marker loci for chickpea (*Cicer arietinum* L.). PLoS One 6(11):e27275

Trapero-Casas A, Jiménez-Díaz RM (1985) Fungal wilt and root rot diseases of chickpea in southern Spain. Phytopathology 75:1146–1151

Tullu A, Muehlbauer FJ, Simon CJ, Mayer MS, Kumar J, Kaiser WJ, Kraft JM (1998) Inheritance and linkage of a gene for resistance to race 4 of fusarium wilt and RAPD markers in chickpea. Euphytica 102(2):227–232

Tullu A, Kaiser WJ, Kraft JM, Muehlbauer FJ (1999) A second gene for resistance to race 4 of fusarium wilt in chickpea and linkage with a RAPD marker. Euphytica 109(1):43–50

Upadhyaya HD, Smithson JB, Haware MP, Kumar J (1983a) Resistance to wilt in chickpea. II. Further evidence for two genes for resistance to race 1. Euphytica 32(3):749–755

Upadhyaya HD, Haware MP, Kumar J, Smithson JB (1983b) Resistance to wilt in chickpea. I. Inheritance of late-wilting in response to race 1. Euphytica 32(2):447–452

Varshney RK, Hiremath PJ, Lekha P et al (2009) A comprehensive resource of drought and salinity responsive ESTs for gene discovery and marker development in chickpea (*Cicer arietinum* L.). BMC Genomics 15:523

Varshney RK, Mohan SM, Gaur PM, Chamarthi SK, Singh VK, Srinivasan S, Swapna N, Sharma M, Pande S, Singh S, Kaur L (2014a) Marker-assisted backcrossing to introgress resistance to fusarium wilt race 1 and Ascochyta blight in C 214, an elite cultivar of chickpea. The Plant Genome 7:1–11

Varshney RK, Mir RR, Bhatia S, Thudi M, Hu Y, Azam S, Zhang Y, Jaganathan D, You FM, Gao J, Riera-Lizarazu O (2014b) Integrated physical, genetic and genome map of chickpea (Cicer arietinum L.). Funct Integr Genomics 14:59–73

Weising K, Kaemmer D, Weigand F, Epplen JT, Kahl G (1992) Oligonucleotide fingerprinting reveals various probe-dependent levels of informativeness in chickpea (*Cicer arietinum* L.). Genome 35(3):436–442

Williams JG, Kubelik AR, Livak KJ, Rafalski JA, Tingey SV (1990) DNA polymorphisms amplified by arbitrary primers are useful as genetic markers. Nucleic Acids Res 18(22): 6531–6535

Winter P, Pfaff T, Udupa SM, Hüttel B, Sharma PC, Sahi S, Arreguin- Espinoza R, Weigand F, Muehlbauer FJ, Kahl G (1999) Characterization and mapping of sequence-tagged microsatellite sites in the chickpea (*Cicer arietinum* L.) genome. Mol Gen Genomics 262:90–101

Winter P, Benko-Iseppon AM, Hüttel B, Ratnaparkhe M, Tullu A, Sonnante G, Pfaff T, Tekeoglu M, Santara D, Sant VJ, Rajesh PN, Kahl G, Muhelbauer FJ (2000) A linkage map of chickpea (*Cicer arietinum* L.) genome based on recombinant inbred lines from a *C. arietinum* × *C. reticulatum* cross: localization of resistance genes for fusarium wilt races 4 and 5. Theor Appl Genet 101:1155–1163

Zhang X, Scheuring CF, Zhang M, Dong JJ, Zhang Y, Huang JJ, Lee MK, Abbo S, Sherman A, Shtienberg D, Chen W (2010) A BAC/BIBAC-based physical map of chickpea, *Cicer arietinum* L. BMC Genomics 11(1):1–10

Dry Root Rot in Chickpea: A Perspective on Disease Resistance Breeding Strategies

4

Shubhashish Ranjan, Rishabh Mirchandani, and Muthappa Senthil-Kumar

Abstract

Chickpea is an essential crop nutritionally rich in protein and grown around the world, generally in rain-fed condition. Dry root rot (DRR) is an emerging and economically devastating disease caused by the chickpea-specific strain *Macrophomina phaseolina*. Environmental conditions such as drought and high temperature aggravate DRR causing significant crop loss. The control of *M. phaseolina* is challenging due to the broad host range of this fungus. Genetic resistance of enhancement of resistance to DRR through breeding is a potential way to prevent crop loss due to the disease. In this chapter, we highlight the importance of breeding strategies for rapidly developing DRR disease-resistant varieties of chickpeas. We also provide a brief overview of the role of next-generation sequencing (NGS) technology and high-throughput phenotyping (HTP) in the next-generation breeding strategy against DRR disease. We suggest that the advancement of sequencing technology and the availability of the high-quality reference genome of chickpeas can facilitate genotyping and mining of the allelic variation among the diverse chickpea population. We also discuss the potential of genome editing integrated with speed breeding to reduce the generation time significantly. Thus, we suggest the combination of genome-wide association study (GWAS) and speed breeding with genome editing can take DRR resistance breeding in chickpea to the next level and have the potential to provide precisely edited chickpeas in a short duration of time.

S. Ranjan · R. Mirchandani · M. Senthil-Kumar (✉)
National Institute of Plant Genome Research, Aruna Asaf Ali Marg, New Delhi, Delhi, India
e-mail: skmuthappa@nipgr.ac.in

U. C. Jha et al. (eds.), *Diseases in Legume Crops*,
https://doi.org/10.1007/978-981-99-3358-7_4

Keywords

Chickpea · Dry root rot · Resistance breeding · High-throughput phenotyping · Genome-wide association study · Speed breeding

4.1 Introduction

Macrophomina phaseolina (Tassi.) Goid is a soil-inhabitant, plant-pathogenic fungus that belongs to the Botryosphaeriaceae family of *Ascomycetes*. This fungus acts as a causal agent of several diseases in more than 500 wild and cultivated plant species (Gupta et al. 2012). These diseases affect the yield of economically important legume crops such as chickpeas, soybean, green gram, and cowpea. *M. phaseolina* causes dry root rot (DRR) disease in chickpea (*Cicer arietinum* L.). It has been reported that specific abiotic stresses, such as drought and heat, aggravate DRR disease symptoms in chickpea (Sinha et al. 2021). Under favorable environmental conditions, up to 100% yield loss can be observed in susceptible cultivars. This pathogen generally attacks the chickpea root system and leads to root architecture damage that can lead to severe yield losses. DRR is an emerging chickpea disease widely spread in major chickpea cultivation regions across the globe (Rai et al. 2022). Further, the changing environmental conditions will favor the geographical spread of the disease (Pandey and Basandrai 2021; Mirchandani et al. 2023).

The control of this fungus is difficult due to its broad host range and prolonged survival in the field. The development of DRR-resistant cultivars of chickpea is one of the important methods to prevent yield loss caused by the disease. DRR phenotyping in the available germplasm is significantly less explored to date. Exploration of chickpea germplasm with high efficiency and accuracy is the minimal requirement for DRR resistance breeding. Efficient utilization of phenomic and genomic tools will be essential in identifying resistant cultivars in the germplasm and developing new cultivars with DRR resistance. Further, techniques such as speed breeding can rapidly advance the generations and significantly reduce the standard breeding period (Samineni et al. 2020). This book chapter aims to highlight the use of HTP and NGS information in identifying associated DRR resistance loci in the chickpea genome and their use in the next-generation breeding strategy to develop DRR disease-resistant varieties.

4.2 DRR Disease Distribution

The disease is particularly prevalent in arid and semiarid agroclimatic conditions worldwide. The total chickpea cultivation area is 13 million hectares (ha), and the total production ranges from approximately 15–16 million tonnes annually. The DRR disease is reported worldwide in major chickpea-growing regions such as Africa, Spain, South Asia, Turkey, the Mediterranean region, and several North

Table 4.1 Examples of chickpea genotypes with contrasting response to DRR

S. No	Genotype name	Disease reaction	Reference
1	JG 62, ICC 1715, JGK 18	Highly susceptible	Chilakala et al. (2022), Talekar et al. (2017)
2	BG212, ICCV 07107, ICCV 07306	Susceptible	Karadi et al. (2021), Talekar et al. (2021)
3	ICCV 08315, ICC 11550, ICC 14395	Moderately susceptible	Talekar et al. (2021)
4	ICC2867, ICC 9023, ICC 14307	Moderately resistance	Talekar et al. (2021)
5	PG 06102, BG 2094	Resistance	Talekar et al. (2021)

American countries. In India, the chickpea area under cultivation is 6.3 million ha, and 85% of the area is under rainfed conditions (FAOSTAT 2019). In mild infection, yield loss can range between 5 and 10%, while in moderate infection, it can be between 30 and 50%, and in severe disease infection, it can be up to 80% (Rai et al. 2022). Changing climatic conditions such as low rainfall and high temperature can elevate the risk of economic yield loss due to DRR. A survey conducted in the chickpea cultivating area of Rajasthan showed that DRR is the major problem in the Churu, Jodhpur, Bikaner, and Jaisalmer districts of Rajasthan. The average disease incidence observed is 9.15% in Rajasthan (Partap and Godara 2022). In contrast, some other regions of the country show higher disease incidence, such as Niwadi (31.5%), middle Gujarat (26%), Kalburgi, and Raichur (30–35%) of Madhya Pradesh, Gujarat, and Karnataka, respectively, (Mirchandani et al. 2023). The disease occurrence varies with soil type or edaphic factors, environmental factors, and cultivated varieties. Table 4.1 represents the name of some reported genotypes against DRR resistance or susceptibility. Figure 4.1 shows the image of DRR infestation in the chickpea field.

4.3 DRR Causal Agent and Disease Cycle

Based on the sequence information of 28s rDNA, *M. phaseolina* is classified under the *Ascomycota* division (Crous et al. 2006). The hyphae are thin-walled, dark to light brown, hyaline, branched, and septate. Branches arise from the parent hyphae, generally at right angles with a constriction at the base. Microsclerotium, a compact mass of fungal mycelium, is light brown (early stage) to dark brown (aging) in color with an oval or spherical shape (Sharma et al. 2015). The fungus reproduces by fragmentation (Sharma and Pande 2013; Ghosh et al. 2013).

The general symptoms associated with DRR disease in chickpea plants are root necrosis, lateral root shedding, yellowing of leaves, and premature drying. The characteristic feature of a DRR-affected field is the presence of irregular dried patches of straw-colored plants. The below-ground symptoms include brownish to black necrotic lesions on lateral and tap roots (Sharma and Pande 2013). Gradual progression of necrosis leads to the complete loss of lateral roots during the later

Fig. 4.1 Dry root rot disease infestation in chickpea field. The arrows represent DRR-infected plants in the field. The photo was taken from a field location at Guntur (India) during rabi 2021

stages of the disease. The taproot may remain intact with plants, but they generally become brittle. Thus, infected plants can be easily uprooted without much force. The premature drying occurs due to blocking in stele by fungal mycelium and microsclerotia growth that reduces the water and nutrient transport to the shoot. Gradual yellowing of leaves from base to top during the vegetative to flowering stage transition period marks the onset of aboveground symptoms. DRR-affected plants remain upright with straw-colored leaves and stem. Healthy chickpea plants become dry only after physiological maturity (90–120 days after sowing (DAS), while DRR-affected plants show premature drying at the reproductive stage (60–80 DAS) (Rai et al. 2022).

DRR disease incidence in the field depends on initial inoculum load, host plant susceptibility, high temperature, and moisture stress in soil. Microsclerotia are present in the soil or on plant debris from the previous cultivated season and act as a source of primary inoculum. The microsclerotia remain dormant but viable in the soil or on plant debris for several years. High soil moisture reduces the survival of microsclerotia, but it can stay in the quiescent stage and be viable for up to 15 years in the soil (Gupta et al. 2012). At the seedling stage (1–10 DAS), microsclerotia can attach to the root and begin epidermal necrosis. Necrotic lesions increase with incubation time and show asymptomatic foliage at the vegetative stage (20–40 DAS). Most pathogen-infected plant tap and lateral roots start to rot, and root loss begins at the reproductive stage (40–60 DAS). The development of symptoms is accelerated under moderate drought stress conditions. The infection period after symptoms appear on foliage is called the active infection period (40–90 DAS) (Rai et al. 2022). However, the abundance of primary inoculum and favorable environmental condition in the field is mainly responsible for the severity of the disease.

4.4 Effect of Abiotic Stresses

Generally, chickpea growth season in India is between November to April. Thus, it encounters terminal drought and heat, i.e., at the reproductive stages of chickpeas (Sharma and Pande 2013; Sinha et al. 2019, 2021) due to rainfed cultivation. In field experiments, Sinha et al. (2019) reported that DRR incidence varied with the severity of drought stress. Higher incidence (40–60%) occurred in severely drought-stressed plots, while pathogen treatment plots with appropriate irrigation had significantly lower disease incidence (0–20%). Under controlled conditions, it was shown that DRR incidence and severity levels increased under drought stress (Sharma and Pande 2013; Sinha et al. 2019). In addition, reduced root water potential caused a rise in the DRR pathogen's lytic enzymatic activity, intensifying the disease (Rai et al. 2022).

In high temperatures, soil-borne necrotrophic pathogens could reproduce, spread widely, and adapt to high-temperature stress better than the host. Studies showing more significant necrosis and colonization in plants at high temperatures indicate a competitive advantage for the pathogen (Desaint et al. 2021). In addition, specific secondary metabolites and enzymes can accumulate in more significant quantities in warm environmental conditions (Rai et al. 2022). The DRR pathogen requires cellulolytic enzymes to lyse the host's cell walls. At temperatures between 15 and 35 °C, the soybean-specific *M. phaseolina* produced the most cellulolytic enzymes in the carboxymethyl cellulose broth medium and less at temperatures lower than 15 °C and higher than 35 °C (Gawade et al. 2018). These studies indicate that infection and colonization of fungus significantly increase with an increase in temperature (Sharath Chandran et al. 2021; Sharma and Pande 2013).

4.5 DRR Resistance Breeding

DRR phenotyping studies have revealed very few resistant genotypes of chickpea. Talekar et al. (2021) screened more than 500 chickpea genotypes in controlled conditions and reported only three resistant and 21 moderately resistant genotypes, while most of the screened genotypes were susceptible. Figure 4.2 depicts the normal distribution curve of a small set of phenotypic data. The data indicates that the distribution is skewed towards susceptibility, i.e., a higher percentage of the genotypes are susceptible. The proportion of resistant genotypes is low at 1.15%, indicating that more germplasm lines should be disease phenotyped and the resistant lines should be explored. The prerequisite for resistance breeding is the availability of resistant pre-breeding material for introgression into elite chickpea cultivars.

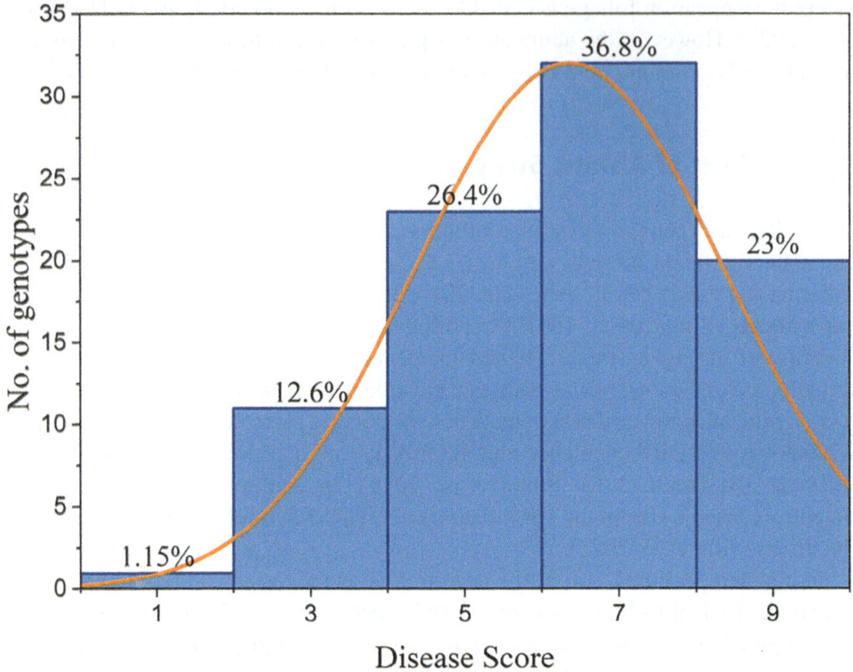

Fig. 4.2 Normal distribution curve of the DRR disease score of a few screened genotypes. The phenotypic data of 88 genotypes (Talekar et al. 2021) is represented here. The percent values indicate the proportion of genotypes with the corresponding disease score in the set of data. The orange line represents the normal distribution curve

Thus, exploring the available large chickpea germplasm is required to discover novel resistant genotypes and for subsequent breeding programs.

Towards DRR resistance breeding, two independent mapping populations have been developed by crossing two contrasting genotypes of chickpeas for DRR. The first mapping population was developed by Talekar et al. (2017). They crossed the highly susceptible genotype L550 and PG 06102, a DRR resistance genotype. The population was comprised of 129 lines. These were phenotyped for DRR resistance by the blotter paper technique. The authors concluded that the resistance to DRR is monogenic. Among the F2:3 mapping lines, 27, 38, and 64 mapping lines show a homozygous resistance, susceptible, and heterozygous disease reaction to DRR, respectively. Thus, the segregating population showed a 1:2:1 Mendelian ratio. Two markers, ICCM0299 and ICCM0120b, were identified in the population for DRR resistance. They reported that the DRR resistance region was between the two markers, and the distance of the DRR resistance gene (named *DRR1*) from ICCM0299 was 7.75 cM and 22.48 cM from the ICCM0120b marker (Talekar et al. 2017).

Karadi et al. (2021) developed a population of 182 recombinant inbred lines (RILs) obtained from the cross between ICCV 08305 (moderately resistant) and BG

212 (susceptible line). The RIL population was developed by the single seed decent technique between the generation advancement from F2 to F9. They phenotyped these lines for DRR resistance by blotter paper technique and used Affymetrix Axiom CicerSNP Array for genotyping the RIL population. A total of 13,110 SNPs were used to construct a linkage genetic map across eight linkage groups with a total length of 1224.11 cM. They identified a minor QTL (qDRR8) on the linkage group CaLG8 for DRR resistance with a phenotypic variance of 6.70% and a LOD score of 3.34. Furthermore, they reported that the QTL is flanked by the markers Ca8_3970986 and Ca8_3904895.

Therefore, the development of these two mapping populations represents only a small region of the chickpea genome or a minor QTL. However, it cannot provide more information about the genetic basis of DRR resistance. Hence, research is required to identify genomic regions conferring DRR disease resistance. Further, these regions will be the pillar of the next-generation resistance breeding strategy for DRR.

4.6 Prospective of Next Generation Breeding in DRR Resistance of Chickpea

4.6.1 Disease Phenotyping

DRR phenotyping includes field sick plot assay, sick pot-based assay, and paper blotter techniques. The rapid way of DRR phenotyping is the blotter paper technique, which requires seedling preparation, fungal culture, and assessment of disease. After incubation, the necrotic lesions and root rot will be observed for disease assessment. Based on the visual observations of infected roots, a score between 1 and 9 will be assigned based on disease reaction from resistant to highly susceptible (Irulappan and Senthil-Kumar 2021). The blotter paper technique is advantageous over sick plot and sick pot assay due to its low time requirement. In addition, the DRR phenotyping for a large number of chickpea germplasm is easy and fast by blotter paper approach relative to other techniques.

HTP is an emerging technology for rapidly analyzing the physical characteristics of many plant individuals to identify genetic variations that may be associated with a specific trait of interest (Song et al. 2021). It is a recent and accurate phenotyping technique that can also be possible with DRR screening techniques. Nondestructive imaging and sensing, including RGB (red, green, and blue), thermal infrared, spectral and hyperspectral, fluorescence, 3D, and computed tomographic imaging by X-ray and MRI (magnetic resonance imaging) techniques, and the use of whole root scanning, will advance the measurement and acquisition of HTP in chickpea phenotyping for DRR. Here we discuss the prospect of using RGB images for HTP in DRR of chickpeas in the greenhouse experiment for the aerial part. Images could be captured between 15 and 30 days of sowing in the greenhouse. The three bands in RGB images are used to compute vegetation indices for further analysis. The images can be processed using a MATLAB-based algorithm to develop a highly accurate disease score of DRR phenotyping (Bari et al. 2023).

Phenotyping for the disease incidence on root trait can be possible by scanning the whole root using a root scanner, and the scanned image is further processed using software like WinRHIZO™ (Regent Instruments Inc.) and RhizoVision Explorer (Seethepalli et al. 2021). These HTP data of disease resistance traits will be further used for linking the genomic information (Song et al. 2021).

4.6.2 Next-Generation Sequencing in Chickpea and Genotyping

NGS and third-generation sequencing technologies have facilitated the development of a high-quality chickpea reference genome (CDC Frontier and ICC4958) and pan-genome. The estimated genome size of chickpea is 738 Mb. The CDC Frontier genome sequence spans 532.29 Mb, which contains 28,269 genes, while the chickpea pan-genome spans 592.58 Mb and contains 29,870 genes (Varshney et al. 2013, 2021). Combining multi-omic assays, large diversity panels, and HTP can bridge the gap between genome-phenome maps (Varshney et al. 2021b).

NGS and third-generation technologies will facilitate efficient allele mining in chickpea. Allele mining is an approach to identifying the new alleles in the genome of cultivars, landraces, and wild relatives. The whole genome survey of the available diversity panel of chickpea (Table 4.2), and high-throughput phenotyping of DRR resistance traits can be associated with the genome marker through GWAS in multiparental populations (Varshney et al. 2021a) (Fig. 4.3). GWAS have been extensively employed to pinpoint the genetic basis for several crop agronomic features. The recent example of GWAS in chickpeas is 429 genotypes from chickpea-growing countries for drought and heat-related stress (Varshney et al. 2019), and 3366 chickpea accessions for yield-related traits were already available (Varshney et al. 2021). GWAS has been conducted in a few legumes under *M. phaseolina* stress.

Table 4.2 Details of sequenced chickpea genotypes for facilitating disease-resistant breeding

S. No.	Number of accessions	Average sequencing depth/ resequencing/genotype by sequencing (GBS)	Genetic variants	References
1.	429 cultivated chickpea	6.8X/resequencing	4.97 million SNPs	Varshney et al. (2019)
2.	3171 cultivated chickpea	10X/ resequencing	3.94 million SNPs	Varshney et al. (2021)
3.	195 wild accessions	10X/resequencing	19.57 million SNPs	Varshney et al. (2021)
4.	100 desi chickpea accession	GBS	44,844 high-quality SNPs	Kujur et al. (2015)

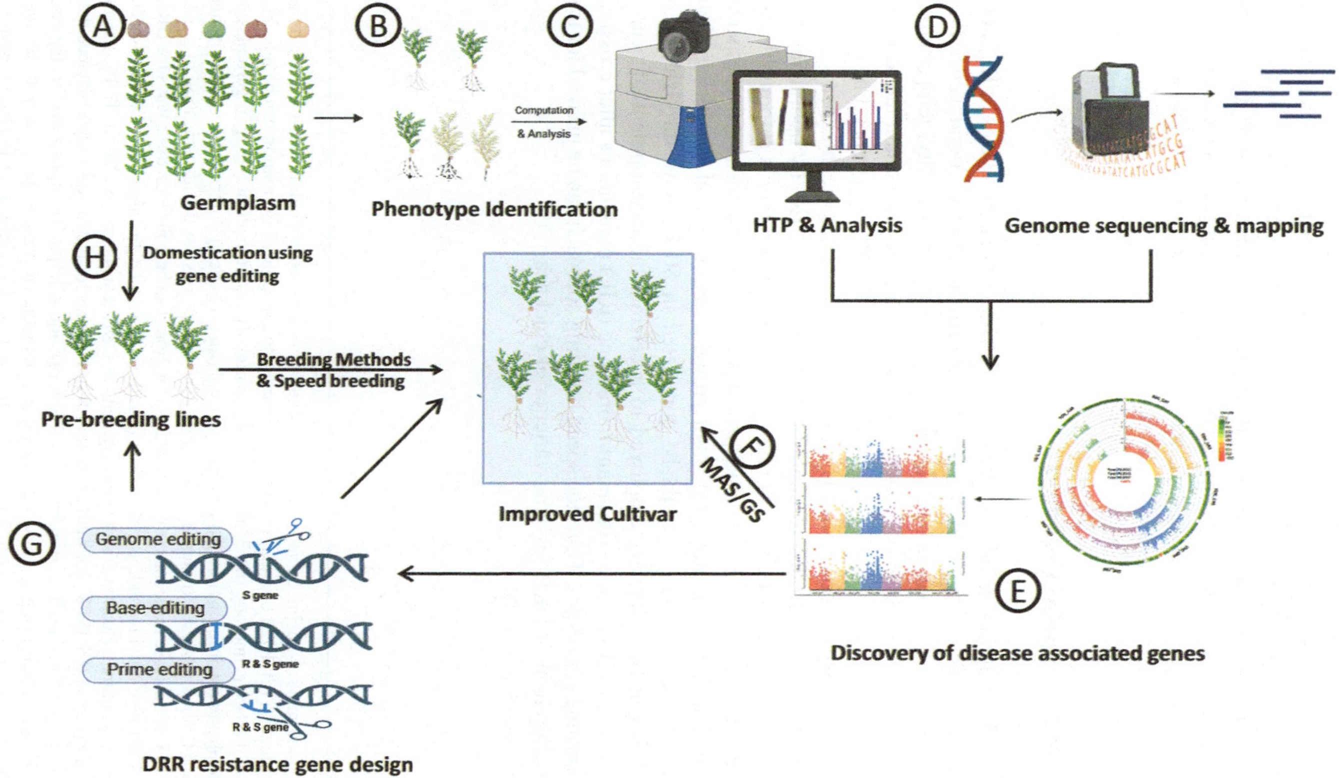

Fig. 4.3 Resistance breeding model for DRR of chickpea. This diagrammatic model depicts the steps involved in DRR resistance breeding in chickpeas. The collection of a large set of chickpea germplasm which represents high genomic diversity (**a**) will be used for DRR phenotyping (**b**), and HTP could be possible

Cosar et al. (Coser et al. 2017) identified 19 linked SNPs by GWAS of soybean against charcoal rot (caused by *M. phaseolina*). Similarly, Muchero et al. (2011) reported eight QTLs in the cowpea RIL population associated with charcoal rot caused by the same fungus. Another powerful method to identify new R genes is resistance gene enrichment sequencing (RanSeq). An understanding of the disease resistance mechanism of the host, the pathogenicity of the fungus, and identification and characterization of the genomic region or disease-contributing alleles can act as a potential pre-breeding material for resistance breeding.

4.6.3 Role of Speed Breeding

Once the GWAS approach has identified the resistance-associated alleles or R genes, speed breeding technology can be utilized. Speed breeding or rapid generation advancement (RGA) involves modulating the photoperiod, humidity, temperature, and the harvesting/germination of immature seeds. Chickpea is a quantitative long-day plant. The chickpea growth period varies from 90 to 160 days, depending on the cultivar and growing conditions. Up to seven generations of chickpea can be completed in a year using this technology (Samineni et al. 2020) by optimizing the life cycle duration. The alteration of photoperiod and temperature is responsible for early flowering and maturation in chickpea. It activates development activities such as germination, leaf expansion, and shoot growth. Samineni et al. (2020) reported that one chickpea generation could be completed in 50–60 days through speed breeding. It can be efficiently employed to accelerate the development of nearly isogeneic lines (NILs) and recombinant inbred line (RIL) populations after crossing two contrasting genotypes. Thus, the speed breeding approach can be utilized for the rapid development of DRR-resistant varieties of chickpea.

Fig. 4.3 (continued) by using cameras and root scanners and their computational analysis to get the traits information (**c**). High efficient genome sequencing and mapping information can be possible by the use of NGS technology (**d**). This can be done parallelly with the phenotyping of germplasm. The DRR phenotype information and chickpea germplasm sequencing information can be further utilized for GWAS analysis to identify associated genomic regions for the identification of genes involved in DRR disease resistance (**e**). Such genomic regions can be utilized in three ways for DRR resistant improvement in chickpea: directly utilized in resistance breeding by marker-assisted selection and genome selection for the development of DRR resistance population (**f**), the use of genome editing tool (CRISPR) for modification of S or R-genes for broad-spectrum resistance that can be used as pre-breeding lines or improved cultivar (**g**), and germplasm can be used for stepwise *de novo* domestication using genome editing for development of pre-breeding material (**h**), and further breeding methods and speed breeding can be used for development of improved cultivar. *Illustration created with* Biorender.com

4.6.4 Role of Genome Editing

A drawback of using specific R gene-mediated resistance to develop cultivars is that resistance could be broken down over time due to the arms race between the host and the pathogen. However, genetic engineering of particular R genes may aid in overcoming this drawback. The receptor regions could be engineered to decrease their specificity and enable them to recognize a broad spectrum of effectors (Segretin et al. 2014). Given that the coding areas of resistance alleles only differ by a small number of nucleotides, CRISPR-mediated homology-directed repair and prime genome editing technology can be utilized to generate new alleles with a broader resistance spectrum (Fig. 4.3) (Deng et al. 2020). An alternate way of improving chickpea resistance against DRR is engineering susceptibility (S) genes. Mutating S-genes typically results in broad-spectrum resistance. However, it has inevitable trade-offs as most S-genes are involved in the host's growth, development, or metabolic functions (Li et al. 2020). Hence, CRISPR-mediated base genome editing can engineer S-genes to produce novel elite alleles that confer broad-spectrum resistance to DRR while potentially alleviating growth and reproductive trade-offs. This has been recently achieved in rice by editing specific SWEET genes (Oliva et al. 2019). Similarly, CRISPR could be used to develop pre-breeding material or cultivars, which can be used in chickpea DRR resistance breeding programs.

4.7 Model for DRR Resistance Breeding

A breeding model for DRR resistance breeding based on the next-generation breeding approach is depicted in Fig. 4.3. An extensive chickpea germplasm can be exploited to explore the available diversity of DRR disease resistance. Phenotyping of this germplasm is required to discover the resistant genotypes. HTP can significantly and efficiently accelerate the analysis of disease resistance traits. The high-throughput advanced genome sequencing platforms or NGS technologies can be used for sequencing and mapping with the reference genome of the chickpea. The phenotypic data and identified single nucleotide polymorphism (SNP) markers can be utilized to conduct a GWAS to identify associated genomic regions or SNPs associated with the studied DRR trait. Once the resistance marker is elucidated, it could act as a source of information for resistance breeding. Alternatively, the associated SNPs or genomic regions can be directly utilized for MAS or genomic selection (GS) to develop new resistant cultivars. The GS approach can predict the genomic estimated breeding values (GEBVs). GEBVs help a breeder to know about the offspring of the crossing program, which serves as parents for the next generation in the breeding cycle. Genomic selection has an advantage over MAS and traditional breeding methods in terms of per annum genetic gain. In contrast, CRISPR-based genome editing can be utilized for broad-spectrum resistance of DRR by engineering the linked genes to recognize a broad range of pathogen effectors. In addition, speed breeding can be employed for RGA. These cutting-edge technologies can help develop DRR-resistant cultivars or improve the chickpea germplasm.

4.8 Conclusion

The number of reported DRR-resistant chickpea cultivars or breeding populations so far is low, only two chickpea breeding populations are known, and a few genotypes reported resistance to DRR. Thus, exploring available chickpea germplasm is required to identify DRR resistance sources. The phenotyping techniques should be robust for the identification of resistance. We highlight that HTP is more feasible and cost-effective and can be utilized for the phenotyping of the germplasm. The DRR disease phenotyping done up to date is insufficient to capture the available germplasm diversity. Currently, more than 3500 chickpea accessions have been sequenced. This information can be exploited for genome-wide analysis if phenome information is available. In the context of DRR, HTP and NGS information could be used to identify genetic variations that confer resistance to the disease in chickpea. The marker-trait association will provide information on the putatively associated locus in the genome for resistance to DRR. Further, the discovered genomic region (s) can be utilized to improve the chickpea germplasms against DRR (Fig. 4.3). Speed breeding and genome editing are cutting-edge technologies that can lead to broad-spectrum resistance and accelerate breeding efforts. Efficient and robust use of the abovementioned tools will be essential in driving the efforts towards breeding for DRR resistance.

Acknowledgment This work was supported by the National Institute of Plant Genome Research Core Funding and the Council of Scientific and Industrial Research (CSIR) junior research fellowship [CSIR File no. – 09/0803(13718)/2022-EMR-I] to S.R. and [CSIR File no. – 09/803 (0177)/2020-EMR-I] to R.M. The authors also thank Mr. Aswin Reddy Chilakala for providing the field image. The authors acknowledge Dr. Prachi Pandey for suggestions on this manuscript.

Conflict of Interest The authors declare that there are no conflicts of interest.

References

Bari MAA, Fonseka D, Stenger J et al (2023) A greenhouse-based high-throughput phenotyping platform for identification and genetic dissection of resistance to Aphanomyces root rot in field pea. Plant Phenom J 6(1). 10.1002/ppj2.v6.1. https://doi.org/10.1002/ppj2.20063

Chilakala AR, Mali KV, Irulappan V et al (2022) Combined drought and heat stress influences the root water relation and determine the dry root rot disease development under field conditions: a study using contrasting chickpea genotypes, vol 13. Front Plant Sci, p 890551. https://doi.org/10.3389/fpls.2022.890551

Coser SM, Chowda Reddy RV, Zhang J et al (2017) Genetic architecture of charcoal rot (*Macrophomina phaseolina*) resistance in soybean revealed using a diverse panel. Front Plant Sci 8:1626. https://doi.org/10.3389/fpls.2017.01626

Crous PW, Slippers B, Wingfield MJ et al (2006) Phylogenetic lineages in the Botryosphaeriaceae. Stud Mycol 55:235–253. https://doi.org/10.3114/sim.55.1.235

Deng Y, Ning Y, Yang D-L et al (2020) Molecular basis of disease resistance and perspectives on breeding strategies for resistance improvement in crops. Mol Plant 13:1402–1419. https://doi.org/10.1016/j.molp.2020.09.018

Desaint H, Aoun N, Deslandes L et al (2021) Fight hard or die trying: when plants face pathogens under heat stress. New Phytol 229:712–734. https://doi.org/10.1111/nph.16965

Food and Agriculture Organization of the United Nations (2019) FAOSTAT statistical database. FAO, Rome

Gawade DB, Perane RR, Deokar CD et al (2018) Effect of physiological factors on production of cellulolytic enzymes by Rhizoctonia bataticola. Indian Phytopathol 71:555–561. https://doi.org/10.1007/s42360-018-0095-y

Ghosh R, Sharma M, Telangre R, Pande S (2013) Occurrence and distribution of chickpea diseases in central and southern parts of India. Am J Plant Sci 04:940–944. https://doi.org/10.4236/ajps.2013.44116

Gupta GK, Sharma SK, Ramteke R (2012) Biology, epidemiology and Management of the Pathogenic Fungus *Macrophomina phaseolina* (Tassi) Goid with special reference to charcoal rot of soybean (Glycine max (L.) Merrill). J Phytopathol 160:167–180. https://doi.org/10.1111/j.1439-0434.2012.01884.x

Irulappan V, Senthil-Kumar M (2021) Dry root rot disease assays in chickpea: a detailed methodology. J Vis Exp. https://doi.org/10.3791/61702

Karadi A, Samineni S, Sajja S et al (2021) Molecular mapping of dry root rot resistance genes in chickpea (Cicer arietinum L.). Euphytica 217:1–13. https://doi.org/10.1007/s10681-021-02854-4

Kujur A, Bajaj D, Upadhyaya HD et al (2015) Employing genome-wide SNP discovery and genotyping strategy to extrapolate the natural allelic diversity and domestication patterns in chickpea. Front Plant Sci 6. https://doi.org/10.3389/fpls.2015.00162

Li W, Deng Y, Ning Y et al (2020) Exploiting broad-spectrum disease resistance in crops: from molecular dissection to breeding. Annu Rev Plant Biol 71:575–603. https://doi.org/10.1146/annurev-arplant-010720-022215

Mirchandani R, Irulappan V, Chilakala AR, Senthil-Kumar M (2023) Dry root rot disease: current status and future implications for chickpea production. Proc Natl Acad Sci India Sect B 1–10

Muchero W, Ehlers JD, Close TJ, Roberts PA (2011) Genic SNP markers and legume synteny reveal candidate genes underlying QTL for *macrophomina phaseolina* resistance and maturity in cowpea [vigna unguiculata (L) walp.]. BMC Genomics 12:8. https://doi.org/10.1186/1471-2164-12-8

Oliva R, Ji C, Atienza-Grande G et al (2019) Broad-spectrum resistance to bacterial blight in rice using genome editing. Nat Biotechnol 37:1344–1350. https://doi.org/10.1038/s41587-019-0267-z

Pandey AK, Basandrai AK (2021) Will *macrophomina phaseolina* spread in legumes due to climate change? A critical review of current knowledge. J Plant Dis Prot 128:9–18. https://doi.org/10.1007/s41348-020-00374-2

Partap M, Godara SL (2022) Occurrence and distribution of dry root rot of chickpea caused by *macrophomina phaseolina* (Tassi) Goid. In arid region of Rajasthan. Legum Res 45(5): 639–645. https://doi.org/10.18805/lr-4429

Rai A, Irulappan V, Senthil-Kumar M (2022) Dry root rot of chickpea: a disease favored by drought. Plant Dis 106:346–356. https://doi.org/10.1094/pdis-07-21-1410-fe

Samineni S, Sen M, Sajja SB, Gaur PM (2020) Rapid generation advance (RGA) in chickpea to produce up to seven generations per year and enable speed breeding. Crop J 8:164–169. https://doi.org/10.1016/j.cj.2019.08.003

Seethepalli A, Dhakal K, Griffiths M et al (2021) RhizoVision explorer: open-source software for root image analysis and measurement standardization. AoB PLANTS 13:plab056. https://doi.org/10.1093/aobpla/plab056

Segretin ME, Pais M, Franceschetti M et al (2014) Single amino acid mutations in the potato immune receptor R3A expand response to *phytophthora* effectors. Mol Plant-Microbe Interact 27:624–637. https://doi.org/10.1094/mpmi-02-14-0040-r

Sharath Chandran US, Tarafdar A, Mahesha HS, Sharma M (2021) Temperature and soil moisture stress modulate the host defense response in chickpea during dry root rot incidence. Front Plant Sci 12:653265. https://doi.org/10.3389/fpls.2021.653265

Sharma M, Pande S (2013) Unravelling effects of temperature and soil moisture stress response on development of dry root rot [*rhizoctonia bataticola* (Taub.)] Butler in chickpea. Am J Plant Sci 04:584–589. https://doi.org/10.4236/ajps.2013.43076

Sharma M, Ghosh R, Pande S (2015) Dry root rot (*rhizoctonia bataticola*(taub.) Butler): an emerging disease of chickpea–where do we stand? Arch Phytopathol Plant Protect 48:797–812. https://doi.org/10.1080/03235408.2016.1140564

Sinha R, Irulappan V, Mohan-Raju B et al (2019) Impact of drought stress on simultaneously occurring pathogen infection in field-grown chickpea. Sci Rep 9:5577. https://doi.org/10.1038/s41598-019-41463-z

Sinha R, Irulappan V, Patil BS et al (2021) Low soil moisture predisposes field-grown chickpea plants to dry root rot disease: evidence from simulation modeling and correlation analysis. Sci Rep 11:6568. https://doi.org/10.1038/s41598-021-85928-6

Song P, Wang J, Guo X et al (2021) High-throughput phenotyping: breaking through the bottleneck in future crop breeding. Crop J 9:633–645. https://doi.org/10.1016/j.cj.2021.03.015

Talekar SC, Lohithaswa HC, Viswanatha KP (2017) Identification of resistant sources and DNA markers linked to genomic region conferring dry root rot resistance in chickpea (*cicer arietinum*L.). Plant Breed 136:161–166. https://doi.org/10.1111/pbr.12448

Talekar SC, Viswanatha KP, Lohithasawa HC (2021) Screening chickpea genotypes for resistance to Rhizoctonia bataticola in controlled conditions. Legum Res 44(1):101–108. https://doi.org/10.18805/lr-4061

Varshney RK, Song C, Saxena RK et al (2013) Draft genome sequence of chickpea (cicer arietinum) provides a resource for trait improvement. Nat Biotechnol 31:240–246. https://doi.org/10.1038/nbt.2491

Varshney RK, Thudi M, Roorkiwal M et al (2019) Resequencing of 429 chickpea accessions from 45 countries provides insights into genome diversity, domestication and agronomic traits. Nat Genet 51:857–864. https://doi.org/10.1038/s41588-019-0401-3

Varshney RK, Roorkiwal M, Sun S et al (2021) A chickpea genetic variation map based on the sequencing of 3,366 genomes. Nature 599:622–627. https://doi.org/10.1038/s41586-021-04066-1

Varshney RK, Bohra A, Yu J et al (2021a) Designing future crops: genomics-assisted breeding comes of age. Trends Plant Sci 26:631–649. https://doi.org/10.1016/j.tplants.2021.03.010

Varshney RK, Bohra A, Roorkiwal M et al (2021b) Fast-forward breeding for a food-secure world. Trends Genet 37:1124–1136. https://doi.org/10.1016/j.tig.2021.08.002

An Overview of Major Bean Diseases and Current Scenario of Common Bean Resistance

5

Sumiah Wani, Qadrul Nisa, Tabia Fayaz, Naziya Nabi, Aasiya Nabi, Irtifa Lateef, Adfar Bashir, Raja Junaid Rashid, Zainab Rashid, Gazala Gulzar, Usma Shafi, Zahoor A. Dar, Ajaz A. Lone, Uday Chand Jha, and Bilal A. Padder

Abstract

Common bean is the most extensively grown legume crop worldwide due to its nutrient composition and inexpensiveness. The biotic and abiotic stresses pose serious constraints to its yield. Among the biotic stresses, fungal diseases are the most devastating followed by bacterial and viral diseases. Resistance stands as the most ecofriendly and viable approach to manage various diseases in common bean. The host resistance genes are short-lived in the co-evolutionary arms race between the pathogens and their hosts. The breeders are thus continuously challenged to explore new resistance sources and introgression in the commercial cultivars. In the modern omics era, new breeding techniques are continuously emerging. The most popular method of selecting resistance candidates, nevertheless, is marker-assisted selection. The chapter summarizes the major bean pathogens, their resistance sources, and the markers that can assist in breeding for the resistance against these pathogens.

Keywords

Common bean · *Phaseolus vulgaris* · Disease resistance · Marker Assisted selection (MAS) · Molecular markers · Plant breeding · Fungal diseases · Bacterial diseases · Viral diseases

S. Wani · Q. Nisa · T. Fayaz · Naziya Nabi · Aasiya Nabi · I. Lateef · A. Bashir · R. J. Rashid · Z. Rashid · G. Gulzar · U. Shafi · Z. A. Dar · A. A. Lone (✉) · B. A. Padder (✉)
Sher-e-Kashmir University of Agriculture Sciences & Technology of Kashmir, Jammu, India
e-mail: ajazlone@skuastkashmir.ac.in; bapadder@rediffmail.com

U. C. Jha
Indian Institute of Pulse Research, Kanpur, India

U. C. Jha et al. (eds.), *Diseases in Legume Crops*,
https://doi.org/10.1007/978-981-99-3358-7_5

5.1 Introduction

Phaseolus vulgaris L., also known as the common bean, is a species of autogamous diploid legume with a relatively small genome (587 Mb) (Schmutz et al. 2014). It was domesticated 8000 years ago in South America and central Mexico (Koinange and Gepts 1992). Common beans were first brought from America to other continents about 500 years ago. Since then, they have become an economically and geographically significant plant, especially in underdeveloped nations where they are a key source of dietary protein (Angioi et al. 2010). Indeed, the common bean is a grain legume that is particularly intriguing for human health and sustainable food production due to its capacity for symbiotic atmospheric nitrogen fixation (Foyer et al. 2016). Plant pathogens significantly reduce the quality of crops and output all around the world. *P. vulgaris* was described by Morales (2006) as "probably the most 'infectible' plant species in the Leguminosae." Fungi, bacteria, and viruses are the most common disease-causing agents. The common disease management strategies are chemicals, certified planting material, and the use of resistant sources. Even if chemical control is successful, the beans are grown by small and marginal farmers who can't afford the skyrocketing cost of the chemicals. Thus, genetic resistance tends to be the most reliable control approach and cheapest alternative in managing bean diseases Murillo et al. 2006. The effective identification of natural resistance was made possible by the enormous genetic advances, which also created new research opportunities to understand the molecular causes of each resistance and its associated phenotype. All of this has aided in the development of plant pathogen resistance in two distinct innate immune systems (Dangl et al. 2013). At the first line of defense, plant pattern recognition receptors (PRRs) identify pathogen-/microbial-associated molecular patterns (MAMPs/PAMPs) and elicit a basal immune response known as PAMP-triggered immunity (PTI) (Dangl et al. 2013). Effector proteins, sometimes referred to as virulence factors, are introduced into host cells by pathogens to block PTI, suppress plant basal defenses, and start effector-triggered susceptibility (ETS). In response, plants recognize pathogen effectors specifically within their cells through dominant resistance genes (R genes). This recognition is part of the plant's innate immunity, i.e., effector-triggered immunity (ETI), in which the effector develops into an avirulence (*Avr*) factor (Dangl et al. 2013). Rapid and effective defenses that confer resistance and restrict the expansion of the invasive pathogen are frequently the outcome of an ETI response. The most frequent response, known as the hypersensitive reaction (HR), results in rapid localized cell death or localized necrotic lesions of infected tissues (Bendahmane et al. 1999). A broad class of intracellular proteins with a nucleotide-binding (NB) domain and a C-terminal leucine rich-repeat (LRR) domain make up the majority of dominant R genes observed in plant-virus interactions (Gouveia et al. 2017). They bestow resistance to a range of diseases, pests, and viruses. Among R-gene-mediated responses, HR is the most prevalent phenotype and consists of local necrotic lesions that prevent infection in infected cells and their surroundings (Heidrich et al. 2011). Aside from dominant R genes, recessive genes control a large proportion of bean disease resistance (Nicaise 2014).

5.2 Fungal Diseases

5.2.1 Bean Anthracnose Resistance in *Phaseolus vulgaris*

Common bean is most severely affected by the disease bean anthracnose, which is inflicted by the fungus *Colletotrichum lindemuthianum* (Kelly and Vallejo 2004). In subtropical and temperate countries where cool, humid conditions are prevalent during the cropping season, the disease causes complete crop failure (Sharma et al. 2008). *C. lindemuthianum* is a member of ascomycetes fungi and follows a hemibiotrophic lifestyle. The fungus lives as a biotroph during the initial stages of infection, obtaining nutrition from living plant tissues through specialized infection structures such as appressorium, vesicles, and primary hyphae. It enters the necrotrophic phase after 72 h of infection and starts to grow secondary hyphae that pierce the host tissue to take up vital nutrients (Mahiya-Farooq Padder et al. 2019). The pathogen affects all the above-ground plant parts, but the most distinctive symptoms are rusty brown lesions that later turn into sunken, dark brown spots with brown to reddish margins on bean pods. The seed and plant residue are the ways that fungi survive during the off-season. The use of disease-free seeds, resistant sources, and chemicals are often the common methods used to combat bean anthracnose. The crop is mostly grown by small and marginal farmers who rely on resistant sources rather than chemicals. Barrus (1911) reported the pathogen's variability, and as of this writing, 298 races have been identified (Padder et al. 2017; Nunes et al., 2021). The constant appearance of new races and the wide range of pathogenic diversity need the development and adoption of durable resistant cultivars.

5.2.2 Bean Anthracnose Resistance Genes, QTLians, and Resistant Sources

In the bean-anthracnose pathosystem illustrating typical gene for gene interaction, 25 significant anthracnose resistance genes have been identified and mapped. Seven out of 11 bean chromosomes contain these genes and are mapped using closely related PCR-based markers. These markers have made it easier to transform them into commercially vulnerable bean cultivars. Meziadi et al. (2016) and Kelly and Bornowski (2018) have creatively detailed the specificities of the main anthracnose resistance genes. The bean anthracnose resistance is conditioned by gene clusters named as *Co*-genes which mainly comprise single, duplicate, or complementary genes, except for recessive *co-8*. Many anthracnose resistance genes, including Co-1 locus, *Co-x*, *Co-1HY*, *Co-1*^{65X}, *Co-1*^{73X}, *Co-Pa*, *Co-14* , and more recently identified *Co-AC* gene, are located in the distal end of Pv01 at a physical position of 50.0–50.5 Mb. There are markers that are closely linked to the Co-1 locus (Nabi et al. 2022), but the InDel marker (Zuiderveen et al. 2016) linked to the *Co-1* locus can be conveniently tracked and will help the breeders introduce the Co-1 locus into cultivars of susceptible beans. On Pv02 a single dominant *Co-u* anthracnose resistance gene and five QTLs are located at Pv02. The Pv03 contains the *Co-17* genes flanked by the NSDU_IND_3_0.0441, an InDel marker, and *Co-13* flanked by the RAPD marker OV20680. The common bean chromosome Pv04 is hub to several resistance genes including the *Co-3* gene and its allelic series. Additionally, anthracnose resistance genes such as *Co-10*, *Co-y*, *Co-z*, *Co-15*, *Co-16*, *Co-RVI*, and *Co-34* are located on Pv04. The *Co-5* and *Co-6* allelic series are mapped to Pv05 and Pv06, respectively. As determined using RIL biparental populations, IAC-UNA × Cal-143 minor alleles have been proved to be durable resistance sources against anthracnose. Additionally, in the nuna bean PHA 1037 from Spain, partial anthracnose resistance genes against races 23 and 1545 were mapped on different bean chromosomes. The pathosystem has been comprehensively reviewed by Nabi et al. (2022).

5.2.3 Angular Leaf Spot (*Pseudocercospora griseola*) Resistance in *Phaseolus vulgaris*

The second most important fungal disease in common beans after anthracnose is angular leaf spot (ALS) caused by *Pseudocercospora griseola* (Sacc.) Crous and Braun, and this disease can curb the production up to 80% when the disease triangle is favored by all the factors (Rava Seijas et al. 1985; Nay et al. 2019). Brazil is the home to common beans as they are the largest producers and consumers, although they find their origin in the American continent. Before 1980s angular leaf spot was not prevalent there, but then in the middle of 1980s, this fungus was contemplated as a major obstruction to bean cultivation in not only Brazil but also eastern and southern Africa and Central America (Aggarwal et al. 2004). Resistance is that considerate option whose importance in bean breeding cannot be ignored. Bean scientists from time to time have reported resistant sources against many diseases, so similarly this disease is not an exception. Common beans have been categorized into two gene pools—Andean and Mesoamerican—based on adaptation and domestication (Broughton et al. 2003). If we want to study host-pathogen interaction, beans are an excellent choice as the pathogen here has been constantly evolving with the host; therefore, the races have been divided into Andean and Mesoamerican (Crous et al. 2006). The same applies to the other two pathogens of beans, namely, *Colletotrichum lindemuthianum* (anthracnose) and *Uromyces appendiculatus* (rust) (Pastor-Corrales and Aime 2004). Being specific for breeding programs concentrated on angular leaf spot, AND 277 cultivar (Andean) is considered as one of the favorites as it has been frequently used from time to time for various studies (Almeida et al. 2021). From the literature we will come across five major loci designated as *Phg* that are imparting ALS resistance. The first single dominant resistance gene *Phg-1* located on chromosome Pv01 is found in AND 277, the Andean cultivar. This locus is surprisingly very closely linked to $Co\text{-}1^4$ on Pv01. The genetic markers flanking these loci are $TGA1.1^{570}$ and $CV542014^{450}$ at 1.3 and 0.7 cM (Gonçalves-Vidigal et al. 2011). Now the second loci such as *Phg-2* found in Mexico 54, the Mesoamerican cultivar is located on chromosome Pv08. The markers for this resistance gene have been found back in 1999. In those days RAPD markers were trending. Therefore, the three RAPDs closely linked to this gene are OPN

02, 0PN 14, and 0PE 04 at 5.9, 6.6, and 11.8 cM, respectively (Sartorato et al. 1999). Recently, Miller et al. (2018) found a g796 marker linked at 3 cM to *Phg-2*. The Ouro Negro, Mesoamerican cultivar harbors *Phg-3* locus on chromosome Pv04 (Correa and Saettler 1987). The eminent bean group, namely, Gonçalves-Vidigal et al. (2013) found that *Phg-3/Co-3*^{4} alleles are closely located at 0.0 cM by a marker g2303 on Pv04. The *Phg-4* resistance gene is reported in G5686, Andean cultivar on chromosome Pv04. This locus is co-segregating with Pv-ag004 at 0.0 cM and it is flanked by 4 M439 and Marker 63 (Mahuku et al. 2009). The last major gene reported to date is *Phg-5* found in two Andean cultivars CAL 143 and G5686 and is located at chromosome Pv10, flanked by GATS11b (33.50 Mb) and IAC137 (4.86 Mb) in CAL 143, respectively (Oblessuc et al. 2012). Apart from these resistance genes, minor QTLs have been mapped on Pv01, Pv03, Pv05, Pv06, Pv08, Pv09, and Pv11. Just now a newly discovered QTL labeled ALS11.1^{AM} was tracked down at the beginning of the Pv11 chromosome. This novel ALS resistance locus was brought into being at distinct plant developmental stages by merging the genome-wide association studies (carioca diversity panel) and linkage mapping (AND 277 X IAC-Milenio) (Almeida et al. 2021). Just now a classical field investigation was aced by Girma et al. (2022) wherein 25 genotypes were analyzed to discern the repercussions to resistance under the hotspot areas of Ethiopia against common bacterial blight and angular leaf spot of beans.

5.2.3.1 Rust

Rust is another important fungal disease affecting common bean which is caused by *Uromyces appendiculatus* F. Strauss. The disease seriously affects the foliar parts of the plant particularly leaves and pods, thus causing considerable damage to the crop and reducing the yield. The peculiar symptoms can be seen initially on the leaves which include dark yellow to brown color pustules. These pustules later appear as fine powder-producing spores (Liebenberg and Pretorius 2010). The pathogen is highly variable and many races of the pathogen have been identified, and therefore disease resistance offers the best possible way to manage the disease (Miklas et al. 1993). Several rust resistance genes have been identified and characterized in the common bean so far. In Mesoamerican genotypes, *Ur 3*, *Ur 5*, *Ur 7,* and *Ur 11* and, while in the Andean gene pool, *Ur 5*, *Ur 6*, *Ur 9*, and *Ur 12* have been identified including some unnamed ones as well (Liebenberg and Pretorius 2010). It is believed that the Mesoamerican gene pool offers a greater spectrum of resistance than the Andean gene pool. In a study conducted by Mienie et al. (2005) which used AFLP markers, five of the seven AFLP fragments were effectively converted to sequence characteristic amplified region (SCAR) markers and were strongly associated with rust resistance. The South African large seed cultivar Kranskop containing the resistance gene *Ur-13* was utilized in this breeding effort. The codominant SCAR markers were found at 1.6 cM from the gene and were obtained from the 405 bp EAACMACC fragment known as KB 126. Similarly in another study conducted by Park et al. (1999), the RAPD marker OA4.1050 was seen to be closely linked with *Ur-9* gene loci at a distance of 8.6 cM. Similarly SI19 a SCAR marker developed by Melotto and Kelly (1998) is a promising marker still used is closely linked to the rust gene *Ur- 5* at a distance of 0 cM. The resistant source utilized was B-190, and Young and Kelly (1997) successfully developed the RAPD marker $OAS19_{350}$ from it which was later converted by Melotto and Kelly into a SCAR marker SI19. Though the previous SCAR markers developed by various workers provide good results and many of them are codominant, still attempts are made to overcome their limitations from time to time. Recently Hurtado-Gonzales et al. (2017) used the fine mapping approach to map the *Ur-3* gene for the development of highly accurate markers. SS68 KASP marker was discovered in this study that was tightly linked to the *Ur-3* gene and produced no false results. This marker was used on a highly diverse panel of 130 common bean cultivars containing all known rust genes validating this KASP marker to be highly efficient for marker-assisted breeding. The use of such markers can result in no erroneous results and a significant reduction in time and labor.

5.2.3.2 Fusarium Wilt

Fusarium oxysporum species are ubiquitous fungal pathogens causing wilt diseases in a broad range of economically important crop species among which common bean occupies a prominent place. Fusarium wilt was first identified on common bean (*Phaseolus vulgaris* L.) in the USA in 1929, and since then this pathogen species has been classified as *Fusarium oxysporum* f. sp. *phaseoli* (Fop) (Xue et al. 2015). The main symptoms of the disease are changes in growth and development such as stunting and complete wilting. This is followed by chlorosis and eventually the plant dies (Batista et al. 2017). Genetic resistance and integrated disease management practices, such as crop rotation and the use of healthy seeds, are two strategies for preventing fusarium wilt.

The host-pathogen interaction was studied in Fop—the common bean pathosystem is very limited except for the few studies carried out on some model plants like *Arabidopsis thaliana* and *Medicago truncatula* (Williams et al. 2016). At least seven different races of this pathogen have been described so far each belonging to a specific geographical area. Also there are reports of some new races like UFV01 and IAC18001 supporting the evolution of this pathogen (Paulino et al. 2021). In common bean the resistance to Fop is reported by major as well as minor genes. The genomic regions linked to Fop traits are located inside or near the candidate genes on Pv01, Pv03, Pv04, Pv05, Pv07, Pv10, and Pv11. Various markers linked to these putative regions have been developed to ease the process of marker-assisted breeding. In germplasm from the Middle American gene pool's common bean races Durango and Mesoamerica, genetic resistance to Fop race 4 has been discovered. A single dominant gene known as *FopA* controls the inheritance of resistance to this strain in some Durango communities. However, in two Durango populations, attempts to find a resistance locus using bulked segregant analysis failed to do so (Cross, 1998). Resistance to fusarium wilt has also been identified in the Andean gene pool of common bean. In populations of snap bean, Ribeiro and Hagedorn (1979) detected a dominant single gene to FOP race 1 and an incompletely dominant gene to race 2. They referred to them as *Fop1* and *Fop2*, respectively. The tight linkage between the putative QTL and U20.750 was observed in a study by Fall et al. in 2001 making this marker a promising candidate for conversion to a sequence characterized amplified region (SCAR) for use in marker-assisted breeding. Therefore, the SCAR marker (U20.750) linked to the QTL was developed, with evaluation in Andean and Mesoamerican germplasm, and the marker had high accuracy in Mesoamerican accessions (Brick et al. 2006). The region 51.50 Mb associated with the significant ss715648096 marker on Pv11 was seen to impart resistance to different races of Fop. The region is also believed to be associated with other important fungal diseases of common bean, such as anthracnose (Biswas and Ghosh 2016). Chen et al. (2019), by using transcriptomic and metabolic studies, reported that differentially expressed genes in Pv03, Pv04, Pv07, Pv08, and Pv11 had important roles in signaling pathways such as ethylene, salicylic acid (SA), and jasmonate.

Several GWAS studies carried out in different areas of the world revealed significant SNPs associated with Fop reactions like ubiquitination, abscisic acid (ABA) production, and phenol production on chromosome Pv01 and Pv05, respectively. Another significant SNP, ss715645397, was found on Pv05 at 0.004 Mb from the Phvul.005G152600 gene (ARM repeat superfamily protein). The Armadillo (ARM) domain has NBS-LRR that has been extensively studied in plants, suggesting a critical role of these repeating peptides in plant cell physiology, plant stress, and plant development (Sharma and Pandey 2016). Association mapping of 133 common bean accessions was performed by Leitão et al. (2020) for race 06 of Fop from Portugal. Significant associations were detected on chromosomes Pv04, Pv05, Pv07, and Pv08 for DSR and AUDPC. They noted that the DART03480 marker on Pv04 was at a small distance of approximately 0.1 Mb from the ss715648681 marker.

5.3 Bacterial Diseases

5.3.1 Halo Blight

Halo blight, a disease of major economic significance that can affect common bean (*Phaseolus vulgaris L.)* production worldwide is caused by the bacterium *Pseudomonas syringae* pv. *phaseolicola* (Psph) (Burkholder 1926; Young et al. 1978). It occurs at higher latitudes in the northern and southern hemispheres and at greater elevations in tropical and subtropical parts of Africa and South America. This seed-transmitted disease is favored by cold and humid conditions. According to Asensio-S-Manzanera et al. (2006), Singh and Schwartz (2010), and Félix-Gastélum et al. (2016), it can result in yield losses of up to 45%. The only effective method for controlling this disease is genetic resistance, which is also necessary for the dependable production of disease-free seed, particularly in bean-producing nations where smallholder farmers depend on seed preserved from a previous crop (Taylor et al. 1996a; Asensio-S-Manzanera et al. 2006; Arnold et al. 2011; Miklas et al. 2014). From the interaction phenotypes between a huge worldwide collection of pathogen isolates and eight distinct bean lines, nine races of Psph were discovered (Taylor et al. 1996a). Tests on a wide variety collection of *P. vulgaris* were conducted using the six most prevalent Psph races (Taylor et al. 1996b). They presented a gene-for-gene model using five major-effect, race-specific plant resistance (R) genes and associated pathogen avirulence (*avr*) genes from the combined data (Taylor et al. 1996b). The positions of major-effect R genes (Pse-1 to Pse-6 *Pse-1* to *Pse-6*) on three *P. vulgaris* chromosomes (Pv02, Pv04, and Pv10), which are effective against one or more Psph races, were later discovered through a genetic study based on linkage mapping (Miklas et al. 2009, 2014). None of these race-specific genes offer protection against the persistent Psph race 6 that continues to endanger bean production all over the world (Taylor et al. 1996a, b; Lamppa et al. 2002; Rico et al. 2003; Félix-Gastélum et al. 2016). Taylor et al. (1996b), on the other hand, recognized bean accession PI150414, an El Salvadorian red dry bean landrace, as a source of quantitative, potentially race nonspecific effective resistance against race 6. This accession was identified by Patel and Walker (1965) as a source of halo blight resistance, and Taylor et al. (1978) found that the resistance was passed down through the recessive or partially dominant expression of a single allele. This resistance has been utilized in the creation of novel cultivars as Wis HBR 72. (Hagedorn et al. 1974), Edmund (Conway et al. 1982), and Kranskop-HR 1. Popular African cultivars have been identified as significant sources of race 6 resistance; the dry bean cultivar CAL 143 from East Africa has been shown to be resistant to a number of diseases, including angular leaf spot (*Phaeoisariopsis griseola*), bean rust (*Uromyces appendiculatus*), and halo blight (Chataika et al. 2011). Preliminary assessments of CAL 143 against the nine Psph races (at the ARC-Grain Crops Institute in Potchefstroom, South Africa) suggest that it possesses quantitative resistance to a number of races, including race 6. This cultivar demonstrated dominant resistance against endemic Psph field isolates (uncharacterized) present in Malawi. Previous research have described minor-effect quantitative trait loci (QTL) for Race 6 resistance. In a RIL population (Xana Cornell 49–242), Trabanco et al. (2014) discovered two minor-effect QTLs (renamed here as HB4.1 and HB6.1) that accounted for 11 and 12% of phenotypic variance. Using an Andean RIL population (PMB0225 PHA1037), González et al. (2016) conducted a multi-environment research to examine the genetic basis of quantitative resistance to nine Psph races in primary and trifoliolate leaf, stem, and pod tissues. They found 1 minor-effect QTL with both epistatic and individual additive effects (explaining 2.64 and 2.04% of phenotypic variance), and 11 minor-effect epistatic QTLs without detectable additive effects (each interaction explaining 8.52% of phenotypic variation). Tock (2017) mapped resistance to Psph race 6 generated from the two common bean lines using high-resolution linkage maps for three recombinant inbred populations in a genome-wide association study (GWAS) of race 6 resistance in an Andean diversity panel of common bean was included as a supplement to this. A single major-effect quantitative trait locus (QTL; HB4.2) on chromosome Pv04, which confers broad-spectrum resistance to eight different races of the pathogen, is the source of race 6 resistance from PI 150414 and maps to that locus. A Rojo CAL 143 population's resistance segregation maps to five chromosomal arms, including HB4.2. On chromosome Pv05 for race 6 resistance, GWAS identified one QTL

Fig. 5.1 Halo blight on bean pods and seed (Courtesy of Howard F. Schwartz, Colorado State University, Bugwood.org)

Fig. 5.2 Halo blight on leaves. (Courtesy of Howard of Howard F. Schwartz, Colorado State University, Bugwood.org)

(HB5.1) with a substantial impact on seed production. The same HB5.1 QTL was discovered in Canadian Wonder PI 150414 and Rojo × CAL 143 populations, which was effective against race 6 but lacks broad resistance (Figs. 5.1 and 5.2).

5.3.2 Bacterial Brown Spot

In the USA, Brazil, and Canada, bacterial brown spot (BBS) disease has been widely characterized (Harveson and Schwartz 2007; Singh and Schwartz 2010). The disease, which is seedborne and brought on by *Pseudomonas syringae* pv. *syringae*, mostly damages the leaves and, to a lesser extent, the pods (Navarro et al. 2007). Additionally, it is particularly severe when beans are cultivated using a monoculture approach (Muedi et al. 2015). In the area where the conditions are conducive, BBS disease has been found to be responsible for up to 55% of yield losses (Muedi et al. 2015; Serfontein 1994). Small water-soaked lesions on leaves and pods may initially indicate symptoms, which may then progress to elliptical, necrotic brown wounds surrounded by a thin yellow-green border portion (Kimani et al. 2005; Muedi et al. 2015). Infected seeds, strong winds, polluted farm equipment, and contaminated soil are some of the sources of infection (Harveson et al. 2015). A significant method of spreading *P. syringae* pv. *syringae* is through infected seed (Kimani et al. 2005). BBS disease thrives in environments with humidity levels of over 95% and temperatures between 28 and 32 °C, which are frequent in South Africa's central and eastern areas, where dry beans are commonly grown on a commercial basis (Muedi et al. 2015).

The most effective, long-lasting strategy to manage the disease is with resistant cultivars (Harveson et al. 2015). In 2003, Jung et al. 2003 discovered RAPD molecular markers connected to BBS resistance quantitative trait loci (QTLs). Navarro et al. (2007) conducted a significant research on the inheritance of BBS resistance. In this study, dry bean-segregating recombinant inbred lines (RIL) were inoculated with *P. syringae* pv. *syringae* seedling stems, and a number of genomic areas located on various linkage groups that were connected to BBS resistance were discovered (Navarro et al. 2007). The Mesoamerican gene pool's small- and medium-seeded cultivars are more resistant to bacterial disease than the Andean gene pool's large-seeded dry and green bean cultivars, which are extremely sensitive to BBS disease (Singh and Schwartz 2010). The discovery of genetic resources resistant to the BSS disease will aid in the creation of cultivars with enhanced BBS resistance, which will be advantageous to farmers. Salegua et al. (2020) tested 5 check cultivars and 415 Andean diversity panel (ADP) dry bean lines for BBS disease resistance in the field in South Africa. According to this study, 45.3% of genotypes were only somewhat resistant to the BBS disease, whereas 17.2% of genotypes were resistant. These genotypes can serve as sources of genetic resistance for the development of improved dry beans in the future. The high variability among genotypes implies better selection criteria based on resistance and yield performance. Genotypes ADP-0592, ADP-0790, ADP-0120, and ADP-0008 were selected for both high disease resistance and high grain yield across three environments. Genotypes ADP-0546, ADP-0630, ADP-0120, and ADP-0279 were selected for both high BBS disease resistance as well as high yield in Warden. Genotypes ADP-0038, ADP-0721, and ADP-0790 were the best performing at Middelburg, whereas genotypes ADP-0120 and ADP-0079 performed better at Potchefstroom. The two QTLs in LGs 2 and 4 were consistently associated with resistance to BSS. Marker-assisted selection for resistance to BBS may improve selection efficiency, due to the low heritabilities of reactions to BBS reported by Antonius (1982).

5.3.3 Bacterial Blight of Bean and Genetic Resistance

Common Bacterial Blight a systemic, seed-transmitted (Aggour et al. 1989) disease caused by *Xanthomonas campestris* pv. *phaseoli* (Smith) Dye (Xcp), attacks beans frequently and severely in the tropics and subtropics (Saettler 1989). The disease is prevalent in areas with warm weather, causing yield reductions of up to 40% (Singh et al. 1999). The most effective way to control CBB is to plant a bacteria-free seed and grow resistant bean cultivars (Zapata et al. 1985). The dearth of available resistance sources complicates the development of cultivars with enhanced resistance to CBB (Beebe and Corrales 1991; Singh and Muñoz 1999).

CBB resistance breeding is a complex process (Miklas et al. 2006; Singh and Miklas 2015; Tugume et al. 2019; Viteri et al. 2014c; Yu et al. 2012). Despite two examples of resistance segregations corresponding to single genes (Adams et al. 1988; Zapata et al. 2011), CBB resistance is primarily quantitative and polygenic (Singh and Miklas 2015; Singh and Schwartz 2010; Yu et al. 2012). Moderate CBB resistance exists in the Mesoamerican gene pool, but no resistance has been reported in the Andean gene pool (Singh and Miklas 2015). To date, the discovery of at least 27 quantitative trait loci (QTLs) for CBB resistance spread across the 11 common bean chromosomes provides a reliable source of resistance. These QTLs do not appear to differentiate Xcf and Xpp, and no significant crossover interactions between strains and common bean genotypes bearing different QTLs have been confirmed so far (Duncan et al. 2011; Mutlu et al. 2008; Opio et al. 1996; Viteri et al. 2014). SAP6, a major QTL derived from the Great Northern landrace cultivar Montana No. 5, has been successfully used to generate CBB-resistant varieties grown in North and South America since the early 2000s (Miklas et al. 2003a, b). In lines derived from crosses between *P. vulgaris* and *P. acutifolius*, two major resistance QTLs have been identified (BC420 and SU91) (Jung et al. 1997; Pedraza García et al. 1997; Shi et al. 2011b; Yu et al. 2000), and analysis of their gene content highlighted several R gene candidates that still require functional validation (Perry et al. 2013; Shi et al. 2011a). To ensure resistance under different conditions and/or in different tissues, a pyramiding strategy was used (Miklas et al. 2000a, b; Mutlu et al. 2005; Viteri and Singh 2014). The bred line VAX6, for example, has high resistance to a wide variety of Xcf and Xpp isolates (Duncan et al. 2011; Mahuku et al. 2006; Singh and Muñoz 1999). Marker-assisted selection of CBB-resistant lines has been facilitated by the development of genetic maps (Bai et al. 1997; Freyre et al. 1998; Meziadi et al. 2016; Shi et al. 2012; Yu et al. 2000) and by the sequencing of Andean and Mesoamerican common bean genomes (Schmutz et al. 2014; Vlasova et al. 2016. However, CBB resistance expression studies are still limited (Cooper 2015; Shi et al. 2011a).

5.3.4 Breeding for Common Blight Resistances and Marker-Assisted Selection

The first common blight-resistant breeding line, XR-235-1-1, derived from an interspecific cross between *P. vulgaris* and *P. coccineus*, was created and registered by Freytag et al. in (1982). A major breakthrough in breeding for common blight resistance occurred when McElroy (1985) developed XAN 159, XAN 160, and XAN 161; Parker (1985) and Scott and Michaels (1992) developed OAC 88–1 (synonymous with BLT 87–13); and Singh and Muñoz (1999) and Singh et al. (2001) developed VAX 1 and VAX 2 interspecific breeding lines with higher levels of resistance derived from *P. vulgaris* × *P. acutifolius* populations. Furthermore, in addition to introgressing resistance from the tepary bean G 40001, Singh and Muñoz (1999) and Singh et al. (2001) pyramided resistance of the tepary bean G 40001 (XA11.4 QTL) and the common bean SAP6 QTL in VAX 1 and VAX 2 (Viteri and

Singh 2014; Viteri et al. 2014b) with that of tepary bean PI 319443 (SU91 QTL, via XAN 263 or XAN 309) that resulted in common bean breeding lines VAX 3, VAX 4, VAX 5, and VAX 6 with even higher levels of pyramided resistance. For example, as noted above VAX 6 exhibited an incompatible response (i.e., resistance) to all 343 pathogenic strains of *X. campestris* pv. *phaseoli* and *X. campestris* pv. *phaseoli* var. *fuscans* from around the world (Mahuku et al. 2006). Five common blight-resistant breeding lines were created by Zapata et al. (2004). Using reciprocal backcross and direct disease screening. A resistant $F^{4:5}$ recombinant inbred line population of the reciprocal crosses between OAC Rex (bb/ss/PP/SuSu) and HR45 (BB/SS/pp/SuSu) was evaluated under artificial field inoculation (Durham et al. 2013) to examine the main and interaction effects of the previously identified CBB resistance quantitative trait loci (QTL) associated with markers BC420 (B) on Pv06, SAP6. While the QTL on Pv06 of HR45 accounted for 37–46% of phenotypic variation in the field when the CBB QTL on Pv08 was present, the effect of the QTL on Pv04 and Pv10 was not significant under field conditions, even when the Pv06 QTL was absent. For CBB severity and the AUDPC broad-sense heritability estimates of CBB resistance and the QTL associated with BC420 were high. This supported the ongoing efforts to pyramid the QTL on Pv06 and Pv08 in the common background, which provides high levels of resistance.

For common blight resistance, only three major QTL effects have been used in marker-assisted selection (MAS). The linked markers SAP6 on Pv10, SU91 on Pv08, and BC420 on Pv06 are equivalent to the QTL. For improved resistance in the market classes of white, beige, black, cream, and gray, BC420 QTL has significant value, but simultaneous selection for SU91 QTL is necessary. Similar to SU91, the SAP6 QTL is frequently discovered in resistant genotypes created using a traditional breeding technique that involves direct pathogen inoculation. O'Boyle et al. (2007) used the SU91 and BC420 QTL for MAS to transfer common blight resistance. Mutlu et al. (2005) developed a common blight-resistant pinto bean breeding line ABCP-8. Similarly, Mutlu et al. (2008) developed a great northern germplasm line ABC-Weihing with pyramided resistance (SAP6 and SU91 QTL) using a combination of direct disease screening and MAS in a backcrossing program.

5.4 Viral Diseases

5.4.1 *Potyviridae*

The potyvirus genus has a notable negative impact on the production of dry and snap beans. Common beans (*Phaseolus vulgaris* L.), as well as a variety of other cultivated and wild legumes, are susceptible to infection by the two most prevalent and damaging viruses in the family *Potyviridae*: bean common mosaic virus (BCMV) and bean common mosaic necrosis virus (BCMNV) (Morales 2006). Yield losses as a result of BCMV and BCMNV can reach 100% (Li et al. 2014). Positive-sense RNA viruses that are monopartite and generate flexuous rod-shaped virions are BCMV and BCMNV (Ivanov et al. 2014). The genomic RNA molecule in the virions is around 10 kb long and has a 30-terminal polyA tail. The virions are about 750 nm long and have a diameter of 11–13 nm (El-Sawy et al. 2013). Two more potyviruses that infect common beans are the bean yellow mosaic virus (BYMV) and clover yellow vein virus (ClYVV). Leaf mosaics are a defining characteristic of these viruses. In reaction to BYMV field infection, 30–40% yield losses have been documented worldwide due to BYMV's widespread distribution of beans. Historically, ClYVV was thought to be the most dangerous BYMV strain. The most common symptoms observed are yellow mosaic, malformation, and reduction of plant size. The virus is spread by aphids and survives the winter in wild legumes. Numerous more potyviruses, such as the soybean mosaic virus (SMV) and the watermelon mosaic virus-2 (WMV-2), have been discovered in common beans. BCMV and BCMNV have the potential to severely damage bean and other leguminous crops. To combat this, common bean cultivars have undergone genetic engineering to be resistant to these viruses (Tracy et al. 1992).

5.4.2 *Geminiviridae*

Begomoviruses and curtoviruses are the two primary genera of the family *Geminiviridae* that contain important viruses that infect beans. A pathogenic plant virus with single-stranded DNA, beet curly top virus (BCTV) is a member of the genus curtoviruses (Strausbaugh et al. 2008). In 1888, it was found in the western USA. An icosahedral capsid with twinned icosahedral capsids surrounds the virus (Horn et al. 2011). The beet leafhopper (*Circulifer tenellus*) spreads the virus, which has a monopartite genome with three viral senses and four complementary open reading frames (ORF). The Worland, Logan, and CFH strains are closely related, and BCTV has been reported to contain a range of strains that are variants of those strains. As a result of the detailed molecular characterization of the genomes published by Stenger and McMahon, the strains are now classified as three distinct viruses in accordance with the species demarcation requirements established by the International Committee on Taxonomy of Viruses. The Logan strain, which is the type species for the genus, received the beet curly top virus (BCTV). Beet mild curly top virus (BMCTV) is what Worland and CHF stand for (BSCTV). Based on the symptoms, it is challenging to differentiate between BCTV, BMCTV, and BSCTV infections in common beans. The common bean is one of BCTV's many talented hosts (Stenger and McMahon 1997; Strausbaugh et al. 2008).

(Singh et al. 2000) Bean golden yellow mosaic virus (BGYMV), tomato yellow leaf curl virus (TYLCV), and bean dwarf mosaic virus (BDMV) are the three main species of the begomovirus genus that are known to cause significant disease in common beans. Whitefly (*Bemisia tabaci*) is the primary vector of BCMV, BGMV, and BGYMV and transmission occurs in a persistent, non-propagative manner (Rosen et al. 2015).

5.4.3 *Secoviridae*

The bean pod mottling virus (BPMV) belongs to the *Secoviridae* subfamily of the genus *Comovirus*. Members of this genus have been identified to cause disease in common bean and other cultivated legume pathogens globally (Giesler et al. 2002). Transmission of BPMV happens mechanically or by insects (Dizadji and Shahraeen 2011). Depending on the strains and host genotypes, sensitive bean genotype symptoms include either severe or minor leaf mottling. Despite being first identified in common bean plants in 1948 (Zaumeyer and Thomas 1948), the BPMV today poses a serious danger to US soybean growing regions, with yield reductions of between 10% and 40% being observed (Ziems et al. 2007). Additionally, co-infection of soybean plants with BPMV and SMV causes synergistic infection, which can impair yield by up to 85%. Other comoviruses that infect common beans include bean rugose mosaic virus (BRMV) and cowpea severe mosaic virus (CpSMV). They are spread mechanically or by insects, just like BPMV. Mild to severe mosaic and leaf deformation are the viral strain's characteristic symptoms. In particular, these viruses cause significant epidemics in improved cultivars with BMCV monogenic dominant resistance in central and southern America.

5.4.4 Viral Resistant Genes and Sources in Common Bean

Numerous investigations have been carried out over the years to find virus resistance genes in common beans, and some have already been applied in plant breeding programs. Some genes were just investigated in preparation for potential loss of effective resistance in the event that new genes were required to replace the succumbed ones. The resistance against viral diseases in common beans is controlled by quantitative trait loci, dominant genes, and recessive genes. The *Ur-5* , *Pse-1*, and *Co-9* gene clusters on chromosome Pv04 contain the major QTLs influencing mosaic resistance (Miklas et al. 1996). According to Mendez-Vigo et al. (2005), the BGYMV QTL located on chromosome Pv04 was observed to cosegregate codominantly with the *Co-9* gene, as deduced from the amplification of various fragment sizes from the linkage of primers for SW12 SCAR. According to Urrea et al. (1996), the only genes found to date that provided resistance against BGYMV, appropriate to MAS, were the SW12.700 SCAR and SR2 SCAR markers associated with Pv04 QTL with the *bgm-1* gene, respectively.

Since the identification of the "I" locus at the terminus of chromosome Pv02, dominant R genes in common beans that protect against potyviruses have received the greatest attention. The "I" locus gives broad-spectrum resistance to at least 10 potyviruses, including the destructive ones like BCMV, BCMNV, BYMV, WMV-2, and ClYVV. The I gene imparts resistance to all BCMV strains. The "I" locus contains two resistance genes, "*Hsw*" and "*Wmv,*" that confer resistance against WMV-2. The genes are present in the genotypes "Black Turtle-1" and "Great Northern 1140." Temperature influences the resistance conferred by the "I" locus. The "I" locus's *Anv* and *Lnv* resistance genes present in genotypes "Iguaçu" and "Pitouco," respectively provide protection against the CpSMV virus. Another R gene, known as R-BPMV, has also been demonstrated to give resistance to BPMV in the BAT93 common bean genotype. The most well-known recessive genes, commonly abbreviated "bc," give viral resistance in common beans (Meziadi et al. 2017).

There are four recessive genes called *bc-u* , *bc-1* (*bc-1* and *bc-12*), and *bc-2 (bc-2* and *bc-22*) that confer resistance to BCMV and BCMNV that is both strain-specific (e.g., *bc-1*, *bc-12*, *bc-2*, and *bc-22*) and nonspecific (bc-3) (Drijfhout et al. 1978). The terminal end of chromosome Pv03 is where the *bc-u* and *bc-1* are located. It has been determined that the *bc-3* belongs to the *eIF4E* gene family and that it is located on chromosome Pv06. In the host group 3 cultivars (such as Olathe, UI 37, and Victor), Strausbaugh et al. (2003) discovered a recessive gene that confers more resistance than *bc-12* to leaf stunting and deformity as well as plant dwarfing brought on by the NL-3 K strain of BCMNV. Additionally, cultivars of the typical host group 3 like Redlands Green Leaf B are vulnerable to NL-3 K. The genes *bc-1* and *bc-u* are related (Strausbaugh et al. 1999). Using a KASP marker developed for the missense single-nucleotide polymorphism (SNP) within *eIF4E*, which is diagnostic for *bc-3*, marker-assisted selection (MAS) of *bc-3* in common bean is carried out (Hart and Griffiths 2013). *Bc-12* is linked to a SCAR marker (SBD5) used for the gene's MAS (Vandemark and Miklas 2005).

As for the genotypic sources of resistance being considered, the great northern, pink, pinto, and red market classes of race Durango cultivars grown in the western USA typically have moderate to high resistance levels against BCTV. Also, small-seeded dry beans from Middle America, such as Porrillo Sintetico, "Tio Canela 75," and "T 39," have shown the highest level of resistance to BCTV. Andean dry beans with large seeds, such as "Cardinal," "Kardinal," "Montcalm," "USWA-33," and "USWA-39, as well as some green beans like "Moncayo" and "Paulista," are extremely resilient (Larsen et al. 2005). Two genes regulate BCTV resistance in various cultivars, including Burtner, Common Red Mexican, and Great Northern "UI 15" (Dean and Schultz, 1947). In addition, one dominant gene (*Bct*) conferring complete resistance to BCTV has been reported in the Andean variety Primo'/ Moncayo (Larsen et al. 2005). The Common Red landrace cultivar and the small white cultivar "Burtners Blightless" are two notable sources that can be linked to the development of CTV resistance in common beans. Early inheritance studies revealed that two genes, one dominant and one recessive, were required for resistance in common beans. Using the marker on linkage group 7 of the *Phaseolus* core map, the *Bct* that confers resistance against BCTV was located. Larsen and Miklas (Larsen et al. 2005) identified a sequence-characterized amplified region (SCAR) marker (SAS8.1550) strongly linked to the original dominant gene Bct that confers resistance to CTV. During screening procedures to evaluate the robustness of SAS8.1550 for marker-assisted selection (MAS) goals, the landrace cv. Jatu Rong (PI 163120 with the corresponding CIAT accession catalog number G122) from India was the only genotype from the Andean gene pool with resistance to CTV that lacked the marker. Bean golden yellow mosaic virus resistance breeding began in 1970, shortly after its significance in the state of Sao Paulo was recognized. Tulmann Neto et al.

(1977) used various treatments and chemicals, such as gamma radiation and ethyl methane sulfonate (EMS), in many commercial bean varieties to create variability for disease resistance. However, none of his experiments resulted in the development of a single resistant cultivar. However, Embrapa Arroz e Feijo released the black bean cultivar Onix in 1980, which demonstrated moderate tolerance to this disease, and Bianchini used the TMD-1 variety as a source of resistance against BGMV in 1999. Embrapa Arroz e Feijo began a recurrent selection program for BGMV tolerance in the 1990s. Compared to most accessions, the starting germplasm was capable of delaying symptoms and causing less severe yellowing and stunting. Despite the efforts of numerous researchers, the program has not, after 25 years, yielded a commercial common bean cultivar that is highly resistant and has sufficient disease tolerance. Independent inheritance studies have shown that disease tolerance is a challenging characteristic to select for using a weighted generation means analysis; Pessoni et al. (1997) showed that plant dwarfing, foliar yellowing, and pod malformation features may all be selected simultaneously. The three qualities were positively linked, and the additive gene action component was substantial.

In addition to the naturally occurring resistance in beans, a synthetic line of common beans that is BGMV- has been created to stop the spread of virus (Bonfim et al. 2007). The BGMV Rep gene, which encodes the viral replication-associated protein (*Rep*) necessary for viral genome replication, was used to create a double-stranded RNA using the biolistic gene insertion technique (Galvez et al. 2014). Transgenic RNAi-mediated virus-resistant beans are genetically modified lines of common beans that silence the BGMV gene and produce durable resistant plants as a result of the double-stranded RNA molecules being inserted (Carvalho et al. 2015). Brazil, where cultivation and consumption were legal, saw success with transgenic beans. Even though this method has significantly improved BGMV resistance levels, it is still technically more challenging to adopt than conventional plant breeding. This is due to two factors: first, the difficulty in obtaining stable transgenic common bean plants; and second, the importance of carefully selecting the viral target gene and having a deep understanding of the viral pathogen. It was unable to create the first transgenic lines of common bean that were BGMV-resistant. Instead of lowering the BGMV-caused sickness, transgenic lines containing the BGMV coat protein, a virulence factor, demonstrated increased viral dissemination and disease signs (Blair and Morales 2008). When there is no natural resistance that may be included in a breeding program, modified plants may be a feasible solution in nations where the growing of transgenic plants is permitted (Meziadi et al. 2017).

References

Adams MW, Kelly JD, Saettler AW (1988) A gene for resistance to common blight (*Xanthomonas campestris* pv. *Phaseoli*). Annu Rep Bean Improv Coop 31:73–74

Aggarwal VD, Pastor-Corrales MA, Chirwa RM, Buruchara RA (2004) Andean beans (*Phaseolus vulgaris* L.) with resistance to the angular leaf spot pathogen (*Phaeoisariopsis griseola*) in southern and eastern Africa. Euphytica 136:201–210. https://doi.org/10.1023/B:EUPH.0000030678.12073.a9

Aggour AR, Coyne DP, Vidaver AK (1989) Comparison of leaf and pod disease reactions of beans (*Phaseolus vulgaris* L.) inoculated by different methods with strains of *Xanthomonas campestris* pv. *phaseoli* (smith) dye. Euphytica 43:143–152

Almeida CP, de Carvalho Paulino JF, Bonfante GFJ, Perseguini JMKC, Santos IL, Gonçalves JGR, Patrício FRA, Taniguti CH, Gesteira G, Garcia AAF, Song Q, Carbonell SAM, Chiorato AF, Benchimol-Reis LL (2021) Angular leaf spot resistance loci associated with different plant growth stages in common bean. Front Plant Sci 12:647043. https://doi.org/10.3389/fpls.2021.647043

Angioi SA, Rau D, Attene G, Nanni L, Bellucci E, Logozzo G, Negri V, Spagnoletti Zeuli PL, Papa R (2010) Beans in Europe: origin and structure of the European landraces of *Phaseolus vulgaris* L. Theor Appl Genet 121(5):829–843

Antonius SH (1982) Identification and inheritance of resistance to bacterial brown spot (*Pseudomonas syringae*) in *Phaseolus vulgaris*, Ph.D. dissertation. University of Wisconsin, Madison, WI

Arnold DL, Lovell HC, Jackson RW, Mansfield JW (2011) *Pseudomonas syringae* pv. *phaseolicola*: from 'has bean' to supermodel. Mol Plant Pathol 12:617–627

Asensio-S-Manzanera MC, Asensio C, Singh SP (2006) Gamete selection for resistance to common and halo bacterial blights in dry bean intergene pool populations. Crop Sci 46:131–135

Bai Y, Michaels TE, Pauls KP (1997) Identification of RAPD markers linked to common bacterial blight resistance genes in *Phaseolus vulgaris* L. Genome 40:544–551

Barrus MF (1911) Variation of cultivars of beans in their susceptibility to anthracnose. Phytopathology 1:190–195

Batista RO, Silva LC, Moura LM, Souza MH, Carneiro PCS, de Souza Carneiro JE (2017) Inheritance of resistance to fusarium wilt in common bean. Euphytica 213(7):1–12

Beebe SE, Corrales MP (1991) Breeding for disease resistance. p. 561–617. A. van Schoonhoven and O. Voysest (ed.) common beans: research for crop improvement. CAB international, Wallingford, UK. Breeding for disease resistance. In: van Schoonhoven A, Voysest O (eds) Common beans: research for crop improvement. CAB International, Wallingford, pp 561–617

Bendahmane A, Kanyuka K, Baulcombe DC (1999) The Rx gene from potato controls separate virus resistance and cell death responses. Plant Cell 11(5):781–791

Biswas K, Ghosh P (2016) Recent advancements and biological management of *Fusarium udum*: a causative agent of pigeonpea wilt. Int J Appl Nat Sci 5(3):57–72

Blair MW, Morales FJ (2008) Geminivirus resistance breeding in common bean. CAB Rev Perspect Agric Vet Sci Nutr Nat Resour 3:1–15. https://doi.org/10.1079/pavsnnr20083089

Bonfim K, Faria JC, Nogueira EO, Mendes ÉA, Aragão FJ (2007) RNAi-mediated resistance to bean golden mosaic virus in genetically engineered common bean (*Phaseolus vulgaris*). Mol Plant-Microbe Interact 20(6):717–726

Brick MA, Byrne PF, Schwartz HF, Ogg JB, Otto K, Fall AL, Gilbert J (2006) Reaction to three races of *Fusarium* wilt in the *Phaseolus vulgaris* core collection. Crop Sci 46(3):1245–1252

Broughton WJ, Hernández G, Blair M, Beebe S, Gepts P, Vanderleyden J (2003) Beans (*Phaseolus* spp.)–model food legumes. Plant Soil 252:55–128. https://doi.org/10.1023/A:1024146710611

Burkholder WH (1926) A new bacterial disease of the bean. Phytopathology 16:915–928

Carvalho JLV, De Oliveira Santos J, Conte C, Pacheco S, Nogueira EOPL, Souza TLPO, Faria JC, Aragão FJL (2015) Comparative analysis of nutritional compositions of transgenic RNAi-mediated virus-resistant bean (event EMBPV051-1) with its non-transgenic counterpart. Transgenic Res 24:813–819. https://doi.org/10.1007/s11248-015-9877-5

Chataika BYE, Bokosi JM, Chirwa RM, Kwapata MB (2011) Inheritance of halo blight resistance in common bean. Afr Crop Sci J 19:325–333

Chen L, Wu Q, He W, He T, Wu Q, Miao Y (2019) Combined de novo transcriptome and metabolome analysis of common bean response to *Fusarium oxysporum* f. sp phaseoli infection. Int J Mol Sci 20(24):6278

Conway J, Hardwick RC, Innes NL, Taylor JD, Walkey DGA (1982) White-seeded beans (*Phaseolus vulgaris*) resistant to halo blight (*Pseudomonas phaseolicola*), to bean common mosaic virus, and to anthracnose (*Colletotrichum lindemuthianum*). J Agric Sci 99:555–560

Cooper DM (2015) Identification and characterization of common bacterial blight resistance genes in the resistant common bean (*Phaseolus vulgaris*) variety OAC rex. University of Guelph, Guelph, ON. http://hdl.handle.net/10214/8952. Accessed 23 Mar 2021

Correa FJ, Saettler AW (1987) Angular leaf spot of red kidney beans in Michigan. Plant Dis 71:915–918. https://doi.org/10.1094/PD-71-0915

Cross H (1998) Resistance to *Fusarium* wilt in races Durango and Mesoamerica of common bean, Doctoral dissertation. Colorado State University

Crous PW, Liebenberg MM, Braun U, Groenewald JZ (2006) Reevaluating the taxonomic status of *Phaeoisariopsis griseola*, the causal agent of angular leaf spot of bean. Stud Mycol 55:163–173. https://doi.org/10.3114/sim.55.1.163

Dangl JL, Horvath DM, Staskawicz BJ (2013) Pivoting the plant immune system from dissection to deployment. Science 341(6147):746–751

Dizadji A, Shahraeen N (2011) Occurrence, distribution and seasonal changes of viruses infecting common bean in northwestern Iran. Arch Phytopathol Plant Protect 44(17):1647–1654

Drijfhout E, Silbernagel MJ, Burke DW (1978) Differentiation of strains of bean common mosaic virus. Neth J Plant Pathol 84:13–26

Duncan RW, Singh SP, Gilbertson RL (2011) Interaction of common bacterial blight bacteria with disease resistance quantitative trait loci in common bean. Phytopathology 101(4):425–435

Durham KM, Xie W, Yu K, Pauls KP, Lee E, Navabi A (2013) Interaction of common bacterial blight quantitative trait loci in a resistant inter-cross population of common bean. Plant Breed 132:658–666. https://doi.org/10.1111/pbr.12103

El-Sawy MA, Mohamed HAE, Elsharkawy MM (2013) Serological and molecular characterisations of the Egyptian isolate of bean common mosaic virus. Arch Phytopathol Plant Protect 47:1–13

Fall AL, Byrne PF, Jung G, Coyne DP, Brick MA, Schwartz HF (2001) Detection and mapping of a major locus for *Fusarium* wilt resistance in common bean. Crop Sci 41(5):1494–1498

Félix-Gastélum R, Maldonado-Mendoza IE, Navarrete-Maya R, Olivas-Peraza NG, Brito-Vega H, Acosta-Gallegos JA (2016) Identification of *Pseudomonas syringae* pv. *phaseolicola* as the causal agent of halo blight in yellow beans in northern Sinaloa, Mexico. Phytoparasitica 44:369–378

Foyer CH, Lam HM, Nguyen HT, Siddique KH, Varshney RK, Colmer TD, Cowling W, Bramley H, Mori TA, Hodgson JM, Cooper JW (2016) Neglecting legumes has compromised human health and sustainable food production. Nature plants 2(8):1–10

Freyre R, Skroch PW, Geffroy V, Adam-Blondon A-F, Shirmohamadali A, Johnson WC (1998) Towards an integrated linkage map of common bean. Development of a core linkage map and alignment of RFLP maps. Theor Appl Genet 97:847–856

Freytag GF, Bassett MJ, Zapata M (1982) Registration of XR-235-1-1 bean germplasm. Crop Sci 22:1268–1269. https://doi.org/10.2135/cropsci1982.0011183X002200060066x

Galvez LC, Banerjee J, Pinar H, Mitra A (2014) Engineered plant virus resistance. Plant Sci 228: 11–25

Giesler LJ, Ghabrial SA, Hunt TE, Hill JH (2002) Bean pod mottle virus: a threat to US soybean production. Plant Dis 86(12):1280–1289

Girma F, Fininsa C, Terefe H, Amsalu B (2022) Evaluation of common bean (*Phaseolus vulgaris*) genotypes for resistance to common bacterial blight and angular leaf spot diseases, and agronomic performances. Heliyon 8:1–16. https://doi.org/10.1016/j.heliyon.2022.e10425

Gonçalves-Vidigal MC, Cruz AS, Garcia A, Kami J, Vidigal Filho PS, Sousa LL, McClean P, Gepts P, Pastor-Corrales MA (2011) Linkage mapping of the *Phg-1* and *Co-14* genes for resistance to angular leaf spot and anthracnose in the common bean cultivar AND 277. Theor Appl Genet 122:893–903

Gonçalves-Vidigal MC, Cruz AS, Lacanallo GF, Vidigal Filho PS, Sousa LL, Pacheco CMNA et al (2013) Cosegregation analysis and mapping of the anthracnose *Co-10* and angular leaf spot *Phg-ON* disease-resistance genes in the common bean cultivar Ouro Negro. Theor Appl Genet 126:2245–2255. https://doi.org/10.1007/s00122-013-2131-8

González AM, Yuste-Lisbona FJ, Godoy L, Fernández-Lozano A, Rodiño AP, De Ron AM (2016) Exploring the quantitative resistance to *Pseudomonas syringae* pv. *phaseolicola* in common bean (*Phaseolus vulgaris* L.). Mol Breed 36:166

Gouveia BC, Calil IP, Machado JPB, Santos AA, Fontes EP (2017) Immune receptors and co-receptors in antiviral innate immunity in plants. Front Microbiol 7:2139

Hagedorn DJ, Walker JC, Rand RE (1974) Wis. HBR 40 and Wis. HBR 72 bean germplasm. Hprt Scoence, pp 402–404

Hart JP, Griffiths PD (2013) A series of *eIF4E* alleles at the *bc-3* locus are associated with recessive resistance to clover yellow vein virus in common bean. Theor Appl Genet 126(11):2849–2863

Harveson RM, Schwartz HF (2007) Bacterial diseases of dry edible beansin the central high plains. Plant Health Prog 10:1093–1094

Harveson RM, Schwartz HF, Urrea CA, Yonts CD (2015) Bacterial wilt of dry-edible beans in the central high plains of the US: past, present, and future. Plant Dis 99:1665–1677

Heidrich K, Wirthmueller L, Tasset C, Pouzet C, Deslandes L, Parker JE (2011) Arabidopsis *EDS1* connects pathogen effector recognition to cell compartment–specific immune responses. Science 334(6061):1401–1404

Horn J, Lauster S, Krenz B, Kraus J, Frischmuth T, Jeske H (2011) Ambivalent effects of defective DNA in beet curly top virus-infected transgenic sugarbeet plants. Virus Res 158(1–2):169–178

Hurtado-Gonzales OP, Valentini G, Gilio TA, Martins AM, Song Q, Pastor-Corrales MA (2017) Fine mapping of *Ur-3*, a historically important rust resistance locus in common bean. G3 Genes Genom Genet 7(2):557–569

Ivanov KI, Eskelin K, Lohmus A, Makinen K (2014) Molecular and cellular mechanisms underlying potyvirus infection. J Gen Virol 95:1415–1429

Jung G, Skroch PW, Coyne DP, Nienhuis J, Arnaud Santana E, Ariyarathne HM, Kaeppler SM, Bassett MJ (1997) Molecular-marker-based genetic analysis of tepary bean-derived common bacterial blight resistance in different developmental stages of common bean. J Am Soc Hortic Sci 122:329–337

Jung G, Ariyarathne HM, Coyne DP, Nienhuis J (2003) Mapping QTL for bacterial brown spot resistance under natural infection in field and seedling stem inoculation in growth chamber in common bean. Crop Sci 43:350e357

Kelly JD, Bornowski N (2018) Marker-assisted breeding for economic traits in common bean. In: Biotechnologies of crop improvement, vol 3. Springer, Cham, pp 211–238. https://doi.org/10.1007/978-3-319-94746-4_10

Kelly JD, Vallejo VA (2004) A comprehensive review of the major genes conditioning resistance to anthracnose in common bean. HortScience 39(6):1196–1207. https://doi.org/10.21273/HORTSCI.39.6.1196

Kimani P, Buruchara R, Ampofo K, Pyndji M, Chirwa R, Kirkby R (2005) Breeding beans for smallholder farmers in eastern, central, and southern Africa: constraints, achievements, and potential, vol 2. PABRA Millennium Workshop, Nairobi, pp 10–11

Koinange EMK, Gepts P (1992) Hybrid weakness in wild *Phaseolus vulgaris* L. J Hered 83 (2):135–139

Lamppa RS, Gross PL, del Río LE (2002) Identification of races of *Pseudomonas syringae* pv. *phaseolicola* present in North Dakota. Phytopathology 92:S139

Larsen RC, Miklas PN, Druffel KL, Wyatt SD (2005) NL-3 K strain is a stable and naturally occurring interspecific recombinant derived from bean common mosaic necrosis virus and bean common mosaic virus. Phytopathology 95(9):1037–1042

Leitão ST, Malosetti M, Song Q, van Eeuwijk F, Rubiales D, Vaz Patto MC (2020) Natural variation in Portuguese common bean germplasm reveals new sources of resistance against

Fusarium oxysporum f. sp. *phaseoli* and resistance-associated candidate genes. Phytopathology 110(3):633–647

Li YQ, Liu ZP, Yang YS, Zhao B, Fan ZF, Wan P (2014) First report of bean common mosaic virus infecting Azuki bean (Vigna angularis Ohwi & Ohashi) in China. Plant Dis 98:1017

Liebenberg MM, Pretorius ZA (2010) 1 common bean rust: pathology and control. Hortic Rev 37:1

Mahiya-Farooq Padder BA, Bhat NN, Shah M, Shikari AB, Awale HE, Kelly JD (2019) Temporal expression of candidate genes at the *Co-1* locus and their interaction with other defense related genes in common bean. Physiol Mol Plant Pathol 108:101424. https://doi.org/10.1016/j.pmpp.2019.101424

Mahuku GS, Jara C, Henriquez MA, Castellanos G, Cuasquer J (2006) Genotypic characterization of the common bean bacterial blight pathogens, *Xanthomonas axonopodis* pv. *phaseoli* and *Xanthomonas axonopodis* pv. *phaseoli* var. fuscans by rep-PCR and PCR-RFLP of the ribosomal genes. J Phytopathol 154:35

Mahuku GS, Iglesias ÁM, Jara C (2009) Genetics of angular leaf spot resistance in the Andean common bean accession G5686 and identification of markers linked to the resistance genes. Euphytica 167:381–396. https://doi.org/10.1007/s10681-009-9897-4

McElroy JB (1985) Breeding for dry beans, *P. vulgaris* L., for common bacterial blight resistance derived from *Phaseolus acutifolius* A. gray. Ph. D. Dissertation. Cornell University, Ithaca, NY

Melotto M, Kelly JD (1998) SCAR markers linked to major disease resistance genes in common bean. Annu Rep Bean Improv Coop 41:64–65

Mendez-Vigo B, Rodriguez-Suarez C, Paneda A, Ferreira JJ, Giraldez R (2005) Molecular markers and allelic relationships of anthracnose resistance gene cluster B4 in common bean. Euphytica 141:237–245

Meziadi C, Richard MM, Derquennes A, Thareau V, Blanchet S, Gratias A, Pflieger S, Geffroy V (2016) Development of molecular markers linked to disease resistance genes in common bean based on whole genome sequence. Plant Sci 242:351–357. https://doi.org/10.1016/j.plantsci.2015.09.006

Meziadi C, Blanchet S, Geffroy V, Pflieger S (2017) Genetic resistance against viruses in *Phaseolus vulgaris* L.: state of the art and future prospects. Plant Sci 265:39–50

Mienie CMS, Liebenberg MM, Pretorius ZA, Miklas PN (2005) SCAR markers linked to the common bean rust resistance gene *Ur-13*. Theor Appl Genet 111(5):972–979

Miklas PN, Stavely JR, Kelly JD (1993) Identification and potential use of a molecular marker for rust resistance in common bean. Theor Appl Genet 85(6):745–749

Miklas PN, Johnson E, Stone V, Beaver JS, Montoya C, Zapata M (1996) Selective mapping of QTL conditioning disease resistance in common bean. Crop Sci 36:1344–1351. https://doi.org/10.2135/cropsci1996.0011183X003600050044x

Miklas PN, Delorme R, Stone V, Daly MJ, Stavely JR, Steadman JR et al (2000a) Bacterial, fungal, and viral disease resistance loci mapped in a recombinant inbred common bean population ('Dorado'/ XAN 176). J Am Soc Hortic Sci 125:476

Miklas PN, Stone V, Daly MJ, Stavely JR, Steadman JR, Bassett MJ, Delorme R, Beaver JS (2000b) Bacterial, fungal, and viral disease resistance loci mapped in a recombinant inbred common bean population 'dorado'/XAN 176. J Am Soc Hortic Sci 125:476–481

Miklas PN, Coyne DP, Grafton KF, Mutlu N, Reiser J, Lindgren DT et al (2003a) A major QTL for common bacterial blight resistance derives from the common bean great northern landrace cultivarMontana no. 5. Euphytica 131:137–146

Miklas PN, Coyne DP, Grafton KF, Mutlu N, Reiser J, Lindgren DT, Singh SP (2003b) A major QTL for common bacterial light resistance derives from the common bean great northern landrace cultivar Montana no. 5. Euphytica 131:137–146. https://doi.org/10.1023/A:1023064814531

Miklas PN, Kelly JD, Beebe SE, Blair MW (2006) Common bean breeding for resistance against biotic and abiotic stresses: from classical to MAS breeding. Euphytica 147(1):105–131

Miklas PN, Fourie D, Wagner J, Larsen RC, Mienie CMS (2009) Tagging and mapping Pse-1 gene for resistance to halo blight in common bean differential cultivar UI-3. Crop Sci 49:41–48

Miklas PN, Fourie D, Trapp J, Davis J, Myers JR (2014) New loci including Pse-6 conferring resistance to halo bacterial blight on chromosome Pv04 in common bean. Crop Sci 54:2099–2108

Miller T, Gepts P, Kimmo S, Arunga E, Luseko AC, Susan N-M et al (2018) Alternative markers linked to the *Phg-2* angular leaf spot resistance locus in common bean using the *Phaseolus* genes marker database. Afr J Biotechnol 17:818–828. https://doi.org/10.5897/ajb2018.16493

Morales FJ (2006) Common beans. In: Loebenstein G, Carr JP (eds) Natural resistance mechanisms of plants to viruses. Springer, Amsterdam, pp 367–382

Muedi HTH, Fourie D, McLaren NW (2015) Reaction of selected common bean germplasm accessions to bacterial brown spot in South Africa. Crop Prot 77:87–93

Murillo IÁ, Falconí E, Marzón N, Peralta IE (2006) Resistance sources for rust, angular leaf spot, and common bacterial blight in common bean for Ecuador. Annu Rep Bean Improv Coop 49: 229–230

Mutlu N, Miklas P, Reiser J, Coyne D (2005) Backcross breeding for improved resistance to common bacterial blight in pinto bean (*Phaseolus vulgaris* L.). Plant Breed 124:282–287

Mutlu N, Vidaver AK, Coyne DP, Steadman JR, Lambrecht PA, Reiser J (2008) Differential pathogenicity of *Xanthomonas campestris* pv. *Phaseoli* and *X. fuscans* subsp. *fuscan*s trains on bean genotypes with common blight resistance. Plant Dis 92:546–554

Nabi A, Lateef I, Nisa Q, Banoo A, Rasool RS, Shah MD, Ahmad M, Padder BA (2022) Phaseolus vulgaris-Colletotrichum lindemuthianum pathosystem in the post-genomic era: an update. Curr Microbiol 79(2):1–13

Navarro F, Skroch P, Jung G, Nienhuis J (2007) Quantitative trait loci associated with bacterial brown spot in *Phaseolus vulgaris* L. Crop Sci 47:1344–1353

Nay MM, Souza TLPO, Raatz B, Mukankusi CM, Pastor-Corrales MA, Abreu AFB et al (2019) A review of angular leaf spot resistance in common bean. Crop Sci 59:1376–1391. https://doi.org/10.2135/cropsci2018.09.0596

Nicaise V (2014) Crop immunity against viruses: outcomes and future challenges. Front Plant Sci 5: 660

Nunes MPBA, Gonçalves-Vidigal MC, Martins VS, Xavier LF, Valentini G, Vaz Bisneta M, Vidigal Filho PS (2021) Relationship of Colletotrichum lindemuthianum races and resistance loci in the Phaseolus vulgaris L. genome. Crop Sci 61(6):3877–3893

O'Boyle PD, Kelly JD, Kirk WW (2007) Use of marker-assisted selection to breed for resistance to common bacterial blight in common bean. J Am Soc Hortic Sci 132:381–386

Oblessuc PR, Baroni RM, Garcia AAF, Chioratto AF, Carbonell SAM, Camargo LEA et al (2012) Mapping of angular leaf spot resistance QTL in common bean (*Phaseolus vulgaris* L.) under different environments. BMC Genet 13:50. https://doi.org/10.1186/1471-2156-13-50

Opio AF, Allen DJ, Teri JM (1996) Pathogenic variation in *Xanthonomas campestris* pv. *phaseoli*, the causal agent of common bacterial blight in *phaseolus* beans. Plant Pathol 45:1126–1133

Padder BA, Sharma PN, Awale H, Kelly JD (2017) *Colletotrichum lindemuthiam* the casual agent of bean anthracnose. J Plant Pathol 99(2):317–330. https://doi.org/10.4454/jpp.v99i2.3867

Park SO, Coyne DP, Bokosi JM, Steadman JR (1999) Molecular markers linked to genes for specific rust resistance and indeterminate growth habit in common bean. Euphytica 105(2): 133–141

Parker JPK (1985) Interspecific transfer of common bacterial blight resistance from *Phaseolus acutifolius* A. gray to *Phaseolus vulgaris* L. M.S. thesis. University of Guelph, Guelph, ON

Pastor-Corrales MA, Aime MC (2004) Differential cultivars and molecular markers segregate isolates of *Uromyces appendiculatus* into two distinct groups that correspond to the gene pools of their common bean hosts. Phytopathology 94:82

Patel PN, Walker JC (1965) Resistance in *Phaseolus* to halo blight. Phytopathology 55:889–894

Paulino JFDC, Almeida CPD, Bueno CJ, Song Q, Fritsche-Neto R, Carbonell SAM et al (2021) Genome-wide association study reveals genomic regions associated with *Fusarium* wilt resistance in common bean. Gene 12(5):765

Pedraza García F, Gallego GJ, Beebe SE, Tohme MJ (1997) Marcadores SCAR y RAPD para la resitencia a la bacteriosis comun (CBB). In: Singh SP, Voysest O (eds) Taller de mejoramiento de frijol para el Siglo XXI: bases para una estrategia para America Latina

Perry G, DiNatale C, Xie W, Navabi A, Reinprecht Y, Crosby W et al (2013) A comparison of the molecular organization of genomic regions associated with resistance to common bacterial blight in two *Phaseolus vulgaris* genotypes. Front Plant Sci 4:318

Pessoni LA, Zimmermann MJO, Faria JC (1997) Genetic control of characters associated with bean gold mosaic geminivirus resistance in *Phaseolus vulgaris* L. Braz J Genet 20:51–58

Rava Seijas CA, Sartorato A, Porto de Carvalho JR (1985) Yield losses in dry bean (*Phaseolus vulgaris* L.) caused by angular leaf spot (*Isariopsis griseola* Sacc.). Annu Rep Bean Improv Coop 28:5–6

Ribeiro RDL, Hagedorn DJ (1979) Screening for resistance to and pathogenic specialization of *Fusarium oxysporum* f. sp. *phaseoli*, the causal agent of bean yellows. Phytopathology 69 (3):272–276

Rico A, López R, Asensio C, Aizpún MT, Asensio-S-Manzanera MC, Murillo J (2003) Nontoxigenic strains of *Pseudomonas syringae* pv. *phaseolicola* are a main cause of halo blight of beans in Spain and escape current detection methods. Phytopathology 93:1553–1559

Rosen R, Kanakala S, Kliot A, Pakkianathan BC, Farich BA, Santana-Magal N, Elimelech M, Kontsedalov S, Lebedev G, Cilia M, Ghanim M (2015) Persistent, circulative transmission of begomoviruses by whitefly vectors. Curr Opin Virol 15:1–8

Saettler AW (1989) Common bacterial blight. In: Bean production problems in the tropics, vol 2. CIAT, Cali, pp 261–283

Salegua V, Rob M, Deidré F, Julia S, Cousin M (2020) Screening Andean diversity panel dry bean lines for resistance to bacterial Brown spot disease under field conditions in South Africa. Plant Dis 104:2509–2514

Sartorato A, Nietsche S, Barros EG, Moreira MA (1999) Inheritance of angular leaf spot resistance and RAPD markers linked to disease resistance gene in common beans. Annu Rep Bean Improv Coop 42:21–22

Schmutz J, McClean PE, Mamidi S, Wu GA, Cannon SB, Grimwood J, Jenkins J, Shu S, Song Q, Chavarro C, Torres-Torres M (2014) A reference genome for common bean and genome-wide analysis of dual domestications. Nat Genet 46(7):707–713

Scott ME, Michaels TE (1992) *Xanthomonas* resistance of *Phaseolus* interspecific cross selections confirmed by field performance. HortScience 27:348–350

Serfontein J (1994) Occurrence of bacterial brown spot of dry beans in the Transvaal province of South Africa. Plant Pathol 43:597–599

Sharma M, Pandey GK (2016) Expansion and function of repeat domain proteins during stress and development in plants. Front Plant Sci 6:1218

Sharma PN, Sharma OP, Padder BA, Kapil R (2008) Yield loss assessment in common bean due to anthracnose (*Colletotrichumlindemuthianum*) under sub-temperate conditions of north–Western Himalayas. Indian Phytopathol 61(3):323–330

Shi C, Chaudhary S, Yu K, Park SJ, Navabi A, McClean PE (2011a) Identification of candidate genes associated with CBB resistance in common bean HR45 (*Phaseolus vulgaris* L.) using cDNA-AFLP. Mol Biol Rep 38:75–81

Shi C, Navabi A, Yu K (2011b) Association mapping of common bacterial blight resistance QTL in Ontario bean breeding populations. BMC Plant Biol 11:52

Shi C, Yu K, Xie W, Perry G, Navabi A, Peter Pauls K (2012) Development of candidate gene markers associated to common bacterial blight resistance in common bean. Theor Appl Genet 125:1525–1537

Singh SP (1999) Production and utilization. In: Common bean improvement in the twenty-first century. Springer, Dordrecht, pp 1–24

Singh SP, Miklas PN (2015) Breeding common bean for resistance to common blight: A review. Crop Sci 55(3):971–984

Singh SP, Muñoz CG (1999) Resistance to common bacterial blight among *Phaseolus* species and common bean improvement. Crop Sci 39:80–89
Singh SP, Schwartz HF (2010) Breeding common bean for resistance to diseases: a review. Crop Sci 50(6):2199–2223
Singh SP, Morales FJ, Miklas PN, Teran H (2000) Selection for bean golden mosaic resistance in intra- and inter-racial bean populations. Crop Sci 40:1565–1572
Singh SP, Muñoz CG, Terán H (2001) Registration of common bacterial blight resistant dry bean germplasm VAX 1, VAX 3, and VAX 4. Crop Sci 41:275–276. https://doi.org/10.2135/cropsci2001.411275x
Stenger DC, McMahon CL (1997) Genotypic diversity of beet curly top virus populations in the western United States. Phytopathology 87(7):737–744
Strausbaugh CA, Myers JR, Forster RL, McClean PE (1999) Bc-1 and bc-u—two loci controlling bean common mosaic virus resistance in common bean are linked. J Am Soc Hortic Sci 124(6): 644–648
Strausbaugh CA, Miklas PN, Singh SP, Myers JR, Forster RL (2003) Genetic characterization of differential reactions among host group 3 common bean cultivars to NL-3 K strain of bean common mosaic necrosis virus. Phytopathology 93(6):683–690
Strausbaugh CA, Wintermantel WM, Gillen AM, Eujayl IA (2008) Curly top survey in the Western United States. Phytopathology 98(11):1212–1217. https://doi.org/10.1094/PHYTO-98-11-1212. PMID 18943410
Taylor JD, Innes NL, Dudley CL, Griffiths WA (1978) Sources and inheritance of resistance to halo-blight of *Phaseolus* beans. Ann Appl Biol 90:101–110
Taylor JD, Teverson DM, Allen DJ, Pastor-Corrales MA (1996a) Identification and origin of races of *Pseudomonas syringae* pv. *phaseolicola* from Africa and other bean growing areas. Plant Pathol 45:469–478
Taylor JD, Teverson DM, Davis JHC (1996b) Sources of resistance to *Pseudomonas syringae pv. phaseolicola* races in *Phaseolus vulgaris*. Plant Pathol 45:479–485
Tock AJ (2017) Applying next-generation sequencing to enable marker-assisted breeding for adaptive traits in common bean (*Phaseolus vulgaris* L.). [Ph.D.thesis] Wellesbourne: University of Warwick72 bean germplasm. HortScience 9:402
Trabanco N, Asensio-Manzanera MC, Pérez-Vega E, Ibeas A, Campa A, Ferreira JJ (2014) Identification of quantitative trait loci involved in the response of common bean to *Pseudomonas syringae* pv. *phaseolicola*. Mol Breed 33:577–588
Tracy SL, Frenkel MJ, Gough KH, Hanna PJ, Shukla DD (1992) Bean yellow mosaic, clover yellow vein, and pea mosaic are distinct potyviruses: evidence from coat protein gene sequences and molecular hybridization involving the 3′ non-coding regions. Arch Virol 122(3–4): 249–261. https://doi.org/10.1007/BF01317187
Tugume JK, Tusiime G, Sekamate AM, Buruchara R, Mukankusi CM (2019) Diversity and interaction of common bacterial blight disease-causing bacteria (*Xanthomonas* spp.) with *Phaseolus vulgaris* L. Crop J 7(1):1–7
Tulmann Neto A, Ando A, Costa AS (1977) Attempts to induce mutants resistant or tolerant to golden mosaic virus in dry beans (*Phaseolus vulgaris* L.). In: induced mutation against plant diseases, IAEA proceedings series. IAEA, New York, pp 281–288
Urrea CA, Miklas PN, Beaver JS, Riley RH (1996) A codominant randomly amplified polymorphic DNA (RAPD) marker useful for indirect selection of BGMV resistance in common bean. J Am Soc Hortic Sci 121(6):1035–1039
Vandemark GJ, Miklas PN (2005) Genotyping common bean for the potyvirus resistance alleles *I* and *bc-1*^{2} with a multiplex real-time polymerase chain reaction assay. Phytopathology 95:499–505
Viteri DM, Singh SP (2014) Response of 21 common beans of diverse origins to two strains of the common bacterial blight pathogen, *Xanthomonas campestris* pv *phaseoli*. Euphytica 200:379–388. https://doi.org/10.1007/s10681-014-1161-x

Viteri DM, Cregan PB, Trapp J, Miklas PN, Singh SP (2014) A new common bacterial blight resistance QTL in VAX 1 common bean and interaction of the new QTL, SAP6, and SU91 with bacterial strains. Crop Sci 54:1598–1608. https://doi.org/10.2135/cropsci2014.01.0008

Viteri DM, Terán H, Asensio-S-Manzanera MC, Asensio C, Porch TG, Miklas PN, Singh SP (2014b) Progress in breeding Andean common bean for resistance to common bacterial blight. Crop Sci 54(5):2084–2092

Viteri DM, Terán H, Asensio-S-Manzanera MC, Asensio C, Porch TG, Miklas PN, Singh SP (2014c) Progress in breeding Andean common bean for resistance to common bacterial blight. Crop Sci 54:2084–2092. https://doi.org/10.2135/cropsci2014.03.0177

Vlasova A, Capella-Gutiérrez S, Rendón-Anaya M, Hernández-Oñate M, Minoche AE, Erb I (2016) Genome and transcriptome analysis of the Mesoamerican common bean and the role of gene duplications in establishing tissue and temporal specialization of genes. Genome Biol 17:32

Williams AH, Sharma M, Thatcher LF, Azam S, Hane JK, Sperschneider J et al (2016) Comparative genomics and prediction of conditionally dispensable sequences in legume–infecting *Fusarium oxysporum* formae speciales facilitates identification of candidate effectors. BMC Genomics 17 (1):1–24

Xue R, Wu J, Zhu Z, Wang L, Wang X, Wang S, Blair MW (2015) Differentially expressed genes in resistant and susceptible common bean (*Phaseolus vulgaris* L.) genotypes in response to *Fusarium oxysporum* f. sp. *phaseoli*. PLoS One 10(6):e0127698

Young RA, Kelly JD (1997) RAPD markers linked to three major anthracnose resistance genes in common bean. Crop Sci 37(3):940–946

Young JM, Dye DW, Wilkie JP (1978) Genus VII *Pseudomonas migula*. In: Young JM, Dye DW, Bradbury JF, Panagopoulos CG, Robbs CF (eds) A proposed nomenclature and classification for plant pathogenic bacteria, New Zeal J Agr Res, vol, vol 21, pp 153–151

Yu K, Park SJ, Poysa V, Gepts P (2000) Integration of simple sequence repeat (SSR) markers into a molecular linkage map of common bean (*Phaseolus vulgaris* L.). J Hered 91:429–434

Yu K, Shi C, Zhang B (2012) Development and application of molecular markers to breed common bean (*Phaseolus vulgaris* L.) for resistance to common bacterial blight (CBB)–current status and future directions. In: Applied Photosynthesis. IntechOpen, Rijeka, p 365

Zapata M, Freytag GF, Wilkinson RE (1985) Evaluation for bacterial blight resistance in beans. Phytopathology 75:1032–1039. https://doi.org/10.1094/Phyto-75-1032

Zapata M, Beaver JS, Porch TG (2011) Dominant gene for common bean resistance to common bacterial blight caused by *Xanthomonas axonopodis* pv. *phaseoli*. Euphytica 179(3):373–382

Zapata M, Freytag G, Wilkinson R (2004) Release of five common bean germplasm lines resistant to common bacterial blight: W-BB-11, W-BB-20-1, W-BB-35, W-BB-52, and W-BB-11-56. J Agric Univ Puerto Rico

Zaumeyer WJ, Thomas HR (1948) Pod mottle, a virus disease of beans. J Agric Res 77:81–86

Ziems AD, Giesler LJ, Graef GL, Redinbaugh MG, Vacha JL, Berry S, Madden LV, Dorrance AE (2007) Response of soybean cultivars to bean pod mottle virus infection. Plant Dis 91(6): 719–726

Zuiderveen GH, Padder BA, Kamfwa K, Song Q, Kelly JD (2016) Genome-wide association study of anthracnose resistance in andean beans (*Phaseolus vulgaris*). PLoS One 11(6):e0156391. https://doi.org/10.1371/journal.pone.0156391

6 Diseases in Cowpea (*Vigna unguiculata* (L.) Walp.): Next Generation Breeding Techniques for Developing Disease-Resistant Cowpea

Sanjeev K. Deshpande, N. M. Kavyashree, Kartar Singh, K. R. Ramya, Namisha Sharma, and Kuldeep Tripathi

Abstract

Next-generation sequencing (NGS) techniques have revolutionized the field of genomics and played a significant role in disease-resistant breeding strategies in cowpea (*Vigna unguiculata*). Cowpea is an important multi-purpose legume crop, widely grown in developing countries, but its production is hampered by various diseases. The application of NGS technologies such as whole-genome sequencing, RNA sequencing, and genotyping-by-sequencing in cowpea breeding programs has enabled to unravel the cowpea genome and identify genetic variations associated with disease resistance. It allows for rapid sequencing and facilitates the identification of specific genomic regions associated with resistance to diseases, such as bacterial blight, fungal pathogens, and viral infections and it will expedite the development of improved cowpea cultivars, ensuring enhanced crop productivity and food security in regions heavily reliant on cowpea. This chapter highlights the economic importance of cowpea and the need to explore

S. K. Deshpande (✉) · N. M. Kavyashree
Department of Genetics and Plant Breeding, University of Agricultural Sciences, Dharwad, Karnataka, India

K. Singh
ICAR-National Bureau of Plant Genetic Resources, RS, Jodhpur, India

K. R. Ramya
Division of Plant Genetic Resources, ICAR-Indian Agricultural Research Institute, New Delhi, India

N. Sharma
Institute of Life Sciences (ILS), Bhubaneswar, Odisha, India

K. Tripathi
ICAR-National Bureau of Plant Genetic Resources, New Delhi, India

U. C. Jha et al. (eds.), *Diseases in Legume Crops*,
https://doi.org/10.1007/978-981-99-3358-7_6

the genetic diversity of cowpea and utilize genetic resources in the breeding of cowpea for disease resistance using traditional and advanced NGS techniques.

6.1 Introduction

Cowpea [*Vigna unguiculata* (L.) Walp.] is one of the most important food and forage legumes in the semiarid tropics that include parts of Asia, Africa, Southern Europe, Southern United States, and Central and South America (Singh 2005 Tripathi et al. 2019). Cowpea is also grown in marginal areas of eastern and southern Africa, especially in Sudan, Somalia, Mozambique, Botswana, and southern Zimbabwe. It is a multipurpose crop, providing food for man and livestock and serving as a valuable and dependable income-generating commodity for farmers and grain traders (Singh et al. 2006). Cowpea is primarily a short-day plant or in some instances, day-neutral (Ehlers and Hall 2011). It is a dicotyledonous belonging to the order Fabaceae, subfamily Faboideae (syn. *Papilionoideae*), tribe Phaseoleae, subtribe Phaseolinae, genus *Vigna*, and section Catiang (Baudoin and Maréchal 2011). *Vigna* is a pantropical and highly variable genus with several species, the number varying from 84 to 184 (Timko et al. 2007).

Cowpea is of major importance to the livelihoods of millions of people in less developed countries of the tropics. It is consumed in many forms. Young leaves, green pods, and green seeds are used as vegetables, and dry seeds are used in various food preparations (Fatokun et al. 1992). With 25% protein (on a dry-weight basis) in its seeds and tender leaves (Carsky et al. 2010), cowpea is a major source of protein, minerals, and vitamins in the daily diets in Africa, and thus it positively influences the health of men, women, and children (Nielson et al. 2010).

According to Langyintuo et al. (2003), some 10 million hectares are under cowpea cultivation worldwide, with the sub-Saharan Africa cowpea belt producing about two-thirds of the annual world yield. The total annual grain production is about 3.7 million tonnes. The second largest production area after Africa is Brazil, where the crop is well suited to the relatively low rainfall and poor soils in the northeastern part of the country. Cowpea is mostly grown as an intercrop with cereals, but little of that harvest reaches regional markets. The most important export market for cowpea in West Africa is Nigeria, simultaneously the world's largest cowpea consumer as well as producer. There is significant cowpea production in the Mediterranean, South Asia, and the southern and southwestern United States.

There are good reasons for the economic importance of cowpea. One of them is the excellent nutrition it offers. The nutritional profile of cowpea grain is similar to that of other pulses with relatively low-fat content and total protein content that is two- to fourfold higher than cereal and tuber crops. Similar to other pulses, the storage proteins in cowpea seeds are rich in the amino acids lysine and tryptophan when compared to cereal grains, but low in methionine and cysteine when compared to animal proteins. Total seed protein content ranges from 23% to 32% of seed weight (Tosti and Negri 2008). Cowpea seeds are also a rich source of minerals and vitamins and among plants have one of the highest contents of folic acid, a B vitamin necessary during pregnancy to prevent birth defects in the brain and spine.

In addition to human consumption, cowpea leaves and stems (stover) is also an important source of high-quality hay for livestock feed (Nielson et al. 2010). Cowpea fodder plays a particularly critical role in feeding animals during the dry season in many parts of the world. Anti-nutritional factors are plants' secondary metabolites which act to reduce food nutrient utilization (Hannah et al. 1976). In cowpea, the predominant anti-nutritional factors found are phytic acid, oxalic acid, and tannins. Thus, foods high in these anti-nutrients should be adequately processed to make them wholesome for consumers. Anti-nutritional factors affect the susceptibility of grains to insect attack (Pant et al. 2007). However, the presence of anti-nutritional factors commonly found in legumes is a major factor limiting the wider food use of these essential tropical plants (Hall et al. 2011). The alarming changes in the climate and global warming have led to a severe effect on agriculture productivity in the long run. Cowpea (*Vigna unguiculata* (L.) Walp.) has a series of disease-causing pathogens constituting the most important biotic stresses of the crop. Although fungal pathogens are more numerous; bacterial, nematode, and viral pathogens also cause serious yield losses. Cowpea is susceptible to many fungal diseases such as damping off in the seedling stage, southern blight, root rot, and leaf spot at vegetative growth. Additional important diseases of adult plants include rust, anthracnose, powdery mildew, and bacterial blight diseases inflicting heavy losses (Emechebe and Lagoke 2002). Many viruses also infect cowpea and are often seed-borne, such as cowpea mosaic virus (CMV) which causes disease symptoms that vary with variety (Singh and Allen 1980). Meanwhile, root-knot nematodes affect the overall growth of the plants (Das et al. 2008). The balance of different diseases varies depending on location. For example, India's production of cowpea is often beset by rust, powdery mildew, and blight as the most common diseases (Raju and Anilkumar 1990). In most cases, Indian farmers grow cowpea during the high humidity of the monsoon season so incidences of these diseases are high; and the management the farmers put into the cowpea fields is minimal because they are grown as a secondary crop compared to a main cash crop of cereal. Advancements in technologies aced the breeding programs with next generation breeding techniques like sequencing, speed breeding, gene editing, and haplotype breeding along with high throughput phenotyping and artificial intelligence (AI) for the breeding of resistant legumes with high yield and high nutrition. In this chapter, the important biotic stress which includes diseases of cowpea and their breeding progress were discussed.

6.2 Diseases in Cowpea

Further division of diseases affecting cowpea is made between those caused by bacterial, fungal, and viral pathogens with nematodes included among these.

6.2.1 Fungal Diseases of Cowpea

6.2.1.1 Anthracnose

Anthracnose is an important disease that accounts for up to 50% yield reduction and was first reported in 1985 (Prasanna 1985) in India. It has been variously advanced and reported as a form of *Colletotrichum lindemuthianum*, *C. gloeosporoides*, *C. dematium*, and recently as *C. destructivum*.

6.2.1.2 Powdery Mildew

Powdery mildew is one of the serious diseases of cowpea and it is spread especially in the southern parts of India, Zambia (Kanniyan et al. 1987), Zimbabwe (Stoffella et al. 1990) United States, Puerto Rico, and other cowpea-growing countries of Latin America where climates with warm days, dry days, and cool nights are common. The causal agent of the disease was identified as *Erysiphe polygoni* (Braun 1987).

6.2.1.3 Rust

Cowpea rust is caused by *Uromyces phaseoli* var. *vignae* (Barclay) which is widespread in all cowpea-growing areas of the world. For management of this disease, some of the promising Indian cowpea lines were identified against *Uromyces phaseoli* var. *vignae*, viz., IC206240, PKB6-2, V-16, EC458483, PKB6-4, EC458480, KBC2, IC402180, 21-2, IC58905, IC249593 (Jayashree et al. 2019).

6.2.1.4 Charcoal Rot (Damping off)

A principal root rot of cowpea is called charcoal rot by the pathogen *Macrophomina phaseolina*. Being widely distributed in tropical and subtropical countries, it is a devastating disease occurring immediately on seedlings and throughout the establishment of the crop. Besides charcoal rot, the pathogen also induces symptoms of dry root rot, wilt, and leaf blight and in adult stages is often referred to as ashy stem blight disease (Singh et al. 1990). Ahmed et al. (2012) reported that the cowpea cultivar ITO4K-217-5 from Nigeria is resistant to *M. phaseolina*.

6.2.1.5 Southern Blight or Stem Rot

Southern blight is a common stem disease of cowpeas worldwide and is caused by the fungal pathogen *Sclerotium rolfsii*. Fery and Dukes (2002) observed that the impact of southern blight on cowpea yield may be more attributable to reduced plant vigor than to plant mortality per se. They reported that the disease can cause dry-seed yield loss of up to 53.4%. Tanimu et al. (2018) evaluated five cowpea varieties (L-25, Ife brown, IT89-KD-374, IT89-KD-434, and IT86-D-715) for resistance to basal stem rot disease and found that three varieties, viz., L-25, IT89-KD, and IT86-D-715, were immune to infection by the pathogen.

6.2.1.6 Fusarium Wilt

Fusarium wilt is thought to be caused by *Fusarium oxysporum* f. sp. *tracheiphilum* and first reported in cowpea from the United States, and in India, it was first reported by Singh and Sinha (1955). The pathogen is soil-borne and probably also seed transmitted. Currently, three races of *F. oxysporum* f. sp. *tracheiphilum* are known: Races 1 and 2 were described in South Carolina. Resistance to the three races has not been well studied in cowpea varieties, so most are considered susceptible.

6.2.1.7 *Cercospora* Leaf Spot

This leaf spot has a widespread distribution and causes leaves to fall off and serious yield losses of up to 40% in cowpea. Two species of *Cercospora*, *C. cruenta* and *C. canescens*, have been associated with leaf spot diseases in cowpea. Unfortunately, no resistance sources for defense against *Cercospora* leaf spot are known in cowpeas.

6.2.2 Bacterial Diseases of Cowpea

6.2.2.1 Bacterial Blight

Bacterial blight has been identified as the most important biotic constraint to cowpea production worldwide. Okechukwu and Florini (2000) have reported yield depressions of 42–71% in pod, 43–68% in seed, and 29–53% in fodder. Bacterial blight is caused by *Xanthomonas campestris* pv. *vignicola*.

6.2.2.2 Viral Diseases of Cowpea

Viral diseases have been considered one of the most important problems in cowpea causing serious reductions in crop productivity. In cowpea, virus infection causes yield loss of up to 70–100% depending on the stage of infection and vector-host plant interaction (Kareem and Taiwo 2007). Worldwide, up to 20 different viruses have been reported to naturally infect cowpea. In India, a total of eight viruses have been identified: namely, alfalfa mosaic virus (AMV, *Bromoviridae*), cowpea chlorotic mottle virus (CCMV, *Bromoviridae*), cowpea mosaic virus (CPMV, *Comoviridae*), cowpea mild mottle virus (CPMMV, *Betaflexiviridae*), and cowpea yellow mosaic virus (CPMV, *Comoviridae*) and Southern bean mosaic virus cowpea strain (SBMV-CS, *Sobemovirus*), as well as two important potyviruses: blackeye cowpea mosaic virus (BLCMV) and cowpea aphid-borne mosaic (CABMV). Each of these viruses can be distinguished only based on molecular diagnostics, serology, transmission tests, and/or symptomatology reactions on diagnostic hosts. CABMV is one of the most damaging viruses in West Africa (Neya et al. 2015). Cowpea severe mosaic virus is found in the Caribbean (Booker et al. 2005).

Excellent sources of resistance to CABMV have been identified among cowpea germplasms which includes cowpea line PI 596353 that was resistant to CABMV (Cisse et al. 1997) and reported an extra early maturing. Bashir and Hampton tested 51 cowpea lines by mechanical inoculation under greenhouse conditions against seven CABMV geographically diverse isolates and identified TVU-410, TVU-1582, and TVU-1593 as immune to all seven isolates.

Recently, apart from studying the genetics of CPMV, the studies have focused on developing this virus as a biotechnology workhorse. This virus has been used as a potent protein expression system (Sainsbury et al. 2010). Further, it has also been used to develop in situ vaccination methods against cancer. In the mice models, CPMV was able to activate pathogen recognition receptors (PRRs) and initiate antitumor immunity (Beiss et al. 2022).

6.3 Breeding for Disease Resistance in Cowpea

The productivity of cowpea is affected by the wide range of diseases which infect the crop. The development of resistant cultivars helps to overcome the problem of yield loss due to disease attacks. As a result, knowledge of the genetic control of resistance helps in accelerating the development of resistant varieties.

6.3.1 Traditional Breeding Approaches

Host plant resistance is a sustainable management option since it is environment-friendly and cost-effective. The cowpea host plant resistance mainly focuses on the development of resistant varieties through traditional breeding and biotechnology approaches.

6.4 Importance of Genetic Resources

The narrow genetic diversity of cowpea is mainly due to its origin in West Africa from a single species and its self-pollinating character. Cowpeas are thought to have evolved from a narrow section of the wild germplasm for the species *V. unguiculata* and related species with only partial gene flow between cultivated and wild types (Boukar et al. 2020). Three subspecies are recognized for the species and correlate with forms of utilization. *V. u.* spp. *cylindrica* is a semiwild type that is sometimes used as forage but not for grain. *V. u.* spp. *sesquipedalis* is a cowpea type selected for long pods in Asia that is used as a vegetable type. Meanwhile, the much more common *V. u.* spp. *unguiculata* is a common cowpea type used as dry grain with some dual purpose as fodder or a source of leaves that can be stewed.

6.5 Genetic Sources in Germplasm Banks

Sources of genes for various traits have been identified by screening of germplasm available in different countries. The main cowpea wild species include *V. vexillata*, *V. spontanea*, *V. tenuis*, *V. protracta*, *V. baoulensis*, and *V. stenophylla* (Boukar et al. 2019). Several researchers have reported on wild accessions having novel resistance against disease stresses, which can be shared and used in breeding programs (Singh 2002; Boukar et al. 2015).

Table 6.1 Cowpea germplasm accessions having resistance to various diseases

Disease	Resistant/tolerant germplasm accessions	References
Fusarium wilt	TVu 109-2, TVu 347, TVu 984, TVu 1000	Singh et al. (1983)
Scab	TVu 853, TVu 1404, TVu 1433	Singh et al. (1983)
Septoria	TVu 456, TVu 483-2, TVu 486, TVu 11761, TVu 112349, TVu 1433	Singh et al. (2002)
Bacterial blight	TVu 347, TVu 410, TVu 483-2	Singh et al. (1983)
BICMV	TVu 2480, TVu 2657, TVu 3433	Bashir (1992)
CABMV	TVu 401, TVu 1582	Bashir (1992)
CPMV	TVu 227, TVu 345, TVu 612	Patel (1982)
CPMoV	TVu 3901	Allen et al. (1982)
Striga and Alectra	TVu 14676, TVu 1070, TVu 1083, TVu 585, TVu 1532, TVu 1537, TVu 1647, TVu 491, B301	Lane et al. (1997)

The main objective of breeding for disease resistance is finding resistant germplasm as listed in Table 6.1. This information is highly useful for the development of resistant varieties through different breeding techniques. Huynh et al. (2017) developed a MAGIC (multi-parent advanced generation intercross) population for cowpea (*Vigna unguiculata* (L.) Walp.) from eight founder parents. These founders were genetically diverse and carried many disease resistance traits.

6.6 Genetics of Resistance

The source of resistance genes and knowledge of their inheritance are two prerequisites for developing varieties resistant to diseases. Resistance to diseases can be governed by both polygenic as well as monogenic inheritance. Disease resistances are much more likely to be under single gene control or simple inheritance, although cowpea R genes are not well studied. The resistant sources for the disease are already available in germplasm resources, and the effective screening helps in the identification of resistant germplasm. Apart from screening in the IITA international collection, EMBRAPA (Brazil), NBPGR (India), and USDA (United States) are examples of large national genebanks with extensive cowpea collections to work on the identification of disease-resistant cowpea germplasm. Along with collaborators in academia, efforts have been made to transfer R genes from wild species to cultivated cowpeas. Pre-breeding as well as elite line breeding, respectively, have been used to widen the genetic base of cowpea and incorporate disease tolerance. The world's largest cowpea germplasm collection (11,000 entries plus) is maintained by IITA, and multiple resistance germplasm has been found in its core collection (Togola et al. 2019). In summary, R-gene transfer has been through conventional breeding as well as the application of molecular techniques, and some resistant lines have been developed for disease resistance. These germplasm resources can be utilized in cowpea breeding programs for the development of further resistant varieties. Resistance sources for many of the soilborne pathogens have been identified, but highly resistant sources are often not available for some of the necrotrophic pathogens like *M. phaseolina*, *S. rolfsii*, and *Rhizoctonia* spp. Moderately resistant sources were identified in 5 of 33 cowpea cultivars (Singh and Lodha 1986) and 4 of 141 cultivars (Sohi and Rawal 1983). Most resistance to root rots in legumes is quantitative and polygenic. Leaf spot resistance to *Cercospora* is controlled by a single dominant gene and its heritability varied from 81 to 97% (Omoigui et al. 2019). Many authors reported different modes of inheritance of rust resistance genes in cowpea. The rust resistance in cowpea was conditioned by dominance genes with additive effects (Rangaiah 1997), recessive genes (Uma and Salimath 2004), or polygenes located at various loci (Uma et al. 2016). The resistance to southern blight is controlled by a dominant single gene (Fery and Dukes 2002).

Viral resistance to Cowpea aphid-borne mosaic virus was governed by more than one recessive gene in eight populations or single recessive genes in another seven populations (Orawu et al. 2013). Barro et al. (2016) observed that resistance to cowpea aphid-borne mosaic virus is governed by two dominant genes, and each parent is contributing a resistant gene. Less discrete results were found by Gumedzoe et al. (1998) in the screening of germplasm for CMV. Since many different viruses infect cowpeas at the same time (Taiwo and Shoyinka 1998) multiple resistance is the key to breeding against this stress. Ogunsola et al. (2010) identified multiple virus resistance to three different virus species in breeding lines IT98K-1092-1 and IT97K-1042-3; however, cowpea varieties with multiple virus resistance are yet to be found. The inheritance of resistance against bacterial blight is quantitative in nature and segregation was decided by the genetic background of parents and modifying factors, however, susceptibility was dominated over resistance and the segregation pattern did not fit into the genetic ratio as reported by Prakash and Shivashankar (1984). Three different types of host reactions were noticed in cowpea during bacterial pustule (*Xanthomonas campestris* pv. *vignae unguiculatae*) infection (Patel 1981) that included non-hypersensitive resistant (R), brown hypersensitive resistant (BHR), and susceptible (S). The inheritance study revealed that BHR reaction was dominated over R and S reaction, and R was recessive to S reaction (Patel 1982). The study indicated that BHR reaction is conditioned by two genes and R reaction is controlled by two or three recessive genes.

6.7 Genomic Resources and Molecular Breeding for Disease Resistance in Cowpea

Implications for crop improvement are provided especially given the availability of genome sequences and multiple molecular markers like RFLPs, SSRs, and SNPs for the plant species. These markers have played a significant role in accelerating various cowpea molecular breeding programs like QTL mapping, marker-assisted selection (MAS), marker-assisted back cross-breeding, and association mapping. Moreover, it also helps to overcome tedious inoculations and screening processes used for the selection of resistant genotypes. Modern biotechnological tools also help to overcome the crossing barrier between wild and cultivated species and help in the deployment of resistant genes from wild species. However, to make use of resistant germplasm knowledge on the inheritance, genomic location and marker association with the resistance genes are requirements. Introgression of disease resistance genes using traditional breeding techniques is complicated and time-consuming. To overcome this problem the best alternative option is to use molecular markers for the identification of resistant individuals in the early generation, which helps in the effective improvement of the breeding procedure (Torres 2010). The bulked segregant analysis reported that a simple sequence repeat (SSR) marker, namely, RB24 differentiates the resistance and susceptibility to *Cercospora* leaf spot (Omoigui et al. 2019). Hence, RB24 marker can be used for marker-assisted selection for this disease. QTLs related to bacterial blight resistance (Agbicodo et al. 2010), *Fusarium* resistance (Pottorff et al. 2012), *Macrophomina* resistance (Muchero et al. 2011), nematode resistance (Huynh et al. 2016), and virus resistance (Gioi et al. 2012) have been reported. Cowpea has a relatively small genome size estimated at 620 Mbp and therefore has been relatively easy to use in genetic mapping studies. Various researchers developed a linkage map of cowpea with different types of molecular and morphological markers. These maps provide resourceful information for various downstream applications including quantitative trait loci (QTL) identification, map-based cloning, diversity analysis, association mapping, and molecular breeding.

Wu et al. (2018) reported one major and two minor quantitative trait loci (QTLs) controlling rust resistance. Major QTLs (named as Ruv1) were mapped to a 12.48 cM interval between the SNP markers 2_01772 and 2_03292 on LG09 which explained 34.8% of the phenotypic variation. The minor QTLs, designated as Ruv2 and Ruv3, were mapped to a 7.01 cM interval on linkage group (LG) 7 and a 6.19 cM interval on LG8, which accounted for 13.4% and 11.9% of the phenotypic variation, respectively. Interval QTL mapping was used to show 98.4% of the variance for the resistance trait mapped in the region of three loci AGB1, VM31, and VM1 covering a genetic interval of 32.1 cM, in which 95% confidence was found for the CYMV resistance. In another study, three QTL against cowpea bacterial blight, namely CoBB-1, CoBB-2, and CoBB-3, were detected in linkage groups LG-3, LG5, and LG-9, respectively, showing that highly potential resistance candidate genes. QTLs such as CoBB-1 and CoBB-2 were reliably confirmed (Agbicodo et al. 2010). Meanwhile, for virus resistance, Gioi et al. (2012) studied the linkage of SSR markers to cowpea yellow mosaic virus (CYMV) by using resistance and susceptible lines of cowpea. Three SSR markers (AG1/AF48383, VM31, and VM1) were linked to resistance in cowpea against CYMV.

6.7.1 Next Generation Breeding Techniques

The first trial method of genetic transformation in cowpea was conducted by Garcia and their team with *Agrobacterium tumefaciens* as the vector with negative results

(Garcia et al. 1987). Genome-wide association studies (GWAS) have allowed researchers to precisely study the genetic basis of agricultural attributes thanks to advanced genotyping and phenotyping systems. By mapping agronomically significant features, quantitative trait loci (QTL) analysis made it possible to analyze how genotype and phenotype are related. Compared to forward genetics (trait to genotype), the reverse genetics (from genotype to phenotype) method often requires less time. Due to the ubiquitous accessibility of sequencing data, scientists may quickly develop reverse genetic approaches to identify gene function which include targeting induced local lesions in genomes (TILLING), EcoTILLING method, which is a modified version of TILLING, and allele mining based on sequencing and the PCR-based allele mining. New approaches to genome-assisted breeding (GAB) include gene cloning and characterization, haplotype-based breeding, allele mining for stress tolerance, and utilizing natural variants (Leng et al. 2017). By making it accessible to enormous plant genomes, recent developments in genome sequencing technology have revolutionized plant breeding and ushered in a new era of genomics (Bassi et al. 2016). Advanced NGS methods allow for the exploration of a wide range of genomic variations that can be connected to clarify complex phenotypes (Unamba et al. 2015). Positional cloning allows for the identification of plant genes using a variety of genomic markers (Bassi et al. 2016). For instance, single-nucleotide polymorphism (SNP), which is affordable, precise, and common in crop genomes, has been extensively utilized to screen thousands of crop germplasm samples (Voss-Fels and Snowdon 2016). A comprehensively annotated crop pan-genome can be built using recent advancements in genomic-assisted breeding (GAB) techniques to restore a species' lost gene repertoire. Pan-genomes offer fresh ways to take advantage of these distinctive genes or genetic variations for breeding program optimization. A greater grasp of the insights into plant phenomics and genetics is possible because of high-throughput phenotyping, which enables high-resolution imaging of thousands of plants (Roitsch et al. 2019). Agriculture is being revolutionized by speed breeding, which can be used to speed up crop breeding processes including cross-breeding, back-crossing, quick gene identification, population mapping, a pyramiding of traits, and the creation of transgenic pipelines (Hickey et al. 2019). Varshney et al. (2021) present the idea of haplotype-based breeding for accelerating the production of designer crops based on these superior haplotypes. This method has many advantages over MABC, which reduces genetic diversity and generates bottleneck effects while taking years and generations to transfer superior genes. By creating intelligent prediction models, next generation AI has recently attracted a lot of interest in plant breeding to address issues with abiotic/biotic stressors, herbicide resistance, crop production, and soil fitness (Muraya et al. 2017). Rapid gene identifications can be made possible by combining speed breeding with genomic and phenomic technologies, which will ultimately speed up crop development operations. Additionally, the incorporation of cutting-edge transdisciplinary breeding platforms can create exciting new opportunities for cultivating crops that are climate-ready and contribute to global food security.

6.8 Genome Sequencing and Next Generation Marker Development

The cowpea genome has been sequenced in various steps with an initial assembly of gene-rich space. Most recently, Munoz-Amatriain et al. (Munoz-Amatriain et al. 2017) and Lonardi et al. (2019) reported whole-genome shotgun (WGS) assemblies, of breeding lineIT97K-499-35. Some of the software commonly used in the molecular breeding of cowpea include "SNP Selector," "KBioConverter," and "Backcross Selector" for the management of genotyping data (Boukar et al. 2015).

6.9 Transgenic Improvement of Cowpea

Transgenic technology also plays a key role in enriching the genetic base of cowpea. It can easily overcome the limitations associated with the cross-compatibility of species. RNAi transgenic lines were developed against the viral disease of the cowpea. Cruz and Aragao (Cruz and Aragão 2014) developed transgenic cowpea lines against cowpea severe mosaic virus (CPSMV) and cowpea aphid-borne mosaic virus (CABMV) through RNAi gene silencing technology. They silenced the proteinase cofactor gene of CPSMV and the coat protein gene of CABMV through RNAi. Out of the ten transgenic lines generated, seven transgenic lines showed milder symptoms while three exhibited enhanced resistance against both viruses. Similarly, Transgenic lines were containing three different intron hairpin (hp) RNAi constructs, containing AC2, AC4, and fusion of AC2 and AC4 (AC2 + AC4) of begomoviruses which codes for transcription activator protein. RNAi transgenic lines were analyzed in T0 and T1 generations. Transgenic lines expressing AC2 hp and (AC2 + AC4) hp RNA showed nearly 100% resistance against MYMIV whereas transgenic lines expressing AC4 hp RNA showed milder symptoms after 5 weeks of infection.

References

Agbicodo EM, Fatokun CA, Bandyopadhyay R, Wydra K, Diop NN, Muchero W, Ehlers JD, Roberts PA, Close TJ, Visser RGF, Van der Linden CG (2010) Identification of markers associated with bacterial blight resistance loci in cowpea [Vigna unguiculata (L.) Walp]. Euphytica 175(2):215–226

Ahmed O, Balogun OS, Shittu BT (2012) Screening of cowpea genotypes for resistance to Macrophomina phaseolina infection using two methods of inoculation. Asian J Plant Pathol 6(1):13–18

Allen DJ, Thottappilly G, Rossel HW (1982) Cowpea mottle virus: field resistance and seed transmission in virus tolerant cowpea Vigna unguiculata. Ann Appl Biol 100:331–336

Barro A, De La Salle TJB, Dieni Z, Kiebre Z, Poda L, Sawadogo M (2016) Inheritance and the allelic relationship of resistance to cowpea aphid borne mosaic virus (CABMV) in two cowpea genotypes, KVX640 and KVX396-4-5-2D, in Burkina Faso. IJCMAS 5(8):285–292

Bashir M (1992) Serological and biological characterization of seed-borne isolates of black eye cowpea mosaic and cowpea aphid borne mosaic potyviruses in Vigna unguiculata (L.) Walp. PhD dissertation. Oregon State University, Corvallis

Bassi FM, Bentley AR, Charmet G, Ortiz R, Crossa J (2016) Breeding schemes for the implementation of genomic selection in wheat (Triticum spp.). Plant Sci 242:23–36

Baudoin JP, Maréchal R (2011) Genetic diversity in *Vigna*. In: Singh SR, Rachie KO (eds) Cowpea research, production and utilization. Wiley, Chichester

Beiss V, Mao C, Fiering SN, Steinmetz NF (2022) Cowpea mosaic virus outperforms other members of the Secoviridae as in situ vaccine for cancer immunotherapy. Mol Pharm 19: 1573–1585

Booker HM, Umaharan P, McDavid CR (2005) Effect of cowpea severe mosaic virus on crop growth characteristics and yield of cowpea. Plant Dis 89:515–520

Boukar O, Fatokun CA, Roberts PA, Abberton M, Huynh BL, Close TJ (2015) Cowpea. In: De Ron AM (ed) Grain legumes. Handbook of plant breeding, vol 10. Springer, New York, pp 219–250

Boukar O, Belko N, Chamarthi S, Togola A, Batieno J, Owusu E, Fatokun C (2019) Cowpea (Vigna unguiculata): genetics, genomics and breeding. Plant Breed 138(4):415–424

Boukar O, Abberton M, Oyatomi O, Togola A, Tripathi L, Fatokun C (2020) Introgression breeding in cowpea [Vigna unguiculata (L.) walp]. Front Plant Sci 11:567425

Braun U (1987) A monograph of the Erysiphales (powdery mildews). Nova Hedwigia Suppl 89: 195–196

Carsky RJ, Vanlauwe B, Lyasse O (2010) Cowpea rotation as a resource management technology for cereal-based systems in the savannas of West Africa. In: Fatokun CA et al (eds) Challenges and opportunities for enhancing sustainable cowpea production. Ibadan, Nigeria, International Institute of Tropical Agriculture, pp 252–266

Cisse N, Ndiaye M, Thiaw S, Hall AE (1997) Registration of "Melakh" cowpea. Crop Sci 37(6): 15–22

Cruz ARR, Aragão FJL (2014) RNAi-based enhanced resistance to cowpea severe mosaic virus and cowpea aphid-borne mosaic virus in transgenic cowpea. Plant Pathol J 63:831–837

Das S, DeMason DA, Ehlers JD, Close TJ, Roberts PA (2008) Histological characterization of root-knot nematode resistance in cowpea and its relation to reactive oxygen species modulation. J Exp Bot 59(6):1305–1313

Ehlers JD, Hall AE (2011) Genotypic classification of cowpea based on responses to heat and photoperiod. Crop Sci 36(3):673–679

Emechebe AM, Lagoke STO (2002) Recent advances in research on cowpea. Challenges and opportunities for enhancing sustainable cowpea production. In: Proceeding of the World Cowpea Conference III. The International Institute of Tropical Agriculture (IITA), Ibadan, p 94

Fatokun CA, Menancio Hautea DI, Danesh D, Young ND (1992) Evidence for orthologus seed weight genes in cowpea and mung bean based on RFLP mapping. Genetics 132(3):841–846

Fery RL, Dukes PD (2002) Southern blight (Sclerotium rolfsii Sacc.) of cowpea: yield-loss estimates and sources of resistance. Crop Protec 21(5):403–408

Garcia JA, Hille J, Vos P, Goldbach R (1987) Transformation of cowpea Vigna unguiculata with a full-length DNA copy of cowpea mosaic virus M-RNA. Plant Sci 48(2):89–98

Gioi TD, Boora KS, Chaudhary K (2012) Identification and characterization of SSR markers linked to yellow mosaic virus resistance genes in cowpea (Vigna unguiculata). Int J Plant Res 2:1–8

Gumedzoe MYD, Rossel HW, Thottappilly G, Asselin A, Huguenot C (1998) Reaction of cowpea (Vigna unguiculata L. Walp.) to six isolates of blackeye cowpea mosaicvirus (BlCMV) and cowpea aphid-borne mosaic virus (CAMV), two potyviruses infecting cowpea in Nigeria. Int J Pest Manag 44(1):11–16

Hall AE, Singh BB, Ehlers JD (2011) Cowpea breeding. Plant Breed Rev 15:215–274

Hannah LC, Ferrero J, Dessauer DW (1976) High methionine lines of cowpea. Trop Grain Legume Bull 4(9)

Hickey LT, Hafeez AN, Robinson H, Jackson SA, Leal-Bertioli SC, Tester M et al (2019) Breeding crops to feed 10 billion. Nat Biotechnol 37:744–754

Huynh BL, Matthews WC, Ehlers JD, Lucas MR, Santos JRP, Ndeve A, Roberts PA (2016) A major QTL corresponding to the Rk locus for resistance to root-knot nematodes in cowpea (Vigna unguiculata L. Walp.). Theor Appl Genet 129:87–95

Huynh BL, Ehlers JD, Maria Munoz-Amatriain M, Lonardi S, Santos JRP, Ndeve A, Roberts PA (2017) A multi-parent advanced generation inter-cross population for genetic analysis of multiple traits in cowpea (Vigna unguiculata L Walp.). BioRxiv 1–24

Jayashree B, Pankaja NS, Sugeetha G, Mahadev J, Suryakanth (2019) Investigations on the rust disease prevalence on cowpea in Mandya district and evaluation of cowpea genotypes for its resistance. Intl J Curr Microbiol App Sci 8(11):1126–1133

Kanniyan J, Greenberg DC, Haciwa HC, Mbewe MN (1987) Screening cowpea for resistance to major diseases in Zambia. Trop Grain Legume Bull 34:23–26

Kareem KT, Taiwo MA (2007) Interactions of viruses in cowpea: effects on growth and yield parameters. Virol J 8(4):15

Lane JA, Moore THM, Child DV et al (1997) Variation in virulence of Striga gesnerioides on cowpea: new sources of crop resistance. In: Singh BB, Mohan Raj DR, Dashiell KE et al (eds) Advances in cowpea research. Co-publication of International Institute of Tropical Agriculture (IITA) and Japan International Research Center for Agricultural Sciences, (JIRCAS)). IITA, Ibadan, pp 225–230

Langyintuo AS, Lowenberg-DeBoer J, Faye M, Lambert D, Ibro G, Moussa B, Kergna A, Kushwaha S, Musa S, Ntoukam G (2003) Cowpea supply and demand in west and Central Africa. Field Crop Res 82:215–231

Leng PF, Lübberstedt T, Xu ML (2017) Genomics-assisted breeding–a revolutionary strategy for crop improvement. J Integr Agri 16:2674–2685

Lonardi S, Muñoz-Amatriaín M, Liang Q, Shu S, Wanamaker SI, Lo S, Tanskanen J, Schulman AH, Zhu T, Luo MC, Alhakami H, Ounit R, Hasan AM, Verdier J, Roberts PA, Santos JRP, Ndeve A, Doležel J, Vrána J, Hokin SA, Farmer AD, Cannon SB, Close TJ (2019) The genome of cowpea (Vigna unguiculata [L.] Walp.). Plant J 98(5):767–782

Muchero W, Ehlers JD, Close TJ, Roberts PA (2011) Genic SNP markers and legume synteny reveal candidate genes underlying QTL for Macrophomina phaseolina resistance and maturity in cowpea [Vigna unguiculata (L) Walp]. BMC Genomics 12:8

Munoz-Amatriain M, Mirebahim H, Xu P, Wanamaker SI, Luo MC, Alhakami H, Close TJ (2017) Genome resources for climate-resilient cowpea, an essential crop for food security. Plant J89: 1042–1054

Muraya MM, Chu J, Zhao Y, Junker A, Klukas C, Reif JC et al (2017) Genetic variation of growth dynamics in maize (Zea mays L.) revealed through automated non-invasive phenotyping. Plant J 89:366–380

Neya BJ, Zida PE, Sereme D, Lund OS, Traore O (2015) Evaluation of yield losses caused by cowpea aphid-borne mosaic virus (CABMV) in 21 cowpea (Vigna unguiculata (L.) Walp.) varieties in Burkina Faso. Pak J Bio Sci 18(7):304–313

Nielson SS, Brandt WE, Singh BB (2010) Genetic variability for nutritional composition and cooking time of improved cowpea lines. Crop Sci 33(2):469–472

Ogunsola KE, Fatokun CA, Boukar O, Ilori CO, Kumar PL (2010) Characterizing genetics of resistance to multiple virus infections in cowpea (Vigna unguiculata L. Walp). In: Abstracts: fifth world cowpea conference: improving livelihoods in the cowpea value chain through advancement of science, September 27–October 1, Saly, Senegal. IITA, Ibadan, pp 26–27

Okechukwu RE, Florini DA (2000) Yield depression in cowpea cultivars infected with Xanthomonas campestris pv. Vignicola in Sudan savanna of Nigeria. Trop Agri Res Ext 3: 98–101

Omoigui LO, Arrey MO, Danmaigona CC, Ekeruo G, Timko MP (2019) Inheritance of resistance to Cercospora leaf spot disease of cowpea [Vigna unguiculata (L.) Walp]. Euphytica 21(5):1–2

Orawu M, Melis R, Laing M, Derera J (2013) Genetic inheritance of resistance to cowpea aphid-borne mosaic virus in cowpea. Euphytica 189:191–201

Pant KC, Chandel KPS, Joshi BS (2007) Analysis of diversity in Indian cowpea genetic resources. Sabro J 14:103–111
Patel PN (1981) Pathogen variability and host resistance to three races of the bacterial pustule pathogen in cowpea. Trop Agric 58:275–280
Patel PN (1982) Genetics of cowpea reactions to two mosaic virus from Tanzania. Phytopathology 72:460–466
Pottorff M, Wanamaker S, Ma YQ, Ehlers JD, Roberts PA, Close TJ (2012) Genetic and physical mapping of candidate genes for resistance to fusarium oxysporum f. sp. tracheiphilum race 3 in cowpea [Vignaunguiculata (L.) Walp]. PLoS One 7:e41600
Prakash CS, Shivashankar G (1984) Inheritance of resistance to bacterial bligh(Xanthomonas campestris pv. Vignicola) in cowpea. Genet Agrar 38:1–10
Prasanna KPR (1985) Seed health testing of cowpea with special reference to anthracnose caused by C. lindemuthianum. Seed Sci Technol 13:821–827
Raju SG, Anilkumar TB (1990) Evaluation of cowpea genotypes for partial resistance to powdery mildew. Euphytica 50:191–195
Rangaiah S (1997) Inheritance of resistance to Uromyces phaseolus in Vigna unguiculata (L.) Walp. Crop Improv 24:251–252
Roitsch T, Cabrera-Bosquet L, Fournier A, Ghamkhar K, Jiménez-Berni J, Pinto F et al (2019) New sensors and data-driven approaches—a path to next generation phenomics. Plant Sci 282:2–10
Sainsbury F, Cañizares MC, Lomonossoff GP (2010) Cowpea mosaic virus: the plant virus-based biotechnology workhorse. Annu Rev Phytopathol 48:437–455
Singh BB (2002) Recent genetic studies in cowpea. In: Fatokun CA, Tarawali SA, Singh BB, Kormawa PM, Tamo M (eds) Challenges and opportunities for enhancing sustainable cowpea production, proceedings of the world cowpea research conference III. International Institute of Tropical Agriculture (IITA), Ibadan, pp 3–13
Singh BB (2005) Cowpea [*Vigna unguiculata* (L.) Walp]. In: Singh RJ, Jauhar PP (eds) Genetic resources, chromosome engineering and crop improvement. CRC Press, Boca Raton, pp 117–162
Singh SR, Allen DJ (1980) Pests, diseases, resistance and protection in cowpea. In: Summerfield RJ, Bunting AH (eds) Advances in legume science. HMSO, London, pp 419–443
Singh S, Lodha S (1986) Varietal resistance of cowpea to Macrophomina phaseolina (Tassi.) Goid. Causing dry root-rot and its control. IJAS 56(8):552–555
Singh RS, Sinha RP (1955) Studies on the wilt disease of cowpea in Uttar Pradesh-occurrence and symptoms of the disease and identity of the causal organism. J Indian Bot Soc 34(4):375–381
Singh SR, Singh BB, Jackai LEN et al (1983) Cowpea research at IITA, Ibadan, Nigeria. Inf Ser 14: 1–20
Singh SK, Nene YL, Reddy MV (1990) Influence of cropping system on M. phaseolina population in soil. Plant Dis 74:814
Singh BB, Ehlers JD, Sharma B et al (2002) Recent progress in cowpea breeding. In: Fatokun CA, Tarawali SA, Singh BB et al (eds) Challenges and opportunities for enhancing sustainable cowpea production. Proceedings of the world cowpea research conference III. International Institute of Tropical Agriculture (IITA), Ibadan, pp 22–40
Singh S, Kundu SS, Negi AS, Singh PN (2006) Cowpea (*Vigna unguiculata*) legume grains as protein source in the ration of growing sheep. Small Ruminant Res 64(3):247–254
Sohi HS, Rawal RD (1983) Field resistance of cowpea varieties to anthracnose (Colletotrichium lindsmuthianum) and stem blight (Macrophomina phaseolina) diseases. IJMPP 13:58–60
Stoffella PJ, Bullock RC, Sonoda RM (1990) Influence of pesticide schedules on growth and yield of cowpea. Proc Interamerican Soc Trop Hort 34:83–87
Taiwo MA, Shoyinka SA (1998) Viruses infecting cowpeas in Africa with special emphasis on the potyviruses. Virus Dis Plants Africa 93:115
Tanimu MU, Mohammed IU, Muhammad A, Kwaifa NM (2018) Response of cowpea varieties to basal stem rot (Sclerotium rolfsii) disease in southern Guinea savanna, Nigeria. Equijost 5(1): 1–8

Timko MP, Ehlers JD, Roberts PA (2007) Cowpea. Genome mapping and molecular breeding in plants. In: Kole C (ed) Pulses, sugar and tuber crops, vol 3. Springer, Berlin, pp 49–67

Togola A, Boukar O, Chamarthi S, Belko N, Tamò M, Oigiangbe N, Ojo J, Ibikunle M, Fatokun C (2019) Evaluation of cowpea mini core accessions for resistance to flower bud thrips Megalurothrips sjostedti Trybom (Thysanoptera: Thripidae). J Appl Entomol 143(6):683–692

Torres AM (2010) Application of molecular markers for breeding disease resistant varieties in crop plants. In: Molecular techniques in crop improvement. Springer, Dordrecht, pp 185–205

Tosti N, Negri V (2008) Efficiency of three PCR-based markers in assessing genetic variation among cowpea (*Vigna unguiculata)* landraces. Genome 45(2):656–660

Tripathi K, Gore PG, Ahlawat SP, Tyagi V, Semwal DP, Gautam NK, Rana JC, Kumar A (2019) Cowpea genetic resources and its utilization: Indian perspective-a review. Legume Res-Int J 42(4):437–446

Uma MS, Salimath PM (2004) Evaluation of segregating populations of cowpea for slow rusting. J Arid Legume 1:181–182

Uma MS, Hegde N, Hittalmani S (2016) Identification of SSR marker associated with rust resistance in cowpea (Vigna unguiculata L.) using bulk segregant analysis. Legume Res 39(1):39–42

Unamba CI, Nag A, Sharma RK (2015) Next generation sequencing technologies: the doorway to the unexplored genomics of non-model plants. Front Plant Sci 6:1074

Varshney RK, Bohra A, Yu J, Graner A, Zhang Q, Sorrells ME (2021) Designing future crops: genomics-assisted breeding comes of age. Trends Plant Sci 26(6):631–649

Voss-Fels K, Snowdon RJ (2016) Understanding and utilizing crop genome diversity via high-resolution genotyping. Plant Biotechnol J 14:1086–1094

Wu X, Wang B, Wu X, Lu Z, Li G, Xu P (2018) SNP marker-based genetic mapping of rust resistance gene in the vegetable cowpea landrace ZN016. Legume Res 41(2):222–225

7 Biotic Stress Resistance in *Vigna mungo* and *Vigna radiata*: A Molecular Perspective

Sandeep Kaur Dhaliwal, Abhishek Pandey, Chayanika Lahkar, Sheetal, and Satinder Kaur

Abstract

Mungbean and blackgram are the two important legumes widely cultivated in the Asian continent for grain, vegetable, green manure, and fodder. Both legumes have a high nutritional value, having essential amino acids, minerals, and vitamins. In spite of good nutritive value, the crops are being neglected and face a major challenge of low productivity due to multiple factors including biotic and abiotic stresses, lack of interest by farmers and consumers, and unfavorable price policies governing the crop. The biotic stresses caused by various organisms including viruses, fungi, bacteria, and insect pests hinder crop productivity potential. An understanding of complex pathogen-host interactions leading to compatible or incompatible response and resistance mechanism is the key to combating these stresses. Identification of novel and diverse sources of resistance is always a preset for improving the durability against diseases and a lot of efforts have been done in these legumes also. But the work pertaining to molecular breeding for biotic stress resistance utilizing advanced omics tools is not at par with mainstream crops. The traditional molecular markers were utilized to understand resistance, but more efforts with cutting-edge technologies are required to accelerate legume breeding so that the appropriate importance of these nutritious legumes can be achieved among major food crops. The chapter would highlight the major biotic concern faced by mungbean and blackgram and efforts made towards incorporating biotic resistance.

S. K. Dhaliwal · A. Pandey · C. Lahkar · Sheetal
Department of Plant Breeding and Genetics, Punjab Agricultural University, Ludhiana, India

S. Kaur (✉)
School of Agricultural Biotechnology, Punjab Agricultural University, Ludhiana, India
e-mail: satinder.biotech@pau.edu

U. C. Jha et al. (eds.), *Diseases in Legume Crops*,
https://doi.org/10.1007/978-981-99-3358-7_7

7.1 Introduction

The *Vigna* species belonging to the subgenus *Ceratotropis* is grown across different parts of the Asian continent. Among the 200 species of *Vigna*, only a few are cultivated around the globe for food purposes. The main species of the genus include *V. radiata* (L.) Wilczek (mungbean), *V. mungo* (L.) Hepper (blackgram), *V. aconitifolia* (Jacq.) (moth bean), *V. angularis* (W) Ohwiand Ohashi (adzuki bean), and *V. umbellata* (Thumb.) Ohwi and Ohashi (rice bean). Mungbean and blackgram are the two main legumes among these which are widely cultivated in the Asian continent as pulse crops (Singh et al. 2022; Dhaliwal et al. 2022). These two legumes are diploid ($2n = 2x = 22$), self-pollinating species which have moderate-sized genomes equal to 0.60 pg/1C (579 Mbp) and 0.59 pg/1C (574 Mbp) (Arumuganathan and Earle 1991), respectively, which are similar with other *Vigna* species also. The crosses between mungbean and blackgram produce successful hybrids in most cases in comparison to their crosses with other *Vigna* species (Singh et al. 2013a). These crops have multipurpose utilization, where seeds, young pods, and sprouts can be used as sources of protein, vitamins, and minerals, and the rest of the plant parts are utilized as green manure and fodder (Tomooka et al. 2002). In addition, 100 g of pulse can provide 5.2–9.8% protein, 1.5–3.8% carbohydrates, 3.8–32.2% fat, and 3.4–14.0% energy of RDA for adolescents and 6.9–12.0% protein, 1.7–4.1% carbohydrates, 5.7–46.8% fat, and 4.2–16.8% energy of RDA for adults. Crops with such nutritional value can aid in curbing malnutrition especially in poor and vegetarian populations across the globe. However, multiple challenges, including biotic (viral, bacterial, fungal, and insects) and abiotic (drought, heat, waterlogging) stresses (Douglas et al. 2020; Gwag et al. 2006) are limiting their productivity potential and interest among farmers. In spite of being regional crops, India is currently importing 3.7 lakh million tonnes of mungbean and blackgram per annum (https://pib.gov.in/Pressreleaseshare.aspx?PRID=1705092). Thus, it is indispensable to employ breeding and biotechnological intervention for coping with the major challenges faced by mungbean and blackgram to unleash their productivity potential. The current chapter highlights the major biotic stresses faced by these two crops, their resistance breeding, and molecular advancements against biotic stresses and digs deeper into the prospects of the stated studies for its practical utilization.

7.2 Origin and Domestication of Mungbean and Blackgram

Cultivated blackgram (*V. mungo*) also known as urdbean, mash, or urd originated and is domesticated in Asia particularly in India as evident from the lineage from its wild progenitor *Vigna mungo* var. *silvestris* (Chandel et al. 1984). Archeological evidence showed the crop domestication occurred about 4500 years ago (Fuller and Harvey 2006). In India, initial records of blackgram were found in Northern Peninsula and Gujarat, where blackgram populations thrive in wild forms (Fuller and Harvey 2006). Blackgram was also introduced to America and Africa by Indian travelers (Vertovec 1994; Jain and Mehra 1980) and cultivated as pulse and green manure famous by the name "woolly pyrol" in America.

One of the Vedas of ancient Indian literature, Yajurveda reported the origin of green gram as early as 7000 BC in India and Central Asia. The archaeological evidence and geographical distribution of wild and weedy types of plants of these species indicated that domestication and cultivation of green gram began in the northwest and far south of India as early as 4000–6000 years ago, and it is likely that green gram is an indigenous crop of India and Central Asia that has been cultivated there since prehistoric times (Vavilov 1926; De Candolle 1886; Singh et al. 1975; Vishnu-Mittre 1974; Fuller and Harvey 2006; Fuller 2007). Green gram is currently distributed throughout South and East Asia, Africa, and America following multiple rounds of domestication and selection (Pratap et al. 2020). Its presumed ancestor, *V. radiata* var. *sublobata*, is native to Northern and Eastern Australia's subtropical and tropical areas and grows as a weed in Southern Africa, Southern and Eastern Asia, and the Pacific islands of Australia and Indonesia.

7.3 Biotic Stresses of Mungbean and Blackgram: Viral Diseases

7.3.1 Yellow Mosaic Disease

Yellow mosaic disease (YMD) in leguminous crops is mainly caused by legumoviruses. It belongs to the genus *Begomoviruses* which is the largest genus of the family *Geminiviridae*, the second largest family of plant viruses causing devastating epidemics. YMDis caused by the four major species of *Begomoviruses*, namely, *mungbean yellow mosaic India virus* (MYMIV), *mungbean yellow mosaic virus* (MYMV), *horsegram yellow mosaic virus* (HYMV), and *dolichos yellow mosaic virus* (DYMV)(Qazi et al. 2007). These 4 viruses are collectively named as legume yellow mosaic viruses (LYMVs), although there are 11 species under the LYMV group, all transmitted by an insect vector whitefly *Bemisia tabaci* (Gennadius) (Hemiptera: Aleyrodidae).

MYMIV is reported to be more established in Northern, Eastern, and Central India (Naimuddin and Akram 2010), and MYMV in more prevalent in the peninsular region of Southern India (Karthikeyan et al. 2004; Malathi and John 2008). In India, the disease has been reported in the states of Punjab, Delhi, Haryana, Uttar Pradesh, Uttrakhand, Himachal Pradesh, Rajasthan, Bihar, Madhya Pradesh, Orissa, Gujarat, Chhattisgarh, Kerala, Tamil Nadu, Andhra Pradesh, and Karnataka (Karthikeyan et al. 2014) posing a major threat for cultivation of mungbean and blackgram.

7.3.2 Symptoms and Crop Loss in YMD

The symptomatology of YMD was first studied by Nariani (1960). The word mosaic in name is due to the mosaic of yellow and green patches on the leaves of infected plants. The disease appears as small yellow spots on the veinlet of young leaves which ultimately cover the entire lamina turning the complete leaf yellow. Two kinds of prominent symptoms, namely, necrotic mottle and yellow mottle, were either caused by different strains of the same virus or due to the altered reactions of various genotypes to the same viral isolate (Nair and Nene 1973). Disease intensity of 25% and above hinders pod formation and reduces yield (Gurha et al. 1982) and crop losses of yield ranged from 10–100% depending upon the stage of infection.

7.3.3 YMD Resistance Mapping

The work on molecular mapping for YMD resistance started with the advent of RFLP markers (Fatokun et al. 1993) followed by SSRs (Gupta et al. 2014) and recent advancements in next generation sequencing identified in SNP markers for the disease (Kang et al. 2014; Jegadeesan et al. 2021; Pootakham et al. 2001). A detailed compilation of different studies involved in mapping different YMD resistance genes is given in Table 7.1.

Until the late 1980s and early 1990s, the mungbean breeders used mutation and hybridization-based strategies to understand and incorporate YMD resistance. Some of the tolerant (MGG 207, RMG 975, GBM 1) moderately resistant (TJM3) and fairly resistant (PM8, PAU 911, GAM 5, KM 2338) varieties have been released/ identified using the traditional approaches (Shanthala et al. 2020). ML1628, one of these kinds of variety was developed by Punjab Agricultural University, Ludhiana having resistance against multiple strains of YMD(Nair et al. 2017). Mutation breeding also played a significant role in developing YMD resistance varieties like NM-92, popular in many South Asian countries and developed jointly by the World Vegetable Center and NIAB (Pakistan).

The genetic nature of the YMD resistance was not simple and has been reported to be governed by a single recessive gene to a single dominant gene to two recessive genes/or complementary recessive genes (Reddy 2009a; Khattak et al. 2000; Sandhu et al. 1985; Pal and Jana 1991; Ammavasai et al. 2004; Dhole and Reddy 2012; Sudha et al. 2013; Saleem et al. 1998; Jain et al. 2013; Mahalingam et al. 2018).

Several attempts have been made by different workers to utilize SSR markers in towards mungbean and blackgram improvement (Kaur et al. 2017). Basak et al. (2005) identified a resistance gene analog (RGA) marker "VMYR1" linked to YMD resistance. Protein prediction by sequencing PCR amplicon showed homology with NB-ARC domains. Selvi et al. (2006) identified 7 RAPD markers of 149 linked with MYMV resistance when tested on resistant and susceptible bulks derived from F_2 populations of ML 267 × CO4. Maiti et al. (2011) identified two RGA-based markers, *CYR1* and *YR4*, in urdbean which also showed significant resistance for mungbean YMD (Kitsanachandee et al. 2013), out of these two markers *CRY1 was* completely linked with YMD resistance. Holeyachi and Savithramma (2013) identified 10 polymorphic RAPD markers of which UBC 499 amplifying 700 bp band was linked to YMD resistance of mungbean line BL 849. Kalaria et al. (2014) did diversity analysis in a set of mungbean germplasm with 200 RAPD and 17 ISSR markers that differentiated the resistant and susceptible genotypes into two separate

Table 7.1 Compilation of different molecular mapping studies involved in YMD resistance mapping in mungbean and blackgram

S. No.	Crop	Trait	Cross	Population	Markers used	Linked markers	Reference
1	Mungbean	YMD	NM10-12-1 × KPS2	F_8 RILs (122)	64 SSR	CEDG100—cp02662, DMB-SSR008 VR113, CEDG166-CEDG304, CEDG100-cp02662, CEDG121—CEDG191	Kitsanachandee et al. (2013)
2	Mungbean	YMD	NM92 × TC1966	200 RILs	356 AFLP and 113 RAPDs	v02a7, mg3pat423, m4pcc585, 9DMB158	Chen et al. (2013)
3	Mungbean	YMD	BM1 × BM6	F2:3 and BC1F1:2	1165 SSR	CEDG275–CEDG006 CEDG041–VES503	Alam et al. (2014)
4	Mungbean	YMD	Sonali × *V. radiata* var. *sublobata*	100 RILs and 93 germplasm lines	213 SSR, STS, SCAR, and RGA	VrD, STSbr1, CEDG044, CEDG228	Singh et al. (2018)
5	Mungbean	YMD	VRMTNAU	108 RILs	538 SNPs	VigSNP_04_32-VigSNP_04_36	Mathivathana et al. (2019)
6	Mungbean	YMD	–	127 genotypes	256 microsatellites	CEDG293 DMB-SSR008 DMB-SSR059	Singh et al. (2020a, b, c)
7	Mungbean	YMD	–	80 genotypes	89 SSR	CP1038	Rohilla et al. (2021)
8	Blackgram	YMD	T9 × mutant T9	F_2 and F_3	40 resistance gene analog (RGA)	RGA 1F-CG/RGA 1R	Basak et al. (2005)
9	Blackgram	YMD	TU 94–2 × *V. mungo* var. *silvestris* i	58 RILs	100 ISSR primers from related species	ISSR8111357	Souframanien and Gopalakrishna (2006)
10	Blackgram	YMD	DPU 88–31	F_2	361 SSR from cowpea	CEDG 180	Gupta et al. (2013)

(continued)

Table 7.1 (continued)

S. No.	Crop	Trait	Cross	Population	Markers used	Linked markers	Reference
11	Blackgram	YMD	759 × T9	F_2	50 SSR	–	Rambabu et al. (2018)
12	Blackgram	MYMIV	KUG 253 ×Mash 114	RILs	SNPs	VM602, VM605, VM610	Dhaliwal et al. (2022)

clusters. Binyamin et al. (2015) identified YMD resistance-based SCAR markers which amplified PCR product in resistant genotypes only.

Kitsanachandee and colleagues (Kitsanachandee et al. 2013) developed the RIL population by crossing YMD-resistant NM10-12-1 with YMD-susceptible KPS2 and identified three QTLs (*qYMIV1*, *qYMIV2*, and *qYMIV3*) for Indian conditions and two QTLs (*qYMIV4* and *qYMIV5*) for Pakistan environment governing YMD resistance. *qYMIV4 and qYMIV1*QTLs were linked between *CEDG100* and *cp02662* SSR markers;, in addition Chen and co-workers (Chen et al. 2013) mapped four QTLs in RILs derived from NM92 × TC1966 of which QTL*MYMIVr9_25* on LG9 was linked with the SSR marker *9DMB158*. Alam et al. (2014) mapped two major QTLs *qMYMIV2.1* and *qMYMIV7.1* linked with *CEDG275* and *CEDG041*SSR markers, respectively, in mapping population derived from BM1 × BM6. In India, Singh et al. (2018) reported two SSR markers named *CEDG228* and *CEDG044* in the RIL population of cross Sonali × *V. radiata* var. *sublobata* linked to YMD resistance. Interspecific hybrids between mungbean and blackgram have been characterized with SSR markers for mapping YMD resistance (Lekhi et al. 2018a, b; Kaur et al. 2017).

Mungbean is ahead of blackgram in utilizing advanced analytical methods and genomic techniques for YMD mapping. LD-based mapping for YMD resistance revealed 2 major MTAs *J01263* and *CEDG220* by Rohilla et al. (2021) from screening of 80 diverse mungbean genotypes in different agroclimatic regions of India. Similarly, Singh et al. (Singh et al. 2020a, b, c) reported three major MTAs *CEDG293*, *DMB-SSR008*, and *DMB-SSR059* associated with MYMIV resistance in mungbean. DMB-SSR008 has also been previously reported by Kitsanachandee et al. (2013). An attempt was made to locate major QTLs regarding MYMIV resistance in mungbean through interspecific crosses with rice bean line VRM (Gg)1 × mungbean TNAU RED (Mathivathana et al. 2019). Through GBS analysis of the RIL population, 5 QTLs have been mapped of which 1 major QTL *qMYMV4_1*in a 1.4 Mb region on chromosome 4 was reported to have 18 candidate genes.

In blackgram less work has been done as compared to mungbean due to its less area and popularity among blackgram growing countries. Initially several attempts have been made to tag the YMD resistance using markers from the crops themselves or with markers from related species. YMD resistance in the mutant of susceptible blackgram cultivar T9 was mapped using resistance gene analog (RGA) primers designed from the functional NBS domains of the resistant gene (Basak et al. 2005) and was used in marker-assisted transfer (Maiti et al. 2011) Souframanien and Gopalakrishna (2006) screened 100 ISSR markers from related species on the RIL population derived from the cross of TU 94-2 × *V. mungo* var. *silvestris* and mapped YMD resistance gene at 6.8 cM of marker *ISSR8111357*. Maiti and co-workers (Maiti et al. 2012) isolated and characterized the CYR1 gene using primer walking. CYR1 protein consisted of 1176 amino acids with coiled-coil (CC) structure at the N-terminal and nucleotide binding site (NBS) at C-terminal and leucine-rich repeats belonging to non-TIR NBS LRR subfamily of resistance genes. Gupta et al. (2013) mapped YMD resistance in the DPU 88-31 cultivar using 361 SSR markers from sister crops of cowpea and mungbean and identified YMD resistance gene linked with CEDG 180 SSR markers at 12.9 cM. A total of 50 SSR markers were amplified in the F_2 population from the cross of LBG 759 × T9, and 1 SSR marker on linkage group 8 was found to be linked with YMD resistance (Rambabu et al. 2018).

The first draft of blackgram genome sequencing was published recently (Pootakham et al. 2021). Subsequently, in the first-ever report of mapping with SNP markers, a major genomic region *qMYMIV6.1.1*, conferring MYMIV resistance was mapped using a sequencing-based *QTL*-seq technique (Dhaliwal et al. 2022). The region was mapped on chromosome 6 of blackgram in the vicinity of three SNPs (*VM602*, *VM605*, *VM610*) harboring interspecific introgression from ricebean.

7.3.4 Transcriptomic Studies for YMD

Advancement in next generation sequencing technology has cut down costs of sequencing DNA to a point so that NGS sequencing is now affordable and the researchers are now switching to functional genomics, including transcriptomic and proteomic approaches to uncover the molecular mechanisms of YMD stress tolerance. Transcriptomics is the quantitative study of all the sets of transcripts inside a cell, and RNA-seq is transcriptome analysis by next generation sequencing (NGS) (Shendure and Ji 2008). The biotic stresses ignite a plethora of genes in the genome of the host which start a cascade of reactions to fight the infection (Fig. 7.1), and a better understanding of the same has been provided by RNA-seq technology. A

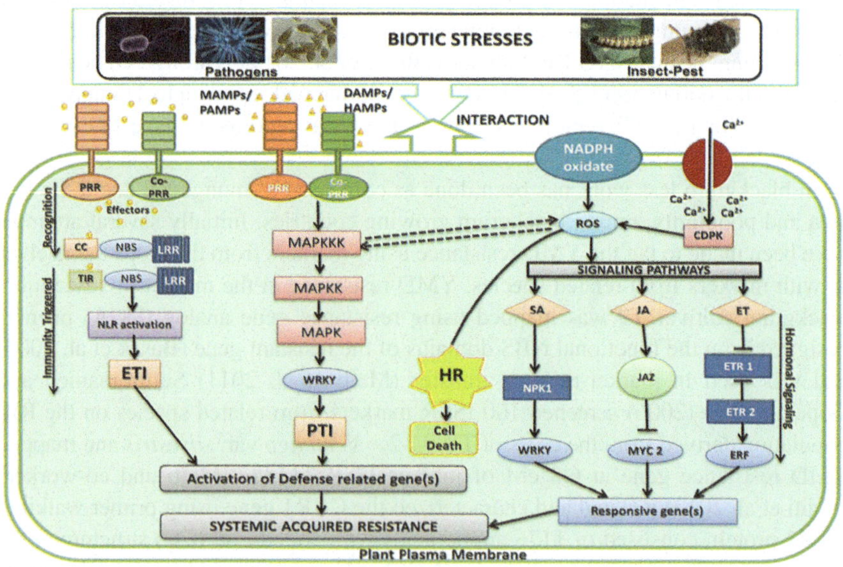

Fig. 7.1 Diagrammatic representation of different pathways describing an overview of the molecular basis of the biotic stress response

compilation of transcriptomics studies done towards YMD resistance in mungbean and blackgram is given in Table 7.2.

In mungbean few studies utilize a RNA-seq-based technique to understand the molecular mechanism of YMD resistance (Dasgupta et al. 2021; Sudha et al. 2022). Dasgupta and colleagues generated transcriptome of resistant (PMR-1) and susceptible (Pusa Vishal) mungbean cultivar and activation of PAMPs along with various signaling cascades and the expression of genes like PR proteins, virus resistance, and R-gene proteins leading to resistance response were identified. Anti-oxidative enzymes peroxidase, (S)-2-hydroxy acid oxidase, and classes of lipoxygenase, along with other enzymes such as O-methyltransferase, D-3-phosphoglycerate dehydrogenase, arogenate dehydratase, cell wall-associated enzymes (chitinase and lignin biosynthesis), were identified to be upregulated with YMD infection.

Sudha and co-workers (Sudha et al. 2022) compared the comparative transcriptome of YMD-resistant VGGRU-1 and YMD-susceptible VRM (Gg) 1 mungbean. Hundreds of differentially expressed genes (DEGs) were reported for both resistant (896) and susceptible (506) genotypes. Through functional annotation, these DEGs were categorized as defense and pathogenesis-related genes.

In blackgram, the contrasting genotypes have been used as appropriate raw material for the discovery of resistance mechanism and candidate genes involved in the pathway for crop improvement (Jasrotia et al. 2017). A transcriptome database, freely accessible at http://webtom.cabgrid.res.in/vmtdb/, was developed by Jasrotia et al. (2017) using existing data of two contrasting genotypes, viz., VM84 (resistant) and T9 (susceptible). The initial studies on transcriptomics for YMD resistance were explored by Kundu et al. (2013)for molecular reactions in the compatible and incompatible interactions between the YMD pathosystem and blackgram. Of the network of different cellular events involved in generating defense response, ascorbate peroxidase, serine/glycine hydroxymethyl transferase, and rubisco activase were the three main hubs with high connectivity. Expressed sequence tag (EST) libraries were prepared by Kundu et al. (2015) from resistant cultivar T9 and susceptible genotype VMR84 of blackgram at different time intervals after inoculation with viruliferous whiteflies. The genes associated with response to viral perturbation included SGT1, HSP90 (R protein complex), PR1 protein, ROS homeostasis (SOD, APOX, TRX, and MET), GST, MAPK, CAM, TS, UBL, PAL, WRKY, RuAc, MADS, ARF, OEC, and CsP. Chakraborty and Bask (Chakraborty and Basak 2018) used suppression subtractive hybridization and identified 145 and 109 differentially regulated transcripts in resistant and susceptible genotypes of blackgram, respectively. The incompatible reaction showed the induction of transcripts present in the jasmonic acid signaling pathway along with the development of a physical barrier during virus invasion and continuance of the reactive oxygen species (ROS) homeostasis. Mitogen-activated protein kinases (MAPK) are an important branch of immunity that transduce signals of pathogen invasion to downstream molecules leading to diverse immune responses. Patel et al. (2017) identified MAPK homolog in the defense signaling pathway in YMD incompatible interaction.

Table 7.2 Compilation of different transcriptomic studies involved in YMD resistance in mungbean and blackgram

S. No.	Crop	Trait	Genotypes	Technique	Sequencing platform	Major candidates	Reference
1	Mungbean	YMD	PMR-1 and (Pusa Vishal)	RNA-seq	Illumina HiSeq 2500	Alpha-dioxygenase, lipase, WRKY, NAC and MYB, RLK, JAZ, LOX and P450	Dasgupta et al. (2021)
2	Mungbean	Fusarium wilt	Zheng8-4 and Zheng8–20	RNA-Seq	Illumina-HiSeq-2500	MADS, MYB, WRKY, AP2, NAC, HD-ZIP gene families, etc.	Chang et al. (2021)
3	Mungbean	YMD	VGGRU-1 and VRM (Gg) 1	Illumina	Illumina-HiSeq-2000	STKs, TIR/CC-NBS-LRR, NB-ARC, RLK, WRKY, MYB, HSP, terpene synthase, galactinol synthase, galactosidase, P450 superfamily	Sudha et al. (2022)
4	Blackgram	YMD	VM84 and T9	EST-sequencing	Sanger sequencing	SGT1, HSP90 (R protein complex), PR1 protein, ROS homeostasis (SOD, APOX, TRX, and MET), GST, MAPK, CAM, TS, UBL, PAL, WRKY, RuAc, MADS, ARF, OEC, and CsP.	Kundu et al. (2015)
5	Blackgram	YMD	VM84	Isolation and cloning of *VmMAPK1*	ABI prism 31,000	VmMAPK1	Patel et al. (2017)
6	Blackgram	YMD	VM1 AND T9	EST SEQUENCING	Sanger sequencing	RUBISCO, PEPC, ribose-5-phosphate isomerase, and aldehyde dehydrogenase	Chakraborty and Basak (2018)
7	Blackgram	YMD	VM84 and T9	RNA-seq	Illumina HiSeq 2000	NB-LRR, WRKY33, ankyrin, argonaute and NAC TF	Kundu et al. (2019)

Hopefully, these studies will help to uncover the molecular mechanism of YMD resistance in blackgram, which can inform future genetic breeding and facilitate the creation of YMD-resistant cultivars.

7.4 Biotic Stresses of Mungbean and Blackgram: Fungal Diseases

The major fungal diseases hampering mungbean and blackgram include ***Cercospora* leaf spot** (CLS) [*Cercospora canescens* (Ellis & G. Martin)], **powdery mildew** [(*Podosphaera fusca* Fr.) U. Braun & Shishkoff, *Erysiphe polygoni* (Vaňha) Weltzien], **anthracnose** [*Colletotrichum acutatum* (J.H. Simmonds), *C. truncatum* (Schwein.) Andrus & Moore, *C. gloeosporioides* (Penz.) Penz. & Sacc)], and **dry root rot** [*Macrophomina phaseolina* (Tassi) Goid]. The other diseases with comparatively less area and severity include **web blight** (*Rhizoctonia solani* Kuhn), **fusarium wilt** [*Fusarium solani* (Mart.) Sacc], and ***Alternaria* leaf spot** [*Alternaria alternata* (Fr.) Keissl] (Ryley and Tatnell 2011; Pandey et al. 2018). The detailed compilation of molecular studies about the mapping of powdery mildew and *Cercospora* leaf spot diseases of mungbean and blackgram are given in Table 7.3.

7.4.1 *Cercospora* Leaf Spot (CLS)

CLS is a major fungal foliar disease affecting mungbean and blackgram production (Duangsong et al. 2016). It is caused by the biotrophic fungi, *Cercospora canescens* (Eliis& Martin), a haploid, hemibiotrophic fungus belonging to the most destructive group of plant pathogens of *Ascomycota*. The disease was first reported in New Delhi, India (Pandey et al. 2009) and is found in all the mungbean and blackgram growing areas of India (Prajapati et al. 2007). Yield losses of 50–70% (Chand et al. 2012) and 23–93% were reported due to CLS in mungbean in India (Batzer et al. 2022).

7.4.1.1 Symptoms and Disease Cycle of CLS

Initially the symptoms appear as water-soaked spots and occur on the upper side of old leaves and progressively spread to the whole plant (Hartman et al. 1993). Disease spread is most rapid at the pod-filling stage, and severe infections cause premature defoliation, delayed maturation, poor pod formation, and small immature seed formation (Shahbaz et al. 2014). The fungus survives on the diseased plant debris as dormant mycelium and produces conidia in plant debris which act as primary inoculum in disease incidence. The favorable conditions for the disease are warm moist weather (20–26 °C) and high rains with 100% relative humidity (Batzer et al. 2022).

Table 7.3 Compilation of different studies involved in the molecular mapping of powdery mildew and *Cercospora* leaf spot resistance in mungbean and blackgram

S. No.	Crop	Trait	Cross	Population	Markers used	Linked markers	Reference
1	Mungbean	*Cercospora* leaf spot	V4718 and KPS1	F_2 and BC_1F_1	SSRs	CEDG117 and VR393	Chankaew et al. (2011)
2	Mungbean	*Cercospora* leaf spot	V4718 and KPS1	F_2 and BC_8F_2	SSRs and Indels	Vr6gCLS085 and VrTAF5_indel	Yundaeng et al. (2021)
3	Mungbean	*Cercospora* leaf spot	CN72 and V4718	$F_{2:9}$ and $F_{2:10}$	ISSR-RGA, ISSR and SSR markers	VR393 and I16274	Tantasawat et al. (2020)
4	Mungbean	*Cercospora* leaf spot	CN72 and V4718	$F_{2:9}$ and $F_{2:10}$ RILs	SSR markers and InDel marker	I16274 and VrTAF5	Papan et al. (2021)
5	Mungbean	Powdery mildew	Berkenand ATF 3640	$F_{2:7}$ and $F_{2:8}$ RILs	RFLP	VrCS65 and VrCS296	Humphry et al. (2003)
6	Mungbean	Powdery mildew	Kamphaeng Saen 1 and VC6468-11-1A	$F_{2:7}$ RILs	SSRs	CEDG282, CEDG191 MB-SSR238, CEDG166	Kasettranan et al. (2010)
7	Mungbean	Powdery mildew	CN72 and V4718	$F_{2:7}$ and $F_{2:8}$ RILs	ISSRs and ISSR-RGAs	I85420 and I42PL229)	Poolsawat et al. (2017)
8	Blackgram	Powdery mildew	–	29 genotypes	MB-SSR238, VrCsSTS1, CEDG191, VrCsSSR1, CEDG166	VrCsSTS1	Samal et al. (2022)
9	Mungbean	Powdery mildew	CN72 and V4758	$F_{2:10}$ RILs	ISSR-RGAs	I41tP379	Tantasawat et al. (2022)

7.4.1.2 Genetics of CLS Resistance

The resistance to CLS has been reported to be controlled by a single dominant gene (Chankaew et al. 2011; Tantasawat et al. 2020) to a single recessive gene (Mishra et al. 1988), to quantitative genetic control (Leabwon and Oupadissakoon 1984; Chankaew et al. 2011). Singh et al. (2017) identified a single dominant gene against CLS from interspecific crosses of susceptible mungbean × resistant blackgram and susceptible mungbean × resistant mungbean.

Chankaew et al. (2011) mapped a CLS resistance QTL, *qCLS* (PVE 65.5–80.5%), with SSR markers, *CEDG117* and *VR393*, in mungbean line V4718. Yundaeng et al. (2021) were able to finely map the *qCLS* at 5.9 cM between markers *Vr6gCLS085* and *VrTAF5_indel* and identified a candidate gene in ~13 Kb genomic region on chromosome 6. Tantasawat et al. (2020) mapped CLS resistance QTL, *qCLSC72V18-1*, in mungbean, explaining the PVE of 79.8% between the SSR marker *VR393* and ISSR marker, *I16274*, at genetic distances of 4 cM and 3 cM, respectively, from the cross CN72 × V4718. Papan et al. (2021) further fine-mapped the QTL *qCLSC72V18-1* using three SSR markers and one InDel marker together with six previously identified markers.

7.4.2 Powdery Mildew (PM)

PM caused by the ascomycete fungi, *Erysiphe polygoni* DC and *Podosphaera fusca* (also known as *P. xanthii*), is a major fungal disease of legumes. *E. polygoni* DC belongs to the *Erysiphales* order of the *Ascomycota* group. It affects all the blackgram cultivating areas and becomes severe in the dry season causing 9.0–50.0% yield loss. In mungbean, a loss of up to 40% has been reported by Humphry et al. (2003) and Khajudparn et al. (2007). An infection at the early growing stage with conducive weather can cause 100% yield reductions (Reddy et al. 1994; Pooja et al. 2018).

7.4.2.1 Symptoms and Disease Cycle of PM

PM generally appears from the early flowering to the pod maturity stage and infects the upper surface of leaves with grayish-white powdery growth with patches gradually covering the lower leaf surfaces (Jyothi 2012). In epidemic form, the fungus covers all parts of the plant adversely affecting the photosynthetic efficiency of the plant leading to forced maturity. This obligate parasite is spread by wind-blown conidia, at a humidity of 50–95% and a temperature of 28–32 °C.

7.4.2.2 Genetics of Resistance

Both qualitative and quantitative PM resistances have been reported in mungbean and blackgram. Single dominant gene (Chaitieng et al. 2002; Khajudparn et al. 2007; Tantasawat et al. 2022), two dominant genes (Reddy et al. 1994; Reddy 2009a, b), single recessive gene (Kaushal and Singh 1989; Kanwade et al. 2019), duplicate recessive genes (Kute et al. 2003), and quantitative genes (Young et al. 1993; Sorajjapinun et al. 2005; Kasettranan et al. 2010; Chankaew et al. 2013) have been reported for PM resistance. Khajudparn et al. (2007) identified dominant resistance genes in each of the three PM resistance mungbean lines (V4718, V4758, and V4785), with each gene nonallelic to other genes. Pulate (2016) reported monogenic dominant resistance against PM in blackgram while Kanwade et al. (2019) identified a single recessive gene for PM resistance in mungbean.

7.4.2.3 Molecular Studies for PM Resistance

An overview of different genes mapped for PM resistance is given in Table 7.3. Molecular studies for mapping PM resistance could be found as early as 1993 when Young et al. mapped genomic regions responsible for PM resistance in resistant mungbean line VC3980A using RFLP markers. The three mapped genomic regions altogether explained 58% of the total PM variation of which two regions were significantly associated with PM resistance at 65-day-old plants, and a third genomic region was associated with resistance at 85-day-old plants.

Using SSR markers, Kasettranan et al. (2010) identified two major PM resistance QTL (*qPMR-1 and qPMR-2*) in mungbean line VC6468-11-1A. SSR markers *CEDG282* and *CEDG191* flanked the *qPMR-1*while SSRs *MB-SSR238* and *CEDG166* flanked *qPMR-2*. Chankaew et al. (2013) also detected a major QTL on linkage group 9 (LG9) and two minor QTLs on LG4 for PM resistance in mungbean line V4718. Further, he also identified two major and one minor QTL on LG6 and LG9 and LG4, respectively, in PM resistance mungbean RUM5.

Poolsawat et al. (2017) identified ISSR marker *I85420* and ISSR-RGA marker *I42PL229* linked to the PM resistance QTL*qPMC72V18-1* at 9 cM and 4 cM, respectively, in mungbean line CN72. Two ISSR-RGA (I27R211 and I27R565) markers were associated with PM resistance in resistant line V4785 of mungbean. They further reported that SSR markers *VR393* and *CEDG084* were linked to major QTL, *qPMC72V18* (PVE 18.72%), for PM resistance, in CN72. Pooja et al. (2018) found SSR marker *MB-SSR238* associated with PM resistance mungbean line TARM1 explaining 11.64% of variations. The diversity analysis (Samal et al. 2022) on 29 blackgram genotypes with five SSR markers *MB-SSR238*, *VrCsSTS1*, *CEDG191*, *VrCsSSR1*, CEDG166, and VrCsSTS1 showed association with PM resistance in different genetic backgrounds. Tantasawat et al. (2022) mapped PM resistance with ISSR-RGA markers *I41tP379* through bulked segregant analysis in mapping population generated from the cross, CN72 (susceptible) × V4758 (resistant) mungbean.

7.4.3 Anthracnose

Anthracnose is a prevalent fungal disease of mungbean and blackgram, widely spread in Asia (Nair et al. 2019) and sub-Saharan Africa (Mbeyagala et al. 2017). This disease of mungbean was first reported in India from the Jorhat district of Assam in 1951 (Majid 1953) and has been found in all major mungbean and blackgram growing states of India. It causes considerable damage to crop plants by reducing seed quality and yield (Sharma et al. 1971); about 30–70% loss was estimated from several mungbean growing areas in India (Shukla et al. 2014; Laxman 2006).

7.4.3.1 Symptoms and Disease Cycle of Anthracnose

Anthracnose in mungbean and blackgram is caused by the fungus, *Colletotrichum truncatum* (Schw.). The pathogen is seed and soil-borne, (Mandal et al. 2015), hemibiotroph that survives in the infected crop residue or seeds (Manandhar and Hartman 1999). The favorable condition for the disease is cool-wet weather of high relative humidity (>90%) with a temperature range of 13–26 °C.

Symptoms of anthracnose disease are circular, black, sunken spots with dark center and bright red-orange margins on leaves and pods. The characteristic symptom of blackgram anthracnose includes horseshoe-shaped brown, sunken necrotic lesions.

7.4.3.2 Anthracnose Resistance

Though several sources of resistance to anthracnose have been reported to be present in blackgram germplasm, nothing is known about the nature of resistance genes and their pattern of inheritance in mungbean. Bindra et al. (2016) found the dominant nature of the anthracnose resistance gene in two resistant urdbean accessions KUG-216 and IPU-05-13. While in mungbean, no such studies on the exploitation of host resistance against anthracnose have been reported so far.

7.4.4 Dry Root Rot (DRR)

DRR, also known as charcoal rot, has become a key emerging yield-limiting disease of mungbean and blackgram in South Asian countries (Latha et al. 2017; Singh et al. 2020a, b, c). It caused 10–48% yield losses in mungbean production in India and Pakistan (Iqbal and Mukhtar 2014). The disease caused a loss in grain yield and protein contents (Indira and Gayatri 2003; Win and Oo 2017).

7.4.4.1 Symptoms and Disease Cycle of DRR

DRR is caused by *Macrophomina phaseolina* (Tassi.) Goid and belongs to the class *Botryosphaeriaceae* of order *Botryosphaeriales*. It is a necrotrophic pathogen and produces dark brown and minute dark, round sclerotia. The fungus survives in facultative parasites and in infected debris. The pathogen is more prevalent at high temperature and prolonged dry season followed by irrigation (Sharma and Pande 2013, Pandey et al. 2020). The nature of resistance to DRR has not been reported so far.

7.4.5 Web Blight

In mungbean, *Rhizoctonia* blight was reported for the first time from the Philippines (Nacien 1924) and later from India for the first time in Kanpur, Uttar Pradesh

(Dwivedi and Saksena 1974). Subsequently the disease has also been reported in blackgram (Saksena 1973; Sharma and Tripathi 2001). The pathogen causes huge losses in yield as it reduces grain yield depending on disease severity in different varieties of mungbean (Singh 2006; Singh et al. 2013a, b).

7.4.5.1 Symptoms and Disease Cycle of Web Blight

The causal organism of web blight is known to be *Rhizoctonia solani* Kuhn. The pathogen has been reported on mungbean under the name *Thanatephorus cucumeris* (frank) Donk (Wang and Yang 1976 of the family *Ceratobasidiaceae*. The colonies of this pathogen are yellowish white to pale brown to dark brown in color (Bal et al. 2019; Rawate et al. 2022).

7.4.6 *Alternaria* Leaf Spot

Alternaria leaf spot is another foliar disease which causes considerable qualitative and quantitative losses in mungbean and blackgram, but not much work has been reported in mungbean and blackgram on this disease (Prathyusha et al. 2021; Meena et al. 2022). *Alternaria* leaf spot was reported from Udaipur, India (Gupta 1970). This disease was reported to cause about 10% yield loss in mungbean (Maheshwari and Krishna 2013) and 50.62–51.29% in blackgram (Ambarish et al. 2021).

7.4.6.1 Symptoms and Disease Cycle of *Alternaria* Leaf Spot

Alternaria leaf spot disease is caused by *Alternaria alternata* (Fr.) Keissl. The fungus produces initially profuse grayish-white mycelium which later becomes dark brown to black. The typical leaf spot symptoms of *Alternaria* first appear on lower leaves at the flower initiation stage. Affected portions in the leaf get separated and fall down resulting in defoliation on severely affected plants during later stages of the crop growth.

7.4.7 Fusarium Wilt

Fusarium wilt is an emerging problem for mungbean and blackgram growers across the world, as it has caused greater than 70% incidence of damage (Zhu et al. 2017). The increased severity is due to favorable environmental conditions of high temperature and humidity (Yin et al. 2016). The incidence of the disease causing yield loss of 15–20% has also been reported from Odisha, India, in blackgram (Biswal et al. 2020).

7.4.7.1 Symptoms and Disease Cycle of Fusarium Wilt

Fusarium wilt is a soil-borne disease caused by *Fusarium oxysporum*. It was first reported in 1950 in China (Tai 1979). Recently, the causal agent of *F. oxysporum* was identified to be a novel forma specialis of *F. oxysporum*, named *F. oxysporum* f. sp. *mungcola* (Fom) (Sun et al. 2017). *F. oxysporum* is mainly confined to vascular tissues and is present both inter- and intracellular in diseased stem or root, particularly near the soil surface. The fungus can infect plants at all growth stages (Tai 1979; Zhu and Duan 2012) and spread upwards along the stem to several inches above ground level and downwards along the tap and lateral roots (Choudhary et al. 2017; Biswal et al. 2020).

7.4.7.2 Resistance Mapping of Fusarium Wilt

Among the molecular mapping for resistance against *Fusarium* wilt, pathotype diversity of *Fusarium oxysporum* causing wilt on mungbean has been reported by Sun et al. (2020). Chang et al. (2021) did the transcriptome analysis of highly resistant (Zheng 8-4) and highly susceptible (Zheng 8-20) mungbean lines to identify putative resistance-related genes for resistance to fusarium wilt infection. Among 3254 DEGs, 24 genes encoding resistance proteins, 22 encoding protein kinases, 20 belonging to transcription factor families, 34 encoding proteins with oxidoreductase activity, 17 involved in stimulation/stress responses, and 54 annotated to pathogen resistance-related pathways were reported.

The molecular studies in fungal diseases, i.e., anthracnose, dry root rot, web blight, and *Alterneria* leaf spot, are scanty. Identification of resistance sources, mapping of putative candidates conferring resistance, and marker-assisted transfer into elite genotypes are proposed strategies for accelerating improvement in mungbean and blackgram.

7.5 Biotic Stress of Mungbean and Blackgram: Bacterial Diseases

Halo blight (*Pseudomonas syringae* pv. *phaseolicola*), bacterial leaf spot (*Xanthomonas campestris* pv. *phaseoli*), and tan spot (*Curtobacterium flaccumfaciens* pv. *flaccumfaciens*) are important bacterial diseases in mungbean and blackgram. Bacterial pathogens are seed-borne and can persist in crop residue. Little work has been done on the screening of mungbean genotypes against bacterial diseases and identifying associated genetic markers (Noble et al. 2019).

7.5.1 Halo Blight

Halo blight of mungbean was first recorded in Queensland in the early 1980s. It caused losses in spring-sown crop with an average yield reduction of 30–50%, and total crop failure may occur in severely infected fields (Sun et al. 2017; Noble et al. 2019).

It is a seed-borne disease caused by *Pseudomonas savastanoi* pv. *phaseolicola. P. s. phaseolicola* can survive between growing seasons on alternative hosts, infected seeds, and infected plant residues. An 18–25 °C of temperature is optimal for the development of the disease. Symptoms include a yellow-green halo surrounding a small dark, water-soaked (shiny) spot on young leaves, which produces necrotic regions in older leaves.

7.5.2 Tan Spot

Tan spot (also known as bacterial scorch and wilt) occurs worldwide and infects several crops including mungbean and blackgram. It is a seed-borne disease caused by the bacterium *Curtobacterium flaccumfaciens* pv. *flaccumfaciens*. Bacterial cells enter plants through the vascular system from infected seeds, and the disease is more commonly seen from the second trifoliate leaf stage onwards. The disease is favored by high temperatures (>30°) and characterized by large, irregular, dry, papery lesions on leaves.

Screening for halo blight and tan spot has been carried out by the Australian breeding program in both controlled (glasshouse) and field conditions to identify useful resistant progenies (Noble et al. 2019).

7.6 Biotic Stresses of Mungbean and Blackgram: Insect Pests

Grain legumes are damaged by more than 150 insect pest species under unprotected conditions in field and storage (Clement et al. 2000; Sharma and Upadhyaya 2016). Mungbean and blackgram are also attacked by several insect pests among which pod borers are major damaging agents with *Helicoverpa armigera* being the most important (Sharma 2001) followed by *Maruca vitrata*—the spotted pod borer (Sharma et al. 1999; Mahalakshmi et al. 2016). Among sap-sucking insects, whitefly (*Bemisia tabaci* Genn.), is a widespread pest followed by *Aphis craccivora* (aphid) and *Empoasca motti* Pruthi (jassids) (Dawoodi et al. 2010). Among defoliators, *Spodopteralitura* (tobacco caterpillar) attacks both mungbean and blackgram in Asia. Bruchids (*Callosobruchus chinensis* L.) and pod-sucking bugs (*Riptortus clavatus* Thunberg) infest all the grain legumes during storage and reproductive stages, respectively, causing worldwide production losses (Wani et al. 2022).

Identifying and characterizing the resistance to insect pest is one of the most challenging tasks due to complex nature of resistance. Moreover, the severity of pest attack is also influenced by environmental factors, preferences of insects, availability of desired food, morphological features of plants, and production of certain defense chemicals or secondary metabolites, all making the breeding for insect pest resistance a herculean task (Mooney et al. 1983). The problem is aggravated as some insect pests play a dual role in damage by acting as vectors for different pathogens. Thus, empowering the pulses with insect pest resistance requires efficient methods for screening with understanding the host × insect × environment interactions (Seram and Devi 2021).

7.6.1 Resistance Sources Against Different Insect Pests

Initial efforts towards achieving insect pest resistance have been made by identifying resistance sources and understanding the mechanism of resistance. The detailed compilation of the different resistance sources identified against different insects in mungbean and blackgram is given in Table 7.4. Different insects pest attacking the mungbean and blackgram and the resistance sources identified are described below.

7.6.2 Bruchid/Cowpea Weevil (Callosobruchus Chinensis and Callosobruchus Maculatus)

Young et al. (1992) identified TC1966, a wild cultivar of mungbean having resistance towards *C. chinensis*. Somta et al. (2008) also identified this resistance in two mungbean cultivars V1128 and V2817. Dixit (2015) identified mungbean line V1123 as having moderate resistance against two bruchid species of *C. chinensis* and *C. maculatus*. V2709 and V2802 varieties of mungbean were also observed to be resistant to both the bruchid species (Samyuktha et al. 2022). *V. mungo* var. *silvestris*, a wild accession of blackgram, was found to have good resistance against *C. maculatus*, and this resistance was successfully introgressed into cultivated blackgram (Souframanien et al. 2010).

7.6.3 Pod Borer (*Helicoverpa armigera*)

Resistance against *H. armigera* has been identified in mungbean lines ML 337, ML 5, M 85-61, and ML 325 while in blackgram, lines CBG 08-011, PLU 54, UH 82-5, IC 8219, and SPS 143 were found have resistance (Soundararajan et al. 2010; Ponnusamy et al. 2014). Somta et al. (2008) observed that the damage to *H. armigera* was lower in lines CGG 08-007 and CGG 08-028 already found resistant to *M. testulalis*. It was also observed that the resistance of TC1966, V2709, V2802, VR1128, and V2817 against *H. armigera* was due to the presence of certain biochemical compounds in the seeds.

7.6.4 Jassid (*Empoasca kerri*) and Stem Fly (*Ophiomyia phaseoli*)

Resistance against the sucking pest *E. kerri* has been identified in Sinkheda 1, Krishna, H 70-3, and UPB 1 accessions of *blackgram* (Dawoodi et al. 2010). They also found resistance against *O. phaseoli* in blackgram lines named Killikullama, 338/3, P 58, Co 4, and Co 5.

Table 7.4 Compilation of different sources for resistance to insect pests in mungbean and blackgram

Mungbean				
S. No.	Insect pest	Scientific name	Resistant source	Reference
1	Bruchid	*Callosobruchus chinensis*	TC1966	Young et al. (1992)
			V1128, V2817	Somta et al. (2008)
		C. chinensis and *C. maculatus*	V2709 and V2802	Samyuktha et al. (2022)
2	Pod borer	*Maruca testulalis*	J1, LM 11, P 526, and P 336	Lakshminarayan et al. (2008)
			ML 337, ML 5, MH 85-61, and ML 325 CGG 08-007 and CGG 08-028	Soundararajan et al. (2010)
		M. vitrata	OBGG109 and BM-4	Kol et al. (2022)
		Helicoverpa armigera		
3	Aphid	*A. craccivora* *A. kerri*	TAM-20, PDM-84-143, and Pusa-105	Devesthali and Joshi (1994)
4	Jassids	*Empoasca kerri*		
5	Sri Lanka weevil	*Myllocerus undecimpustulatus*		
6	Stem fly	*Ophiomyia centrosematis*	Co 3	Devesthali and Joshi (1994)
BLACKGRAM				
S. No.	Insect pest	Scientific name	Resistant source	Reference
1	Bruchid	*C. maculatus*	*V. mungo* var. *silvestris*	Souframanien et al. (2010)
			TC2210	Somta et al. (2019)
			TU68	Subramaniyan et al. (2021)
2	Pod borer	*H. armigera*	CBG 08-011 and PLU 54, UH 82-5, IC 8219, and SPS 143	Soundararajan et al. (2010)
				Ponnusamy et al. (2014)
3	Jassid	*Empoasca kerri*	Sinkheda 1, Krishna, H 70-3 and UPB 1	Dawoodi et al. (2010)
4	Stem fly	*Ophiomyia phaseoli*	Killikullama, 338/3, P 58, Co 4 and Co 5	
5	Pink pod borer	*Cydia ptychora*	SKNU-03-03	
6	Aphid	*Aphis craccivora*	MBM-390-94-Y and MBM-07-Y2	Souframanien et al. (2010)

7.6.5 Aphid (*Aphis craccivora*)

Aphid resistance was identified by Souframanien et al. (2010) in two blackgram cultivars MBM-390-94-Y and MBM-07-Y2.

7.6.6 Molecular Studies for Insect Pest Resistance

The identification and development of molecular markers and other genomic resources have facilitated marker-based identification of insect pest resistance in mungbean and blackgram (Kumar et al. 2011). The detailed compilation of the different insect pest resistance genes mapped in various studies in mungbean and blackgram is given in Table 7.5.

Different studies reported the quantitative type of nature of insect pest resistance in the two legume crops of mungbean and blackgram. Young et al. (1992), Chotechung et al. (2016), and Souframanien et al. (2010) found a major *Br* locus in C618, TC1966 of mungbean, TC2210 in blackgram and in *V. mungo* var. *silvestris*.

In green gram, RFLP markers were used to map bruchid resistance in wild mungbean, TC1966 onto LG 8 (Young et al. 1992), and the marker was further used for marker-assisted introgression of this resistance (Yang et al. 1998). Further Kaga and Ishimoto (1998) found a positive correlation of bruchid resistance of TC1966 with insecticidal cyclopeptide alkaloids involved in insect avoidance. A number of studies have confirmed the association of the *Br l*ocus with an EST-SSR marker *DMB-SSR158* in the C618 mungbean cultivar. This locus encodes *VrPGIP2*, a polygalacturonase inhibitor to bruchid (Chotechung et al. 2016; Chen et al. 2021). Resistance to bruchids is governed by a major dominant gene, *Br* on chromosome 8 in TC1966, V2709BG, and V2802BG varieties of mungbean (Young et al. 1992; Mahato et al. 2015). Although wild mungbean TC1966 and derived purelines V2709BG and V2802BG have been found completely immune to *C. chinensis* and *C. macualtus*, to date, no cultivated mungbean or blackgram is found free from the attack of bruchids.

Souframanien et al. (2010) developed 104 RIL lines by crossing the wild *V. mungo* var. *silvestris* conferring resistance to *C. maculatus* with bruchid susceptible cultivar TU 94-2. Markers including 86 RAPDs, 47 SSRs, 41 ISSRs, and 254 AFLPs used to map QTLs associated with *C. maculatus* and QTLs contributing to adult emergence, developmental period, and developmental stage resistance were mapped.

Hong et al. (2015) developed an F_2 mapping population from the cross of Sunhwa (susceptible) × Jangan (resistance) for pod-sucking bug (*Riptortus clavatus*) and bruchid (*C chinensis*), and resistance was mapped using 118 SSR and 190 RAPD markers. One QTL conferring bruchid resistance flanked by markers MB87 and COPU11 and another QTL flanked by markers RP and COPU06 imparting resistance to both bruchid and pod-sucking bug were mapped.

Table 7.5 Compilation of different molecular mapping studies for insect pest resistance in mungbean and blackgram

S. No.	Crop	Trait	Cross	Population	Markers used	Linked markers	References
1	Blackgram	CMRAE1	*V. mungo* var. *silvestris* × TU-94-2	104 RILs	428 SSR	EACT/MCTC-8-EACT/ MCAT-5 CEDG086-CEDG154	Souframanien et al. (2010)
		CMRDP1	*V. mungo* var. *silvestris* × TU-94-2	104 RILs	86 RAPD+ 47 SSR+ 41 ISSR + 254 AFLP	CEDG133-CEDG149 EACG/MCTA-15-EAGG/MCTA-1 OPL14-1300-OPI20-600 OPR1-1380-EACA/ MCAT-12 EAAG/MCTA-1-UBC827-1800 EACG/MCTA-14-EAAG/MCAA-4	
2	Mungbean	Br	Sunhwa × Jangan	460 F_2	118 SSR + 190 RAPD	MB87-COPU0	Hong et al. (2015)
3	Mungbean	Bean bug	Sunhwa × Jangan	460 F_2	118SSR + 190 RAPD	RP-COPU06	Hong et al. (2015)
4	Blackgram	PDS AUDPS	BC48 × TC2210	150 RILs	3675 SNPs	Marker14881-Marker9514	Somta et al. (2019)
5	Blackgram	AE	TU68 × MDU1	108 F_2	625 SSR	CEDG020-CEDG067	Subramaniyan et al. (2021)
		SD	TU68 × MDU1	108 F_2	625 SSR	CEDG020-CEDG067	
		DT	TU68 × MDU1	108 F_2	625 SSR	CEDG302-GMES1248	
6	Mungbean	Br	C618 × Zhonglv 5	182 F_7	17 SSR	Vr04-246-Vr04-483	Chen et al. (2021)

In blackgram, 625 SSR markers were used for mapping the QTLs associated with the total number of adult emergence (AE), percentage of seed damage (SD), and developmental time (DT) against *C. maculatus* on mapping population of 108 F_2 individuals from the cross, TU68 (resistant) × MDU1 (susceptible) (Subramaniyan et al. 2021). They identified three QTLs associated with AE, SD, and DT in linkage groups 5 and 8. The QTLs for AD and SD, namely, *qbr_AE@50DAI* and *qbr_SD@50DAI*, respectively, were mapped on linkage group 5 flanked by CEDG020 and CEDG067 markers, while SSR markers CEDG302 and GMES1248 flanking the QTL for DT, *qbr_Dev.T*, were mapped on linkage group 8.

Chen et al. (2021) mapped bruchid (*C chinensis*) resistance QTL flanked by markers *Vr04–221* and *Vr04–604* on chromosome 8 spanning a physical region of 111 Kb using RILs population from the cross Zhonglv 5 (S) × C618 (R). A total of 11 genes involved in resistance were identified of which *Vr04g00919* was reported as the candidate gene involved in brucid resistance.

Most of the cultivated varieties of mungbean and blackgram are susceptible to insect pests. Efficient screening methods combined with the use of a high-throughput marker system are required to develop the cultivar's resistance to insect pests.

7.6.7 Insect Pest Resistance

The deployment of high-throughput genomic sources for the deployment of resistance in the two important legume crops promises genetic gains. The genomic-assisted breeding (GAB) along with high throughput genotyping/phenotyping, sequencing platforms, and high-density linkage/QTL maps has helped in speeding up the genetic improvement of major pulses, which lead to the rapid development of improved cultivars with higher yield, wider adaptability, and enhanced stress resistance (Bohra et al. 2014).

In blackgram, Somta et al. (2019) mapped *C. maculatus* resistance genes on a 160.2 Kb region on linkage group 6 with SLAF sequencing associated encoding lectin receptor kinase (LecRK) and chitinase genes imparting resistance to bruchid in the wild accession, TC2210.

Chen et al. (2021), through RNA sequencing in mungbean, identified 11 candidate genes in a physical interval of 111 kb governing resistance to bruchids. The identified region encodes for serine/threonine-protein kinase, mitogen-activated protein (MAPK), cellulose synthase, and lipid-transfer protein. Fine mapping this region led to the identification of six highly polymorphic SNPs, and eight high-throughput SNPs were identified in the amino acid sequence of Vr04g00919 gene encoding for polygalacturonase inhibition.

7.7 Future Prospects of Biotic Stress Resistance

The conventional breeding methods coupled with biotechnological interventions have the potential for combating biotic stresses and to meet the ever-growing need of increasing legume production. Precise phenotypic screening remains challenging due to the complex nature and uneven distribution of insect pests, and quantitative inheritance of resistance in some diseases poses another challenge. The screening is also affected when the insect pest infesting the host is a vector for a certain disease making it difficult to accurately score and collect data under natural epiphytotic conditions, e.g., for viral diseases. The application of MAS for the genetic improvement of the cultivated gene pool has been limited to a few biotic stresses in grain legumes. The conventional molecular marker studies have led to the identification of important genes/QTLs governing major biotic stresses, but in view of screening diverse germplasm sets, high-throughout genotyping can play an important role. In recent times, genomic-assisted breeding strategies involving sequence analysis in combination with genetic markers have been established as time and cost-effective for the identification, mapping, and transfer of candidate genes conferring resistance to targeted traits. In a nutshell, it is high time to enhance genetic improvement in important legumes by incorporating genes against biotic stresses, ultimately increasing productivity potential to meet the nutritive demands by using conventional breeding and high-throughput molecular techniques.

References

Alam AKM, Somta P, Srinives P (2014) Identification and confirmation of quantitative trait loci controlling resistance to mungbean yellow mosaic disease in mungbean [*Vigna radiata* (L.) Wilczek]. Mol breed 34:1497–1506

Ambarish KV, Kumar AP, Adinarayana M, Madhumathi T (2021) Spatial distribution and characterization of *Alternaria alternata* causing leaf spot in blackgram. Int J Ecol Environ Sci 3(1): 77–80

Ammavasai S, Phogat D S, Solanki I S, (2004) Inheritance of resistance to mungbean yellow mosaic virus (MYMV) in green gram [Vigna radiata (L.) Wilczek]

Arumuganathan K, Earle ED (1991) Nuclear DNA content of some important plant species. Plant Mol Biol Rep 9:211–215

Bal SS, Kumar S, Mishra IOP, Panda PK, Panigrahi RK, Prusti AM (2019) Evaluation of mungbean and urdbean genotypes against web blight disease under natural field condition. Int J Chem Stud 7(5):3374–3377

Basak J, Kundagrami S, Ghose TK et al (2005) Development of yellow mosaic virus (YMV) resistance linked DNA marker in *Vigna mungo* from populations segregating for YMV-reaction. Mol Breeding 14:375–383. https://doi.org/10.1007/s11032-005-0238-6

Batzer JC, Singh A, Rairdin A, Chiteri K, Mueller DS (2022) Mungbean: a preview of disease management challenges for an alternative US cash crop. J Integr Pest Manag 13(1):4

Bindra SH, Mittal RK, Sood VK, Sharma PN (2016) Inheritance of resistance in urdbean (*Vigna mungo*) to anthracnose caused by *Colletotrichum truncatum*. Indian Phytopathol 69:311–313

Binyamin R, Khan MA, Khan NA, Khan AI (2015) Application of SCAR markers linked with mungbean yellow mosaic virus disease-resistance gene in Pakistan mungbean germplasm. Genet Mol Res 14:2825–2830

Biswal K, Sahu KC, Ranasingh N, Moharana RL (2020) *Fusarium oxysporum* causing black gram wilt a new threat in the Western undulated zones of Odisha. Int J Curr Microbiol App Sci 9(06): 3027–3031

Bohra A, Pandey MK, Jha UC, Singh B, Singh IP, Datta D, Chaturvedi SK, Nadarajan N, Varshney RK (2014) Genomics-assisted breeding in four major pulse crops of developing countries: present status and prospects. Theor Appl Genet 127:1263–1291

Chaitieng B, Kaga A, Han OK, Wang X, Wongkaew S, Laosuwan P et al (2002) Mapping a new source of resistance to powdery mildew in mungbean. Plant Breed 121:521–525

Chakraborty N, Basak J (2018) Comparative transcriptome profiling of a resistant vs. susceptible *Vigna mungo* cultivar in response to Mungbean yellow mosaic India virus infection reveals new insight into MYMIV resistance. Curr Plant Biol 15:8–24

Chand R, Singh V, Pal C, Kumar P, Kumar M (2012) First report of a new pathogenic variant of Cercospora canescens on mungbean (*Vigna radiata*) from India. New Dis Rep 26(1):6

Chandel KPS, Lester RN, Starling RJ (1984) The wild ancestors of urid and mung beans (Vigna mungo (L.) Hepper and V. radiata (L.) Wilczek). Bot J Linn 89(1):85–96. https://doi.org/10.1111/j.1095-8339.1984.tb01002.x

Chang Y, Sun F, Sun S, Wang L, Wu J, Zhu Z (2021) Transcriptome analysis of resistance to fusarium wilt in mung bean (*Vigna radiata* L.). Front Plant Sci 12:679629

Chankaew S, Somta P, Sorajjapinun W, Srinives P (2011) Quantitative trait loci mapping of Cercospora leaf spot resistance in mungbean, *Vigna radiata* (L.) Wilczek. Mol Breed 28(2): 255–264

Chankaew S, Somta P, Isemura T, Tomooka N, Kaga A, Vaughan DA, Srinives P (2013) Quantitative trait locus mapping reveals conservation of major and minor loci for powdery mildew resistance in four sources of resistance in mungbean [*Vigna radiata* (L.) Wilczek]. MolBreed 32(1):121–130

Chen HM, Ku HM, Schafleitner R, Bains TS, George Kuo C, Liu CA, Nair RM (2013) The major quantitative trait locus for mungbean yellow mosaic Indian virus resistance is tightly linked in repulsion phase to the major bruchid resistance locus in a cross between mungbean [*Vigna radiata* (L.) Wilczek] and its wild relative *Vigna radiata* ssp. sublobata. Euphytica 19:205–216

Chen H, Hu L, Wang L, Wang S, Cheng X (2021) Identification of a candidate gene for bruchid resistance by combining fine mapping and transcriptome profiling in mung bean (*Vigna radiata* L.)

Chotechung S, Somta P, Chen J, Yimram T, Chen X, Srinives P (2016) A gene encoding a polygalacturonase-inhibiting protein (PGIP) is a candidate gene for bruchid (Coleoptera: *bruchidae*) resistance in mungbean (*Vigna radiata*). Theor Appl Genet 129:1673–1683

Choudhary A, Ashraf S, Musheer N (2017) Screening of phytoextracts to control of *fusarium oxysporum* f. sp. Vigni incitant of mungbean (*Vigna radiata*) wilt. Int J Acad Res Dev 2:1181–1184

Clement SL, Wightman JA, Hardie DC, Bailey P, Baker G, McDonald G (2000) Opportunities for integrated management of insect pests of grain legumes. In: Linking research and marketing opportunities for pulses in the 21st century, pp 467–480

Dasgupta U, Mishra GP, Dikshit HK, Mishra DC, Bosamia T, Roy A et al (2021) Comparative RNA-Seq analysis unfolds a complex regulatory network imparting yellow mosaic disease resistance in mungbean [Vigna radiata (L.) R. Wilczek]. Plos One 16:e0244593

Dawoodi JT, Parsana GJ, Jethva DM et al (2010) Screening of blackgram varieties for resistance against pink pod borer, *Cydia ptychora* (Meyrick). Legume Res 33:54–56

De Candolle A (1886) Origin of cultivated plants. Kejen Paul Trench and Co., London

Devesthali S, Joshi M (1994) Infestation and varietal preference of insect pests in green gram. Indian Agric 38:263–272

Dhaliwal SK, Gill RK, Sharma A (2022) A large-effect QTL introgressed from ricebean imparts resistance to Mungbean yellow mosaic India virus in blackgram (*Vigna mungo* (L.) Hepper). Theor Appl Genet 135:4495. https://doi.org/10.1007/s00122-022-04234-5

Dhole VJ, Reddy KS (2012) Genetic analysis of resistance to mungbean yellow moSsaic virus in mungbean (*Vigna radiata*). Plant Breed 131:414–417

Dixit GP (2015) All India coordinated research project on chickpea. Project Coordinator's report. www.Aicrpchickpea.res.in/pdf_files/pc_report2014_190915.pdf

Douglas C, Pratap A, Rao BH, Manu B, Dubey S, Singh P (2020) Breeding progress and future challenges: abiotic stresses. In: Nair RM, Schafleitner R, Lee SH (eds) The mungbean genome. springer, Cham, pp 81–96

Duangsong U, Kaewwongwal A, Somta P, Chankaew S, Srinives P (2016) Identification of a major QTL for resistance to Cercospora leaf spot disease in cowpea (*Vign aunguiculata* (L.) Walp.) revealed common genomic region with that for the resistance to angular leaf spot in common bean (*Phaseolus vulgaris* L.). Euphytica 209:199–207

Dwivedi RP, Saksena HK (1974) Occurance of web blight caused by *Thanatephorus cucumeris* on mung bean. Indian J Farm Sci 2:100

Fatokun CA, Danesh D, Young ND, Stewart EL (1993) Molecular taxonomic relationships in the genus *Vigna* based on RFLP analysis. Theor Appl Genet 86(1):97–104. https://doi.org/10.1007/BF00223813

Fuller DQ (2007) Contrasting patterns in crop domestication and domestication rates: recent archaeobotanical insights from the Old World. Ann Bot 100:903–924

Fuller DQ, Harvey EL (2006) The archaeobotany of Indian pulses: identification, processing and evidence for cultivation. Environ Archaeol 11:219–246

Gupta BM (1970) New disease on mungbean from Udaipur, Rajasthan. India Plant Dis Rep 54:453

Gupta S, Gupta DS, Anjum TK, Pratap A, Kumar J (2013) Inheritance and molecular tagging of MYMIV resistance gene in blackgram (Vigna mungo L. Hepper). Euphytica 193:27–37

Gupta SK, Bansal R, Gopalakrishna T (2014) Development and characterization of genic SSR markers for mungbean (*Vigna radiata* (L.) Wilczek). Euphytica 195:245–258

Gurha SN, Misra DP, Kamthan KP (1982) Studies on some aspects of yellow mosaic disease of black gram (*Vigna mungo* (L.) Hepper). Madras Agri J 69:435–438

Gwag JG, Chung JW, Chung HK, Lee JH, Ma KH, Dixit A (2006) Characterization of new microsatellite markers in mungbean, (*Vigna radiata* L.). Mol Ecol Notes 6:1132–1134. https://doi.org/10.1111/j.1471-8286.2006.01461.x

Hartman GL, Wang TC, Kim D (1993) Field evaluation of mungbeans for resistance to Cercospora leaf spot and powdery mildew. Int J Pest Manag 39(4):418–421

Holeyachi P, Savithramma DL (2013) Identification of RAPD markers linked to MYMV resistance in mungbean (*Vigna radiata* (L). Wilczek). Bioscan 8:1409–1411

Hong MG, Kim KH, Ku JH, Jeong JK, Seo MJ, Park CH et al (2015) Inheritance and quantitative trait loci analysis of resistance genes to bruchid and bean bug in mungbean (*Vigna radiata* L. Wilczek). Plant breed. Biotech 3(1):39–46

Humphry ME, Magner T, McIntyre CL, Aitken EA, Liu CJ (2003) Identification of a major locus conferring resistance to powdery mildew (*Erysiphe polygoni* DC) in mungbean (*Vigna radiata* L. Wilczek) by QTL analysis. Genome 46(5):738–744

Indira N, Gayatri S (2003) Management of blackgram root rot caused by *Macrophomina phaseolina* by antagonistic microorganisms. Madras Agric J 90:490–494

Iqbal U, Mukhtar T (2014) Morphological and pathogenic variability among *Macrophomina phaseolina* isolates associated with mungbean (*Vigna radiata* L.) Wilczek from Pakistan. Sci world J 2014:1–9

Jain HK, Mehra KL (1980) Evolution adaptation relationships and uses of the species of vigna cultivated in India. In: Summerfield RJ (ed) Advances in legume science. Royal Botanic Garden Kew, Richmond Surrey, pp 459–468

Jain R, Lavanya GR, Reddy PA, Babu GS (2013) Genetic inheritance of yellow mosaic virus resistance in mungbean [*Vigna radiata* (L.) Wilczek]. Biosci Trends 6:305–306

Jasrotia RS, Iquebal MA, Yadav PK, Kumar N, Jaiswal S, Angadi UB, Kumar D (2017) Development of transcriptome based web genomic resources of yellow mosaic disease in Vigna mungo. Physiol Mol Biol Plants 23:767–777

Jegadeesan S, Raizada A, Dhanasekar P (2021) Draft genome sequence of the pulse crop blackgram [*Vigna mungo* (L.) Hepper] reveals potential R-genes. Sci Rep 11:11247

Jyothi D (2012) Epidemiology and management of powdery mildew of greengram caused by *Erysiphe polygoni* DC. M Sc.(Agri.) Thesis

Kaga A, Ishimoto M (1998) Genetic localization of a bruchid resistance gene and its relationship to insecticidal cyclopeptide alkaloids, the vignatic acids, in mungbean (*Vigna radiata* L. Wilczek). Mol Gen Genet MGG 258:378–384

Kalaria RK, Chauhan DIGVIJAY, Mahatma MK, Mahatma LALIT (2014) Identification of RAPD and ISSR makers for resistance against Mungbean yellow mosaic virus in mungbean (*Vigna radiata* L.) under South Gujarat agroclimatic condition of India. Bioscan 9:1177–1182

Kang Y, Kim S, Kim M (2014) Genome sequence of mungbean and insights into evolution within *Vigna* species. Nat Commun 5:5443

Kanwade DG, Kute NS, Bedis MR, Pawar VY, Deshmukh GP, Jaybhaye CP, Giri MD (2019) Inheritance of powdery mildew resistance in mung bean.[*Vigna radiata* (L.) R. Wilczek]. J Pharmacogn Phytochem 8(3):3642–3644

Karthikeyan AS, Vanitharani R, Balaji V, Anuradha S, Thillaichidambaram P, Shivaprasad PV, Parameswari C, Balamani V, Samathan M, Veluthambi K (2004) Analysis of an isolate of Mungbean yellow mosaic virus (MYMV) with a highly variable DNA-B component. Arch Virol 149:1643–1652

Karthikeyan A, Shobhana VG, Sudha M, Raveendran M, Senthil N, Pandiyan M, Nagarajan P (2014) Mungbean yellow mosaic virus (MYMV): a threat to green gram (*Vigna radiata*) production in Asia. Int J Pest Mgmt 60:314

Kasettranan W, Somta P, Srinives P (2010) Mapping of quantitative trait loci controlling powdery mildew resistance in mungbean (*Vigna radiata* (L.) Wilczek). J Crop Sci Biotechnol 13(3): 155–161

Kaur S, Gill RK, Bains TS, Kaur M, Thakur S (2017) Comparative assessment of SSR markers derived from different sources in genetic diversity analysis of *Vigna* genotypes. Agric Res J 54(4):462–468

Kaushal RP, Singh BM (1989) Genetics of resistance in black gram to powdery mildew caused by *Erysiphe polygoni*. Indian J Agric Sci 59(10):684–685

Khajudparn P, Wongkaew S, Thipyapong P (2007) Mungbean powdery mildew resistance. Identification of genes for resistance to powdery mildew in mungbean. In 8th African crop science society conference, El-Minia, Egypt. African crop science society, Kanana, pp 743–745

Khattak GSS, Haq MA, Ashraf M, Elahi T (2000) Genetics of Mungbean yellow mosaic virus (MYMV) in mungbean (*Vigna radiata* L.) Wilczek. J Genet Breed 54:237–243

Kitsanachandee R, Somta P, Chatchawankanphanich O, Akhtar KP, Shah TM, Nair RM, Bains TS, Sirari A, Kaur L, Srinives P (2013) Detection of quantitative trait loci for mungbean yellow mosaic India virus (MYMIV) resistance in mungbean (*Vigna radiata* (L.) Wilczek) in India and Pakistan. Breed Sci 63:367–373

Kol U, Singh V, Chandrakar G (2022) Screening of germplasm of mungbean against pod borer complex. J Curr Res Food Sci 3:37–40

Kumar J, Choudhary AK, Solanki RK, Pratap A (2011) Towards marker-assisted selection in pulses: a review. Plant Breed 130:297–313

Kundu S, Chakraborty D, Kundu A, Pal A (2013) Proteomics approach combined with biochemical attributes to elucidate compatible and incompatible plant-virus interactions between Vigna mungo and Mungbean yellow mosaic India virus. Proteome Sci 11(1):15

Kundu A, Patel A, Paul S, Amita P (2015) Transcript dynamics at early stages of molecular interactions of MYMIV with resistant and susceptible genotypes of the leguminous host, *Vigna mungo*. PLOS ONE 10(4):e0124687

Kundu A, Singh PK, Dey A, Ganguli S, Pal A (2019) Complex molecular mechanisms underlying MYMIV-resistance in *Vigna mungo* revealed by comparative transcriptome profiling. Sci Rep 9(1):8858

Kute NS, Deshmukh RB, Patil JV (2003) Inheritance resistance to powdery mildew (*Erysiphe polygoni* D.C) in mungbean (*Vigna radiata* (L.) Wilczeck). Res Legume 26(1):24–27

Lakshminarayan S, Singh PS, Mishra DS (2008) Relationship between whitefly population, YMV disease and morphological parameters of green gram germplasm. Environ Ecol 26:978–982

Latha P, Karthikeyan M, Rajeswari E (2017) Development of bioformulations for the management of blackgram dry root rot caused by *Rhizoctonia bataticola* (Taub Butler). Adv Res 9(4):1–2

Laxman R (2006) Studies on leaf spot of greengram caused by *Colletotrichum truncatum* (Schw.) Andrus and Moore. M. Sc. (Agri.) Thesis, University eof Agricultural Sciences, Dharwad, Karnataka, India

Leabwon U, Oupadissakoon S (1984) Inheritance of resistance to Cercospora leaf spot in mungbean. Agric Nat Res 18:14–19

Lekhi P, Gill RK, Kaur S, Bains TS (2018a) Generation of interspecific hybrids for introgression of mungbean yellow mosaic virus resistance in *Vigna radiata* (L.) Wilczek. Legume Res 41:526–531

Lekhi P, Gill RK, Kunal KS (2018b) Identification of molecular marker linked to mungbean yellow mosaic virus (MYMV) resistance in *Vigna radiata* (L.) Wilczek. Electron. J Plant Breed 9(3): 839–845

Mahalakshmi MS, Sreekanth M, Adinarayana M, Reni YP, Rao YK, Narayana E (2016) Incidence, bionomics and management of spotted pod borer [*Maruca vitrata* (Geyer)] in major pulse crops in India-a review. Agric Rev 37:19–26

Mahalingam A, Satya VK, Manivannan N, Narayanan SL, Sathya P (2018) Inheritance of mungbean yellow mosaic virus disease resistance in greengram [*Vigna radiata* (L.) Wilczek]. Int J Curr Microbiol App Sci 7:880–885

Mahato S, Jena M, Mohanty P, Behera C, Senapati N, Dash GB et al (2015) Bruchid resistance in mung bean: an overview. e-planet 13(2):10–15

Maheshwari SK, Krishna H (2013) Field efficacy of fungicides and bio-agents against *Alternaria* leaf spot of mungbean. Ann Plant Prot Sci 21:364–367

Maiti S, Basak J, Kundagrami S, Kundu A, Pal A (2011) Molecular marker-assisted genotyping of mungbean yellow mosaic India virus resistant germplasms of mungbean and urdbean. Mol Biotechnol 47:95–104

Maiti S, Paul S, Pal A (2012) Isolation, characterization, and structure analysis of a non-TIR-NBS-LRR encoding candidate gene from MYMIV-resistant Vigna mungo. Mol Biotechnol 52(3): 217–233

Majid S (1953) Annual report of Department of Agriculture, Assam for the year 1951-1952 and 1952-1953. The Grow More Food Campaign. 2: 112

Malathi VG, John P (2008) Gemiviruses infecting legumes. In: Govind P, Rao P, Kumar PL, Holguin-Pena RJ (eds) Characterization, diagnosis & management of plant viruses. Stadium Press LLC, Texas, pp 97–123

Manandhar JB, Hartman GL (1999) Anthracnose. In: Hartman GL, Sinclair JB, Rupe JC (eds) Compendium of soybean diseases. American Phytopathological Society, St. Paul, MN, pp 13–14

Mandal R, Narender S, Mehta SK, Khirbat SK (2015) Influence of environmental conditions on the incidence of anthracnose of mungbean. Ann Agri Bio Res 20(2):184–187

Mathivathana MK, Murukarthick J, Karthikeyan A, Jang W, Dhasarathan M, Jagadeeshselvam N, Sudha M, Vanniarajan C, Karthikeyan G, Yang TJ, Raveendran M (2019) Detection of QTLs associated with mungbean yellow mosaic virus (MYMV) resistance using the interspecific cross of Vigna radiata × Vigna umbellata. J Appl Genet 60:255–268

Mbeyagala KE, Amayo R, Obuo JP, Pandey AK, War AR, Nair RM (2017) A manual for mungbean (greengram) production in Uganda. National Agricultural Research Organization (NARO) Entebbe 32

Meena NK, Trivedi A, Kumar S, Chand S (2022) Assessment of disease severity influence by Alternaria leaf spot and powdery mildew in organic mungbean (*Vigna radiata* L.). The Pharma Innov J 11(2):964–966

Mishra SP, Asthana AN, Yadav L (1988) Inheritance of Cercospora leaf spot resistance in mung bean, Vigna radiata (L.) Wilczek. Plant Breed 100(3):228–229

Mooney HA, Field C, Gulmon SL, Rundel P, Kruger FJ (1983) Photosynthetic characteristic of south African sclerophylls. Oecologia 58:398–401

Nacien CC (1924) Studied on *Rhizoctonia* blight of beans. Philipp Agric 8:315–321

Naimuddin K, Akram M (2010) Detection of mixed infection of begomoviruses in cowpea and their molecular characterization based on CP gene sequences. J Food Legume 23:191–195

Nair NG, Nene YL (1973) Studies on yellow mosaic of urd bean (*Phaseolus mungo* L.) caused by mungbean yellow mosaic virus I transmission studies. Indian J Farm Sci 1:109–110

Nair RM, Gotz M, Winter S, Giri RR, Boddepalli VN, Sirari A, Kenyon L (2017) Identification of mungbean lines with tolerance or resistance to yellow mosaic in fields in India where different begomovirus species and different *Bemisia tabaci* cryptic species predominate. Eur J Plant Pathol 149:349–365

Nair RM, Pandey AK, War AR, Hanumantharao B, Shwe T, Alam A, Pratap A, Malik SR, Karimi R, Mbeyagala EK, Douglas CA, Rane J, Schafleitner R (2019) Biotic and abiotic constraints in Mungbean production—Progress in genetic improvement. Front Plant Sci 10: 1340

Nariani TK (1960) Yellow mosaic of mung (*Phaseolus aureus*). Indian Phytopath 13:24–29

Noble TJ, Young AJ, Douglas CA, Williams B, Mundree S (2019) Diagnosis and management of halo blight in Australian mungbeans: a review. Crop Pasture Sci 70(3):195–203

Pal AK, Jana PK (1991) Effect of phosphorus, rhizobium inoculation and irrigation on growth, yield and water-use efficiency of summer greengram (*Phaseolus radiatus*). Indian J Agron 36: 536–540

Pandey S, Sharma M, Kumari S, Gaur PM, Chen W, Kaur L, Macleod W, Basandrai AK, Basandrai D, Bakr A, Sandhu JS, Tripathi HS, Gowda CLL (2009) Integrated foliar diseases management of legumes. In: Ali M et al (eds) Grain legumes: genetic improvement, management and trade. Indian Society of Pulses Research and Development, Kanpur, pp 143–161

Pandey AK, Burlakoti RR, Kenyon L, Nair RM (2018) Perspectives and challenges for sustainable management of fungal diseases of mungbean [*Vigna radiata* (L.) R. Wilczek var. radiata]: a review. Front Environ Sci 6:53

Pandey AK, Burlakoti RR, Rathore A, Nair RM (2020) Morphological and molecular characterization of *Macrophomina phaseolina* isolated from three legume crops and evaluation of mungbean genotypes for resistance to dry root rot. Crop Prot 127:104962

Papan P, Chueakhunthod W, Poolsawat O, Arsakit K, Tharapreuksapong A, Tantasawat PA (2021) Validation of molecular markers linked to Cercospora leaf spot disease resistance in mungbean (*Vigna radiata* [L.] Wilczek). SABRAO J Breed Genet 1:53(4), 749

Patel A, Dey N, Chaudhuri S, Pal A (2017) Molecular and biochemical characterization of a *Vigna mungo* MAP kinase associated with Mungbean yellow mosaic India virus infection and deciphering its role in restricting the virus multiplication. Plant Sci 262:127–140

Ponnusamy D, Pratap A, Singh SK, Gupta S (2014) Evaluation of screening methods for bruchid beetle (*Callosobruchus chinensis*) resistance in greengram (*Vigna radiata*) and blackgram (*Vigna mungo*) genotypes and influence of seed physical characteristics on its infestation. VEGETOS 27:60–67

Pooja SK, Sarkale AP, Bhat S (2018) Validation of molecular marker associated with powdery mildew resistance in mungbean. Indian J Genet Plant Breed 78(03):386–388

Poolsawat O, Kativat C, Arsakit K, Tantasawat PA (2017) Identification of quantitative trait loci associated with powdery mildew resistance in mungbean using ISSR and ISSR-RGA markers. Mol Breed 37(12):1–2

Pootakham W, Nawae W, Naktang C, Sonthirod C, Yoocha T, Kongkachana W et al (2021) A Chromosome-scale assembly of the black gram (Vigna mungo) Genome. Mol Ecol Resour 21:238–250. https://doi.org/10.1111/1755-0998.13243

Prajapati RK, Choudhary RG, Viswadhar (2007) Integrated management strategies on pulse. In: Ahmed S, Narain U (eds) Eco friendly management of plant disease. Daya Publishers, New Delhi, pp 277–287

Pratap A, Douglas C, Prajapati U, Kumari G, War AR, Tomar R, Dubey S (2020) Breeding progress and future challenges: biotic stresses. In: The mungbean genome. Springer, Cham, pp 55–80

Prathyusha D, Adinarayana M, Manoj Kumar V, Ambarish KV (2021) Status and distribution of leaf spot of blackgram incited by *Alternaria alternata* in Krishna and Guntur districts of Andhra Pradesh. Biolog Forum 13(1):479–483

Pulate SC (2016) Inheritance of powdery mildew resistance, grain yield and yield components in black gram (*Vigna mungo* (L.) Hepper) Ph.D. Thesis, Mahatma Phule Krishi Vidyapeeth, Rahuri (M.S.) India

Qazi J, Ilyas M, Mansoor S, Briddon RW (2007) Legume yellow mosaic viruses: genetically isolated Begomoviruses. Mol Pl Path 8:343–348

Rambabu E, Anuradha C, Sridhar V, Reddy SS (2018) Identification of molecular markers linked to yellow mosaic virus resistance in blackgram (Vigna mungo (L.) Hepper). Int J Curr Microbiol App Sci 7(2):3810–3817

Rawate R, Kaushal A, Awadhiya GK, Patel DP (2022) Survey and management of web blight disease of mung bean. Pharma Innov J 11(8):477–485

Reddy KS (2009a) A new mutant for yellow mosaic virus resistance in Mungbean (*Vigna radiata* (L.) Wilczek) variety SML-668 by recurrent gamma-ray irradiation. In: Induced plant mutation in the genomics era. IEAE, Mumbai, pp 361–362

Reddy KS (2009b) Identification and inheritance of a new gene for powdery mildew resistance in mungbean (*Vigna radiata* L. Wilczek). Plant Breed 128:521–523

Reddy KS, Pawar SE, Bhatia CR (1994) Inheritance of powdery mildew (*Erysiphepolygoni* DC) resistance in mungbean (*Vigna radiata* L. Wilczek). TheorAppl Genet 88(8):945–948

Rohilla V, Yadav RK, Poonia A, Sheoran R, Kumari G, Shanmugavadivel PS, Pratap A (2021) Association mapping for yield attributing traits and yellow mosaic disease resistance in mung bean [*Vigna radiata* (L.) Wilczek]. Front plant Sci 12:749439

Ryley MJ, Tatnell JR (2011) Management of the major foliar diseases of mungbeans and peanuts in Australia. In proceedings of the 4th Asian conference on plant pathology and the 18th biennial Australasian plant pathology society conference (ACPP/APPS 2011) (p. 61). Australasian Plant Pathology Society

Saksena HK (1973) Banded blight disease of paddy in North India, Proceedings, International Rice Research Conference, International IRRI, LosBanos

Saleem M, Haris WAA, Malik IA (1998) Inheritance of yellow mosaic virus in mungbean (*Vigna radiata* L. Wilczek). Pakistan J Phytopathol 10:30–32

Samal KC, Ramakrishna B, Sahoo JP, Behera L (2022) Molecular characterization and validation of micro-satellite markers linked to powdery mildew disease resistance in mini Core germplasm of Urdbean [*Vigna mungo* (L.) Hepper]. Res Biotica 4(2):62–68

Samyuktha SM, Malarvizhi D, Mariyammal I, Karthikeyan A, Seram D, Dhasarathan M et al (2022) The hunt for Mungbean (*Vigna radiata* (L.) Wilczek) genotypes and breeding lines resistance to south Indian Bruchid strain. Agric 12:1050

Sandhu TS, Brar JS, Sandhu SS, Verma MM (1985) Inheritance of resistance to mungbean yellow mosaic virus in green gram. J Res Punjab Agr Univ

Selvi R, Muthiah AR, Manivannan N, Raveendran TS, Manickam A, Samiyappan R (2006) Tagging of RAPD marker for MYMV resistance in mungbean (*Vigna radiata* (L.) Wilczek). Asian J Plant Sci 5(2):277–280

Seram D, Devi YK (2021) Insect resistance screening techniques and scoring for important horticultural crops. Plant Cell Biotechnol Mol Biol 22:155–162

Shahbaz MU, Iqbal MA, Rafiq M, Batool A, Kamran M (2014) Efficacy of different protective fungicides against Cercospora leaf spot of mungbean (*Vigna radiata* l. Wilczek). Pak J Phytopathol 26(2):187–191

Shanthala J, Savithramma DL, Gazala P, Jambagi BK, Desai SKP (2020) Genomics-assisted breeding green gram (*Vigna radiata* (L.) Wilczek) for accelerating genetic gain. In: Accelerated Plant Breeding, vol 3. Springer, Cham, pp 143–171

Sharma HC (2001) Crop protection compendium: *Helicoverpa armigera*. In: Electronic compendium for crop protection. Commonwealth Agricultural Bureaux International, Wallingford. www.cabi.org/c&c/about/contributors-p-s1

Sharma M, Pande S (2013) Unravelling effects of temperature and soil moisture stress response on development of DRR [*Rhizoctonia bataticola* (Taub.)] butler in chickpea. Am J Plant Sci 4:584–589

Sharma J, Tripathi HS (2001) Host range of *Rhizoctonia solani* Kuhn, the causal agent of web blight of urdbean *Vigna mungo* (L.) Hepper. J Mycol Pl Patho 131:81–82

Sharma S, Upadhyaya HD (2016) Pre-breeding to expand primary genepool through introgression of genes from wild *Cajanus* species for pigeonpea improvement. Legume Perspect 11:17–20

Sharma HC, Khare MN, Joshi LK, Kumar SM (1971) Efficacy of fungicides in the control of diseases of *kharif* pulses mung and urid. In All India Workshop on *Kharif* Pulses. 2

Sharma HC, Saxena KB, Bhagwat VR (1999) The legume pod borer, *Maruca vitrata*: bionomics and management. Information bulletin 55, international crops research institute for the SemiArid tropics (ICRISAT), Patancheru

Shendure J, Ji H (2008) Next-generation DNA sequencing. Nat Biotechnol 26:1135–1115

Shukla V, Baghel S, Maravi K, Singh SK (2014) Yield loss assessment in mungbean [*Vigna radiata* (L.) Wilczek] caused by anthracnose [*Colletotrichum truncatum* (schw.) Andrus and Moore]. Bioscan (Ranchi) 9:1233–1235

Singh A (2006) *Rhizoctonia solani* in rice-wheat cropping system. Ph.D. Thesis, Department of Plant Pathology, G.B. Pant University of Agriculture and Technology, Pantnagar, U.S. Nagar

Singh HB, Joshi BS, Chandel KPS, Pant KC, Saxena RK (1975) Genetic diversity in some Asiatic Phaseolus species and its conservation. Indian J Genet Plant Breed, pp 52–57

Singh I, Sandhu JS, Gupta SK, Singh S (2013a) Introgression of productivity and other desirable traits from ricebean (*Vigna umbellata*) into blackgram (*Vigna mungo*). Plant Breed 132:401. https://doi.org/10.1111/pbr.12068

Singh J, Mishra KK, Singh AK (2013b) Current status of web blight of mung bean. Asian J Soil Sci 8(2):495–504

Singh DK, Vaishampayan A, Singh MN, Singh SK, Vennela PR, Varsha G, Jaiswal DK, Vaibhav S (2017) Inheritance studies on Cercospora leaf spot (CLS) resistance in mung and urdbean. Vegetos 30(Special Issue 2):388–391

Singh N, Mallick J, Sagolsem D, Mandal N, Bhattacharyya S (2018) Mapping of molecular markers linked with MYMIV and yield attributing traits in mungbean. Indian J Genet Plant Breed 78: 118–126

Singh CM, Pratap A, Gupta S, Biradar RS, Singh NP (2020a) Association mapping for mungbean yellow mosaic India virus resistance in mungbean (*Vigna radiata* L. Wilczek). 3Biotech 10:1–12

Singh J, Mishra KK, Chaubey AK (2020b) Important diseases of green gram (*Vigna radiata* L. Wilczek) and their management. In: Diseases of Field crops: diagnosis and management. Apple Academic Press, New Jersey, pp 61–89

Singh L, Kumar A, Kaur S, Gill RK (2020c) Genetic variability, heritability and genetic advance as mean for yield and yield contributing traits in urdbean (*Vigna mungo* L. Hepper). Int J Curr Microbiol Appl Sci 9:2778–2786

Singh L, Dhillon GS, Kaur S, Dhaliwal SK, Kaur A, Malik P, Kumar A, Gill RK, Kaur S (2022) Genome-wide association study for yield and yield-related traits in diverse Blackgram panel (*Vigna mungo* L. Hepper) Reveals Novel Putative Alleles for Future Breeding Programs. Front Genet 13:849016. https://doi.org/10.3389/fgene.2022.849016

Somta C, Somta P, Tomooka N, Ooi PC, Vaughan DA, Srinives P (2008) Characterization of new sources of mungbean (*Vigna radiata* (L.) Wilczek) resistance to bruchids, *Callosobruchus* spp. (Coleoptera: *Bruchidae*). J Stored Prod Res 44:316–321

Somta P, Chen J, Yundaeng C, Yuan X, Yimram T, Tomooka N, Chen X (2019) Development of an SNP-based high-density linkage map and QTL analysis for bruchid (*Callosobruchus maculatus* F.) resistance in black gram (*Vigna mungo* (L.) Hepper). Sci Rep 9(1):1–9

Sorajjapinun W, Rewthongchum S, Koizumi M, Srinives P (2005) Quantitative inheritance of resistance to powdery mildew disease in mungbean (*Vigna radiata* (L.) Wilczek). SABRAO J Breed Genet 37(2):91

Souframanien J, Gopalakrishna T (2006) ISSR and SCAR markers linked to the mungbean yellow mosaic virus (MYMV) resistance gene in blackgram [Vigna mungo (L.) Hepper]. Plant Breed 125:619–622

Souframanien J, Gupta SK, Gopalakrishna T (2010) Identification of quantitative trait loci for bruchid (*Callosobruchus maculatus*) resistance in black gram [*Vigna mungo* (L.) Hepper]. Euphytica 176(3):349–356

Soundararajan RP, Chitra N, Ramasamy M (2010) Host plant resistance to insect-pests of urdbean and mungbean. In: National workshop on paradigm shifts in research on crop resistance to pests, Annamalai University, Annamalai Nagar, Mar p 4–5

Subramaniyan R, Narayana M, Krishnamoorthy I, Natarajan G, Gandhi K (2021) Mapping and mining of major genomic regions conferring resistance to Bruchine (*Callosobruchus maculatus*) in blackgram (*Vigna mungo*). Plant Breed 140(5):896–906

Sudha M, Karthikeyan A, Anusuya P, Ganesh NM, Pandiyan M, Senthil N, Raveendran M, Nagarajan P, Angappan K (2013) Inheritance of resistance to mungbean yellow mosaic virus (MYMV) in inter and intra specific crosses of mungbean (*Vigna radiata*). Am J Plant Sci 4:1924

Sudha M, Karthikeyan A, Madhumitha B, Veera Ranjani R, Kanimoli Mathivathana M, Dhasarathan M, Murukarthick J, Samu Shihabdeen MN, Eraivan Arutkani Aiyanathan K, Pandiyan M, Senthil N (2022) Dynamic transcriptome profiling of Mungbean genotypes unveil the genes respond to the infection of Mungbean yellow mosaic virus. Pathogens 11:190

Sun S, Zhi Y, Zhu Z, Jin J, Duan C, Wu X, Xiaoming W (2017) An emerging disease caused by pseudomonas syringae pv. Phaseolicola threatens mung bean production in China. Plant Dis 101:95–102

Sun S, Zhu L, Sun F, Duan C, Zhu Z (2020) Pathotype diversity of *fusarium oxysporum* f. sp. mungcola causing wilt on mungbean (*Vigna radiata*). Crop Pasture Sci 71(10):873–883

Tai FL (1979) Sylloge fungorum sinicorum. Beijing, Science Press, Academia Sinica

Tantasawat PA, Poolsawat O, Arsakit K, Papan P (2020) Identification of ISSR, ISSR-RGA and SSR markers associated with Cercospora leaf spot resistance gene in mungbean. Intl J Agric Biol 23:447–453

Tantasawat PA, Poolsawat O, Kativat C, Arsakit K, Papan P, Chueakhunthod W, Pookhamsak P (2022) Inheritance and identification of ISSR-RGA markers associated with powdery mildew resistance in mungbean for marker-assisted breeding. Chilean J Agric Res 82(1):3–9

Tomooka N, Maxted C, Thavarasook AHM, Jayasuriya (2002) Two new species, sectional designations and new combinations in *Vigna* subgenus *Ceratotropis* (piper) Verdcourt (Leguminosae). Kew Bull 57:613–624

Vavilov NI (1926) Studies on the origin of cultivated plants. Bull Appl Bot 26:248

Vertovec S (1994) "Official" and "Popular" Hinduism in the Caribbean: historical and contemporary trends in surinam. Trinidad Guyana Contrib Indian Soc 28:123–147

Vishnu-Mittre B (1974) Palaeobotanical evidence in India. In: Hutchinson J (ed) Evolutionary studies in world crops: diversity and change in the Indian sub-continent. Cambridge University Press, Cambridge, pp 3–30

Wang TC, Yang CY (1976) Mungbean leaf blight, a new disease of mungbean. Plant Prot Bull 17: 8–9

Wani AA, Bhat FA, Nisa RT, Wani TA, Raja W, Jan SK et al (2022) Bean angular leaf spot (*Phaeoisariopsis griseola*): epidemiology and yield loss assessment. Pharm Innov J 11(2): 291–297

Win KT, Oo AZ (2017) Salt-stress-induced changes in protein profiles in two blackgram (*Vigna mungo* L.) varieties differing salinity tolerance. Adv Plant Agric Res 7(1):00239

Yang TJ, Kim DH, Kuo GC, Kumar L, Yong ND, Park HG (1998) RFLP marker-assisted selection in backcross breeding for introgression of the bruchid resistance gene in mungbean. Korean J Breed 30:8–15

Yin X, Olesen JE, Wang M, Öztürk I, Zhang H, Chen F (2016) Impacts and adaptation of the cropping systems to climate change in the northeast farming region of China. Eur J Agron 78: 60–72

Young ND, Kumar L, Menancio-Hautea D, Danesh D, Talekar NS, Shanmugasundarum S, Kim DH (1992) RFLP mapping of a major bruchid resistance gene in mungbean (*Vigna radiata.* L Wilczek). Theor App Genet 84(7):839–844

Young ND, Danesh D, Menancio-Hautea D, Kumar L (1993) Mapping oligogenic resistance to powdery mildew in mungbean with RFLPs. Theor Appl Genet 8:243–249

Yundaeng C, Somta P, Chen J, Yuan X, Chankaew S, Chen X (2021) Fine mapping of QTL conferring Cercospora leaf spot disease resistance in mungbean revealed TAF5 as candidate gene for the resistance. Theor App Genet 134(2):701–714

Zhu ZD, Duan CX (2012) Identification and control of diseases and pests in mung bean (China agricultural science and technology press: Beijing) [in Chinese] pp 17–19

Zhu L, Sun SL, Sun FF, Duan CX, Zhu ZD (2017) The method for evaluation of mung bean resistance to fusarium wilt. Zhiwu Yichuan Ziyuan Xuebao 18:696–703

Disease Resistance an Essential for Better Adaptability and Production of Faba Bean in India (*Vicia faba L.*)

8

Neha Tiwari, Surendra Barpete, Tapan Kumar, D. R. Saxena, and Fouad Maalouf

Abstract

A popular leguminous crop, faba bean (*Vicia faba* L.), is rich in micronutrients and cool-tolerant. It is cultivated across several countries, including Australia, China, Ethiopia, France, and United Kingdom. The primary purpose of its cultivation is to provide digestible and valuable protein-rich food for human consumption and animal feed. This crop also improves soil quality and contributes to sustainable agriculture by fixing atmospheric nitrogen in symbiosis with Rhizobium. Fababean cultivated area is around 2.57 million-hectare and production 5.4 million tonnes across 38 countries around the globe. In India, it is grown as a minor crop mainly as a garden crop. Faba beans' adaptability and productivity are severely limited by diseases and pests. Severe diseases of faba bean caused by fungi, viruses, nematodes, and parasitic weeds have an effect on seed yields. Genetics and genomics research on fababean genotypes provides the opportunity to develop disease-tolerant, high-yielding genotypes. Disease-resistant varieties could be generated through the use of resistant/tolerant sources in breeding programs. In order to successfully apply biotechnology to disease resistance, a complete understanding of fababean biology is required. In this chapter, we discuss the major biotic stresses facing the crop and how they can be managed to improve the adaptability and productivity.

N. Tiwari (✉) · S. Barpete · T. Kumar
International Centre for Agricultural Research in the Dry Areas (ICARDA), Amlaha, Madhya Pradesh, India

D. R. Saxena
Rafi Ahmed Kidwai College of Agriculture, Sehore, Madhya Pradesh, India

F. Maalouf
International Centre for Agricultural Research in the Dry Areas (ICARDA), Rabat, Morocco

U. C. Jha et al. (eds.), *Diseases in Legume Crops*,
https://doi.org/10.1007/978-981-99-3358-7_8

Keywords

Fababean · Biotic stress · Disease management · Disease resistance

8.1 Introduction

Faba bean (*Vicia faba* L.) known as broad bean, horse bean and field bean is a micronutrient-rich and cool-tolerant leguminous crop grown widely in Australia, China, Ethiopia, France and the United Kingdom (Labba et al. 2021). It is primarily grown as digestible and protein-rich food for human consumption and as a forage for animals. In addition to being consumed as a green vegetable (whole pods), faba beans can be consumed as split seeds as well. Faba bean also improves soil quality and helps to achieve sustainable agriculture by fixing atmospheric nitrogen (70–223 kg N/ha) symbiotically with *Rhizobium* (soil bacteria) (Dhull et al. 2022). Through nitrogen fixation, faba beans improve soil fertility when grown in rotation with cereal crops. Nitrogen fixation by faba bean and other legume crops reduces the requirement of extensive use of chemical fertilizers that directly help the protecting arable soil and water quality (Stagnari et al. 2017), faba bean global cultivated area of around 2.57 million hectare and production of 5.4 million tonnes across 38 countries around the globe (FAOSTAT 2020). The crop is minor and primarily cultivated as a garden crop in Assam, Bihar, Madhya Pradesh, Manipur, Nagaland, Rajasthan, Uttar Pradesh and West Bengal (Singh et al. 2012a, b) in India. Indian plains grow it mostly during rabi, though it is successfully grown in mountainous and hilly areas during Kharif season also (Singh and Bhatt 2012; Tiwari et al. 2021). The total cultivated area is declining because of the unreliable and poor seed yield primarily because of the crop's susceptibility and vulnerability to major pests and the diseases (Sillero et al. 2009). The major damage to the crop is due to a number of soil-borne pathogenic fungi in association with root-rot complexes, viruses, nematodes, and competitive parasitic weeds which majorly affect the seed yields (Sillero et al. 2009; Stoddard et al. 2006). Faba beans are susceptible to a number of diseases, including *Ascochyta* blight (*Ascochyta fabae*), rust (*Uromyces fabae*), chocolate spot (*Botrytis fabae* and *B. cinerea*), cercospora leaf spot (*Cercospora zonata*), alternaria leaf blight (*Alternaria* spp.) and downy mildew (*Peronospora viciae*) (Stoddard et al. 2010; Singh et al. 2012b). Breeding disease-resistant crops is more suitable and cost-effective means of controlling the insects and pests damage (Stoddard et al. 2010; Tekalign et al. 2017). However, molecular marker-based approaches as if marker-assisted selection (MAS) are still underway due to polygenic resistance and have to exploit its potential fully (Khater et al. 2022). Whereas phenotypic expression of resistance in other leguminous crops such as soybean and common bean exists complete monogenic resistance, Marker-assisted selection is efficiently used in these legume crops (Garcia-Villalba et al. 2008; Maalouf et al. 2019).

Although in faba bean, genomic resources are significantly progressing and many genetic maps are available for important traits (Ellwood et al. 2008a, b; Webb et al.

2016; Satovic et al. 2019). However, the marker density (genetic mapping) is relatively large distance and enables accurate prediction of desired quantitative trait loci (QTLs); therefore, associate molecular mapping hampered the accurate and effective use of MAS in the faba bean breeding programme. However, few studies on SNP markers with traits of interest identified resistance to *Ascochyta* and broomrape (Kaur et al. 2014; Ocana-Moral et al. 2017). The efficiency of MAS in faba bean progress would be increased by adopting next-generation sequencing (NGS) and integration of functional genomics and comparative mapping. The productivity and adaptability are highly influenced and limited by major diseases and hampered faba bean acceptance in the farmer's field, especially in the new area where it was not cultivated earlier. Therefore, the successful application of modern biotechnology and molecular tools has developed disease resistance cultivars of faba bean. The large genome size (13 Gb) also hampered the goal (Cooper et al. 2017). However, significant progress has been made in the next-generation sequencing (NGS) technologies which contributed to the generation of large volumes of sequences that have facilitated the revelation of single-nucleotide polymorphisms (SNPs) associated with key breeding traits either through genome-wide association studies (GWAS) or biparental mapping, but more research is needed to understand their full potential (Gnanasambandam et al. 2012; Carrillo-Perdomo et al. 2020), although appropriate progress has also been attempted for tissue culture and agrobacterium-mediated genetic transformation in faba bean (Gantait and Mukherjee 2021). As a complement to conventional and MAS breeding, this type of approach is a highly useful for amelioration of faba beans breeding programme. However, genomic and biotechnological applications are slow to progress because of the difficulties and relatively low investment in genetic, genomic, and disease interactions for breeding faba bean resistance to disease. The focus of this chapter will be on current and future strategies for breeding disease-resistant faba beans.

8.2 Production

Based on the production of faba bean, it raked ranked fifth crop world average production in the last decade (Bangar and Kajla 2022). The faba bean global production has been increasing every year due to its main producer, namely, China, Ethiopia, the United Kingdom, Australia and France. However, its production is affected by import-export demand, processing and domestic prices of the countries which consume large quantities for nutritional security. The world production of faba beans is 5.4 million metric tonnes, and the cultivated area is around 2.57 million hectares across 38 countries, which represented an increase of around 25% compared with 4.3 million metric tonnes per million hectares (FAOSTAT 2020). In the production scenario at the regional level, Asia is leading with a total faba bean production of 33% at the global level. Asia, Europe, and Africa are making their presence for production, respectively, at 29% and 27%, the leading countries that produce and export faba beans. Ethiopia produced almost 50% of the total global production and is the second-largest producer of faba beans after China. However, the European Union, the United Kingdom and France are among the leading producers. Following Australia, in terms of faba bean exports in 2019 (about 30% of total exports), countries like the United Kingdom, Lithuania and Latvia were the top exporters, while the top four countries for faba bean imports globally were Germany, Norway, Saudi Arabia and France (FAOSTAT 2020).

8.3 Uses

Faba bean is emerging as a quality plant protein source that fulfils the demand for nutritious and healthy foods with the potential to meet the growing global populations (Martineau-Cote et al. 2022). In addition to the high protein content, it is also rich in amino acids, phenolic compounds, bioactive peptides and L-DOPA with many health-enhancing properties. Faba bean bioactive peptides released after gastrointestinal digestion have shown properties as antidiabetic, antioxidant, cholesterol-lowering, antihypertensive and anti-inflammatory effects showing strong potential as a functional food that can be helpful for preventing the incidences of non-communicable diseases. Faba beans-based products are nutrient-rich and popular in many countries. The isolates of faba bean protein or starch have been utilized in many products, i.e. including pasta, spaghetti, bread, tofu and yoghurt (Gimenez et al. 2013; Rosa-Sibakov et al. 2016; Sozer et al. 2019; Jiang et al. 2020). Additionally in beef patties, faba beans are also used to partially substitute the meat and fat (Sulaiman et al. 2018).

8.4 Understanding the Pathogen

Pathogens consistently are the most limiting components for the cultivation of faba bean among the many constraints. Many diseases target faba bean crop (Singh et al. 2012b). Important fungal and bacterial diseases are:

8.4.1 Fungal Disease

Chocolate spot (*Botrytis fabae* and *B. cinerea*), Anthracnose (*Colletotrichum lindemuthianum*), Sclerotinia stem rot (*Sclerotiniatri foliorum, S. sclerotiorum*), rust (*Uromyces viciae-fabae*) and *Ascochyta* blight (*Ascochyta fabae*) are the important fungal diseases of faba bean.

8.4.1.1 Chocolate Spot

Chocolate spot disease is caused by *Botrytis fabae* and *Botrytis cinerea* and occurs everywhere in the faba bean growing areas. It has been reported from around the world like Syria, Europe, Middle East and North Africa. The disease is also reported in Morocco, Libya, Ethiopia, England, Spain, Norway, Germany, Scotland, Tunisia,

Algeria, Canada, North and South America, Russia, Japan, China and Australia (Hanounik and Hawtin 1982; Singh et al. 2012b).

Symptoms: All the parts of plants are affected, and symptoms can be seen on leaves, stems, flowers as well as in the pods. A thin layer of reddish-brown spots appears on the leaves, stems and flowers at the initial stage of the disease. On leaves and stems, initial grey spots developed with red-brown margin and grey centre. These spots later on turn into chocolate in colour. These spots coalesce at later the stage and form irregular larger dark-brown lesions on the entire leaf surface. Under favourable conditions, severe defoliation occurs and flowers drop, and affected tissues show necrosis that leads to the death of the plant. In susceptible varieties, the infection often results in severe crop lodging.

Chocolate spot disease is caused by *Botrytis fabae* and *Botrytis cinerea*. Sporulation of primary injury may cause chocolate spot epiphytotic under suitable conditions. At more than 90% humidity and temperature between 15 and 20 °C will trigger the disease (Harrison 1984). Abundant sporulation of the pathogen can be found on blackened tissue at the aggressive stage. *B. fabae* and *B. cinerea* on faba beans are frequently mistaken for one another, but the separation between the two pathogens is always possible. The pathogen *B. fabae* is more virulent with larger conidia size, shorter conidiophores and smaller sclerotia (Leach 1955; Mansfield and Widdowson 1973; Harrison 1984; Lee et al. 2020) than *B. cinere*a due to is its high pathogenicity. Reduced vigour of the crop due to factors like excessive weeds, deficiency of phosphorus and potassium and water logging makes the plants susceptible to chocolate spots. Injuries of leaves due to insect damage or due to wheel tracks also make the crop more susceptible to the disease.

Disease cycle: *B. fabae* can survive on the weeds and crop debris from the diseased faba bean plant for more than one season. Primary infection is initiated early in the season by wind dispersing conidiospores into new crops. Despite being capable of traveling long distances, conidiospores generally fall within a few hundred meters of their source. The infection is spread throughout a crop when spores are blown or splashed on other leaves and plants that have dropped onto the soil surface. In warm and relative humid weather conditions, the disease spreads quickly. Development of stem infection during late stage can cause crop to lodge.

Management: The seeding rates should be low, and the adjustment of date of sowing should be considered to avoid long humid weather conditions (Wilson 1937; Hanounik and Hawtin 1982), removal of plant debris harbouring hyphae or sclerotia of *B. fabae* (Harrison 1979; Hanounik and Hawtin 1982) from the field, the use of healthy seed and wider row spacing reduce the severity of the disease. Faba bean-wheat intercropping is a good method to control chocolate spot disease and increase the faba bean yield (Guo et al. 2020). The use of vinclozolin as foliar spray could control and increase the yield (Hanounik 1981). Carbendazim and mancozeb are fungicide treatment for better management of this disease. In few reports, *Penicillium citrinum* and *P. cyclopium* were reported to prevent the spread of disease by suppressing the *B. fabae* spores. In recent years, the use of nanoparticles has been explored in the management of charcoal spots of faba bean. According to Ahmad (Ahmad et al. 2017), treatment with 100 ppm of silver nanoparticles resulted in the highest reduction in the disease intensity of chocolate spots (52.94%), followed by treatment with 80 ppm of chitosan nanoparticles (50.59%). The antifungal activity of biosynthesized zinc oxide nanoparticles (ZnO NPs) suggests that they could be employed as a fungicide to protect faba bean against *B. cinerea* (Issam et al. 2021). In the early 1980s, three faba bean lines, viz., BPL 1179, 1196 and 710, were identified as reliable sources of resistance to *B. fabae* by Hanounik and Hawtin (1982), Hanounik and Viha (1986) and Hanounik and Robertson (1988). The highest level of resistance was found in the ICARDA lines, ILB-4726, ILB-938 and BPL-710 (Beyene et al. 2018).

8.4.1.2 Anthracnose Disease

Anthracnose is caused by the hemibiotrophic fungus *Colletotrichum lindemuthianum* (Sacc. & Magn.) Scrib. is a major disease of faba bean and does more losses in the temperate region compared to the tropics. In the North Indian hills, the disease appears in mid-June and does the maximum damages from the starting of August month to September. Crop losses up to 100% if contaminated seeds are planted for disease development under favourable weather conditions.

Symptoms: Angular reddish-brown patches on the leaves are the most important symptoms to identify. There are elliptical lesions that form on stems and petioles. The fungus is often present in infected pods which have brown lesions. If the infected seeds are planted, then the seedlings will die earlier than emergence. On the seedlings, lesions with pink masses of spores in the centre are common. The fungus destroys both the cotyledons and damages the stem if conditions are humid. The conidia develop singly at free hyphal ends or on stromatoid masses in acervuli. During the infection period, *C. lindemuthianum* initially produces biotropic primary hyphae with larger diameters and are entirely intracellular. They are followed by inter- or intracellular, necrotrophic secondary hyphae that are narrower.

The secondary hyphae are not surrounded by an extracellular matrix. The fungus produces perithecia that range in size from 120 to 240 μm in diameter. Asci, filiform paraphyses and hyaline are present in perithecia. Eight ascospores are present in each ascus, and they can either be allantoid (6.5 x 20 m) or ellipsoid (4 x 10 m) in shape (Muimba-Kankolongo 2018).

Disease Cycle: The fungus is a seed- and soil-borne in nature. Although it can survive in the soil on the decaying remains of disease plants, it cannot survive for very long on its own when separated from the debris. If the plant debris are not well buried in the soil, the fungus can persist for up to 2 years in the regions with cool climates. There is a significant correlation between the amount of debris left on the soil surface and the infected pods from anthracnose disease in the field (Ntahimpera et al. 1997). In the seed, the fungal pathogen of anthracnose can survive as long as the seed is viable. Since the amount of secondary inoculum developed will be proportional to the amount of primary inoculum, the percentage of infected seed is crucial for determining the severity of anthracnose during the growth season (Arraya Fernandez et al. 1987). First lesions form and produce spore masses on the cotyledons when infected seeds are planted or from the soil-borne inoculum. These initiate the disease development in the crop (Muimba-Kankolongo 2018).

Management: The use of healthy seeds, seed treatment, foliar fungicide spraying, rotation with non-host crops, tillage techniques and the use of resistant varieties are a few management practices. The infected plant parts of faba bean either should be removed completely or deeply buried (Ntahimpera et al. 1997). If the disease takes on a serious form, the crop should be sprayed with 2 kg/ha with suggested fungicides such as zineb or mancozeb. Benlate, Ziram, Vitavax, Ferbam and lime sulphur were suggested by Sindhan and Bose (1981) for foliar applications in the given order. Bavistin and Vitavax fungicides were recommended for seed treatment. Spraying mancozeb at 0.25% at intervals of 10–15 days following the onset of symptoms efficiently manages the disease and increases yield as well.

8.4.1.3 Sclerotinia Stem Rot

Sclerotinia spp. cause disease in pulses, flower crops and vegetables. The disease is reported from different parts of the world and has very wide host range attacking over 60 families and 350 plant species. It affects seedlings, mature plants and harvested seeds. The faba bean crop damage is varying depending on upon the weather, host susceptibility and infection type.

Symptoms: A water-soaked region appearing on any aerial plant component is the disease's most prominent and typical early symptom. A white fluffy mycelial (white mould) growth appears 1–3 days after this watery soft rot symptom.

Under favorable temperatures of 20 to 24 °C, the white cottony growth spreads rapidly and kills a plant within 4–10 days. Black sclerotia developed on the infected tissues. When the dry position of the stem is opened, sclerotia may be seen filling the pith. They can also form on the surface of the stem as a white mycelial web adhering to the host surface. In the pods, white mycelium develops on the pod's surface causing necrosis, and later fungus enters in pods and starts rotting seeds (Lithourgidis et al. 2004).

Disease cycle: The pathogen can persist in the form of sclerotia close to the soil surface, in crop debris or as an admixture with the seed. The forthcoming crop becomes infected by ascospores produced after germination of the sclerotia. If cattle are fed contaminated feed, the disease may potentially spread through animal manure. In the northern part of India, during the middle of winter, the disease outbreaks by ascospores from germinating sclerotia occur. Winter is the time of year when apothecia are most commonly formed since the conditions are ideal for sclerotial germination in terms of temperature, sunlight and moisture (Lithourgidis et al. 2004).

Management: A seed must be taken which is free from sclerotia and seed infection. Sclerotia are primarily carried with seed and are removed using the flotation process. Seed treatment with fungicides as captan, PCNB, thiabendazole or fludioxonil can inactivate seed-borne mycelium. The fungicide treatment with captan + PCNB + thiabendazole totally inhibits mycelial growth from seed (Mueller et al. 2002). The primary source of infection for the crop is soil-borne inoculum. It is effective yet impractical to treat soil with chemicals like cyanamides. Sugha (2001) found that pre-sowing seed treatments with carbendazim (10 kg/ha) combined with a 1:1 mixture of carbendazim+ thiram (2.5 g/kg) and 0.1% carbendazim after flowering began helps to prevent white rot in peas by 97%. Destruction of sclerotia-load-bearing plant parts by burning is necessary. It helps in destroying most of the sclerotia, and those survive have less germination capability. Deep plough of soil is also recommended to destroy sclerotia, and deep-buried sclerotia fail to produce apothecia. *Trichoderma harzianum* treatment seeds with mycelial preparation and application at the rate of 200 g/m^2 is found to control stem rot (Sharma et al. 1999). In order to control the white rot in the case of peas, soil application as well as seed treatment with *T. harzianum* and *T. viride* have produced good results. *S. sclerotium* has been treated with fungicides through seed treatment, fumigation, soil drenching and foliar spraying in a variety of crops.

Sprays of ziram, ferbam or systemic benzimidazole group fungicides (such carbendazim) have shown to be beneficial in controlling the disease spread. Studies have proven that the sclerotinia can be effectively treated with tetrasodium thiocarbamate as well as dicarboximide fungicides such procymidone, vinclozolin and iprodione. In order to prevent ascospore infection of plant tissues, systemic fungicides are applied as foliar application. It only works when chemicals are applied to tissues that are senescent, like flower petals.

8.4.1.4 Rust

Rust is a widely distributed disease of faba bean in the world, mainly caused by *Uromyces viciae* fabae pers. Schroet. This is more severe in humid tropical and subtropical areas (Guyot 1975; Hebblethwaite 1983). The disease is also reported from all over West Asia and North Africa or WANA region (Hawtin and Stewart 1979). The yield loss will be up to 45% if severe infections occur and appear as black-brown masses or pustules usually surrounded by a pale halo (Bekhit et al. 1970; Mohamed 1982).

Symptoms: Rusty red pustules surrounded by a light yellow halo appear late in the crop season and can cause up to more than 20% losses in the crop production (Bekhit et al. 1970; Mohamed 1982). These losses may extent up to 45%, in severe infections which appear as black-brown masses or pustules usually surrounded by a pale halo. This is the telial stage of the fungus that contains resting spores which allow survival over summer.

Disease cycle: The fungus mainly survives on plant debris of previous infected crops. Spores are blown to another or new crops through the wind. The volunteer faba bean plays an important role in the early development of infection. Later, spores are released from pustules and spreads through wind, insects and from farm machinery to other plants and crops. Severe infections can cause premature defoliation and reduction of leaf area, small seed size and resulting as yield losses. Yield losses are much higher in mixed infection with chocolate spot. The disease occurrence is high if the weather is humid and rainy conditions.

Management: Agricultural practices such as crop rotation with a non-host crop, burning or deep plowing of crop debris, weeding and proper plant spacing can reduce the pathogen's inoculum. Several workers have reported that destruction of infected plant debris and crop rotation with non-host crops plays an important role in minimizing the chances of pathogen survival and avoided primary infections in the field (Conner and Bernier 1981). Clean and contamination free seeds are also recommended for the disease management.

Control measures should be taken before the disease established to minimize the yield losses. Many fungicides give good disease control of rust. Mancozeb (0.2%), calixin (0.2%) and bayleton (0.05%) fungicides found very effective against the pathogen. As soon as the disease is found in the field, the first spray is applied, followed by three subsequent sprays at 10 days intervals. All of the triazole fungicides are reported to give excellent control when used 72 hours after inoculation (Gupta and Shyam 2000). Tridemorph, chlorothalonil, zineb and mancozeb sprays have all been effective in controlling the disease.

Rust occurs in the late season; therefore, fungicide treatment is not economical. When rust takes place with *Ascochyta* at the same time, a foliar spray of Mancozeb is beneficial to control the disease. Mancozeb spray is found to decrease *Ascochyta* and chocolate spot in faba beans by reducing the risk of infections on the pod and seed (Mansour et al. 1975). Consequently, this minimizes blemishes on the seed and the spread of *Ascochyta* disease. There have been many reports of rust-resistant faba bean lines. Faba bean lines BPL1179, 261, 710, 8, 406, 417 and 484 were found resistant, and some L82009s, L82007s, L82011s and L82010s were found to be resistant to rust and chocolate spots as well (ICARDA 1987; Maalouf et al. 2016).

8.4.1.5 *Ascochyta* Blight

It is a severe disease of faba bean and is scattered throughout the world, and its severity differs from crop and season. Up to 30% loss of yield can occur if the season is favourable for the disease.

Symptoms: The symptoms of the disease appear on leaves, stems and pods and create confusion with the early stages of the chocolate spot. On both sides of the affected leaves, circular, dark-brown spots appear. During the disease development, lesions increase and change to dark grey colour. The lesions on the leaf may become black and necrotic with numerous pin-head pycnidia of the fungus developing under moist conditions (Matthews and Marcellos 2003).

Lesions tend to be elongated in the stem and are typically covered with scattered pycnidia. These lesions girdle the affected part of the stem, which may split, break and lodge the plant. The lesions on the pods have a pale centre and dark margin covered by pycnidia. Infected seeds are having brown stains on the seed coat. Additionally, *Ascochyta* blight is always confused with some minor diseases like *Alternaria* and *Cercospora* leaf spots. At the podding stage, an infection on the pod level occurs during cool, humid weather, which results in seed infection and seed discoloration (Maalouf et al. 2016).

Disease Cycle: The pathogen survives in the seed as well as on diseased plant debris left in the field as pycnidia (Singh et al. 2012b). However, seeds affected from disease do not germinate and serve as a substrate for the growth of fungus under. According to Singh and Pal (1995), infected seeds rarely result in infected seedlings. When crop debris is buried in snow and frozen for an extended amount of time in temperate climates, pseudothecia with ascospores are released. Moisture is necessary for pycnidia to survive in debris, regardless of temperature. When the soil surface's relative humidity is 0–3%, pycnidia can survive for more than 2 years at temperatures between 10 and 350C (Vishunawat and Chaube 1986). The primary inoculum multiplies near the base of the stem after the fungus develops from germination of the seed or from infected soil, and it subsequently spreads to aerial parts. Additionally, insects, contact between leaves, conidia disseminated by rain-drop splashes in windy conditions, and animal movement around the field all contribute to the disease spread.

Management: Seed-borne pathogens are externally and internally transmitted, so the only preventive measure is to use seed from healthy crops that is clean and certified. Additionally, it is encouraged that farmers avoid using seeds that are discoloured, especially those that are more than 25% discoloured, as this could significantly lower their faba bean's grain yield. In the absence of resistant cultivars, seed treatment is an additional important control measure. It is suggested to treat the seeds with copper sulphate, thiram, benomyl or calixin M. In order to control the disease in the field, Tripathi et al. (1987) reported that 2.5 g/kg of seeds should be treated with carbendazim + thiram in a 1:3 ratio, followed by three sprays of carbendazim (0.5 kg/ha) at 10 days interval when the disease initially appears. Sprays of chlorothalonil, mancozeb, zineb, ferbam, maneb and daconil are recommended in the case of a mild attack or if it persists and spreads throughout the faba bean crop. Mancozeb has a 30-day withholding period for fungicides in grain prior to harvest, but chlorothalonil has a 7-day withholding period. This is why chlorothalonil is thought to be a preferable alternative. Singh and Singh (1990) suggest that for suppressing secondary infection, foliar sprays of Bravo (1500 mL/ha) and calixin M (900 g/ha) are considerably superior to other fungicides, followed by Rovral (750 g/ha) and Hexacap (900 g/ha) in 300L of water.

8.4.2 Bacterial Disease

Halo blight and bacterial brown spot are bacterial diseases affecting the faba bean crop production.

8.4.2.1 Halo Blight

This is a serious seed-borne disease of beans worldwide that can reduce the yield of beans and the quality of the beans. There are two races of phaseolicola, differentiated by their pathogenicity on *Phaseolus vulgaris* cultivars, and they have been reported from the United States and United Kingdom (Akhavan et al. 2013; Tock et al. 2017).

Studies conducted at CIAT (International Center for Tropical Agriculture) in 1986 reported the existence of a third race of the bacterium. In India, the disease was first reported in Delhi (Patel and Jindal 1972) and later in Maharashtra (Shirsat et al. 1976).

Symptoms: Infected seeds produce rotted cotyledons and chlorotic primary leaves. The water-soaked angular spots appear as tiny pinpricks on the lower surface of the leaf later becoming tan-brown. Halos do not develop under relatively higher temperature conditions above 21 °C. The spots eventually turn brown and are surrounded by a broad yellow-green halo when temperatures are optimum. Later, these spots become reddish brown to brown and dry. A non-host-specific toxin known as phaseolotoxin is responsible for the infection (Lelliott and Stead 1987; Prosen et al. 1993). During bacterial growth in a minimal medium between 18 and 20 °C, it is produced ex planta but is not detected at 28 °C. In the arginine biosynthesis pathway, phaseolotoxin targets the enzyme ornithine carbamoyltransferase (OCTase), which converts ornithine and carbamoyl phosphate to citrulline (Aguilera et al. 2012).

Cool and wet conditions favour the pod infection resulting in circular to oval, dark green, water-soaked, "greasy" lesions up to 9 mm in diameter. There may be crusty bacterial ooze on the surface of pod lesions as pods mature and turn yellow. Developing seeds may be shriveled or discoloured if lesions extend to the pod surface. The bacteria *P. syringae* pv. phaseolicola survive both as a parasite or saprophyte on plant tissue. It also produces a phytotoxin called phaseolotoxin. All species of beans are hosts of the bacterium, but the most susceptible types are snap, red kidney, cranberry and yellow-eye field beans. It is possible for the bacteria to survive for 6 to 18 months in plant refuse, on or above the soil surface, or in bean cull piles within a field, even under dry conditions. Contaminated seeds are the main source of primary inoculum in areas where bacteria in debris cannot survive in extreme temperatures. The bacterium is seed transmitted externally or internally in *Phaseolus vulgaris* and probably all other hosts.

Disease Cycle: During rainstorms, the moist wind particles can carry bacteria that can move many miles. Wet and cool conditions (18–22 °C) promote the development of diseases. Primary infection is due to emerging seedling that comes into contact with infected plant material. A layer of available water is necessary for infection. The bacteria enter in the plant via natural openings and create wounds than the bacterium can spread through the xylem.

Management: Disease-free seeds should be used. Detection of this pathogen in the seed is essential for effective control of the disease. Seeds can be grown in areas, where humidity and rainfall are normally too low for the bacteria to infect plants. Seed treatment with streptomycin reduces surface contamination. Soaking seeds for 1–5 min (Taylor and Dudley 1977) in 100 PPM streptomycin solution or dipping in sodium hypochlorite (1–2% available chlorine) reduce both surface infection and contamination by infected dust or debris. Handling plants wet with dew or rain should be avoided. Collection and burning of all infected plant debris as soon as after harvest should be done. Three-year rotations should be adopted in disease-prone areas. The cultivation of soybeans, cowpeas and other plants affected by this disease should be avoided during the rotation. Due to the appearance of new races of the pathogen, efforts to develop bean varieties resistant to the disease with acceptable characteristics were not very successful. However, the cultivation of bean varieties is recommended which are resistant.

8.4.2.2 Bacterial Brown Spot

The bacterial disease significantly causes economic losses to faba bean growers. It occurs in all the bean-growing areas. Due to its severe infection, the spots may coalesce and destroy the plant surface, and the infected plants appear blighted (Hirano and Upper 1983).

Symptoms: At each and every stage of the host's growth, symptoms appear. There are a few small, oval necrotic lesions found on the leaves. Around the lesions, a narrow yellow-green tissue zone may be seen. Leaves seem tattered as a result of small lesions coalescing and their centres falling out. Dark brown, smaller lesions that develop on pods lead to pod malformation by inciting cessation of growth of nearby tissue (Singh et al. 2012b).

The bacterium *Pseudomonas syringae* pv. Syringae cause the disease. The primary sources of infection are plant debris and weed hosts, which help in surviving the pathogen. Weather that is favourable to disease development includes somewhat warm temperatures, clouds and humidity. Plants damaged by unfavourable weather are more prone to infection. Infected tissue frequently releases large amounts of bacteria in humid or wet conditions, which is spread preferably by windblown rain to fresh tissues or plants and initiates new infections.

Disease Cycle: The disease development requires high humidity and temperature between 12 and 20 °C. These humid and cooler conditions along with heavy dew or fog provide favourable condition for bacterial brown spot. There are several routes through which bacteria gain access to a plant, including natural leaf openings (stomata and hydathodes) and wounds caused by hail, insects, blowing soil particles or cultivator injuries. Rust and the bacterial brown spot are frequently detected combined in the same lesion because the brown spot pathogen frequently infects the leaves through rust pustules. The bacteria ooze from the surface of diseased cotyledons (when the seedling sprouts and emerges) to the neighbouring plants and enter the stomata to infect the plant. The disease symptoms will start appearing within 2 to 5 days after the penetration of pathogen (Singh et al. 2012b).

Management: To control the bacterial disease, a combination of control measures is required. Sowing of healthy seeds with crop rotation with non-host crop is useful in checking the pathogen multiplication. The use of plant protection chemicals has been generally less useful to control the fungal diseases. The use of copper compounds as foliar sprays can give the good results. To control diseases like leaf spots, brown spots and blights use of Bordeaux mixture, fixed coppers and cupric hydroxide are commonly used.

8.5 Modern Approaches

Disease resistance has been successfully achieved through modern breeding methods that focus on the most cost-effective means of prevention with high impact. It is well known that complete monogenic resistance exists in a wide range of other crops and that it can be used effectively in marker-assisted selection (MAS) (Miklas et al. 2006; Garcia-Villalba et al. 2008). Since faba bean genomic resources are poorly developed, the genetic basis of resistance is quite unknown. It is relatively difficult to utilize quantitative trait loci for MAS due to a relatively large gap between their associated molecular markers and the QTL itself. Therefore in faba bean, it is quite challenging to identify tightly linked markers and the exact location of QTLs due to the little gnomic resources developed for faba bean and the low saturation of the genetic map holding putative QTLs (Torres et al. 2010; Rispail et al. 2010).

Marker-assisted selection (MAS) in breeding programmes is the most promising for the development of new cultivars. Many legume crops have been improved genetically through MAS, including common bean, soybean and pea; however, species like faba bean still remains in a developmental phase. It is more important to pay attention to the genes that govern simple traits rather than traits regulated by multiple genes. Recently, markers associated with a gene controlling plant growth patterns or impacting the nutritional quality of seeds like vicines/convicine, tannins and other nutrient content have also been reported, which may boost the selection of new cultivars with high-nutrition and disease resistance (Maalouf et al. 2021). Therefore, it is important for researchers to develop the molecular markers for many other highly sought-after traits as they are challenging to breed conventionally, such as frost, drought or disease tolerance in the near future. New candidate genes and selectable markers for MAS required for important traits will be identified by using comparative genomics and synteny investigations across the closely related legumes, along with mapping of resistance gene analogues (RGAs) (Torres et al. 2010).

8.5.1 Genetic and Genomic Resources

In the early 1980s, ICARDA identified the effective source of resistance to *Ascochyta* blight and chocolate spot disease (Robertson 1984; Hanounik and Robertson 1989) which was used by national partners to develop lines with high yields and good levels of resistance. Later researchers at the Ethiopian Institute of Agricultural Research (EIAR) released several varieties resistant to chocolate spot and high-yielding cultivars that were derived by transferring the resistance source from ICARDA germplasm into the locally adapted varieties helped to improve resistance to a specific disease. Some of the varieties known for partial resistance to chocolate spot are 'Moti' (ILB 4432 × Kuse-2-27-33), 'Walki' (ILB 4615 × Bulga 70), Obsie (ILB 4427 × CS20DK) and 'Gebelcho' (ILB 4726 × 'Tesfa') (Maalouf et al. 2016, 2019). Efforts are currently being made to develop multiple disease-resistant lines of faba beans (Maalouf et al. 2019), which are currently being used in the ICARDA breeding programme to develop cultivars with multiple disease resistance to different target environments.

8.5.2 Genome Mapping in Faba Bean

In 2010, the release of 5000 expressed sequence tags (ESTs) from the developing embryos of the broad bean variety "Windsor" represent the first significant contribution to the faba bean transcriptome (Ray and Georges 2010). The high transcriptome coverage produced came from Illumina sequencing of a library of mixed tissues ameliorate with embryo cells (Zhang et al. 2015). Construction of a comparative genetic map is an important discovery that is the foundation of a genetic map composed completely of sequence-based markers (Ellwood et al. 2008a, b), with the related model plants (e.g. Medicago) to easily trace. Bi-parental populations were mainly used for genetic studies and identification of QTLs. According to a study conducted by Satovic et al. (2013), the genetic linkage consensus map was aligned, which covers a high number of loci through merging three RIL populations' genetic maps. Recently by using the Goettingen Winter Bean MAGIC panel, markers for frost tolerance were identified (Sallam and Martsch 2015). In another study, two new resistance genes Uvf-2 and Uvf-3 found resistant against a range of faba bean rust pathotypes which can be used for gene pyrimidine through MAS (Ijaz et al. 2021).

8.5.3 Genome Editing

Genetic modification allows scientists to study the function of a gene by altering its expression through knockouts and targeted mutations. This process is also used for research into the generation of specific phenotypes, based on knowledge about how a particular gene functions. The first report of stable germline transformation of Vf using in vitro regeneration of *Agrobacterium*-infiltrated internode stem segments was described by Bottinger et al. (2001). The aim of reducing generation time utilizing tissue-culture based embryo rescue has received some attention in the absence of a reliable and effective transformation approach, with some degree of success (Mobini et al. 2020). A number of researchers have been investigating effective transformation methods. A renewed interest in genetic transformation could potentially be stimulated by new insights into allelic variations and the more advanced biotechnology opportunities provided by quickly developing genome editing technologies (Gaj et al. 2013; Maalouf et al. 2019). Recently, the first whole genome assembly of *B. fabae* was published and will help the researchers to understand the mechanisms of disease in better way (Lee et al. 2020).

8.6 Conclusion

Diseases are a major limitation to the adaptability and productivity of faba beans. A complete understanding of faba bean biology and the underlying resistance mechanisms is necessary for the successful application of biotechnology to disease resistance in faba beans. The large genome size of faba beans has made achieving success in this area much more difficult than with other crops. Even with significant advancements in genetic modification and tissue culture, the faba bean still lags behind other crops in terms of biotechnological advancements.

Faba bean is a minor legume crop grown primarily in the eastern states of India. While few studies have conclusively shown that genotypes from the ICARDA faba bean breeding programme may play an important role in helping new varieties adapt and thrive in India under the different biotic and abiotic stresses. Genetic research has made notable advancements in recent years as a result of increased access to genetic data. This includes improvement of cultivation techniques, which has led to more diverse cultivars being created, as well as advances in genomic sequencing technology. The genetics and genomics research community targeting faba bean genotypes provide the opportunity to choose resilient, disease-tolerant genotypes with high yields. Utilizing these tolerant sources in national breeding programmes could help to generate disease-resistant varieties. Though it has numerous reasons to benefit society as well as farmers, more research needs to be conducted before this knowledge can truly take hold.

Acknowledgements The CGIAR's Research Program on Grain Legumes supported the research and is greatly acknowledged by the authors.

References

Aguilera S, De la Torre-Zavala S, Hernández-Flores JL, Murillo J, Bravo J, Alvarez-Morales A (2012) Expression of the gene for resistance to Phaseolotoxin (argK) depends on the activity of genes phtABC in pseudomonas syringae pv. Phaseolicola. PLoS One 7(10):e46815

Ahmad A, Wei Y, Ullah S, Shah SI, Nasir F, Shah A, Iqbal Z, Tahir K, Khan UA, Yuan Q (2017) Synthesis of phytochemicals-stabilized gold nanoparticles and their biological activities against bacteria and Leishmania. Microb Pathog 110:304–312

Akhavan A, Bahar M, Askarian H, Lak MR, Nazemi A, Zamani Z (2013) Bean common bacterial blight: pathogen epiphytic life and effect of irrigation practices. Springerplus 2(1):41

Arraya Fernandez CM, Dhingra OD, Kushalappa AC (1987) Influence of primary inoculum on bean anthracnose prevalence. Seed Sci Technol 15:45

Bangar SP, Kajla P (2022) Introduction: global status and production of Faba bean. In: Punia Bangar S, Bala Dhull S (eds) Faba bean: chemistry, properties and functionality. Springer, Cham, pp 1–15

Bekhit MR, Rizk Z, Mansour K, Abdel-Moneim A, Kamel B, Boshra S (1970) Study of the effect of spraying with some fungicides at different dates and different intervals on the control of chocolate spot and rust of field beans. Agr Res Rev Cairo 48:37–63

Beyene AT, Derera J, Sibiya J (2018) Genetic variability of faba bean genotypes for chocolate spot (Botrytis fabae) resistance and yield. Euphytica 214:132

Bottinger P, Steinmetz A, Schieder O, Pickardt T (2001) Agrobacterium-mediated transformation of Vicia faba. Mol Breed 8:243–254

Carrillo-Perdomo E et al (2020) Development of new genetic resources for faba bean (Vicia faba L.) breeding through the discovery of gene-based SNP markers and the construction of a high-density consensus map. Sci Rep 10:1–14

Conner RL, Bernier CC (1981) Host range of Uromyces viciae fabae. Phytopathology 72:687–689

Cooper JW et al (2017) Enhancing faba bean (Vicia faba L.) genome resources. J Exp Bot 68:1941–1953

Dhull SB, Kidwai MK, Noor R, Chawla P, Rose PK (2022) A review of nutritional profile and processing of faba bean (Vicia faba L.). Legume Sci 4(3):e129

Ellwood SR et al (2008b) Construction of a comparative genetic map in faba bean (Vicia faba L.); conservation of genome structure with Lens culinaris. BMC Genomics 9:1–1

Ellwood SR, Phan HT, Jordan M, Hane J, Torres AM, Avila CM et al (2008a) Construction of a comparative genetic map in faba bean (Vicia faba L.); conservation of genome structure with Lens culinaris. BMC Genomics 9(1):380:380

FAOSTAT (2020). Food and agriculture Organization of the United Nations. www.Faostat.fao.org Accessed 18 Nov 2022

Gaj T, Gersbach CA, Barbas ICF (2013) ZFN, TALEN, and CRISPR/Cas-based methods for genome engineering. Trends Biotechnol 31:397–405

Gantait S, Mukherjee E (2021) Tissue culture-based genetic improvement of fava bean (Vicia faba L.): analysis on previous achievements and future perspectives. Appl Microbiol Biotechnol 105(18):6531–6546

Garcia-Villalba R, Leon C, Dinelli G, Segura-Carretero A, Fernandez-Gutierrez A, Garcia-Canas V, Cifuentes A (2008) Comparative metabolomic study of transgenic versus conventional soybean using capillary electrophoresis-timeof-flight mass spectrometry. J Chromatogr A 1195:164–173

Gimenez MA, Gonzalez RJ, Wagner J, Torres R, Lobo MO, Samman NC (2013) Effect of extrusion conditions on physicochemical and sensorial properties of corn-broad beans (Vicia faba) spaghetti type pasta. Food Chem 136:538–545

Gnanasambandam A et al (2012) Impact of molecular technologies on faba bean (Vicia faba L.) breeding strategies. Agronomie 2:132–166

Guo Z, Dong Y, Dong K, Zhu J, Ma L (2020) Effects of nitrogen management and intercropping on faba bean chocolate spot disease development. Crop Prot 127:104972

Gupta SK, Shyam KR (2000) Post-infection activity of ergosterol biosynthesis inhibiting fungicides against pea rust. J Mycol Plant Pathol 30:414–419

Guyot AL (1975) Les Ureudinees (ou rouilles des vegetaux) III Uromyces. Encycl Mycol 29:505–554

Hanounik SB (1981) Influence of "Ronilan" on the severity of chocolate spot and yield of faba bean. FABIS Newsl 3:50–51

Hanounik SB, Hawtin GC (1982) Screening for resistance to chocolate spot caused by Botrytis fabae. In: Hawtin GC, Webb C (eds) Faba Bean Improvement. Martinus Nijhoff Publishers, The Hague, The Netherlands, pp 243–250

Hanounik SB, Robertson LD (1988) New sources resistance in Vicia faba to chocolate spot caused by Botrytis fabae. Plant Dis 72:596–698

Hanounik SB, Robertson LD (1989) Resistance in Vicia faba germplasm to blight caused by Ascochyta fabae. Plant Dis 73:202–205

Hanounik SB, Viha N (1986) Horizontal and vertical resistance in Vicia faba to chocolate spot caused by Botrytis fabae. Plant Dis 70:770–773

Harrison JG (1979) Overwintering of Botrytis fabae. Trans Br Mycol Soc 72:389–394

Harrison JG (1984) Effect of humidity on infection of field bean leaves by Botrytis fabae and germination of conidia. Trans Br Mycol Soc 82:245–248

Hawtin GC, Stewart R (1979) The development and problems of faba bean (Viciafaba) in West Asia and North Africa. FABIS Newsl 1:7–9

Hebblethwaite PD (1983) The Faba bean. Butterworths, London, p 573

Hirano SS, Upper CD (1983) Ecology and epidemiology of foliar bacterial plant pathogens. Annu Rev Phytopathol 21:243–269

ICARDA (1987) Faba bean pathology Progress report 1986–1987. Food Legume Improvement Program, ICARDA, Aleppo, Syria

Ijaz U, Sudheesh S, Kaur S, Sadeque A, Bariana H, Bansal U, Adhikari K (2021) Mapping of two new rust resistance genes Uvf-2 and Uvf-3 in Faba bean. Agronomy 11:1370

Issam N, Naceur D, Nechi G, Maatalah S, Zribi K, Mhadhbi H (2021) Green synthesised ZnO nanoparticles mediated by Olea europaea leaf extract and their antifungal activity against Botrytis cinerea infecting faba bean plants. Arch Phytopathol Plant Protect 54:1083–1105

Jiang ZQ, Wang J, Stoddard F, Salovaara H, Sontag-Strohm T (2020) Preparation and characterization of emulsion gels from whole faba bean flour. FoodReview 9:755

Kaur S et al (2014) SNP discovery and high-density genetic mapping in faba bean (Vicia faba L.) permits identification of QTLs for ascochyta blight resistance. Plant Sci 217:47–55
Khater LA, Maalouf F, Jighly JQ, Alsamman AM, Rubiales D, Rispail N et al (2022) Genomic regions associated with herbicide tolerance in a worldwide faba bean (Vicia faba L.) collection. Sci Rep 12:158
Labba ICM, Frokiaer H, Sandberg AS (2021) Nutritional and antinutritional composition of fava bean (Viciafaba L., var. minor) cultivars. Food Res Int 140:110038
Leach R (1955) Recent observations of the botrytis infection beans. Trans Br Mycol Soc 38:171
Lee RC, Farfan-Caceres LM, Debler JW, Syme RA (2020) Characterization of growth morphology and pathology, and draft genome sequencing of Botrytis fabae, the causal organism of chocolate spot of Faba bean (Vicia faba L.). Front Microbiol 11:217
Lelliott RA, Stead DE (1987) Methods for the diag-nosis of bacterial diseases of plants. Blackwell Scientific Publications, Oxford. pp216
Lithourgidis AS, Roupakias DG, Damalas CA (2004) Evaluation of faba beans for resistance to sclerotinia stem rotcaused by Sclerotinia trifoliorum. Phytoprotection 85:89–94
Maalouf F, Ahmed S, Bishaw Z (2021) Faba bean. In: Pratap A, Gupta S (eds) The Beans and the Peas. Woodhead Publishing, pp 105–131
Maalouf F, Hu J, O'Sullivan DM, Zong X, Hamwieh A, Kumar S, Baum M (2019) Breeding and genomics status in faba bean (Vicia faba). Plant Breed 138:465–473
Maalouf F, Ahmed S, Shaaban K, Bassam B, Nawar F, Singh M, Amri A (2016) New faba bean germplasm with multiple resistances to Ascochyta blight, chocolate spot and rust diseases. Euphytica 211(2):157–167
Mansfield JW, Widdowson DA (1973) The metabolism of wyerone acid (aphytoalexin from Vicia faba L.) by Botrytis fabae and B. cinerea. Physiol Plant Pathol 3:393–404
Mansour K, Risky Z, Fouad W (1975) Studies on the control of chocolate spot and rust of horse beans. Agr Res Rev Cairo 53:89–95
Martineau-Cote D, Achouri A, Karboune S, L'Hocine L (2022) Faba bean: an untapped source of quality plant proteins and bioactives. Nutrients 14(8):1541
Matthews P, Marcellos H (2003) Faba bean, 2nd edn. NSW Agriculture, Australia, p 25
Miklas PN, Kelly JD, Beebe SE, Blair MW (2006) Common bean breeding for resistance against biotic and abiotic stresses: from classical to MAS breeding. Euphytica 147:105–131
Mobini S, Khazaei H, Warkentin TD, Vandenberg A (2020) Shortening the generation cycle in faba bean (Vicia faba) by application of cytokinin and cold stress to assist speed breeding. Plant Breed 139(6):1181–1189
Mohamed HAR (1982) Major disease problems of faba beans in Egypt. In: Hawtin GC, Webb C (eds) Faba bean improvement. Maainus Nijhoff Publishers, The Hague, The Netherlands, pp 213–225
Mueller DS, Hartman GL, Pederson WL (2002) Effect of crop rotation and tillage system on Sclerotinia stem rot of soybean. Can J Plant Pathol 24:450
Muimba-Kankolongo A (2018) Leguminous Crops. In: Muimba-Kankolongo A (ed) Food crop production by smallholder farmers in southern Africa. Academic Press, pp 173–203
Ntahimpera N, Dillard HR, Cobb AC, Seem RC (1997) Influence of tillage practices on anthracnose development and distribution in dry bean. Plant Dis 81:71
Ocana-Moral S, Gutierrez N, Torres AM, Madrid E (2017) Saturation mapping of regions determining resistance to Ascochyta blight and broomrape in faba bean using transcriptome-based SNP genotyping. Theor Appl Genet 130:2271–2282
Patel PN, Jindal JK (1972) Bacterial leaf spot and halo blight diseases of mung bean (Phaseolus aureus) and other legumes in India. Indian Phytopathol 25(4):517–525
Prosen D, Hatziloukas E, Schaad NW, Panopoulos NJ (1993) Specific detection pseudomonas syringae DNA in bean seed by PCR-based amplification of a phaseolotoxin gene region. Phytopathology 83:965–970
Ray H, Georges F (2010) A genomic approach to nutritional, pharmacological and genetic issues of faba bean (Vicia faba): prospects for genetic modifications. GM Crops 1(2):99–106

Rispail N, Kalo P, Kiss GB, Ellis THN, Gallardo K, Thompson RD, Prats E, Larrainzar E, Ladrera R, Gonzalez EM, Arrese-Igor C, Ferguson BJ, Gresshoff PM, Rubiales D (2010) Model legumes contribute to faba bean breeding. Field Crop Res 115(3):253–269

Robertson LD (1984) A note on the I.L.B. source of Botrytis fabae resistance. In: Chapman GP, Tarawali SA (eds) Systems for cytogenetic analysis in Vicia Faba L. Martinus Nijhoff Dr W, Junk Publishers, p 80

Rosa-Sibakov N, Heinio RL, Cassan D, Holopainen-Mantila U, Micard V, Lantto R, Sozer N (2016) Effect of bioprocessing and fractionation on the structural, textural and sensory properties of gluten-free faba bean pasta. LWT-Food Sci Technol 67:27–36

Sallam A, Martsch R (2015) Association mapping for frost tolerance using multi-parent advanced generation inter-cross (MAGIC) population in faba bean (Vicia faba). Genetica 143(4):501–514

Satovic Z, Avila CM, Cruz-Izquierdo S, Diaz-Ruiz R, Garcia-Ruiz GM, Palomino C et al (2013) A reference consensus genetic map for molecular markers and economically important traits in faba bean (Vicia faba L.). BMC Genomics 14(1):932

Satovic Z et al (2019) A reference consensus genetic map for molecular markers and economically important traits in faba bean (Vicia faba L.). BMC Genomics 14:1–15

Sharma SK, Verma BR, Sharma BK (1999) Biocontrol of Sclerotinia sclerotiorum causing stem rot of chickpea. Indian Phytopathol 52:44

Shirsat AM, Kale UV, Pawar NB (1976) Incidence of halo blight caused by pseudomonas phaseolicola (Burk.) Dowson on green gram (Phaseolus aureus) in Maharashtra state. J Maharashtra Agric Univ 1:314

Sillero JC, Villegas-Fernandez AM, Thomas J, Rojas-Molina MM, Emeran AA, Fernandez-Aparicio M, Rubiales D (2009) Faba bean breeding for disease resistance. Field Crop Res 115:297–307

Sindhan GS, Bose SK (1981) Evaluation of fungicides against anthracnose of F French bean caused by Colletotrichum lindemuthianum. Indian Phytopath 34:325–329

Singh AK, Bhat BP, Sundaram PK, Chndra N, Bharati RC, Patel SK (2012a) Faba bean (Vicia faba L.) phenology and performance in response to its seed size class and planting depth. Int J Agric Stat Sci 8:97–109

Singh AK, Bhatt BP (2012) Faba bean: unique germplasm explored and identified. Hort Flora Res Spectrum 1:267–269

Singh D, Sangle UR, Kumar B, Tripathi HS, Singh KP, Gupta AK (2012b) Integrated disease management of Faba bean (Vicia faba L.). In: Singh AK, Bhatt BP (eds) Faba bean (Vicia faba L.): a potential leguminous crop of India, Patna, pp 279–301

Singh G, Singh M (1990) Chemical control of Ascochyta blight of chickpea. Indian Phytopathol 43: 59

Singh P, Pal M (1995) Seed transmission of Ascochyta rabiei, the causal organism of chickpea blight. Indian Phytopathol 48:279

Sozer N, Melama L, Silbir S, Rizzello CG, Flander L, Poutanen K (2019) Lactic acid fermentation as a pre-treatment process for faba bean flour and its effect on textural, structural and nutritional properties of protein-enriched gluten-free faba bean breads. Food Rev 8:431

Stagnari F, Maggio A, Galieni A, Pisante M (2017) Multiple benefits of legumes for agriculture sustainability: an overview. Chem Biol Technol Agric 4:1–13

Stoddard F, Balko C, Erskine W, Khan HR, Link W, Sarker A (2006) Screening techniques and sources of resistance to abiotic stresses in cool season food legumes. Euphytica 147:167–186

Stoddard FL, Nicholas AH, Rubiales D, Thomas J, Villegas-Fernandez AM (2010) Integrated pest management in faba bean. Field Crop Res 115:308–318

Sugha SK (2001) Management of white rot of pea by fungicides. J Mycol Plant Pathol 31:335

Sulaiman N, Orfila C, Ho P, Maycock J (2018) Viciafaba: a cheap and sustainable source of protein and its application in beef products. Proc Nutr Soc 77(OCE4):E137

Taylor JD, Dudley CL (1977) Seed treatment for the control of halo blight of beans (pseudomonas syringae). Ann Appl Biol 85:223–232

Tekalign A, Sibiya J, Derera J, Fikre A (2017) Analysis of genotype_ environment interaction and stability for grain yield and chocolate spot (Botrytis fabae) disease resistance in faba bean (Vicia faba). Aust J Crop Sci 11:1228–1235

Tiwari N, Kumar T, Saxena DR, Swain N, Maalouf M, Kemal SA, Sarker A (2021) Evaluation of disease resistant and high yielding faba bean germplasm in India. J Genet 100:34

Tock AJ, Fourie D, Walley PG, Holub EB, Soler A, Cichy KA, Pastor-Corrales MA, Song Q, Porch TG, Hart JP, Vasconcellos RCC, Vicente JG, Barker GC, Miklas PN (2017) Genome-wide linkage and association mapping of halo blight resistance in common bean to race 6 of the globally important bacterial pathogen. Front Plant Sci 8:1170

Torres AM, Avila CM, Gutierrez N, Palomino C, Moreno MT, Cubero JI (2010) Marker-assisted selection in faba bean (Vicia faba L.). Field Crop Res 115(3):243–252

Tripathi HS, Singh RS, Chaube HS (1987) Effect of fertilizers on the severity of Ascochyta blight of chickpea. Indian Phytopathol 40:561

Vishunawat K, Chaube HS (1986) Survival of Ascochyta rabiei in gram seed. Indian J Mycol Plant Pathol 16:183

Webb A et al (2016) A SNP-based consensus genetic map for synteny-based trait targeting in faba bean (Vicia faba L.). Plant Biotechnol J 14:177–185

Wilson AR (1937) The chocolate spot disease of beans (Vicia faba L.) caused by Botrytis cinerea Pers M. Appl Biol 24:258–288

Zhang HM, Wheeler S, Xia X, Radchuk R, Weber H, Offler CE, Patrick JW (2015) Differential transcriptional networks associated with key phases of ingrowth wall construction in trans-differentiating epidermal transfer cells of Vicia faba cotyledons. BMC Plant Biol 15(1):103

9 Next-Generation Crop Breeding Approaches for Improving Disease Resistance in Groundnut (*Arachis hypogaea* L.)

Kiranmayee Bangaru, Anurag Mathew, Rachana Bagudam, Ankush Purushottam Wankhade, Ashutosh Purohit, Esnart Nyirenda Yohane, Seltene Abady, Dnyaneshwar Bandu Deshmukh, and Janila Pasupuleti

Abstract

Breeding programs across the globe focus on improving the groundnut/peanut yield, quality, and resistance to abiotic and biotic stresses. Diseases are widespread across the groundnut growing regions reducing the pod yield and impairing the kernel and haulm quality and are therefore economically important. Development of disease-tolerant/resistant cultivars for major groundnut diseases was achieved through breeding that used disease-resistant sources or its derivative as one of the parents, and "disease screening" to advance the selected breeding populations to next generation based on the disease reaction. Therefore, a reliable and repeatable disease screening protocol is critical to make progress for improvement of disease resistance. However, owing to the changing climatic scenario, and with challenges in disease screening, it may be difficult to improve genetic resistance for diseases with breeding methods that rely only on disease screening. Genomic tools enable selection of the disease-resistant/moderately resistant phenotype by the use of DNA markers thus circumvent the need to screen large number of "selection candidates." Moreover, with the recent advances in -omics technologies and sequencing of peanut genomes, it is possible

K. Bangaru · A. Mathew · R. Bagudam · A. P. Wankhade · A. Purohit · D. B. Deshmukh · J. Pasupuleti (✉)
International Crops Research Institute for the Semi-Arid Tropics (ICRISAT), Hyderabad, India
e-mail: Janila.Pasupuleti@icrisat.org

E. N. Yohane
Department of Agricultural Research Services, Lilongwe, Malawi

S. Abady
School of Plant Sciences, Haramaya University, Dire Dawa, Ethiopia

U. C. Jha et al. (eds.), *Diseases in Legume Crops*,
https://doi.org/10.1007/978-981-99-3358-7_9

to identify the genomic regions governing resistance to diseases. This chapter describes the major diseases of groundnut, genetics, and sources of host resistance and the new breeding technologies that can be implemented in breeding disease resistance which is critical in attenuating disease-incurred damage and progress toward making groundnut varieties more resilient to disease outbreaks.

Keywords

Groundnut · Biotic stress · Host-pathogen interaction · Omics · Haplotype

9.1 Introduction

Groundnut or peanut (*Arachis hypogaea* L.) is a leguminous food and oilseed crop of global importance, and Africa and Asia account for >90% groundnut area. With a total cultivation area of 32.7 million hectares (Mha) and a production of 53.9 million tons (Mt), groundnut is grown in 112 different nations (FAO 2021). It is an allotetraploid with a chromosome number of $2n = 4 \times (\text{AABB}) = 40$. Kernels are a rich source of energy (564 kcal from 100 g kernels), oil (48–50%), protein (25–28%), dietary fiber vitamins, minerals, and antioxidants (Janila et al. 2013). It supplies minerals like K, Na, Ca, Mn, Fe, and Zn, (Özcan and Seven 2003) as well as biologically active ingredients such as arginine, resveratrol, phytosterols, and flavonoids (Higgs 2003). Groundnut oil is one of the preferred cooking oils in several countries, such as India, China, Nigeria, etc., and globally, groundnut is mostly utilized in the food industry to make peanut butter and confectionery products, as well as for direct consumption as boiled, salted, and roasted nuts (Variath and Janila 2017).

Among the biotic stresses, diseases are economically important production-limiting factors to groundnut production, and they particularly impede the livelihood of small-holder farmers of Africa and Asia. The foliar fungal diseases, early and late leaf spots (ELS & LLS), and rust are important diseases all over the world. An outbreak of ELS can reduce the pod yield by 70% (Zongo et al. 2017), whereas LLS and rust can reduce by 50–70% (Wankhade et al. 2021). Aflatoxin-producing fungi, *Aspergillus flavus* and *Aspergillus parasiticus*, can infect the groundnut pods or kernels or both, at field and storage conditions (Variath and Janila 2017). Aflatoxin is a carcinogenic secondary metabolite produced by the fungus, and aflatoxin contamination in groundnut is major consumer health and trade concern, globally as aflatoxin-contaminated food and fodder could imperil the life of humans and animals. Among the soil-borne diseases of groundnut, *Sclerotium* stem rot is an important disease which can reduce the yield by 50–80% in the infested field (Agmon et al. 2022). Groundnut rosette disease (GRD) is an important disease across the growing regions in Africa. Peanut smut caused by *Thecaphora frezii*, which started as an emerging disease in South America, is now a threat to Argentine peanut production (Rago et al. 2017). Another disease, *Cylindrocladium* black rot (CBR) of peanut caused by *Cylindrocladium parasiticum* is an important disease in the USA (Coffelt and Garren 1982), and in 2010, this disease was reported in China (Gai et al. 2012). It is important to keep the track of the emerging diseases because of their ability to spread through seed and soil, and the destructive potential of the pathogens may pose a serious threat to global peanut production.

The important groundnut diseases are fungal diseases such as early leaf spot (ELS), late leaf spot (LLS), rust, aflatoxin disease, pod rot, *Sclerotinia* blight, *Cylindrocladium* black rot (CBR), and *Sclerotium* stem rot and viral diseases such as peanut bud necrosis virus (PBNV), peanut mottle virus (PMV), groundnut rosette disease (GRD), tomato spotted wilt virus (TSWV), peanut stripe virus (PStV), tobacco streak virus (TSV), Indian peanut clump virus (IPCV), and bacterial wilt caused by a bacterium. The diseases caused by fungi, bacteria, and viruses are described in Table 9.1.

Chemical measures of control of diseases increase the groundnut production cost by 10% (Coffelt and Porter 1986) and also pose a threat to the environment. Therefore, host-plant resistance is the best-bet approach to reduce the yield and quality losses caused by the diseases as it is environmentally sustainable and cost-efficient. Considering the economic importance, moderate resistance/resistance to LLS, ELS, rust for both Asia and Africa, and GRD for Africa are prioritized as "must-have"` traits in the breeding pipeline at International Crops Research Institute for the Semi-Arid Tropics (ICRISAT). Groundnut varieties with moderate resistance to LLS and resistance to rust have been developed using efficient screening for diseases under artificially inoculated conditions (Janila et al. 2016b). However, disease nurseries require huge resources, time, may be season, or location specific; consequently, the progress in improving disease resistance is slow, with low rate of genetic gains (Janila et al. 2016b).

Recently, molecular markers are used in groundnut to breed varieties with resistance to rust (Pasupuleti et al. 2016) and late leaf spot (Holbrook et al. 2022) and for both LLS and rust (Deshmukh et al. 2020; Shasidhar et al. 2020; Rajarathinam et al. 2022). QTLs for late leaf spot disease resistance have been identified in peanut involving wild diploid species (Bertioli et al. 2009) and derivatives of the wild species for both LLS and rust resistance (Gowda et al. 2002). Furthermore, nematode resistance genes have been introduced from a wild species (*A. cardenasii*) into an elite peanut variety (Simpson et al. 2003). Mapping populations, such as RIL (recombinant inbred lines), NAM (nested-association mapping) (Holbrook et al. 2013; Pandey et al. 2016), and MAGIC (multi-parent advanced generation intercross) (Wankhade et al. 2021; Guo et al. 2018), have been successfully developed and applied in groundnut for identifying the QTLs/genomic regions governing disease resistance. The development of advanced genetic resources, such as a transcriptome map (Clevenger et al. 2016) and a high-density genotyping array (Clevenger et al. 2017), has enabled researchers to develop new breeding approaches for use in breeding disease-resistant cultivars in groundnut. In this chapter, we describe the major groundnut diseases, their phenotypic screening methods, mechanisms of disease resistance, −omic technologies, and the new breeding approaches that can be applied to improve resistance to diseases in groundnut.

Table 9.1 Major diseases of groundnut, causal organism, and their symptoms

Pathogen type	Disease	Causal organism	Symptoms	References
Fungi	Early leaf spot	*Cercospora arachidicola*	Presence of subcircular, dark brown lesions with yellow hallow on the upper leaflet surface	Subrahmanyam et al. (1992)
	Late leaf spot	*Phaeoisariopsis personata*	Presence of circular lesions, darker than ELS, on the lower leaflet surface, leaf defoliation under severe infection is observed	Subrahmanyam et al. (1992)
	Rust	*Puccinia arachidis*	Presence of orange-colored pustules on the lower leaflet surface	Subrahmanyam et al. (1992)
	Stem rot	*Sclerotium rolfsii*	Yellowing and wilting of the lateral branch or the whole plant, when the main stem is attacked; leaves of infected branches turn chlorotic and then to brown as they rapidly dry out; presence of white mycelium sheaths of fungus around the infected plants near the soil surface	Bera et al. (2014)
	Cylindrocladium black rot	*Cylindrocladium crotalariae*	Presence of yellow, chlorotic, blighted leaves and cause wilting of plants, rotting of taproots, and hypocotyls. Presence of reddish orange spherical fruiting bodies on lesions present in plant	Subrahmanyam et al. (1992)
	Aflatoxin contamination	*Aspergillus flavus*	The fungus causes the kernels it attacks into a dry, shrivelled brown or black mass covered in yellow or greenish yellow spores. In the	Subrahmanyam et al. (1992)

(continued)

Table 9.1 (continued)

Pathogen type	Disease	Causal organism	Symptoms	References
			emerging seedlings, the radicle and hypocotyl become infected and rapidly decay. When the strain causing seedling disease produces aflatoxin, the adult plants may be severely stunted and have chlorotic or pale green leaves with vein-clearing of leaflets and have pointed tips	
	Collar rot	*Aspergillus niger*	Pre- and postemergence symptoms are visible. Preemergence symptoms include covering of the seed with black masses of conidia, giving a sooty appearance, the seeds fail to germinate. Postemergence, the seedlings are affected at the collar region causing yellowing of lower leaves, slow death due to blighting of the shoot, finally leading to death of the crown	Subrahmanyam et al. (1992)
Virus	Groundnut rosette disease	*Peanut rosette assistor luteovirus, peanut rosette umbravirus, and a satellite RNA*	Cause extreme stunting, due to shortening of the internodes and reduced leaf size; the two predominant symptoms are chlorotic and green rosette plants	Waliyar et al. (2007)
	Peanut bud necrosis disease	*Peanut bud necrosis virus*	Necrosis at growing buds, shorted internodes, petiole	Kesmala et al. (2006)

(continued)

Table 9.1 (continued)

Pathogen type	Disease	Causal organism	Symptoms	References
			bending at upper leaves, stunting, and proliferation of axillary branches	
	Tomato spotted wilt virus	*Tomato spotted wilt virus*	Presence of concentric ringspots, chlorotic spots on leaflets, stunting of all aboveground parts, reddish discoloration, and cracking on seed coat	Culbreath et al. (2003)
	Peanut stripe virus	*Peanut stripe virus*	Presence of intermittent stripes and green bands along the lateral veins of the leaves; stunted growth	Bera et al. (2022)
Bacteria	Bacterial wilt	*Ralstonia solanacearum*	Presence of slight drooping or curling of leaves; plants appear to dry and bend at the tip and eventually turn brown, wither, and die; discoloration of roots and pods is observed	Subrahmanyam et al. (1992)

9.2 Phenotyping Tools for Screening the Disease Resistance Reaction in Groundnut

The essential components for disease screening include multiplication of the pathogen, selection of proper phenotyping technique and time, time or crop stage for inoculum application, maintenance of optimum temperature and moisture conditions, and, eventually, evaluating the disease parameters (Kasundra and Kamdar 2016). An efficient screening tool is an essential requirement to identify resistance sources and use them in breeding program in exercising selection decisions. While the methods of screening for most the diseases in groundnut are reliable and repeatable, the screening for preharvest aflatoxin contamination (PAC) is prone to sampling errors, and consequently the repeatability is poor. Disease hot-spots (LLS, ELS, rust, GRD) (Chaudhari et al. 2019; Wankhade et al. 2021; Chapu et al. 2022) or sick fields (Bacterial wilt, Sclerotium rot) are also used for disease screening in groundnut but suffer from uniform disease pressure over years and across the field. Recently, image-based screening methods are also being developed for disease screening.

9.2.1 Early Leaf Spot, Late Leaf Spot, and Rust

Effective screening methods are available for screening for ELS, LLS, and rust both under field and controlled conditions. An infector row technique developed at ICRISAT (Subrahmanyam et al. 1995) is a widely used screening technique for LLS (Pasupuleti et al. 2013; Wankhade et al. 2021), rust (Pasupuleti et al. 2016;) and ELS (Zongo et al. 2017) in field conditions. Under controlled conditions, detached leaf assay, a rapid screening technique for LLS can be used to screen large number of entries in a short time (Foster et al. 1980; Pasupuleti et al. 2013; Deshmukh et al. 2020; Wankhade et al. 2021). Chapu et al. (2022) used normalized difference vegetation index (NDVI), red-green-blue (RGB), and color space indices (CSI), as high-throughput phenotyping measures for LLS.

9.2.2 Stem Rot

Screening for the stem rot disease is possible with artificial inoculation under sick plot and greenhouse conditions. Kasundra and Kamdar (2016) have described a screening procedure for stem rot disease under sick plot conditions. Under artificially inoculated sick field conditions, the number of infected and dead plants are counted at 30 DAP (days after planting), 60 DAP, 90 DAP to assess the disease severity in terms of percent infection and percent mortality, respectively. Based on percent mortality, genotypes are categorized into highly resistant (<10% mortality), resistant (10–19% mortality), moderately resistant (20–29% mortality), and susceptible (>30% mortality) (Kasundra and Kamdar 2016). Shokes et al. (1998) have designated a scale of 1–5 for disease severity assessment under glasshouse conditions. The scale is denoted by 1 = healthy plants (resistant) to 5 = > 50% of stems wilted or dead (highly susceptible). At ICRISAT, a laboratory method of screening for the disease, using oxalic acid assay, has been standardized (unpublished data).

9.2.3 *Cylindrocladium* Black Rot (CBR)

Under field conditions, CBR disease screening is conducted by artificially inoculating the groundnut plants with the microsclerotia of the pathogen, and recommended cultural practices are followed with regular irrigations. Two recommended methods of screening groundnut genotypes for CBR resistance were reported by Hammons et al. (1981). The first approach, which is particularly suited to the laboratory, makes use of a sterile environment, and second approach involves raising seedlings in non-sterile environmental chambers.

9.2.4 Aflatoxin Contamination

In groundnut, screening for aflatoxin contamination is done at both field and laboratory levels. At ICRISAT, a three-step screening process was adopted based on the targeted site for infection, which includes preharvest infection to the pods, seed coat mediated in vitro seed colonization and aflatoxin production in the cotyledons. After field screening, pods are to be collected and shelled, and then kernels are cultured in petri plates in the laboratory to test for preharvest seed infection. The infected seeds are evaluated for mycelial growth, green color, and colonies formed using the scale of 0–10 which ranges from 0 (no growth, green color, or fluffy colonies) to 10 (dense mycelium on all quarters, dark green color, or all fluffy colonies) (Xue et al. 2003). In addition to the *Aspergillus* infection, measuring the postharvest aflatoxin contamination is also important to screen the genotypes for tolerance. The aflatoxin contamination in the cotyledons is tested using enzyme-linked immunosorbent assay (ELISA) procedure. ICRISAT developed an indirect and direct competitive ELISA, which involves the separation of unreacted toxin in liquid phase from the bound toxin in solid phase (Waliyar and Sudini 2012).

With the current method of aflatoxin estimation, biased results are expected, owing to the sampling and analysis errors. Sampling is the most important step of aflatoxin estimation as it accounts for 90% of the total variability, while the sampling procedure accounts for 10% (Whitaker et al. 1993). Huge variability is observed between the replications for the same genotype, owing to small sample size. The true aflatoxin concentration of a population cannot be estimated with absolute accuracy due to the variability among the replicated samples of that population (Whitaker et al. 2004). Whitaker et al. (2004) stated that, by increasing the plot length of the standard deviation among the plot, aflatoxin values could be reduced, thus, reducing the variability among the aflatoxin values with large plot sizes.

9.2.5 Collar Rot

Screening for collar rot disease resistance can be carried out like stem rot disease screening by mass multiplication of the fungus on the sorghum grains followed by artificial inoculation to the groundnut plants. Regular irrigation must be ensured to maintain sufficient soil moisture level to create conducive environment for the fungus. Percent disease incidence is recorded for the assessment of the disease. Based on the disease incidence, the genotypes can be divided into four categories: (resistant, 1 to 10%), (moderately resistant, 11 to 20%), (susceptible, 21 to 30%), and (highly susceptible, >30%) (Kumari et al. 2016).

9.2.6 Groundnut Rosette Disease, Peanut Bud Necrosis Disease, Tomato Spotted Wilt Virus, and Peanut Stripe Virus

Disease assessment is based on percent disease incidence and disease severity. The percent disease incidence accounts for the number of plants infected at a given time and is recorded at 10 days interval, starting from 30 days after sowing (DAS) to a week before harvest (Kasno 1988; Waliyar et al. 2007; Bera et al. 2014).

$$\text{Percent disease incidence} = \frac{\text{Number of disease infected plants}}{\text{Total number of plants}} \times 100$$

Disease incidence multiplied with duration of days from 30 DAS to date of observation measures the disease progress by obtaining AUDPC score (Gopal et al. 2004).

AUDPC $= \sum [Y_{i+1} + Y_i] / 2[T_{i+1} + T_i]$ where Y_{i+1} = apparent incidence at the ith observation, T_i = time (days) at the ith observation, and n = total number of observations.

Disease severity scores are rated on a scale of 1 to 5 based on GRD (Essandoh et al. 2022), PBND (Gopal et al. 2004), TSWV (Garcia et al. 2000), and PStV symptoms. Infector row technique has also been found to be effective for screening PStV disease resistance (Kasno 1988). As PStV is endemic to certain locations, screening at disease hotspot can be utilized as an effective technique (Middleton et al. 1988). Diagnostic assays such as triple antibody sandwich enzyme-linked immunosorbent assay (TAS-ELISA) for GRD (Waliyar et al. 2007) and direct antigen coating enzyme-linked immunosorbent assay (DAC-ELISA) for PBND (Pensuk et al. 2004) and PStV (Hobbs et al. 1987) are used to confirm the presence or lack of viral agents during genotype evaluation.

9.2.7 Bacterial Wilt

Field evaluation for identification of resistant sources against bacterial wilt is mostly done in wilt sick plots or disease nursery. Test entries are planted in replications along with susceptible checks at regular intervals to maintain uniform disease pressure. Test entries can be categorized as immune (no wilting symptoms), highly resistant (more than 90% survival rate), and moderately resistant (more than 80% survival rate) based on the plant survival percentage at harvest stage (Mehan et al. 1994). Artificial inoculation with pure culture under greenhouse conditions can be done to ensure resistant reaction found in the field.

9.3 Host-Plant Resistance, Genetics of Resistance, and Resistant Sources

9.3.1 Early Leaf Spot, Late Leaf Spot, and Rust

Disease resistant genotypes are distinguished by leaves with thick epidermis, low stomatal index, smaller size of the stomata, high palisade index, and the presence of higher amounts of secondary metabolites which were effective in defense mechanism (Basra et al. 1985; Kaur and Dhillon 1990). The principal metabolites which are high in concentration in the LLS-resistant genotypes are sugars, malic acid, and citric acid. In addition, ribonic acid, cinnamic acid, flavanols, phenols, peroxidase (POD), and polyphenol oxidase (PPO) enzymes are also present (Jyosthna et al. 2004; Sukand and Kulkarni 2006; AL Harde et al. 2019 & Mahatma et al. 2021). Differentially expressed genes (DEGs)/R genes which regulate the resistance mechanism for leaf spot diseases (Dang et al. 2019, 2021; Gong et al. 2020) and rust (Rathod et al. 2020b) were identified.

Genetic studies on ELS, LLS, and rust resistance conducted before 20 years suggest that resistance to these diseases is complex and polygenic in nature and probably controlled by several recessive genes (Sharief et al. 1978; Nevill 1982; Green and Wynne 1986; Motagi 2001; Dwivedi et al. 2002). Furthermore, additive genetic variance seems to contribute predominantly to the resistance (Kornegay et al. 1980; Hamid et al. 1981; Anderson et al. 1986; Jogloy et al. 1987). Resistant sources for ELS, LLS, and rust are reported in both cultivated and wild species (Table 9.2). QTLs/linked genes were identified and mapped for diseases, namely, rust (Sujay et al. 2012; Kolekar et al. 2016; Ahmad et al. 2020), LLS (Sujay et al. 2012; Kolekar et al. 2016; Chu et al. 2019; Desmae et al. 2019; Ahmad et al. 2020), and ELS (Han et al. 2018; Chu et al. 2019). Using a QTL-*Seq* technique, Pandey et al. (Pandey et al. 2017b) revealed 25 candidate genes for LLS resistance and 9 candidate genes for rust resistance. Shirasawa et al. (2018) employed a method based on next-generation sequencing (NGS) called double-digest restriction site-associated DNA sequencing (ddRAD-Seq) to identify genetic loci for LLS and rust resistance.

9.3.2 Stem Rot

In some plants, whose stem is not well developed, the leaf anatomy such as cuticle thickness, stomatal morphology, and trichome structure caters to resistance. Tang et al. (2015) reported that the stomatal characteristics such as density of stomata, length/width ratio, and their position define resistance in some *Cyperus* species. Formation of tyloses (Sujitkumar 2015) determines the resistance to *S. rolfsii*. Biochemical resistance is achieved through the release of certain enzymes (Chen et al. 2000), enhanced activity of POD, PPO, chitinase, and β-1, 3-glucanase (Nandi et al. 2013) and distinguished metabolic pattern of sugars in *S. rolfsii* inoculated groundnut plants (Mahatma et al. 2018). Kajal kumar Biswas and Chitreswar Sen (Biswas and Sen 2000) reported role of microbial biological control agents, i.e.,

Table 9.2 Resistant sources identified for groundnut diseases reported in the literature

Disease	Sources of resistance	References
Early leaf spot	PI 276325, 33,820, 262,797; NRCGCS-77, NRCGCS-85, NRCGS-86; *A. stenosperma*, *A. diogoi*; *A. cardenasii*, *A. chacoense*, *A. appressipila*, *A. pusila*, *A. kempff-mercadoi*, *A. batizocoi*, *A. hagenbeckii*, *A. glabrata*, and *A. repens*	Nigam (2014); Motagi et al. (2022); Foster et al. (1981)
Late leaf spot	PI 276325, 262,141, 338,280, PI 262797, *A. diogoi*, *A. cardenasii*, *A. glabrata*, *A. stenosperma*, *A. repens*, *A. appressipila*, *A. paraguariensis*, *A. villosulicarpa*, *A. hagenbeckii*, *A. chacoense*, *A. batizocoi*, *A. duranensis*, *A. correntina*, *A. villosa*, *A. pusila*, *A. kempff-mercadoi*; ICGV 99001, 99,004; GPBD 4; NRCGCS-77, NRCGCS-85, NRCGS-86; VRI Gn 5	Nigam (2014); Singh et al. (2003); Gowda et al. (2002); Motagi et al. (2022); Vindhiyavarman and Mohammed (2001); Subrahmanyam et al. (1982a, b)
Rust	PI 298639, 338,312, 219,823, 263,133, 331,194, 338,280, 262,141, 276,235, 210,554, *A. batizocoi*, *A. correntina*, *A. cardenasii*, *A. duranensis*, *A. diogoi*, *A. pusilla*, *A. villosa*; *A. chacoense*, *A. stenosperma*, *A. repens*, *A. appressipila*, *A. hagenbeckii*, *A. glabrata*, *A. correntina*, *A. villosa*, *A. kempff-mercadoi*, *A. paraguariensis*, *A. villosulicarpa*, *A. spegazzinii*; ICGV 99003, 99,005; GPBD 4; NRCGCS-77, NRCGCS-85, NRCGS-86; VRI Gn 5, ICGV 87354; AB-ICGS76-7-1, AB-ICGS76-18-4, AB-ICGS76-40-6	Nigam (2014); Singh et al. (2003); Gowda et al. (2002); Motagi et al. (2022); Vindhiyavarman and Mohammed (2001); Reddy et al. (2001); Kumari et al. (2014); Abdou et al. (1974), Subrahmanyam et al. (1982a, b); (1985a, b)
Stem rot	NC 2; NC 3033; CS-319, 21, 86, 222, Florida-07, GBFDS-272, Georgia-07 W, CS-19, Georgia-03 L, Dh 8; ICGV 86590; ICGV 87157	Cook (1981); Beute et al. (1976), Bera et al. (2016); Reddy et al. (1993); Nigam et al. (1992)
Cylindrocladium black rot	*A. duranensis*; *A. valida*, *A. williamsii*, *A. cruziana*, *A. batizocoi*, and *A. correntina*; NC 3033; "NC 8C," "NC 10C," "NC 12C," and "Perry"; Georgia Greener, Georgia-06G, Georgia-07 W, Georgia-02C, and Carver; Tifton-8	Kochert et al. (1996); Tallury et al. (2014); Beute et al. (1976); Wynne and Beute (1983); Branch and Brenneman (2012); Coffelt and Garren (1982)

(continued)

Table 9.2 (continued)

Disease	Sources of resistance	References
Collar rot	RG-425, CSNG-19-1, SNG-69, GG-21, and RG-559-3; U4-47-7; C421 and C1780; GG 2	Kumari et al. (2016); Aulakh and Sandhu (1970); Dasgupta and Raj (1997); Nathawat et al. (2014)
In vitro seed colonization	PI 337394 F and PI 337409; Ah 78,223, U 4-47-7, Var 27, Faizpur, and Monir 240-30; Zhonghua 6; ICG 13212, ICG 11560, ICG 8131, and ICG 14875	Mixon and Rogers (1973); Mehan (1989); Liao et al. (2009); Liang et al. (2009); Thakur et al. (2000)
Preharvest aflatoxin contamination (PAC)	ICGs 13,603, 1415, 14,630, 3584, 5195, 6703 and 6888	Waliyar et al. (2016)
Aflatoxin contamination	ICG 1326, ICG 3263, ICG 3336, ICG 3700, ICG 4749, and ICG 7633; ICGV 87084, ICGV 87094, and ICGV 87110; J11, ICG 9610, ICG 1323, ICG 10094, ICG 9407, ICG 3263, ICG 4749, ICG 1859, and ICG 7633; ICG 12625, ICG 4750; Zh.h0551 and Zh.h2150; ICG 13212, ICG 11560, ICG 8131 and ICG 14875; PI 468319, PI 468200, PI 262133, PI 262141, and PI 475997 *A. cardenasil* and *A. duranensis*	Mehan et al. (1991); Waliyar et al. (1994); Nigam et al. (2009); Yu et al. (2019); Thakur et al. (2000); Xue et al. (2004); Nigam et al. (1991a)
Groundnut rosette disease	69-101, RMP 12, RMP 40, and RG 1, KH149 A, KH 241C, KH 241 D, CN94C, QH 243C, and ICGV-SM 90704; ICG 12991; RMP 40, RMP91, and RG1 Hybrid from an interspecific cross between *A. hypogaea* and *A. chacoense*; *A. repens*, *A. glabarata*	Van der Merwe et al. (2001); Subrahmanyam et al. (2000); Waliyar et al. (2007); Subrahmanyam et al. (1985a)
Peanut bud necrosis disease	ICG 8199, 8956, 11,552, 11,553, 11,555, 8132, 8189; ICGV 87160; ICGV 86590; ICGV 86699; ICGV 87157; ICGS 11; ICGV 87141	Nigam (2014); Reddy et al. (1992); Reddy et al. (1993); Reddy et al. (1996); Nigam et al. (1992); Nigam et al. (1991c); Nigam et al. (1991b)
Tomato spotted wilt virus	PI 262794, 33,864, 468,141, 468,144; Georgia Green; Georgia Bold; Georgia Hi-O/L peanut; *A. cardenasii*, *A. diogoi*, *A. correntina*, and *A. pusilla*	Nigam (2014); Branch (1996); Branch (1998); Branch (2000); Subrahmanyam et al. (1985a, b)
Peanut stripe virus	PI 468141, PI 468142, PI 468144, PI 468345; PI 475998, PI 476012, PI 476013, PI 276235; PI 262801, PI 262794, PI 210555-1, GKP 9530-31; PI 468174, PI 468363, PI 468366; PI 476004, PI 468170, PI 468176; Huayu 16 and Huayu 17	Culver et al. (1987); Prasada Rao et al. (1989); Xu and Zhang (1987); Li and Qiu (2000)

(continued)

Table 9.2 (continued)

Disease	Sources of resistance	References
Bacterial wilt	ICG 1703, 1705, 7893, 7894 11,325 5272, 5273, 5276, 1609, 5313, 7343, 7968, and 8666; Schwarz 21, Banteng, Gajah, Kidang and Macan, Pelanduk, Tapi and Tupai, Guiyou 28, El Hua 5, Lu Hua 3 and Zhonghua 2, Zhonghua 4, Tianfu 11, Zhonghua 6, and Zhonghua 21; ICGV 87165	Singh et al. (1997); Bera et al. (2022); Moss et al. (1997)

Trichoderma harzianum (three isolates, viz., T8, T10, and T2), which induced resistance without directly interacting with the pathogen.

Resistance to soil-borne diseases including stem rot is governed by polygenes controlled by additive gene effects (Fry 1982; Vander Plank 1968; Bera et al. 2016). Resistance to stem rot is heritable; thus, it can be improved in the further generations (Smith et al. 1989). Some of the resistant sources are given in Table 9.2, and a stem rot-resistant variety, Bailey, was released in the USA for commercial cultivation. QTLs governing resistance to stem rot were reported (Dodia et al. 2019; Luo et al. 2020b; Cui et al. 2020; Agmon et al. 2022).

9.3.3 Cylindrocladium Black Rot

Combining phenological suppression as an escape mechanism with metabolic resistance could possibly lead to resistant reactions for CBR (Shew et al. 1987). Formation of an additional effective periderm is one of the resistance mechanisms to CBR (Beute 1980). Susceptible peanut cultivars depict more breachments on the tap root and are less effective in formation of additional periderms than the resistant lines (Harris and Beute 1980). Sources of resistance to CBR in both wild and cultivated species are reported (Table 9.2). Heritability studies revealed that resistance is quantitative, with only additive genetic effects involved in the inheritance of resistance (Hadley et al. 1979; Hammons et al. 1981; Green et al. 1983), and is influenced by morphophysiological traits. Physiological resistance is governed by cytoplasmic factors (Coffelt and Porter 1982). Nigam (2014) stated that the broad-sense heritability ranges from 48 to 65%, and narrow sense heritability is around 51.7% for CBR disease resistance.

9.3.4 Aflatoxin Contamination

The mechanisms of resistance to aflatoxin contamination in groundnut include resistance against infection in the pod wall, seed invasion, seed coat colonization, and aflatoxin production in cotyledons (Soni et al. 2020a). The first point of contact between the host and pathogen is pod shell, and resistance is due to the structure of pod shell (Upadhyaya et al. 2002). The pathogen must first break through the pod wall during infection in order to get to the cotyledons, where they obtain nutrients and produce aflatoxin. *A. flavus* can enter the pod through cracked pod shells in drought-prone areas as there is less moisture in the pod or soil, which increases aflatoxin contamination (Girdthai et al. 2010). Aflatoxin levels in injured shells are higher than in undamaged shells (Sudhakar et al. 2007). The fungal pathogen can enter through the seed coat when the developing pods in the soil are damaged by insects, extreme temperatures, or drought. The seed coat, or testa, acts as the second barrier to the pathogen. This resistance is due to the palisade layer thickness, density, wax layers, and lack of cracks and cavities in the seed coat (Nigam 2014). Liang et al. (2003) emphasized the role of wax and cutin layers in groundnut seed coat as a physical barrier for imparting resistance against the fungal pathogen. Biochemical resistance is achieved due to the presence of tannins (Sanders and Mixon 1978) and 5–7-dimethoxyisoflavone (Turner et al. 1975); they have been identified as important inhibitors of *A. flavus* infection. The genetics, structural, biochemical, and molecular mechanisms (genes and proteins) and the genomic regions governing aflatoxin resistance have been detailed by Soni et al. (2020a).

Preharvest infection, seed coat-mediated colonization, and production of aflatoxin mechanisms are all independently inherited. Occurrence of reciprocal differences and additive gene effects is reported for seed coat resistance. Gene effects are primarily nonadditive for aflatoxin production (Nigam 2014). QTLs have been reported for aflatoxin (Lei et al. 2005, 2006; Yu et al. 2019; Khan et al. 2020; Ding et al. 2022).

9.3.5 Groundnut Rosette Disease

The mechanism of resistance involves limitations on virus movement, creation of the sat RNA that causes symptoms (Ntare et al. 2002), and resistance to aphids, the transmitters of the virus. These factors may cause the resistance mechanisms to differ from one variety to the other. It is hypothesized that under field conditions, aphid resistance is responsible for ICG 12991's low rosette incidence, whereas resistance to the virus is responsible for ICGV-SM 90704's and RG 1's low rosette incidence (Van der Merwe et al. 2001). However, it does not imply that the genotypes that are resistant to aphids are also resistant to the disease as they fail to show resistance to the virus. Some resistant varieties for GRD are released in Africa for commercial cultivation. Resistance to GRD is a must-have trait for the cultivars to be released in Africa.

Resistance to GRD has been reported to be under the control of two independent recessive genes (Nigam and Bock 1990; Olorunju et al. 1992; Ntare et al. 2002). The existence of genetic variability for resistance to GRD, with more significant additive gene action, has been observed (Kayondo et al. 2014). High broad-sense heritability has been reported for resistance to GRD (Kayondo et al. 2014). Limited studies were conducted on QTL identification for GRD resistance (Essandoh 2021).

9.3.6 Peanut Bud Necrosis Disease, Tomato Spotted Wilt Virus, and Peanut Stripe Virus

Host plants are categorized as susceptible, tolerant, resistant, and immune to viral infection in groundnut (Walkey 1985). Susceptible host supports rapid virus infection, multiplication, and systemic spread within host cell. The tolerant plants show only mild symptoms without marked effect on plant growth and vigor or yield. A resistant host, however, does not readily allow virus infection, proliferation, or spread. Immune plants do not have surface receptors that would allow for viral particle adhesion, preventing virus penetration into host cells.

Pathogen-derived resistance (PDR), a mechanism of resistance against PBNV and TSWV, involves the expression of viral coat protein encoded by nucleocapsid gene, replicase protein, movement protein, and resistance genes (R genes). Plants expressing the coat protein of one virus will restrict the entry or replication of other viruses. The resistance induced by the R genes is systemic and is effective against different types of viruses. RNA interference (RNAi) is used as an antiviral mechanism that will degrade the pathogenic RNA, and resistance induced by RNAi is virus-specific systemic resistance (Anderson et al. 2018). These mechanisms have shown significant impact in plant disease resistance against viruses.

Pensuk et al. (2004) reported that PBND resistance is a polygenic trait, as it is governed by multiple genes. Genetic analysis studies have identified that resistance to PBND is due to additive inheritance and the absence of dominance and epistatic gene effects (Jadhav et al. 2019). Nigam (2014) reported significant general combining ability and specific combining abilities and transgressive segregation for TSWV resistance. With limited efforts to understand the genetics behind PStV resistance, significant general combining ability in two Virginia-type parents NC Ac 2821 and ICGS 4 has been reported by Anderson et al. (1990). Partially dominant mono-gene "*Pst*" from a cultivated variety of soybean has been reported to confer resistance against PStV (Choi et al. 1989). QTLs governing resistance to TSWV (Qin et al. 2012; Agarwal et al. 2019; Gaurav et al. 2019; Zhang et al. 2019) and PBND (Jadhav et al. 2019; Jasani et al. 2021) have been reported.

9.3.7 Bacterial Wilt

Groundnut exhibits host plant resistance against bacterial wilt pathogen through a longer latent period of infection. Late maturing Virginia runner types are comparatively resistant to bacterial wilt than early maturing Spanish and Valencia types. Several morphological features are associated with bacterial wilt resistance, especially the root architecture and number of root nodules (Liao et al. 1994). Cultivars with long lateral root system and fewer root nodules exhibit resistance reaction compared to long main root with a greater number of nodules. Hypersensitive reaction-mediated partial wilting symptom has been conferred as a defense reaction of the host plant where resistant cultivars have a longer wilting period. Several biochemical components like gallic acid, catechin, p-coumaric acid, polyphenol oxidase, and esterase govern resistance (Duan et al. 1994).

Sources of resistance for bacterial wilt are used in breeding to develop and release bacterial wilt resistant varieties in China, Vietnam and Indonesia, where this is an economically important disease. For bacterial wilt, three pairs of major genes and few minor genes govern resistance with partial dominance (Liao et al. 1986), whereas, there are reports of resistance reaction being a recessive trait (Wang et al. 1985), in case of tomato, bacterial wilt resistance has been found to be governed by both dominant and recessive genes (Scott et al. 1988; Messiaen et al. 1991). QTLs were reported for resistance to bacterial wilt (Zhao et al. 2016; Luo et al. 2020a).

9.4 Omics Resources

9.4.1 Genomic Resources

The diploid progenitors of peanut, *A. duranensis* and *A. ipaensis*, have their genome sequenced (Bertioli et al. 2016), and recently the tetraploid genome is sequenced independently by two groups (Bertioli et al. 2019; Chen et al. 2019a). Accessibility to the complete genome sequence is aiding in breeding of groundnut varieties with desired traits. The availability of low-, mid-, and high-density SNP genotyping assays for groundnut has accelerated the breeding process. The low-density assay containing 10–100 polymorphic SNPs can be used for quality control (QC) for testing purity of the parents and genotypes in the breeding programs (Pandey et al. 2020). Groundnut mid-density panel comprises of 2500 SNP markers distributed across the genome, and it is easily accessible (https://excellenceinbreeding.org/toolbox/services/groundnut-mid-density-genotyping-services). The marker panel comprises 20 quality control (QC) and 72 associated markers for 8 key traits, namely, leaf rust resistance, late leaf spot resistance, net blotch resistance, high oleic acid, seed weight, shelling percentage, fresh seed dormancy, and blanchability, hence, can be used in disease resistance breeding. The 58 k (Axiom_Arachis) and 48 k high density SNP arrays were developed and validated independently at ICRISAT (Pandey et al. 2017) and University of Georgia (Clevenger et al. 2017) from diverse groundnut genotypes including both tetraploids and diploids. These high-density genotyping arrays are now often employed in trait mapping and genetic diversity studies. Thus, implementing genomic technologies into the breeding process, combined with higher precision in yield trialling and phenotyping, will increase efficiency and genetic gain for the release of better groundnut varieties (Desmae et al. 2019).

9.4.2 Transcriptomics

With the availability of the reference and cultivated genomes for groundnut (Bertioli et al. 2016; Chen et al. 2016) along with pathogen, genome sequencing revealed novel genes and complex genomic architecture. Additionally, transcriptomics studies have substantially improved our understanding of gene expression and the pathways that are enriched in the host-pathogen interaction which controls the groundnut disease-related mechanism. Insights into the molecular understanding of resistance mechanisms have now been considerably accelerated by modern methods like sequencing-based RNA-*seq* for early leaf spot (Rathod et al. 2020a), late leaf spot (Gangurde et al. 2021), rust (Rathod et al. 2020b), stem rot (Jogi et al. 2016; Bosamia et al. 2020), preharvest aflatoxin contamination (Clevenger et al. 2016; Soni et al. 2021), and aflatoxin production (Wang et al. 2016; Korani et al. 2018; Soni et al. 2020b; Cui et al. 2022) in groundnut. Details of transcriptomic studies for groundnut diseases are provided in Table 9.3. These investigations have improved our knowledge of genetic controls and the molecular mechanisms underlying host-pathogen interactions, paving the way for the development of new strategies for breeding disease-resistant groundnut varieties.

9.4.3 Proteomics

Identification of genes and proteins responsible for stress tolerance and disease resistance is required to increase crop productivity. Proteomics is a powerful tool for understanding the dynamics of proteins expressed by genomes in response to various environmental stresses and biological processes (Graves and Haystead 2002; Jamshidi Goharrizi et al. 2020). Plant-pathogen interactions rely heavily on changes in proteome composition and protein activity (Elmore et al. 2021; Jamshidi Goharrizi et al. 2020). Proteome profiling, during pathogen infection, can identify specific proteins and associated biological pathways that contribute to disease resistance and susceptibility (Bhatnagar-Mathur et al. 2021).

In groundnut, proteomics approaches have been previously used to understand the development of seed (Kang et al. 2007), gynophore (Zhao et al. 2015), leaf (Katam et al. 2010), allergens (Porterfield et al. 2009), and drought tolerance (Kottapalli et al. 2013). With respect to disease resistance studies, proteomics has been applied to identify and reveal some differentially expressed proteins (DEPs) associated with peanut preharvest aflatoxin contamination (Wang et al. 2013; Zhao et al. 2019). Nontarget proteomics, based on 2D gel electrophoresis and liquid chromatography-mass spectroscopy (LC-MS), was used to explain the mechanism of resistance against *A. flavus* (Ouakhssase et al. 2019; Bhatnagar-Mathur et al. 2021). Bhatnagar et al. (Bhatnagar-Mathur et al. 2021) revealed that several resistance proteins associated with secondary metabolic pathways were strongly induced

Table 9.3 Transcriptome-based studies for identification of key genes and associated pathways underlying disease resistance mechanisms in groundnut

Disease	Key genes/transcription factors/pathways identified	Function	Reference
Early leaf spot (ELS)	Thaumatin, glutathione peroxidase, and cinnamyl alcohol dehydrogenase	Defence-related genes	Rathod et al. (2020a)
Late leaf spot (LLS)	RPP13-like protein and nucleotide binding site leucine-rich repeat (NBS-LRR)	Disease-related genes	Gangurde et al. (2021)
	Ethylene-responsive factor (ERF) and ethylene-responsive nuclear protein (ERF2), and early responsive dehydration gene (ERD)	Causes leaf defoliation	
	Antibiotic biosynthesis, flavonoid biosynthesis, and phenylpropanoid biosynthesis pathways	Triggered against infection by pathogen causing LLS	
Rust	Pathogenesis-related (PR), thaumatin, and F-box proteins	Defense-related proteins	Rathod et al. (2020b)
	Ethylene-responsive transcription factor	Regulates the expression of defense-related genes	
	Chitinase, cytochrome P450, glutathione S-transferase, and NBS-LRR	Plant defense mechanism	
Stem rot	Nucleotide-binding site leucine-rich repeat (NBS-LRR), thaumatin-like proteins, glutathionine S-transferases, polygalacturonase-inhibiting proteins, and resveratrol synthase	Defense-related genes	Jogi et al. (2016)
	Pathogenesis-related genes (PR)-3, 4	Anti-fungal activity	
	WRKY transcription factor	Induces systemic acquired resistance (SAR) by the activation of the jasmonic acid defense signaling pathway	Bosamia et al. (2020)
	Toll/Interleukin1 receptor-nucleotide-binding site leucine-rich repeat (TIR-NBS-LRR), dirigent proteins, CC-NBS-LRR, LRR, and NB-ARC domain protein	Defense related genes	

(continued)

Table 9.3 (continued)

Disease	Key genes/transcription factors/pathways identified	Function	Reference
Preharvest aflatoxin contamination (PAC)	ABR1	Ethylene-responsive transcription factor and repressor of ABA signaling	Clevenger et al. (2016)
	Pathogenesis related-2	Stress and defense-responsive gene	
	Fatty acid biosynthesis, flavonoid biosynthesis, seed lineolate gene expression, chalcone synthase, 9 s-LOX, resveratrol synthase, and glutathione S-transferase	Defense response genes	Soni et al. (2021)
	ABA-responsive ABR17	Co-regulates the genes of ABA-responsive elements during drought stress	
Aflatoxin contamination	Pathogenesis-related-1,2,5; NBS-LRR genes	Defense-related genes	Wang et al. (2016)
	WRKY	Stress regulative transcription factor	Korani et al. (2018)
	Toll/interleukin1 receptor-nucleotide-binding site leucine-rich repeat (TIR-NBS-LRR)	Highly conserved disease-resistant genes in plants	
	Ethylene-responsive factors	Transcriptionally regulates jasmonate signaling pathway	
	Transcription factors like ARF, DBB, MYB, NAC, and C2H2	Jasmonic acid, salicylic acid, and phenyl propanoid biosynthesis pathways	Soni et al. (2020b)
	1-aminocyclopropane-1-carboxylate oxidase (ACO1)	Ethylene signaling pathway	Cui et al. (2022)
	Pathogenesis-related proteins (PR10), serine/threonine kinase (STK), pentatricopeptide repeat (PPR) protein	Disease resistance proteins	
	Mitogen-activated protein kinase (MAPK)	Protein kinase superfamily protein	
	Pattern recognition receptors (PRRs)	Pattern-triggered immunity (PTI) response	
	Cytochrome P450	Disease resistance gene	

in the resistant genotypes. Kumar and Kirti (2015) have reported 233 differentially expressed genes through proteomic analyses of resistant host responses in *Arachis diogoi* against late leaf spot pathogen. Advanced omics technology like proteomics is sparsely explored in the peanut disease resistance breeding programs, and only a few reports are available. To cull out the underlying genes and proteins governing peanut disease resistance, the application of proteomic approaches to peanut disease resistance studies is essential.

9.4.4 Metabolomics

Metabolomics deals with expression of metabolites and their changes, along with their interaction with plant phenotypic traits during stress. Hence, studying the metabolomics of legumes in response to biotic stress will help in understanding the pathways related to stress response. Studies have been reported on metabolite profiling across legumes under biotic stresses: chickpea infected with *Fusarium oxysporum* (Narula et al. 2020); pea infected with *Didymella pinodes* and *R. solani*; and soybean infected with *Aspergillus* infection (Makhumbila et al. 2022).

Metabolite profiling was also done in groundnut using gas chromatography-mass spectrometry (GC-MS), against fungal diseases – stem rot caused by *Sclerotium rolfsii* (Mahatma et al. 2018); aflatoxin produced by *Aspergillus flavus* (Sharma et al. 2021); and late leaf spot caused by *Cercospora personata* (Mahatma et al. 2021). Those metabolites having higher accumulation in resistant genotypes can be used as biomarkers in identifying resistant germplasm, which are ultimately involved in defence mechanism against the pathogens. Breeding process could be accelerated and made more effective by combining metabolomics, a metabolite-based selection approach, along with other omics approaches. As a result, metabolomics approaches will be very helpful to identify metabolite markers for resistance as well as various molecular pathways involved in response to environmental stimulus.

9.5 Next Crop Breeding Approaches for Disease Resistance

9.5.1 Marker-Assisted Breeding (MAB)

The application of genomic tools to groundnut breeding programs is quite recent. Since then, groundnut genomics has been progressing, with significant achievements with respect to marker development, genetic and phenotypic mapping, and genome sequencing. These developments have led to greater understanding of the groundnut genome, resulting in identification of genes for traits of interest and the incorporation of marker-assisted breeding for selected traits (Desmae et al. 2019). The practice of applying genomic tools in groundnut breeding began with focus toward specific traits. Mapping populations had been developed for disease resistance studies, for economically important diseases (Janila et al. 2016a).

Among the different genomic tools, marker-assisted breeding has been used extensively either as marker-assisted selection (MAS) or marker-assisted backcrossing (MABC). In groundnut, the first root-knot nematode-resistant groundnut variety, NemaTAM, was successfully bred using the MAS method and released in the USA (Simpson et al. 2003). Marker-assisted backcrossing (MABC) has been the most successful approach for trait improvement and gene pyramiding (Janila et al. 2016b). The first successful application of the MABC scheme was for combining foliar disease (rust) resistance with early maturity in groundnut at ICRISAT. Varshney et al. (2014) introduced the rust-resistant QTL into three early maturing elite groundnut cultivars, namely, ICGV 91114, JL 24, and TAG 24, using GPBD 4 as the donor parent, employing the markers flanking the rust QTL which were earlier identified by Khedikar et al. (2010) and Sujay et al. (2012). In the similar direction, MABC has been applied to improve popular groundnut cultivars (GJG 9, GG 20, and GJGHPS 1) (Shasidhar et al. 2020), Kadiri 6 (Deshmukh et al. 2020) for foliar disease resistance and high oleic acid content, to improve LLS and rust resistance in TMV2 (Ramakrishnan et al. 2020; Rajarathinam et al. 2022). A 10-SNP panel encompassing related SNPs for high oleic acid and foliar disease resistance has been established, and high-throughput genotyping initiative has so far genotyped more than 55,000 groundnut breeding lines (HTPG) (Pandey et al. 2020).

9.5.2 Transgenics

Genetic transformation is a complementing method to crop breeding, particularly for the traits that are not present in peanut germplasm. Resistance through transgenics can therefore be used as a strategy to break down the gene transfer barriers for high agricultural productivity and nutritional quality, which eventually results in increased crop yield. In groundnut, protocols for transformation are well established, and transgenics have been characterized for resistance to various diseases under controlled and field conditions. The first successful transformation and plant regeneration in groundnut, using microprojectile bombardment technique, was reported in the USA, 1993 (Nigam 2014). Under biotic stress, more attention has been paid to transgenic research on viral diseases than on fungi, using various techniques, such as coat protein-mediated resistance and RNAi-mediated resistance.

Agrobacterium-mediated transformation and particle bombardment techniques were used to transfer nucleocapsid gene (np), encoding for viral coat protein for developing resistance against viral diseases—PBND (Sharma and Anjaiah 2000; Rao et al. 2013; Swamy et al. 2015); TSWV (Li et al. 1997; Magbanua et al. 2000), PStV (Higgins et al. 2004), and PSND (Mallikarjuna et al. 2016). Several genes, i.e., glucanase, chitinase, SniOLP, and Rs-AFP2 genes for early leaf spot (ELS), late leaf spot (LLS); chitinase gene for rust and stilbene synthase, glucanase, chitinase, mod 1, anionic peroxidase, synthetic peptide D4E1, LOX 1; and nonheme chloroperoxidase genes for *Aspergillus flavus* infection, were used for development of fungal disease resistance by genetic transformation in groundnut (Prasad et al. 2011). Various studies have focused on developing peanut cultivars resistant to *A. flavus*. Increased resistance to *A. flavus* was found in transgenic peanuts that expressed the rice chitinase (Prasad et al. 2012), tobacco glucanase (Sundaresha et al. 2010), lipoxygenase (Ozias-Akins et al. 1999), and soyabean *loxl* (Bhatnagar-Mathur et al. 2015) genes. Bhatnagar-Mathur et al. (2015) have described the achievements made through genetic transformations for developing peanuts resistant to *A. flavus*. These genes slowed the progression of the disease, increased resistance, and reduced the frequency of disease in transgenic plants.

9.5.3 Gene Editing

With the advent of genome editing tools like mega-nucleases, zinc-finger nucleases (ZFN), transcription activation-like effector nucleases (TALENs), and CRISPR-Cas9 (clustered regularly interspaced short palindromic repeats)-Cas9 (CRISPR-associated protein), targeted modification of genomic region is now feasible in various crop species (Chen and Gao 2013). To confer disease resistance in various crops, several "S" (susceptible) genes have been knocked out to produce transgenic and transgene-free mutants. There is a scope of targeting nutritional immunity of pathogens which can limit a diverse set of pathogens to establish themselves in the host plants (Hood and Skaar 2012; Ren et al. 2016). Genome editing has been deployed successfully in peanut to identify genetic factors associated with nitrogen-fixing bacterial symbiosis (Sinharoy and DasGupta 2009), reduction in growth of *Aspergillus flavus* (Arias et al. 2015), and reduction in allergen production (Dodo et al. 2008). Therefore, genome editing tools have potential to develop disease-resistant cultivars in the future.

9.5.4 Speed Breeding

Breeding a new crop variety takes over a decade, with 3–6 years spent on seasonal generation advancements (depending on number of generations in a year, and it can be maximum of 2 under field conditions), to arrive at the elite lines that go for testing and then released as varieties. Speed breeding is the rapid generation advancement method successfully applied in crops like wheat, barley, and chickpea (Hickey et al. 2009, 2012; Watson et al. 2018). This concept was used by National Aeronautics and Space Administration (NASA) to shorten the life cycle and accelerate crop growth in groundnut (Rowell et al. 1999). O'Connor et al. (2013) demonstrated that groundnut could produce four generations per year using speed breeding technique. The plants were grown in controlled environmental conditions with 24 h of light intensity and temperature range of 28–32 °C for the rapid development of a population from F_2 to F_5 generations.

Recent work by the wheat breeding team at the University of Queensland has shown that speed breeding techniques combined with high-throughput phenotypic tests were used for quick introgression of traits like grain dormancy, rust, and yellow spot in spring wheat (Hickey et al. 2010, 2012; Richard et al. 2015; Dinglasan et al. 2016; Riaz and Hickey 2017) and multiple disease resistance (leaf rust, net, and spot blotch) in barley (Hickey et al. 2017). Using this approach, genes for target traits such as disease resistance are rapidly transferred into elite cultivars. The speed breeding protocol is standardized at ICRISAT (unpublished data), and the speed breeding techniques can be employed for multiple disease resistance in groundnut breeding.

9.5.5 Haplotype-Based Breeding (HBB)

Haplotypes are SNPs or other markers or variants on the same chromosome that are inherited together with little chance of contemporary recombination, i.e., recombination during the most recent generation of meiosis (Stram and Seshan 2012). Biallelic nature, the presence of rare alleles, and abundant levels of linkage drag are the major disadvantages of SNPs, which sometimes limit the discovery of candidate genomic regions (Bhat et al. 2021). On the other hand, haplotypes overcome the SNP disadvantages and offer more resolution to reach candidate genomic regions (Qian et al. 2017). The recent availability of third-generation sequencing (TGS) platforms has made it possible to construct haplotypes in less time and at a lower cost than second-generation sequencing (Maestri et al. 2020). The TGS, such as genotyping by sequencing (GBS) (Pandey et al. 2016) and SNP arrays (Wankhade et al. 2023), have been used in QTL mapping and GWAS for disease resistance in groundnut. An analysis called haplo-pheno analysis, which offers identification of superior haplotypes, has been used in rice (Abbai et al. 2019; Chen et al. 2019b) and soybean (Guan et al. 2014). This haplotype-based breeding strategy could significantly contribute to improving disease resistance in groundnut. The identified superior haplotypes can be introgressed from cultivar to cultivar using haplotype-specific molecular markers.

9.6 Future Perspectives

Groundnut production is affected by fungal, bacterial, and viral diseases leading to significant yield and quality losses. The climate change effect can further exacerbate the losses caused by diseases through the emergence of new strains of pathogens. Yet another challenge is the spread of diseases such as peanut smut, CBR, and others to the new groundnut growing regions. So far, world over, progress has been made in improving disease resistance in groundnut using the "disease screening" protocols (Janila et al. 2016b), and marker-assisted breeding (MAB) approach that use DNA markers linked to disease resistance to select the desired phenotype is also used (Janila et al. 2016b; Varshney et al. 2014; Kolekar et al. 2017; Shasidhar et al. 2020; Burow et al. 2013; Ramakrishnan et al. 2020; Yeri et al. 2014). However, disease resistance breeding using "disease screening" is often time-consuming and resources demanding. Consequently, the breeding progress for disease resistance is slow. Application of genomic tools in groundnut is lagging compared to other crops due to its large genome size and the complexity of the tetraploid groundnut genome (Zhang et al. 2017). In future, it is essential to enhance disease resistance in groundnut for multiple diseases using efficient and cost-effective breeding methods that use cutting-edge technologies. The following way-forward is suggested:

1. High-throughput phenotyping platform (HTPP) is a viable option to increase genetic gain for disease resistance (Mahlein 2016). HTPP enables screening of a large number of "selection candidates." For a given number of "selected candidates," the selection intensity will be high when they are selected from large number of "selection candidates" thus increasing the genetic gain for disease resistance. Image-based technologies that are under development can be viable in future HTPP platforms.
2. Speed breeding that involved advancing generation under controlled environment of photoperiod, light intensity, temperature, and relative humidity enables to take more number of cycles per year resulting in increased rate of genetic gain. Speed breeding is cost-effective tools that can be deployed in the breeding programs effectively.
3. Understanding host-resistance mechanism: The available -omics technologies such as genomics, transcriptomics, proteomics, and metabolomics have immense potential in carving out the mechanisms of host-pathogen interactions and plant defences against the pathogens and thus enable identification of target genome segments and/or candidate genes governing disease resistance.
4. Identify QTLs governing host resistance: Genetic populations such as recombinant inbred lines (RIL), nested association mapping (NAM), multi-parent advanced generation intercross (MAGIC), and association panels will be valuable to map QTLs governing disease resistance.
5. Identification of superior haplotypes for the candidate genes governing disease resistance and the accessions carrying these superior haplotypes is now possible with available phenomic and genomic technologies.
6. Marker-assisted breeding and gene pyramiding: DNA markers can be used to select the desired phenotype conferred by the candidate gene or QTLs. Breeding schemas that integrate MAB and speed breeding will increase the rate of genetic gain for resistance to diseases. Stacking of resistance for more than one disease is possible with the use of DNA markers.
7. Gene editing. The immediate application of gene editing tools will be for discovery of candidate genes for disease resistance. Gene editing offers development of transgene-free nongenetically modified disease-resistant cultivars. However, information on host genome sequence and the target gene is required for the successful application of genome editing (Ali et al. 2022).
8. The breeding programs must develop and implement improved "breeding schema" that employs speed breeding, singe-seed descent of generation advancement, genomics-assisted breeding, reliable phenotyping to assess the disease reaction of the genotype-based selected progenies, and multi-environment testing to increase the rate of genetic gain for disease resistance in groundnut.

References

Abbai R, Singh VK, Nachimuthu VV, Sinha P, Selvaraj R, Vipparla AK, Singh AK, Singh UM, Varshney RK, Kumar A (2019) Haplotype analysis of key genes governing grain yield and quality traits across 3 K RG panel reveals scope for the development of tailor-made rice with enhanced genetic gains. Plant Biotechnol J 17(8):1612–1622

Abdou YAM, Gregory WC, Cooper WE (1974) Sources and nature of resistance to *Cercospora arachidicola Hori* and *Cercosporodium personatum* (Beck & Curtis) Deighton in *Arachis* species. Peanut Sci 13:6–11

Agarwal G, Clevenger J, Kale SM, Wang H, Pandey MK, Choudhary D, Yuan M, Wang X, Culbreath AK, Holbrook CC, Liu X, Varshney RK, Guo B (2019) A recombination bin-map identified a major QTL for resistance to tomato spotted wilt virus in peanut (*Arachis hypogaea* L.). Sci Rep 9:18246

Agmon S, Kunta S, Dafny Yelin M, Moy J, Ibdah M, Harel A, Rabinovitch O, Levy Y, Hovav R (2022) Mapping of stem rot resistance in peanut indicates significant effect for plant architecture locus. Crop Sci 62(6):2197–2211

Ahmad S, Nawade B, Sangh C, Mishra GP, Bosamia TC, Kumar N, Dobaria JR, Gajera HP (2020) Identification of novel QTLs for late leaf spot resistance and validation of a major rust QTL in peanut (*Arachis hypogaea* L.). 3 Biotech 10:1–3

Harde AL, Perane RR, Shinde VS, Sakore GD (2019) Biochemical defense mechanism in groundnut genotypes against rust caused by *Puccinia arachidis* Speg. Int J Chem Stud 7:834–841

Ali Q, Yu C, Hussain A, Ali M, Ahmar S, Sohail MA, Riaz M, Ashraf MF, Abdalmegeed D, Wang X, Imran M, Manghwar H, Zhou L (2022) Genome engineering Technology for Durable Disease Resistance: recent Progress and future outlooks for sustainable agriculture. Front Plant Sci 13:860281

Anderson EJ, Ali S, Byamukama E, Yen Y, Nepal MP (2018) Disease resistance mechanisms in plants. Gene 9:1–10

Anderson WF, Patanothai A, Wynne JC, Gibbons RW (1990) Assessment of a diallele cross for multiple foliar pest resistance in peanut the international crops research for the semi-arid tropics (ICRI-SAT), Hyderabad. India Oleagineux 45(8–9):373–378

Anderson WF, Wynne JC, Green CC (1986) Potential for incorporation of early and late leaf spot resistance in peanut. Plant Breed 97:163–170

Arias RS, Dang PM, Sobolev VS (2015) RNAi-mediated control of aflatoxins in peanut: method to analyze mycotoxin production and transgene expression in the peanut/aspergillus pathosystem. J Vis Exp 106:e53398

Aulakh KS, Sandhu RS (1970) Reactions of groundnut varieties against *aspergillus Niger*. Plant Dis Rep 54:337

Basra RK, Kaur S, Dhillon M (1985) Anatomical and biochemical studies of the resistance and susceptibility of groundnut varieties to *Cercospora* leaf spot. Ann Biol 1:7–12

Bera SK, Kamdar JH, Kasundra SV, Thirumalaisamy PP (2016) Identification of groundnut genotypes and wild species resistant to stem rot using an efficient field screening technique. Electron J Plant Breed 7:61–70

Bera SK, Kasundra SV, Kamdar JH, BC A, Lal C, Thirumalasmy PP, Dash P, Maurya AK (2014) Variable response of interspecific breeding lines of groundnut to *Sclerotium rolfsii* infection under field and laboratory conditions. Electron J Plant Breed 5:22–29

Bera SK, Rani K, Kamdar JH, Pandey MK, Desmae H, Holbrook CC, Burow MD, Manivannan N, Bhat RS, Jasani MD, Bera SS (2022) Genomic designing for biotic stress resistant Peanut in genomic designing for biotic stress resistant oilseed crops. Springer International Publishing, Cham, pp 137–214

Bertioli DJ, Cannon SB, Froenicke L, Huang G, Farmer AD, Cannon EK, Liu X, Gao D, Clevenger J, Dash S, Ren L, Moretzsohn MC, Shirasawa K, Huang W, Vidigal B, Abernathy B, Chu Y, Niederhuth CE, Umale P, Araujo AC, Kozik A, Kim KD, Burow MD, Varshney RK, Wang X, Zhang X, Barkley N, Guimaraes PM, Isobe S, Guo B, Liao B, Stalker

HT, Schmitz RJ, Scheffler BE, Leal-Bertioli SC, Xun X, Jackson SA, Michelmore R, Ozias-Akins P (2016) The genome sequences of *Arachis duranensis* and *Arachis ipaensis*, the diploid ancestors of cultivated peanut. Nat Genet 48(4):438–446

Bertioli DJ, Jenkins J, Clevenger J, Dudchenko O, Gao D, Seijo G, Samoluk SS (2019) The genome sequence of segmental allotetraploid peanut *Arachis hypogaea*. Nat Genet 51:877

Bertioli S, José AC, Alves-Freitas DM, Moretzsohn MC, Guimarães PM, Nielen S, Vidigal BS, Pereira RW, Pike J, Fávero AP, Parniske M (2009) Identification of candidate genome regions controlling disease resistance in *Arachis*. BMC Plant Biol 9:1–2

Beute MK (1980) Cylindrocladium black rot (CBR) disease of peanut (*Arachis hypogaea*). In: *International Workshop on Groundnuts*, p s

Beute MK, Wynne JC, Emery DA (1976) Registration of NC 3033 Peanut germplasm 1 (Reg. No. GP 9). Crop Sci 16:881–887

Bhat JA, Yu D, Bohra A, Ganie SA, Varshney RK (2021) Features and applications of haplotypes in crop breeding. Commun Biol 4:1–12

Bhatnagar-Mathur P, Sunkara S, Bhatnagar-Panwar M, Waliyar F, Sharma KK (2015) Biotechnological advances for combating *Aspergillus flavus* and aflatoxin contamination in crops. Plant Sci 234:119–132

Bhatnagar-Mathur P, Yogendra K, Parankusam S, Sanivarapu H, Prasad K, Lingampali SB, Sharma KK (2021) Comparative proteomics provide insights on the basis of resistance to *aspergillus flavus* infection and aflatoxin production in peanut (*Arachis hypogea* L.). J Plant Interact 16(1): 494–509

Bosamia TC, Dodia SM, Mishra GP, Ahmad S, Joshi B, Thirumalaisamy PP, Kumar N, Rathnakumar AL, Sangh C, Kumar A, Thankappan R (2020) Unraveling the mechanisms of resistance to *Sclerotium rolfsii* in peanut (*Arachis hypogaea* L.) using comparative RNA-Seq analysis of resistant and susceptible genotypes. PLoS One 15(8):e0236823

Branch WD (1996) Registration of 'Georgia Green' peanut. Crop Sci 36:806

Branch WD (1998) Registration of 'Georgia Bold' peanut. Crop Sci 38:895–896

Branch WD (2000) Registration of Georgia hi-O/L peanut. Crop Sci 40:1823–1824

Branch WD, Brenneman TB (2012) New sources of *Cylindrocladium* black rot resistance among runner-type Peanut cultivars. Peanut Sci 39:38–42

Burow MD, Leal-Bertioli SC, Simpson CE, Ozias-Akins P, Chu Y (2013) Marker-assisted selection for biotic stress resistance in peanut. In: Varshney RK, Tuberosa R (eds) Translational genomics for crop breeding, volume I: biotic stresses, 1st edn. Wiley, New York, pp 125–150. (Chapter 13)

Chapu I, Okello DK, Okello RCO, Odong TL, Sarkar S, Balota M (2022) Exploration of alternative approaches to phenotyping of late leaf spot and groundnut rosette virus disease for groundnut breeding. Front Plant Sci 13:912332

Chaudhari S, Khare D, Patil SC, Sundravadana S, Variath MT, Sudini HK, Manohar SS, Bhat RS, Pasupuleti J (2019) Genotype × environment studies on resistance to late leaf spot and rust in genomic selection training population of peanut (*Arachis hypogaea* L.). front. Plant Sci 10:1338

Chen CQ, Belanger RR, Benhamou N, Paulitz TC (2000) Defense enzymes induced in cucumber roots by treatment with plant growth-promoting rhizobacteria (PGPR) and *Pythium aphanidermatum*. Physiol Mol Plant Pathol 56:13–23

Chen K, Gao C (2013) Targeted genome modification technologies and their applications in crop improvements. Plant Cell Rep 33:575–583

Chen X, Li H, Pandey MK, Yang Q, Wang X, Garg V, Li H, Chi X, Doddamani D, Hong Y, Upadhyaya HD, Guo H, Khan AW, Zhu F, Zhang X, Pan L, Pierce GJ, Zhou G, Krishnamohan KAVS, Chen M, Zhong N, Agarwal G, Li S, Chitikineni A, Zhang G, Sharma S, Chen N, Liu N, Janila P, Li S, Wang M, Wang T, Sun J, Li X, Li C, Wang M, Yu L, Wen S, Singh S, Yang Z, Zhao J, Zhang C, Yu Y, Bi J, Zhang X, Liu Z, Paterson AH, Wang S, Liang X, Varshney RK, Yu S (2016) Draft genome of the peanut A-genome progenitor (*Arachis duranensis*) provides insights into geocarpy, oil biosynthesis, and allergens. Proc Natl Acad Sci U S A 113:6785–6790

Chen X, Lu Q, Liu H, Zhang J, Hong Y, Lan H, Li H, Wang J, Liu H, Li H, Pandey MK, Zhang Z, Zhou G, Yu J, Zhang G, Yuan J, Li X, Wen S, Meng F, Yu S, Wang X, Siddique KHM, Liu Z-J, Paterson AH, Varshney RK, Liang X (2019a) Sequencing of cultivated peanut, *Arachis hypogaea*, yields insights into genome evolution and oil improvement. Mol Plant 12(7): 920–934

Chen K, Zhang Q, Wang CC, Liu ZX, Jiang YJ, Zhai LY, Zheng TQ (2019b) Genetic dissection of seedling vigour in a diverse panel from the 3,000 Rice (*Oryza sativa* L.) Genome Project. *Sci Rep* 9:4804

Choi SH, Green SK, Lee DR, Yoon JY (1989) Incompletely dominant single resistance gene for peanut stripe virus in soybean line AGS 129. Euphytica 40:193–196

Chu Y, Chee P, Culbreath A, Isleib TG, Holbrook CC, Ozias-Akins P (2019) Major QTLs for resistance to early and late leaf spot diseases are identified on chromosomes 3 and 5 in Peanut (*Arachis hypogaea* L.). Front Plant Sci 10:883

Clevenger J, Chu Y, Chavarro C, Agarwal G, Bertioli DJ, Leal-Bertioli SC, Pandey MK, Vaughn J, Abernathy B, Barkley NA, Hovav R (2017) Genome-wide SNP genotyping resolves signatures of selection and tetrasomic recombination in peanut. Mol Plant 10:309–322

Clevenger J, Chu Y, Scheffler B, Ozias-Akins P (2016) A developmental transcriptome map for allotetraploid *Arachis hypogaea*. Front Plant Sci 7:1446

Coffelt TA, Garren KH (1982) Screening for resistance to *Cylindrocladium* black rot in peanuts (*Arachis hypogaea* L.). Peanut Sci 9:1–5

Coffelt TA, Porter DM (1982) Screening peanuts for resistance to *Sclerotinia* blight. Plant Dis 66: 385–387

Coffelt TA, Porter DM (1986) Field screening of reciprocal Chico x Florigiant peanut populations for resistance to leafspot in Virginia. Peanut Sci 13:57–60

Cook M (1981) Susceptibility of peanut leaves to *Cercosporidium personatum*. Phytopathology 71: 787–791

Cui M, Han S, Wang D, Haider MS, Guo J, Zhao Q, Du P, Sun Z, Qi F, Zheng Z, Huang B (2022) Gene co-expression network analysis of the comparative transcriptome identifies hub genes associated with resistance to *Aspergillus flavus* L. in cultivated peanut (*Arachis hypogaea* L.). Front Plant Sci 13:899177

Cui R, Clevenger J, Chu Y, Brenneman T, Isleib TG, Holbrook CC, Ozias-Akins P (2020) Quantitative trait loci sequencing-derived molecular markers for selection of stem rot resistance in peanut. Crop Sci 60:2008–2018

Culbreath AK, Todd JW, Brown SL (2003) Epidemiology and management of tomato spotted wilt in peanut. Annu Rev Phytopathol 41:53–75

Culver JN, Sherwood JL, Melouk HA (1987) Resistance to peanut stripe virus in *Arachis* germplasm. Plant Dis 71:1080–1082

Dang PM, Lamb MC, Bowen KL, Chen CY (2019) Identification of expressed R-genes associated with leaf spot diseases in cultivated peanut. Mol Biol Rep 46:225–239

Dang PM, Lamb MC, Chen CY (2021) Association of differentially expressed R-gene candidates with leaf spot resistance in peanut (*Arachis hypogaea* L.). Mol Biol Rep 48:323–334

Dasgupta S, Raj SK (1997) Screening of groundnut germplasm for resistance to seed rot and collar rot diseases caused by *Aspergillus niger*. Int J Arachis Newsl 17:11–13

Deshmukh DB, Marathi B, Sudini HK, Variath MT, Chaudhari S, Manohar SS, Rani CV, Pandey MK, Pasupuleti J (2020) Combining high oleic acid trait and resistance to late leaf spot and rust diseases in groundnut (*Arachis hypogaea* L.). Front Genet 11:514

Desmae H, Janila P, Okori P, Pandey MK, Motagi BN, Monyo E, Mponda O, Okello D, Sako D, Echeckwu C, Oteng-Frimpong R (2019) Genetics, genomics and breeding of groundnut (*Arachis hypogaea* L.). Plant Breed 138:425–444

Ding Y, Qiu X, Luo H, Huang L, Guo J, Yu B, Sudini H, Pandey M, Kang Y, Liu N, Zhou X (2022) Comprehensive evaluation of Chinese peanut mini-mini core collection and QTL mapping for aflatoxin resistance. BMC Plant Biol 22(1):207

Dinglasan E, Godwin I, Mortlock M, Hickey L (2016) Resistance to yellow spot in wheat grown under accelerated growth conditions. Euphytica 209:693–707

Dodia SM, Joshi B, Gangurde S, Thirumalaisamy PP, Mishra GP, Kumar N, Soni P, Rathnakumar AL, Dobariya JR, Sangh C, Chitikineni A, Chanda SV, Pandey MK, Varshney RK, Radhakrishnan T (2019) Genotyping-by-sequencing based genetic mapping reveals large number of epistatic interactions for stem rot resistance in groundnut. Theor Appl Genet 132(4): 1001–1016

Dodo HW, Konan KN, Chen FC, Egnin M, Viquez OM (2008) Alleviating peanut allergy using genetic engineering: the silencing of the immunodominant allergen Ara h 2 leads to its significant reduction and a decrease in peanut allergenicity. Plant Biotechnol J 6:135–145

Duan NX, Peng Z, Tan YJ, Jiang HF (1994). Preliminary studies on the biochemical mechanisms of resistance to Pseudomonas solanacearum in groundnut

Dwivedi SL, Pande S, Rao JN, Nigam SN (2002) Components of resistance to late leaf spot and rust among interspecific derivatives and their significance in a foliar disease resistance breeding in groundnut (*Arachis hypogaea* L.). Euphytica 125:81–88

Elmore JM, Griffin BD, Walley JW (2021) Advances in functional proteomics to study plant-pathogen interactions. Curr Opin Plant Biol 1(63):102061

Essandoh DA, Odong T, Okello DK, Fonceka D, Nguepjop J, Sambou A, Ballén-Taborda C, Chavarro C, Bertioli DJ, Leal-Bertioli SC (2022) Quantitative trait analysis shows the potential for alleles from the wild species *Arachis batizocoi* and *A. duranensis* to improve groundnut disease resistance and yield in East Africa. Agron J 12:2202

Essandoh DA (2021) Performance of interspecific derivatives of *Arachis* species in Uganda and identification of quantitative trait loci associated with resistance to groundnut rosette and late leaf spot. (Doctoral dissertation,. Makerere University

FAO (2021) Food and Agriculture Organization of the United Nations-Statistic Division https://www.fao.org/faostat/en/#data

Foster DJ, Stalker HT, Wynne JC, Beute MK (1981) Resistance of *Arachis hypogaea* L. and wild relatives to *Cercospora arachidicola* Hori. Oieagineux 36:139–143

Foster DJ, Wynne JC, Beute MK (1980) Evaluation of detached leaf culture for screening peanuts for leaf spot resistance. Peanut Sci 7:98–100

Fry WE (1982) Principles of plant disease management. Academic Press, New York, p 378

Gai Y, Deng Q, Pan R, Chen X, Deng M (2012) First report of *Cylindrocladium* black rot of peanut caused by *Cylindrocladium parasiticum* (teleomorph *Calonectria ilicicola*) in Jiangxi Province. China Plant Dis 96:586

Gangurde SS, Nayak SN, Joshi P, Purohit S, Sudini HK, Chitikineni A, Hong Y, Guo B, Chen X, Pandey MK, Varshney RK (2021) Comparative transcriptome analysis identified candidate genes for late leaf spot resistance and cause of defoliation in groundnut. Int J Mol Sci 22(9):4491

Garcia LE, Brandenburg RL, Bailey JE (2000) Incidence of tomato spotted wilt virus (*Bunyaviridae*) and tobacco thrips in Virginia-type peanuts in North Carolina. Plant Dis 84: 459–464

Gaurav A, Josh C, Hui W, Divya C, Mei Y, Xingjun W, Corley Jr HC, Xin L, Baozhu G (2019) A recombination bin-map facilitates identification of major QTL on chromosome A01 and potential candidate genes for resistance to tomato spotted wilt virus in peanut (*Arachis hypogaea* L.)

Girdthai T, Jogloy S, Vorasoot N, Akkasaeng C, Wongkaew S, Holbrook CC, Patanothai A (2010) Associations between physiological traits for drought tolerance and aflatoxin contamination in peanut genotypes under terminal drought. Plant Breed 129:693–699

Gong L, Han S, Yuan M, Ma X, Hagan A, He G (2020) Transcriptomic analyses reveal the expression and regulation of genes associated with resistance to early leaf spot in peanut. BMC Res Notes 13:1–6

Gopal K, Jagadeswar R, Babu GP, Upadhyaya HD (2004) Sources of resistance to bud necrosis disease in groundnut. Pathology 24:1–10

Gowda MV, Motagi BN, Naidu GK, Diddimani SB, Sheshagiri R (2002) GPBD 4: a spanish bunch groundnut genotype resistant to rust and late leaf spot

Graves PR, Haystead TA (2002) Molecular biologist's guide to proteomics. Microbiol Mol Biol Rev 66:39–63

Green CC, Wynne JC (1986) Diallel and generation means analyses for the components of resistance to *Cercospora arachidicola* in peanut. Theor Appl Genet 73:228–235

Green CC, Wynne JC, Beute MK (1983) Genetic variability and heritability in early generation crosses of large-seeded Virginia-type peanuts with lines resistant to *Cylindrocladium* black rot. Peanut Sci 10:47–51

Guan R, Qu Y, Guo Y, Yu L, Liu Y (2014) Salinity tolerance in soybean is modulated by natural variation in GmSALT3. Plant J 80:937–950

Guo B, Wang H, Yaduru S, Pandey MK, Zhao C, Culbreath AK, Holbrook CC, Wang X, Varshney RK (2018) Development of a next-generation, multi-parent advanced generation intercross (MAGIC), fine-mapping population for advancing genetics and genomics studies in peanut. In: Proceedings of international conference on advances in *Arachis* genomics & biotechnology (AAGB), Senegal

Hadley BA, Beute MK, Wynne JC (1979) Heritability of *Cylindrocladium* black rot resistance in peanut. Peanut Sci 6:51–54

Hamid MA, Isleib TG, Wynne JC, Green CC (1981) Combining ability analysis of *Cercospora* leaf spot resistance and agronomic traits in *Arachis hypogaea* L. Oleagineux 36:605–612

Hammons RO, Bell DK, Sobers EK (1981) Evaluating peanuts for resistance to *Cylindrocladium* black rot. Peanut Sci 8:117–120

Han S, Yuan M, Clevenger JP, Li C, Hagan A, Zhang X, Chen C, He G (2018) A SNP-based linkage map revealed QTLs for resistance to early and late leaf spot diseases in peanut (*Arachis hypogaea* L.). front. Plant Sci 9:1012

Harris NE, Beute MK (1980) Histological responses of peanut germplasm resistant and susceptible to *Cylindrocladium crotalariae* in relationship to inoculum density. (Doctoral dissertation,. North Carolina State University

Hickey LT, Dieters MJ, DeLacy IH, Christopher MJ, Kravchuk OY, Banks PM (2010) Screening for grain dormancy in segregating generations of dormant× non-dormant crosses in white-grained wheat (*Triticum aestivum* L.). Euphytica 172(2):183–195

Hickey LT, Dieters MJ, DeLacy IH, Kravchuk OY, Mares DJ, Banks PM (2009) Grain dormancy in fixed lines of white-grained wheat (*Triticum aestivum* L.) grown under controlled environmental conditions. Euphytica 168(3):303–310

Hickey LT, Germán SE, Pereyra SA, Diaz JE, Ziems LA, Fowler RA, Platz GJ, Franckowiak JD, Dieters MJ (2017) Speed breeding for multiple disease resistance in barley. Euphytica 213(3): 1–4

Hickey LT, Wilkinson PM, Knight CR, Godwin ID, Kravchuk OY, Aitken EAB, Bansal UK, Bariana HS, DeLacy IH, Dieters MJ (2012) Rapid phenotyping for adult-plant resistance to stripe rust in wheat. Plant Breed 131:54–61

Higgins CM, Hall RM, Mitter N, Cruickshank A, Dietzgen RG (2004) Peanut stripe potyvirus resistance in peanut (*Arachis hypogaea* L.) plants carrying viral coat protein gene sequences. Transgenic Res 13:59–67

Higgs J (2003) The beneficial role of peanuts in the diet-part 2. Nutr Food Sci 33(2):56–64

Hobbs H, Reddy D, Rajeshwari R, Reddy A (1987) Use of direct antigen coating and protein a coating ELISA procedures. Plant Dis 71:747–749

Holbrook CC, Isleib TG, Ozias-Akins P, Chu Y, Knapp SJ, Tillman B, Guo B, Gill R, Burow MD (2013) Development and phenotyping of recombinant inbred line (RIL) populations for peanut (*Arachis hypogaea*). Peanut Sci 40:89–94

Holbrook CC, Ozias-Akins P, Chu Y, Lamon S, Bertioli DJ, Leal-Bertioli SC, Culbreath AK, Godoy I (2022) Registration of TifGP-3 and TifGP-4 peanut germplasm lines. J Plant Regist 16(1):120–123

Hood MI, Skaar EP (2012) Nutritional immunity: transition metals at the pathogen-host interface. Nat Rev Microbiol 10:525–537

Jadhav Y, Manohar SS, Sunkad G, Kannalli VP, Pandey MK, Variath MT, Yaduru S, Kona P, Varshney RK, Pasupuleti J (2019) Genomic regions associated with resistance to peanut bud necrosis disease (PBND) in a recombinant inbred line (RIL) population. Plant Breed 138:748–760

Jamshidi Goharrizi K, Fatehi F, Nazari M, Salehi F, Maleki M (2020) Assessment of changes in the content of sulforaphane and expression levels of CYP79F1 and myrosinase genes and proteomic profile of *Lepidium draba* plant under water-deficit stress induced by polyethylene glycol. Acta Physiol Plant 42:1–18

Janila P, Nigam SN, Pandey MK, Nagesh P, Varshney RK (2013) Groundnut improvement: use of genetic and genomic tools. Front Plant Sci 4:23

Janila P, Pandey MK, Manohar SS, Variath MT, Nallathambi P, Nadaf HL, Sudini H, Varshney RK (2016a) Foliar fungal disease-resistant introgression lines of groundnut (*Arachis hypogaea* L.) record higher pod and haulm yield in multilocation testing. Plant Breed 135:355–366

Janila P, Variath MT, Pandey MK, Desmae H, Motagi B, Okori P, Manohar SS, Rathnakumar AL, Radhakrishnan T, Liao B, Varshney RK (2016b) Genomic tools in groundnut breeding program: status and perspectives. Front Plant Sci 7:289

Jasani MD, Kamdar JH, Bera S, Sunkad G, Bera SK (2021) Novel and stable major QTLs conferring resistance to peanut bud necrosis disease and identification of resistant high yielding peanut breeding lines. Euphytica 217(6):1–3

Jogi A, Kerry JW, Brenneman TB, Leebens-Mack JH, Gold SE (2016) Identification of genes differentially expressed during early interactions between the stem rot fungus (*Sclerotium rolfsii*) and peanut (*Arachis hypogaea* L.) cultivars with increasing disease resistance levels. Microbiol Res 184:1–2

Jogloy S, Wynne JC, Beute MK (1987) Inheritance of late leaf spot resistance and agronomic traits in peanut. Peanut Sci 14:86–90

Jyosthna MK, Reddy NE, Chalam TV, Reddy GLK (2004) Morphological and biochemical characterization of *Phaeoisariopsis personata* resistant and susceptible cultivars of groundnut (*Arachis hypogaea* L.). Plant Pathol Bull 13:243–250

Biswas K, Sen C (2000) Management of stem rot of groundnut caused by *Sclerotium rolfsii* through *Trichoderma harzianum*. Indian Phytopath 53:290–295

Kang I-H, Srivastava P, Ozias-Akins P, Gallo M (2007) Temporal and spatial expression of the major allergens in developing and germinating peanut seed. Plant Physiol 144:836–845

Kasno A (1988) Screening, breeding, and maintenance of groundnut varieties resistant to peanut stripe virus. In: 1. Meeting to coordinate research on Peanut stripe virus disease of groundnut. ICRISAT, Malang (Indonesia)

Kasundra S, Kamdar J (2016) Phenotyping technique for stem rot disease in groundnut under field conditions

Katam R, Basha SM, Suravajhala P, Pechan T (2010) Analysis of peanut leaf proteome. J Proteome Res 9:2236–2254

Kaur J, Dhillon M (1990) Structural features of resistance in leaves of groundnut in relation to *Cerosporidium* leaf spot. Plant Dis Res 5:70–72

Kayondo SI, Rubaihayo PR, Ntare BR, Gibson PT, Edema R, Ozimati A, Okello DK (2014) Genetics of resistance to groundnut rosette virus disease. Afr Crop Sci J 22:21–30

Kesmala T, Jogloy S, Wongkaew S, Akkasaeng C, Patanothai A (2006) Evaluation of ten Peanut genotypes for resistance to Peanut bud necrosis virus (PBNV). Evaluation 28(3):460

Khan SA, Chen H, Deng Y, Chen Y, Zhang C, Cai T, Ali N, Mamadou G, Xie D, Guo B, Varshney RK, Zhuang W (2020) High-density SNP map facilitates fine mapping of QTLs and candidate genes discovery for *aspergillus flavus* resistance in peanut (*Arachis hypogaea* L.). Theor Appl Genet 133:2239–2257

Khedikar YP, Gowda MVC, Sarvamangala C, Patgar KV, Upadhyaya HD, Varshney RK (2010) A QTL study on late leaf spot and rust revealed one major QTL for molecular breeding for rust resistance in groundnut (*Arachis hypogaea* L.). Theor Appl Genet 121:971–984

Kochert G, Stalker HT, Gimenes M, Galgaro L, Moore K (1996) RFLP and cytogenetic evidence for the progenitor species of allotetraploid cultivated peanut (*Arachis hypogaea* L.). Am J Bot 83:1282–1291

Kolekar RM, Sujay V, Shirasawa K (2016) QTL mapping for late leaf spot and rust resistance using an improved genetic map and extensive phenotypic data on a recombinant inbred line population in peanut (*Arachis hypogaea* L.). Euphytica 209:147–156

Kolekar RM, Sukruth M, Shirasawa K, Nadaf HL, Motagi BN, Lingaraju S, Patil PV, Bhat RS (2017) Marker-assisted backcrossing to develop foliar disease-resistant genotypes in TMV 2 variety of peanut (*Arachis hypogaea* L.). Plant Breed 136(6):948–953

Korani W, Chu Y, Holbrook CC, Ozias-Akins P (2018) Insight into genes regulating postharvest aflatoxin contamination of tetraploid peanut from transcriptional profiling. Genetics 209(1): 143–156

Kornegay JL, Beute MK, Wynne JC (1980) Inheritance of resistance to *Cercospora arachidicola* and *Cercosporidium personatum* in six Viginia-type peanut lines. Peanut Sci 7:4–9

Kottapalli KR, ZabetMoghaddam M, Rowland D, Faircloth W, Mirzaei M, Haynes PA, Payton P (2013) Shotgun label-free quantitative proteomics of water-deficit-stressed midmature peanut (*Arachis hypogaea L.*) seed. J Proteome Res 12:5048–5057

Kumar D, Kirti PB (2015) Transcriptomic and proteomic analyses of resistant host responses in *Arachis diogoi* challenged with late leaf spot pathogen, *Phaeoisariopsis Personata*. PLoS One 10:0117559

Kumari M, Singh M, Godika S, Choudhary S, Sharma J (2016) Effect of different fungicides, plant extracts on incidence and varietal screening against collar rot of groundnut (*Arachis hypogaea* L.) caused by *aspergillus Niger* van Tiegham. Int Quarterly J Life Sci 11(4):2835–2839

Kumari V, Gowda M, Tasiwal V, Pandey MK, Bhat RS, Mallikarjuna N, Upadhyaya HD, Varshney RK (2014) Diversification of primary gene pool through introgression of resistance to foliar diseases from synthetic amphidiploids to cultivated groundnut (*Arachis hypogaea* L.). Crop J 2: 110–119

Lei Y, Liao BS, Wang SY, Li D, Jiang HF (2005) Identification of AFLP markers for resistance to seed invasion by *aspergillus flavus* in peanut (*Arachis hypogaea* L.). Acta Agron Sin 31:1349–1353

Lei Y, Liao B-S, Wang S-Y, Zhang Y-B, Li D, Jiang H-F (2006) A SCAR marker for resistance to *aspergillus flavus* in peanut (*Arachis hypogaea* L.). Hereditas 28(9):1107–1111

Li Z, Qiu Q (2000) Huayu 16: a new high-yielding, improved quality groundnut cultivar with wide adaptability for northern China. Int Arachis Newsl 20:31–32

Li Z, Jarret RL, Demski JW (1997) Engineered resistance to tomato spotted wilt virus in transgenic peanut expressing the viral nucleocapsid gene. Transgenic Res 6:297–305

Liang X, Zhou G, Hong Y, Chen X, Liu H, Li S (2009) Overview of research progress on peanut (*Arachis hypogaea* L.) host resistance to aflatoxin contamination and genomics at the Guangdong academy of agricultural sciences. Peanut Sci 36:29–34

Liang XQ, Zhou GY, Pan RZ (2003) Study on the relationship of wax and cutin layers in peanut seeds and resistance to invasion and aflatoxin production by *aspergillus flavus*. J Trop Subtrop Bot 11:11–14

Liao B, Zhuang W, Tang R, Zhang X, Shan S, Jiang H, Huang J (2009) Peanut aflatoxin and genomics research in China: progress and perspectives. Peanut Sci 36:21–28

Liao BS, Duan NX, Wang YY, Sun DR, Mehan VK (1994) Host-plant resistance to groundnut bacterial wilt: genetic diversity and enhancement. In: Mehan VK, MeDonald D (eds) Groundnut bacterial wilt in Asia: proceedings of the third working group meeting, Wuhan, China. ICRISAT, Patancheru, India, pp 91–96

Liao BS, Li WR, Sun DR (1986) A study on inheritance of resistance to *pseudomonas solanacearum* E.F. Smith in *Arachis hypogaea* L. Oil Crops of China 3:1–8

Luo HY, Pandey MK, Zhi Y, Zhang H, Xu SL, Guo JB, Wu B, Chen HW, Ren XP, Zhou XJ, Chen YN, Chen WG, Huang L, Liu N, Sudini HK, Varshney RK, Lei Y, Liao BS, Jiang HF (2020a)

Discovery of two novel and adjacent QTLs on chromosome B02 controlling resistance against bacterial wilt in peanut variety Zhonghua 6. Theor Appl Genet 133:1133–1148

Luo Z, Cui R, Chavarro C, Tseng YC, Zhou H, Peng Z, Chu Y, Yang X, Lopez Y, Tillman B, Dufault N, Brenneman T, Isleib TG, Holbrook C, Ozias-Akins P, Wang J et al (2020b) Mapping quantitative trait loci (QTLs) and estimating the epistasis controlling stem rot resistance in cultivated peanut (*Arachis hypogaea* L.). Theor Appl Genet 133(4):1201–1212

Maestri S, Maturo MG, Cosentino E, Marcolungo L, Iadarola B, Fortunati E, Rossato M, Delledonne M (2020) A long-read sequencing approach for direct haplotype phasing in clinical settings. Int J Mol Sci 21(23):9177

Magbanua ZV, Wilde HD, Roberts JK, Chowdhury K, Abad J, Moyer JW, Wetzstein HY, Parrott WA (2000) Field resistance to tomato spotted wilt virus in transgenic peanut (*Arachis hypogaea* L.) expressing an antisense nucleocapsid gene sequence. Mol Breed 6:227–236

Mahatma MK, Thawait LK, Jadon KS, Thirumalaisamy PP, Bishi SK, Jadav JK, Khatediya N, Golakiya BA (2018) Metabolic profiles of groundnut (*Arachis hypogaea* L.) genotypes differing in *Sclerotium rolfsii* reaction. Eur J Plant Pathol 151:463–474

Mahatma MK, Thawait LK, Jadon KS, Thirumalaisamy PP, Bishi SK, Rathod KJ, Verma A, Kumar N, Golakiya BA (2021) Metabolic profiling for dissection of late leaf spot disease resistance mechanism in groundnut. Physiol Mol Biol Plants 27(5):1027–1041

Mahlein A-K (2016) Plant disease detection by imaging sensors—parallels and specific demands for precision agriculture and plant phenotyping. Plant Dis 100:241–251

Makhumbila P, Rauwane M, Muedi H, Figlan S (2022) Metabolome profiling: a breeding prediction tool for legume performance under biotic stress conditions. Plan Theory 11:1–10

Mallikarjuna G, Rao TS, Kirti PB (2016) Genetic engineering for peanut improvement: current status and prospects. Plant Cell Tissue Organ Cult 125:399–416

Mehan VK (1989) Screening groundnuts for resistance to seed invasion by *aspergillus flavus* and to aflatoxin production. In: McDonald M, Mehan VK, Hall SD (eds) Proceeding of international work-shop on aftatoxin contamination of groundnut. International Crops Research Institute for the Semi-Arid Tropics, Patancheru, pp 323–334

Mehan VK, Amadou B, McDonald D, Renard JL, Rao RN, Jayanthi S (1991) Field screening of groundnuts for resistance to seed infection by *aspergillus flavus*. Oleagineux 46:109–118

Mehan VK, Boshou LS, Tan YJ, Robinson-Smith A, McDonald D, Hayward AC (1994) Bacterial wilt of groundnut. Information bulletin no. 35. International crops research Institute for the Semi-Arid Tropics, Patancheru, India

Messiaen CM, Blancard D, Rouxel F, Lafon R (1991) Les maladies des Plantes Maraicheres, 3rd edn. Institut National de la Recherche Agronomique, Paris, France

Middleton KJ, Saleh N, Horn N, Reddy DVR (1988) Production of peanut stripe virus-free seed and screening groundnuts for resistance to peanut stripe virus. In: *1. Meeting to coordinate research on Peanut stripe virus disease of groundnut*. ICRISAT, Malang (Indonesia)

Mixon AC, Rogers KM (1973) Peanut accessions resistant to seed infection by *aspergillus flavus*. Agron J 65:560–562

Moss JP, Singh AK, Reddy LJ, Nigam SN, Subrahmanyam P, McDonald D, Reddy AG (1997) Registration of ICGV 87165 peanut germplasm line with multiple resistance. Crop Sci 37:1028

Motagi BN (2001) Genetic analysis of resistance to late leaf spot and rust Vis-à-Vis productivity in groundnut (*Arachis hypogaea* L.). Dissertation, University of Agriculture Sciences, Dharwad, India

Motagi BN, Bhat RS, Pujer S (2022) Genetic enhancement of groundnut: current status and future prospects. Accelerat Plant Breed 4:63–110

Nandi S, Dutta S, Mondal A, Adhikari A, Nath R, Chattopadhaya A, Chaudhuri S (2013) Biochemical responses during the pathogenesis of *Sclerotium rolfsii* on cowpea. Afr J Biotechnol 12(25)

Narula K, Elagamey E, Abdellatef MA, Sinha A, Ghosh S, Chakraborty N, Chakraborty S (2020) Chitosan-triggered immunity to *fusarium* in chickpea is associated with changes in the plant extracellular matrix architecture, stomatal closure and remodeling of the plant metabolome and proteome. Plant J 103(2):561–583

Nathawat BDS, Patel DS, Singh RP, Mahendra P (2014) Effect of different plant age on incidence and varietal screening against of collar rot in groundnut. Biosci Trends 7(7):580–581

Nevill DJ (1982) Inheritance of resistance to *Cercosporidium personatum* in groundnuts: a genetic model and its implications for selection. Oleagineux 37:355–362

Nigam SN, Bock KR (1990) Inheritance of resistance to groundnut rosette virus in groundnut (*Arachis hypogea* L.). Ann Appl Biol 117:553–560

Nigam SN, Dwivedi SL, Gibbons RW (1991a) Peanut breeding: constraints, achievements and future possibilities. Plant Breed 61:1127–1136

Nigam SN, Dwivedi SL, Rao YL, Gibbons RW (1991b) Registration of 'ICGV 87141'peanut. Crop Sci 31:1096

Nigam SN, Dwivedi SL, Rao YL, Gibbons RW (1991c) Registration of ICGS 11 peanut cultivar. Crop Sci 30:960

Nigam SN, Reddy LJ, Subrahmanyam P, Reddy AG, McDonald D, Gibbons RW (1992) Registration of ICGV 87157, an elite peanut germplasm with multiple resistance to diseases. Crop Sci 32:837

Nigam SN, Waliyar F, Aruna R, Reddy SV, Kumar PL, Craufurd PQ, Diallo AT, Ntare BR, Upadhyaya HD (2009) Breeding peanut for resistance to aflatoxin contamination at ICRISAT. Peanut Sci 36:42–49

Nigam SN (2014) Groundnut at a glance, p 121

Ntare BR, Subrahmanyam P, Waliyarc F (2002) Progress in combatting groundnut rosette disease in sub-Saharan Africa. Plant virology in sub-Saharan Africa 285–293

O'Connor DJ, Wright GC, Dieters MJ, George DL, Hunter MN, Tatnell JR, Fleischfresser DB (2013) Development and application of speed breeding technologies in a commercial peanut breeding program. Peanut Sci 40:107–114

Olorunju PE, Kuhn CW, Demski JW, Misari SM, Ansa OA (1992) Inheritance of resistance in peanut to mixed infections of groundnut rosette virus (GRV) and groundnut rosette assistor virus and a single infection of GRV. Plant Dis 76:95–100

Ouakhssase A, Chahid A, Choubbane H, Aitmazirt A, Addi EA (2019) Optimization and validation of a liquid chromatography/tandem mass spectrometry (LC-MS/MS) method for the determination of aflatoxins in maize. Heliyon 5:01565

Özcan M, Seven S (2003) Physical and chemical analysis and fatty acid composition of peanut, peanut oil and peanut butter from Çom and NC-7 cultivars. Grasas Aceites 54(1):12–18

Ozias-Akins P, Gill R, Yang H, Lynch R (1999) Genetic engineering of peanut: progress with Bt, peroxidase, peptidyl MIM D4E1, and lipoxygenase. In: Proceedings of the USDA-ARS 1999 Aflatoxin Elimination Workshop, pp 69–70

Pandey MK, Agarwal G, Kale SM, Clevenger J, Nayak SN, Sriswathi M, Chitikineni A, Chavarro C, Chen X, Upadhyaya HD, Vishwakarma MK (2017) Development and evaluation of a high density genotyping 'Axiom_Arachis' array with 58 K SNPs for accelerating genetics and breeding in groundnut. Sci Rep 7(1):1–10

Pandey MK, Khan AW, Singh VK, Vishwakarma MK, Shasidhar Y, Kumar V, Garg V, Bhat RS, Chitikineni A, Janila P (2017b) QTL-seq approach identified genomic regions and diagnostic markers for rust and late leaf spot resistance in groundnut (*Arachis hypogaea* L.). Plant Biotechnol J 15:927–941

Pandey MK, Pandey AK, Kumar R, Nwosu V, Guo B, Wright G, Bhat RS, Chen X, Bera SK, Yuan M, Jiang H, Faye I, Radhakrishnan T, Wang X, Liang X, Liao B, Zhang X, Varshney RK, Zhuang W (2020) Translational genomics for achieving higher genetic gains in groundnut. Theor Appl Genet 133:1679–1702

Pandey MK, Roorkiwal M, Singh VK, Ramalingam A, Kudapa H, Thudi M, Chitikineni A, Rathore A, Varshney RK (2016) Emerging genomic tools for legume breeding: current status and future prospects. Front Plant Sci 7:45510

Pasupuleti J, Pandey MK, Manohar SS, Variath MT, Nallathambi P, Nadaf HL, Sudini H, Varshney RK (2016) Foliar fungal disease-resistant introgression lines of groundnut (*Arachis hypogaea* L.) record higher pod and haulm yield in multilocation testing. Plant Breed 135:355–366

Pasupuleti J, Ramaiah V, Rathore A, Rupakula A, Reddy RK, Waliyar F, Nigam SN (2013) Genetic analysis of resistance to late leaf spot in interspecific groundnuts. Euphytica 193:13–25

Pensuk V, Jogloy S, Wongkaew S, Patanothai A (2004) Generation means analysis of resistance to peanut bud necrosis caused by peanut bud necrosis tospovirus in peanut. Plant Breed 123:90–92

Porterfield H, Murray K, Schlichting D, Chen X, Hansen K, Duncan M, Dreskin S (2009) Effector activity of peanut allergens: a critical role for Ara h 2, Ara h 6, and their variants. Clin Exp Allergy 39:1099–1108

Prasad K, Bhatnagar-Mathur P, Narasu ML, Waliyar F, Sharma KK (2011) Transgenic approaches for improving fungal disease resistance in groundnut. Technol Spectr Rev 5(1):1–0

Prasad K, Bhatnagar-Mathur P, Waliyar F, Sharma KK (2012) Overexpression of a chitinase gene in transgenic peanut confers enhanced resistance to major soil borne and foliar fungal pathogens. J Plant Biochem Biotechnol 22:222–233

Prasada Rao RDVJ, Reddy AS, Chakravarthy SK, Sastry KS, Reddy DVR, Ramanath, Moses JP (1989) Peanut stripe virus research in India. II. Peanut stripe coordinator meeting held at ICRISAT, Hyderabad, India;456(2)

Qian L, Hickey LT, Stahl A, Werner CR, Hayes B, Snowdon RJ, Voss-Fels KP (2017) Exploring and harnessing haplotype diversity to improve yield stability in crops. Front Plant Sci 8:1534

Qin H, Feng S, Chen C, Guo Y, Knapp S, Culbreath A, He G, Wang ML, Zhang X, Holbrook CC, Ozias-Akins P (2012) An integrated genetic linkage map of cultivated peanut (*Arachis hypogaea* L.) constructed from two RIL populations. Theor Appl Genet 124:653–664

Rago AM, Cazón LI, Paredes JA, Molina JPE, Conforto EC, Bisonard EM, Oddino C (2017) Peanut smut: from an emerging disease to an actual threat to argentine peanut production. Plant Dis 101:400–408

Rajarathinam P, Palanisamy G, Narayana M, Alagirisamy M (2022) Marker assisted backcross to introgress late leaf spot and rust resistance in groundnut (*Arachis hypogaea* L.). Mol Biol Rep 31:1–9

Ramakrishnan P, Manivannan N, Mothilal A, Mahaingam L, Prabhu R, Gopikrishnan P (2020) Marker assisted introgression of QTL region to improve late leaf spot and rust resistance in elite and popular variety of groundnut (*Arachis hypogaea* L.) cv TMV 2. Australas. Plant Pathol 105: 105

Rao SC, Bhatnagar-Mathur P, Kumar PL, Reddy AS, Sharma KK (2013) Pathogen-derived resistance using a viral nucleocapsid gene confers only partial non-durable protection in peanut against peanut bud necrosis virus. Arch Virol 158:133–143

Rathod V, Hamid R, Tomar RS, Padhiyar S, Kheni J, Thirumalaisamy P, Munshi NS (2020a) Peanut (*Arachis hypogaea*) transcriptome revealed the molecular interactions of the defense mechanism in response to early leaf spot fungi (*Cercospora arachidicola*). Plant Gene 23: 100243

Rathod V, Hamid R, Tomar RS, Patel R, Padhiyar S, Kheni J, Thirumalaisamy PP, Munshi NS (2020b) Comparative RNA-Seq profiling of a resistant and susceptible peanut (*Arachis hypogaea*) genotypes in response to leaf rust infection caused by *Puccinia arachidis*. 3 Biotech 10(6):1–5

Reddy LJ, Nigam SN, Moss JP, Singh AK, Subrahmanyam P, McDonald D, Reddy AG (1996) Registration of ICGV 86699 peanut germplasm line with multiple disease and insect resistance. Crop Sci 36:821

Reddy LJ, Nigam SN, Rao NR, Reddy NS (2001) Registration of ICGV 87354 peanut germplasm with drought tolerance and rust resistance. Crop Sci 41:274

Reddy LJ, Nigam SN, Subrahmanyam P, Reddy AG, McDonald D, Gibbons RW, Pentaiah V (1992) Registration of 'ICGV 87160′ peanut. Crop Sci 32(4):1075

Reddy LJ, Nigam SN, Subrahmanyam P, Reddy AG, McDonald D, Gibbons RW, Pentaiah V (1993) Registration of 'ICGV 86590′ peanut cultivar. Crop Sci 33:357–358

Ren W, Wang K, Liao P, Yang G, Zhao Y, Zhou Y (2016) The regulation of innate immunity by nutritional factors. Biomed Res Int 2016:5138706

Riaz A, Hickey LT (2017) Rapid phenotyping adult plant resistance to stem rust in wheat grown under controlled conditions. In: Wheat rust diseases. Humana Press, New York, NY, pp 183–196

Richard CA, Hickey LT, Fletcher S, Jennings R, Chenu K, Christopher JT (2015) High-throughput phenotyping of seminal root traits in wheat. Plant Methods 11(1):1–1

Rowell T, Mortley DG, Loretan PA, Bonsi CK, Hill WA (1999) Continuous daily light period and temperature influence peanut yield in nutrient film technique. Crop Sci 39(4):1111–1114

Sanders TH, Mixon AC (1978) Effect of peanut tannins on percent seed colonization and *in vitro* growth by *aspergillus parasiticus*. Mycopathologia 66:169–173

Scott JW, Somodi GC, Jones JB (1988) Bacterial spot resistance is not associated with bacterial wilt resistance in tomato. Proc Fla State Hort Soc 101:390–392

Sharief Y, Rawlings JO, Gregory WC (1978) Estimates of leaf spot resistance in three interspecific hybrids of *Arachis*. Euphytica 27:741–751

Sharma S, Choudhary B, Yadav S, Mishra A, Mishra VK, Chand R, Chen C, Pandey SP (2021) Metabolite profiling identified pipecolic acid as an important component of peanut seed resistance against *aspergillus flavus* infection. J Hazard Mater 404:124155

Sharma KK, Anjaiah V (2000) An efficient method for the production of transgenic plants of peanut (*Arachis hypogaea* L.) through *agrobacterium tumefaciens*-mediated genetic transformation. Plant Sci 159:7–19

Shasidhar Y, Variath MT, Vishwakarma MK, Manohar SS, Gangurde SS, Sriswathi M, Sudini HK, Dobariya KL, Bera SK, Radhakrishnan T, Pandey MK, Janila P, Varshney RK (2020) Improvement of three Indian popular groundnut varieties for foliar disease resistance and high oleic acid using SSR markers and SNP array in marker-assisted backcrossing. Crop J 8:1–15

Shew BB, Wynne JC, Beute MK (1987) Field, microplot and green house evaluation of resistance to *Sclerotium rolfsii* in groundnut. Plant Dis 71:188–192

Shirasawa K, Bhat RS, Khedikar YP, Sujay V, Kolekar RM, Yeri SB, Sukruth M, Cholin S, Byregowda A, Pandey MK, Varshney RK, Gowda MVC (2018) Sequencing analysis of genetic loci for resistance for late leaf spot and rust in peanut (*Arachis hypogaea* L.). front. Plant Sci 9: 1727

Shokes FM, Weber Z, Gorbet DW, Pudelko HA, Taczanowski M (1998) Evaluation of peanut genotypes for resistance to southern stem rot using an agar disk technique. Peanut Sci 25:12–17

Simpson CE, Starr JL, Church GT et al (2003) Registration of 'NemaTAM' peanut. (registrations of cultivars). Crop Sci 43(4):1561–1562

Singh AK, Dwivedi SL, Pande S, Moss JP, Nigam SN et al (2003) Registration of rust and late leaf spot resistant peanut germplasm lines. Crop Sci 43:440–441

Singh AK, Mehan VK, Nigam SN (1997) Sources of resistance to groundnut fungal and bacterial disease: an update and appraisal. Information bulletin, no. 50. ICRISAT, Patencheru 502324, Andhra Pradesh, India, p 48

Sinharoy S, DasGupta M (2009) RNA interference highlights the role of CCaMK in dissemination of endosymbionts in the Aeschynomeneae legume *Arachis*. Mol Plant-Microbe Interact 22: 1466–1475

Smith OD, Boswell TE, Grichar WJ, Simpson CE (1989) Reaction of select peanut (*Arachis hypogaea* L.) lines to southern stem rot and *Pythium* pod rot under varied disease pressure. Peanut Sci 16:9–14

Soni P, Gangurde SS, Ortega-Beltran A, Kumar R, Parmar S, Sudini HK, Lei Y, Ni X, Huai D, Fountain JC, Njoroge S (2020a) Functional biology and molecular mechanisms of host-pathogen interactions for aflatoxin contamination in groundnut (*Arachis hypogaea* L.) and maize (*Zea mays* L.). Front Microbiol 11:227

Soni P, Nayak SN, Kumar R, Pandey MK, Singh N, Sudini HK, Bajaj P, Fountain JC, Singam P, Hong Y, Chen X (2020b) Transcriptome analysis identified coordinated control of key pathways regulating cellular physiology and metabolism upon *aspergillus flavus* infection resulting in reduced aflatoxin production in groundnut. J Fungus 6(4):370

Soni P, Pandey AK, Nayak SN, Pandey MK, Tolani P, Pandey S, Sudini HK, Bajaj P, Fountain JC, Singam P, Guo B (2021) Global transcriptome profiling identified transcription factors, biological process, and associated pathways for pre-harvest aflatoxin contamination in groundnut. J Fungus 7(6):413

Stram DO, Seshan VE (2012) Multi-SNP haplotype analysis methods for association analysis. Methods Mol Biol 850:423–452

Subrahmanyam P, Ghanekar AM, Nolt BL, Reddy DVR, McDonald D (1985a) Resistance to peanut diseases in wild Arachis species. In: Proceedings of international workshop cytogenet. *Arachis*, 31 Oct.–Nov. 1983. ICRISAT, Patancheru, India, pp 49–55

Subrahmanyam P, McDonald D, Gibbons RW, Nigam SN, Nevill DJ (1982b) Resistance to rust and late leafspot diseases in some genotypes of *Arachis hypogaea*. Peanut Sci 9:6–10

Subrahmanyam P, McDonald D, Gibbons RW, Reddy LJ (1985b) Peanut rust: a major threat to peanut production in the semiarid tropics. Plant Dis 69:813–819

Subrahmanyam P, McDonald D, Gillons RW (1982a) Variation in *Cercosporidium personatum* symptoms on certain cultivars of *Arachis hypogaea*. Oleagineux 37:63–67

Subrahmanyam P, McDonald D, Waliyar F, Reddy LJ, Nigam SN, Gibbons RW, Rao VR, Singh AK, Pande S, Reddy PM, Rao PS (1995) Screening methods and sources of resistance to rust and late leaf spot of groundnut. Information Bulletin no 47

Subrahmanyam P, Van der Merwe PJ, Reddy LJ, Chiyembekeza AJ, Kimmins FM, Naidu RA (2000) Identification of elite short-duration, rosette resistant lines in world germplasm collections. Int Arachis Newsl 20:46–50

Subrahmanyam P, Wongkaew S, Reddy DVR, Demski JW, McDonald D, Sharma SB, Smith DH (1992) Field diagnosis of groundnut diseases. ICRISAT information bulletin no. 36, Patancheru, AP

Sudhakar P, Latha P, Babitha M, Reddy PV, Naidu PH (2007) Relationship of drought tolerance traits with aflatoxin contamination in groundnut. Indian J Plant Physiol 12:261

Sujay V, Gowda MVC, Pandey MK, Bhat RS, Khedikar YP, Nadaf HL, Gautami B, Sarvamangala C, Lingaraju S, Radhakrishan T, Knapp SJ, Varshney RK (2012) Quantitative trait locus analysis and construction of consensus genetic map for foliar disease resistance based on two recombinant inbred line populations in cultivated groundnut (*Arachis hypogaea* L.). Mol Breed 32(2):773–788

Sujitkumar B (2015) Biochemical and molecular basis of innate and *Pseudomonas fluorescens* induced stem rot tolerance in groundnut (*Arachis hypogaea* L.). M.Sc (Biochemistry) thesis. Junagadh Agric Univ, Junagadh, India

Sukand G, Kulkarni SR (2006) Studies on structural and biochemical mechanism of resistance in groundnut to *Puccinia arachidis*. Indian Phytopathol 59:323–328

Sundaresha S, Kumar AM, Rohini S, Math S, Keshamma E, Chandrashekar S, Udayakumar M (2010) Enhanced protection against two major fungal pathogens of groundnut, *Cercospora arachidicola* and *aspergillus flavus* in transgenic groundnut over-expressing a tobacco β 1–3 glucanase. Eur J Plant Pathol 126:497–508

Swamy KM, Patil MS, Bhat RS, Bydgi AS (2015) Development of coat protein mediated resistance in Peanut plants against Peanut bud necrosis virus. J Mycol Plant Pathol 45(2):156–165

Tallury SP, Hollowell JE, Isleib TG, Stalker HT (2014) Greenhouse evaluation of section *Arachis* wild species for *Sclerotinia* blight and *Cylindrocladium* black rot resistance. Peanut Sci 41:17–24

Tang W, Kuang J, Qiang S (2015) The pathogenicity of *Sclerotium rolfsii* on *Cyperus difformis* and its potential host specificity among the genus Cyperus. J Plant Pathol Microbiol 6(3):S3-002

Thakur RP, Rao VP, Reddy SV, Ferguson M (2000) Evaluation of wild Arachis germplasm accessions for in vitro seed colonization and anatoxin production by *aspergillus flavus*. Int Arachis Newsl 20:44–46

Turner RB, Lindsey DL, Davis DD, Bishop RD (1975) Isolation and identification of 5,7-Dimethoxyisoflavone, an inhibitor of *aspergillus flavus* from peanuts. Mycopathologia 57:39

Upadhyaya HD, Nigam SN, Thakur RP (2002) Genetic enhancement for resistance to aflatoxin contamination in groundnut. In: Summary proceedings of the seventh ICRISAT regional groundnut meeting for Western and Central Africa, Cotonu, Benin, 6–8 December 2000. International Crops Research Institute for the Semi-Arid Tropics, Patancheru, India, pp 29–36

Van der Merwe PJA, Subrahmanyam P, Kimmins FM, Willekens J (2001) Mechanisms of resistance to groundnut rosette. Int Arachis Newsl 21:43–46

Vander Plank JE (1968) Disease resistance in plants. Academic Press, New York

Variath MT, Janila P (2017) Economic and academic importance of peanut. In: The peanut genome, vol 7, p 26

Varshney RK, Pandey MK, Pasupuleti J, Nigam SN, Sudini H, Gowda MVC, Sriswathi M, Radhakrishnan T, Manohar SS, Nagesh P (2014) Marker-assisted introgression of a QTL region to improve rust resistance in three elite and popular varieties of peanut (*Arachis hypogaea* L.). Theor Appl Genet 127:1771–1781

Vindhiyavarman P, Mohammed SE (2001) Release of foliar disease resistant groundnut cultivar VRI Gn 5 in Tamil Nadu. India Int Arachis Newsl 21:16–17

Waliyar F, Hassan H, Bonkoungou S (1994) Sources of resistance to *aspergillus flavus* and aflatoxin contamination in groundnut genotypes in West Africa. Plant Dis 78:704–708

Waliyar F, Kumar K, Diallo M, Traore A, Mangala UN, Upadhyaya HD, Sudini H (2016) Resistance to pre-harvest aflatoxin contamination in ICRISAT's groundnut mini core collection. Eur J Plant Pathol 145:901–913

Waliyar F, Kumar PL, Ntare BR, Monyo E, Nigam SN, Reddy AS, Osiru M, Diallo AT (2007) A century of research on groundnut rosette disease and its management. Information Bulletin no. 75

Waliyar F, Sudini H (2012) ELISA: an inexpensive and highly precise tools for estimation of aflatoxins. International Crops Research Institute for the Semi-Arid Tropics

Walkey DGA (1985) Control through resistant cultivars. In: Applied virology. Wiley, New York, pp 234–259

Wang H, Lei Y, Wan L, Yan L, Lv J, Dai X, Ren X, Guo W, Jiang H, Liao B (2016) Comparative transcript profiling of resistant and susceptible peanut post-harvest seeds in response to aflatoxin production by *aspergillus flavus*. BMC Plant Biol 16(1):1–6

Wang YY, Wang CH, Xia XM (1985) A preliminary study on inheritance of resistance to bacterial wilt in peanut. Oil Crops of China 4:15–17

Wang T, Chen XP, Li HF, Liu HY, Hong YB, Yang QL, Chi XY, Yang Z, Yu SL, Li L (2013) Transcriptome identification of the resistance-associated genes (RAGs) to *aspergillus flavus* infection in pre-harvested peanut (*Arachis hypogaea*). Funct Plant Biol 40:292–303

Wankhade AP, Chimote VP, Viswanatha KP, Yadaru S, Deshmukh DB, Gattu S, Sudini HK, Deshmukh MP, Shinde VS, Vemula AK, Pasupuleti J (2023) Genome-wide association mapping for LLS resistance in a MAGIC population of groundnut (*Arachis hypogaea* L.). Theor Appl Genet 136(3):43–43

Wankhade AP, Kadirimangalam SR, Viswanatha KP, Deshmukh MP, Shinde VS, Deshmukh DB, Pasupuleti J (2021) Variability and trait association studies for late leaf spot resistance in a groundnut MAGIC population. Agron J 11(11):2193

Watson A, Ghosh S, Williams MJ, Cuddy WS, Simmonds J, Rey MD, Hatta MAM (2018) Speed breeding is a powerful tool to accelerate crop research and breeding. Nat Plants 4:23–29

Whitaker TB, Dorner JW, Giesbrecht FG, Slate AB (2004) Variability among aflatoxin test results on runner peanuts harvested from small field plots. Peanut Sci 31(1):59–63

Whitaker TB, Dowell FE, Hagler WM, Giesbrecht FG, Wu J (1993) Variability associated with sampling, sample preparation, and chemically testing farmers stock peanuts for aflatoxin. J Assoc Off Anal Chem Int 77:107–116

Wynne JC, Beute MK (1983) Registration of NC 8C peanut (Reg. No. 27). Crop Sci 23:183–184

Xu ZY, Zhang ZY (1987) Test of wild species of peanut for resistance to peanut mild mottle virus (in Chinese). Oil Crops China 4:91–93

Xue HQ, Isleib TG, Payne GA, Wilson RF, Novitzky WP, O'Brian G (2003) Comparison of aflatoxin production in normal and high-oleic backcross-derived peanut lines. Plant Dis 87: 1360–1365

Xue HQ, Isleib TG, Stalker HT, Payne GA, OBrian G (2004) Evaluation of *Arachis* species and interspecific tetraploid lines for resistance to aflatoxin production by *aspergillus flavus*. Peanut Sci 31:134–141

Yeri SB, Kolekar RM, Motagi BN, Nadaf HL, Lingaraju S, Gowda MVC, Bhat RS (2014) Development of late leaf spot and rust tolerant genotypes from TMV 2 and JL 24 by marker assisted backcross breeding in groundnut AAGB 2014—7th international conference on advances in *Arachis* through Genomics & Biotechnology. Georgia, USA, p 45

Yu B, Huai D, Huang L, Kang Y, Ren X, Chen Y, Zhou X, Luo H, Liu N, Chen W, Lei Y, Pandey MK, Sudini H, Varshney RK, Liao B, Jiang H (2019) Identification of genomic regions and diagnostic markers for resistance to aflatoxin contamination in peanut (*Arachis hypogaea* L.). BMC Genet 20(1):32

Zhang X, Zhang J, He X, Wang Y, Ma X, Yin D (2017) Genome-wide association study of major agronomic traits to domestication in peanut. Front Plant Sci 8:1611

Zhang H, Chu Y, Dang P, Tang Y, Li J, Jiang T, Clevenger JP, Ozias-Akins P, Holbrook C, Wang ML, Jacobson A (2019) Identification of potential genes for resistance to tomato spotted wilt and leaf spots in peanut (*Arachis hypogaea* L.) through GWAS analysis

Zhao Y, Zhang C, Chen H, Yuan M, Nipper R, Prakash CS, Zhuang W, He G (2016) QTL mapping for bacterial wilt resistance in peanut (*Arachis hypogaea* L.). Mol Breed 36:13

Zhao C, Zhao S, Hou L, Xia H, Wang J, Li C, Li A, Li T, Zhang X, Wang X (2015) Proteomics analysis reveals differentially activated pathways that operate in peanut gynophores at different developmental stages. BMC Plant Biol 15:1–12

Zhao X, Li C, Yan C, Wang J, Yuan C, Zhang H, Shan S (2019) Transcriptome and proteome analyses of resistant preharvest peanut seed coat in response to *aspergillus flavus* infection. Electron J Biotechnol 39:82–90

Zongo A, Nana AT, Sawadogo M, Konate AK, Sankara P, Ntare BR, Desmae H (2017) Variability and correlations among groundnut populations for early leaf spot, pod yield, and agronomic traits. Agron J 7:52

Disease Resistance Breeding in *Lathyrus sativus* L.

10

Surendra Barpete, Arpita Das, Parisa Pourali Kahriz, Mahsa Pourali Kahriz, K. M. Khawar, Xu Qunale, Kuldeep Tripathi, A. K. Parihar, and Shiv Kumar

Abstract

Grass pea (*Lathyrus sativus* L.) is a climate-resilient and cool-season legume crop with wide range of genetic diversity in the genus that is valuable sources for agronomical traits including biotic stresses. Grass pea is a rich source of protein and calories and provides food and nutritional security to many low-income communities of different underdeveloped regions of the world. It is a hardy and climate smart crop under the changing scenario of climate change and option for agroecosystems. Stigma of neurotoxin (β-ODAP) major constraint in grass pea production is known as β-N-oxalyl-l-α,β-diaminopropionic acid causing neurolathyrism in humans and animals. Grass pea is better tolerance to variable levels of abiotic and biotic stresses as compared to other legumes crops, although it is well-adapted to a number of biotic stresses but still considerable yield losses (approx. 15–25%) due to biotic stress. Due to underutilized and neglected crop, negligible genetic resources utilized and developed grass pea genotype resistant

S. Barpete (✉) · S. Kumar
ICARDA-Food Legume Research Platform, Amlaha, Madhya Pradesh, India

A. Das
Bidhan Chandra Krishi Viswavidyalaya, Mohanpur, Nadia, West Bengal, India

P. P. Kahriz · M. P. Kahriz · K. M. Khawar
Department of Field Crops, Faculty of Agriculture, Ankara University, Ankara, Turkey

X. Qunale
College of Life Sciences, Northwest A&F University, Yangling, Shaanxi, China

K. Tripathi
Indian Council of Agricultural Research-National Bureau of Plant Genetic Resources, New Delhi, India

A. K. Parihar
ICAR-Indian Institute of Pulses Research, Kanpur, India

U. C. Jha et al. (eds.), *Diseases in Legume Crops*,
https://doi.org/10.1007/978-981-99-3358-7_10

against diseases and pests. This chapter reviews the present status of economic importance, genetic transformation, and genetic and genomic resources for developing biotic stress-resistant grass pea genotypes.

Keywords

Grass pea · Biotic stresses · Diseases · *Lathyrus sativus* L. · Pathogens

10.1 Introduction

Several phytopathogens from fungus bacteria to viruses and viroids cause diseases in the *Lathyrus* species. Their damages range from a few to catastrophic. Phytopathogens are very difficult to control as they are dependent on genotype, space, and prevailing environmental conditions, and sometimes these pathogens evolve resistance (Mitra et al. 2019). Therefore, the diseases are minimized by inhibition of their virulence and the introduction of genetic diversity of the systems of agriculture.

The studies on breeding, agronomic studies, and the use of grass pea had never been so important as it has been today (Campbell et al. 1994; Girma and Korbu 2012; Kumar et al. 2013; Dixit et al. 2016; Das et al. 2021; Parihar et al. 2022). There is a rapid increase in the breeding of grass pea in the last 30 years, and efforts are there for breeding more improved and low-ODAP varieties for animal and human consumption (Banerjee et al. 2022; Sellami et al. 2022; Tripathi et al. 2022). Although grass pea cultivation has reduced in the current decade, grass pea breeding has gradually started receiving more attention from scientists' community for its potential as a functional food to promote homoarginine content like nutraceuticals and others with time (Banerjee et al. 2022). Generally, the seeds of grass pea are sold without identification of cultivar or variety in the markets; however, the seeds of identified origin and registered varieties in limited quantities could be obtained from the research centers and gene banks globally (Hammer et al. 2019; Santos et al. 2021a; Parihar et al. 2022; Kock 2022). With the rapid increase in the human population, basic food requirements have changed and increased (Kopittke et al. 2019). The breeders are after every crop plant that could be of use and could be utilized as food, feed, or otherwise in breeding programs (Chiurugwi et al. 2019). Early farmers were bestowed with a natural wealth of a large number of weeds and forage plants that could be utilized as natural pastures and grasslands (Baxter et al. 2022; Lanfranco et al. 2022). These (pastures and grasslands) are invaded and encroached by farmlands with the gradual extinction of native grasslands (Palit and DeKeyser 2022; de Souza et al. 2022). The farmers have been forced to rely upon cultivated legumes and seeded grasses to compensate for these losses to meet their forage and fodder needs. With this awareness, the farmers, plant breeders, biologists, medical scientists, and policy-makers are gradually becoming more conscious of the value of grass pea. The α and β-ODAP isomers, respectively, exist in a ratio of 5:95 in grass pea. However, β-isomer is believed to be more toxic. Both tend to modify under different conditions of heat treatments (Yan et al. 2006). β-ODAP is produced under drought or water stress (Cocks et al. 2000), excessive iron in soil, and zinc deficiency (Lambein et al. 2019). The breeders are taking interest to breed new varieties for economic grass pea free of β-ODAP plant for use in human food and animal feed production the world over (Duguma and Janssens 2021; Tittonell et al. 2020; Barpete et al. 2021, 2022). They are also devising ways to make use of grass pea in soil conservation and find ways to use it as a medicinal plant (Lan et al. 2013). This recognition and interest are increasing day after day, after the passing of the Plant Variety Protection Act 1970 in the USA and similar laws in many other countries. Many private seed companies/ entrepreneurs have started taking interest in forage plants including grass pea (Adebola 2019; Verma 2021; Gonçalves et al. 2022; Sahu and Amin 2022). These efforts have added and increased the speed of grass breeding research tremendously. Farming of forage and fodder including grass pea is occupying a prominent position gradually (Sharma et al. 2021). Many scientists and organization group with different backgrounds, nationalities, and origins are working individually or integratively in understanding the problems related to grass pea (Gonçalves et al. 2022).

A large number of Lathyrus species pave the way for developing low-ODAP grass pea varieties through breeding efforts. However, domestication and breeding of grass pea require several generations of hybridization and selection (Kahriz et al. 2022). The breeders have collected a large amount of germplasm stored in the gene banks of several countries to increase its utilization compared to the past studies, when grass pea was considered a taboo crop with limited utilization (Eastwood et al. 2022). Breeding of grass pea is very new if we compare it with wheat, maize, cotton, etc. which are cultivated for thousands of years before the start of common era-BCE (Lambein et al. 2019). Grass pea contains a neurotoxin β-ODAP that is a causal agent of neurolathyrism in human beings, cattle, and poultry (Ramya et al. 2021), if consumed for several months (Llorens et al. 2011). Moreover, the phytopathogens neurotoxins in the plant itself could hamper food supplies and security that besides β-ODAP-related problems in grass pea for human and animal consumption. The breeders need to pay more attention to the elimination of the ODAP in future cultivars (Parihar et al. 2022).

10.2 Economic Importance of the Crop

Grass pea is a popular legume and pulse crop of economic importance for Ethiopia, Pakistan, Turkey, Iran, Greece, Spain, Nepal, India, Bangladesh, China, Mediterranean and Central European countries of Europe (Campbell et al. 1994; Campbell 1997; Yadav and Mehta 1995), and Australia where it is naturalized (Cocks et al. 2000). Species in the genus *Lathyrus* are popularly used as underutilized sources of proteins in African, South Asian, temperate Mediterranean countries in the Northern hemisphere and Australia as food and feed crops (*L. sativus*, *L. cicera*, *L. clymenum*, or *L. ochrus*) green manure (Sarpaki and Jones 1990; Campbell 1997). Ornamental plants in gardens and landscaping, nitrogen fixers in soil, soil stabilizers and conservation, and most importantly as a medicinal plant (Lan et al. 2013) could be counted as an important plant for commercially (Campbell 1997; Kenicer et al. 2005; Skiba et al. 2007). Most of the species in the *Lathyrus* genus are mesophytes, grow in forests, roadsides, or field margins as annuals, biennials, or perennial plants with the climbing habit (Kenicer et al. 2005; Skiba et al. 2007) and are very resistant to drought and heat, whereas *L. sativus* (grass pea), *L. cicera*, and *L. odoratus* are the significantly important plant species and are used as food and feed crops since time immemorial. Kislev (1989) maintains cultivation of grass pea started about 6000 years BCE and was the first cultivated and domesticated crop in the Middle East and Europe. Some estimates suggest that India is the largest producer of grass pea crops in the world (Skiba et al. 2007, Das et al. 2021). It is closely followed by Bangladesh and Ethiopia (Gautam et al. 1997; Granati et al. 2003; Tripathi et al. 2022). All germplasm of grass pea have a significant amount of ODAP that cause neurolathyrism; therefore, many countries in the world discourage its cultivation. Even then, it is cultivated in Turkey, South Asian, and Sub-Saharan African countries (Campbell 1997; Enneking 2011; Kumar et al. 2011; Hillocks and Maruthi 2012).

Almost all species of the genus *Lathyrus* contain β-ODAP in variable amounts that are notoriously famous for neurolathyrism of forelimbs, if overconsumed (3—6 months) by poultry, animals, and human beings as a sole diet (Lambein and Kuo 2009). However, breeding of grass pea is still considered an important topic by international research centers like ICARDA with considerable efforts to reduce β-ODAP or nitriles (Llorens et al. 2011; Kumar et al. 2021) and since the last 50 years with the introduction of several low β-ODAP cultivars (Kumar et al. 2011; Sengupta et al. 2021). They have high resistance against both biotic and abiotic stress (Lambein et al. 1994; Zhang et al. 2003) and are produced as both food and feed in South Asia, Mediterranean countries, Europe, and North Africa. Moreover, the roots of the grass pea plant produce nodules that fix atmospheric nitrogen ($\geq$ 125 N kg/year) through symbiotic relationships and increase soil fertility. It is assumed that β-ODAP acts as a carrier of zinc ions that scavenges hydroxyl radicals (Lambein et al. 1994; Gongke et al. 2001), enabling the protection of plants under high intensity of light during photosynthesis (Zhang et al. 2003) and help plants to resist oxidative stress.

Most researchers agree that *L. sativus* seeds could be partially detoxified depending on the methods of cooking, treatment with alkaline water solutions, etc. (Kuo et al. 2000; Barpete et al. 2021). Other important species in the genus include *L. cicera* used as a forage and fodder plant (Kislev 1989; White et al. 2002), and *L. odoratus*, *L. rotundifolius*, *L. latifolius*, *L. grandiflorus*, *L. tingitanus*, *L. vernus*, *L. chloranthus*, and *L. belinensis* are also popularly used as cut flowers, and as garden decoration in rock gardens, and landscaping (Kahriz et al. 2022). Grass pea seeds are rarely used in ethnomedicinal systems. However, recently β-ODAP (dencichine) has been approved by the Chinese and US patent offices for its applications in hemostatic and neuroprotective characteristics (Zhao 2012, Compositions and Methods for Treating Haemorrhagic Condition. 2011). The seed decoction of *L. aphaca* is used to treat and heal the wounds in the Khyber

Pakhtunkhwa province of Pakistan in local ethnomedicinal systems (Sher et al. 2015).

10.3 Advances and Challenges for Exploitation of Sources of Resistance in Grass pea Breeding and Major Achievement

Grass pea is considered as resilient and less affected by biotic cues in comparison to other food legumes (Das et al. 2021). However, an array of biotic stresses has been reported in the crop which causes hindrance in global grass pea production, and up to 15–25% yield reduction was reported (Table 10.1).

10.3.1 Powdery Mildew (*Erysiphe pisi* Syn. *E. polygoni*)

Powdery mildew caused by *Erysiphe pisi* syn. *E. polygoni* is a serious menace toward successful grass pea cultivation (Campbell et al. 1994; Vaz Patto and Rubiales 2014). *E. pisi* is obligate biotroph or holoparasitic in nature and complete its life cycle exploiting a suitable host. Pathogenesis phase of this pathogen involves attachment of spore followed by germination, aspersorium formation, and its penetration through epidermal cell wall. In subsequent stages, the pathogen creates haustoria inside of living plant cells, which reroute the host's metabolism to serve the needs of the infection (Glawe 2008). Resistance breeding is a holistic way to overcome this issue (Vaz Patto et al. 2006b; Sharma et al. 2022), although the introgression of resistance genes into well-adapted cultivars creates a menace with high rates of resistance breakdown. However, very meager research findings are available regarding powdery mildew resistance in grass pea, and the disease reactions were not thoroughly examined. Among the cultivated gene pool, some of the identified accessions (RLK-1, RLK-281, RLK-617, RPL-26, RLK-273-1, RLK-273-3, JRL-6, and JRL-41) detected with powdery mildew resistance (Lal et al. 1986; Pandey et al. 1997; Asthana and Dixit 1998). Grass pea lines with intermediate resistance to powdery mildew have been reported from Syria and India (Campbell et al. 1994; Robertson and Abd El-Moneim 1996; Asthana and Dixit 1998). Vaz Patto and associates decisively screened both *L. sativus* and *L. cicera* germplasm against powdery mildew both at natural and artificial epiphytic situation (Vaz Patto et al. 2006a, 2007). Under growth chamber conditions, both species showed in most cases a compatible reaction of high infection rate with no macroscopically visible hypersensitivity. However, disease severity varied significantly among accessions, and low disease severity values were far more frequent in *L. sativus* than *L. cicera* (Vaz Patto et al. 2006a, 2007). Disease severity was more in case of *L. sativus* accession in comparison to *L. cicera*. Under growth chamber and field conditions, partially resistant accessions with low disease severity with virulent infection type strain of powdery mildew have been found. Mixed disease reactions were much more common on *L. sativus* in comparison to *L. cicera* in the

Table 10.1 Major biotic stresses and sources of resistance in grass pea

S. No.	Trait	Status	Sources of resistance	Reported country	References
1	Powdery mildew (*Erysiphe* sp.)	Major disease	IPLy2-10, RPLK-26, RL-41, RLS-2, RPLK-26 and RL-21, LS8246, RLK-1, RLK-281, RLK-617, RPL-26, RLK-273-1, RLK-273-3, JRL-6, and JRL-41	India	Narsinghani and Kumar (1979), Lal et al. (1986), Asthana and Dixit (1998), Pandey et al. (1997), Sastri (2008)
			L. cicera	Spain	Vaz Patto et al. (2006a), Vaz Patto et al. (2007)
2	Downy mildew (*Peronospora* sp.)	Major disease	RLS-1, RLS-2, JRS-115, JRL-43, and JRL-16	India	Lal et al. (1986)
			L. cicera	India; Canada	Asthana and Dixit (1998), Campbell (1997)
3	Ascochyta blight (*Ascochyta pisi* and *Mycosphaerella pinodes*)	Major disease	ATC 80878	Australia	Skiba et al. (2004a, b), Gurung et al. (2002)
4	Rust (*Uromyces pisi*)	Major disease	BG-15744 and BG-23505 (*L. sativus*: partially resistance), *L. cicera*	Spain	Vaz Patto and Rubiales (2009, 2014)
5	Fusarium wilt (*Fusarium oxysporum f.* sp. *pisi*)	Minor disease causing damage to some extent	ACC273 and SITNICA	Portugal	Sampaio et al. (2021a)
6	Thrips (*Caliothrips indicus*)	Noxious pest in southeast Asia and Ethiopia	RLK-1, RLK-281, RLK-617, RLK-26, RLK-273-1, RLK-273-3, JRL-6 and JRL-41	India	Asthana and Dixit (1998); Pandey et al. (1997), Lal et al. (1986)
7	Bruchid resistance (*Callosobruchus* sp.)	Major stored grain pest	cv. Rodos (*L. cicera*)	Greece	Tsialtas et al. (2020)
8	Cyst nematode (*Heterodera ciceri*)	Major pest of Mediterranean region	IFLA 347	ICARDA	Di Vito et al. (2001)

9	Root knot nematode (*Meloidogyne artiellia*)	Major pest of Mediterranean region	cv. PI 236481 and UT2921 (*L. latifolius*), cv. PI 358879 (*L. sylvestris*), cv. PI 440462 (*L. hirsutus*)	Washington, USA	Rumbaugh and Griffin (1992)
10	Broomrape (*Orobanche crenata*)	Major plant parasitic weed of Mediterranean region	*L. ochrus*, *L. clymenum*, and *L. choranthus*, IG-64782 and IG-65197 (*L. sativus*)	Spain	Linke et al. (1993), Sillero et al. (2005)
	O. foetida		*L. annuus*, *L. tingitanus*, and *L. pseudocicera*	Morocco and Tunisia	Abdallah et al. (2021)

case of the powdery mildew inoculation (Vaz Patto et al. 2006a, 2007). Qualitative resistance mechanism associated with *E. pisi* involved collapse of sporelings shortly after germination, followed by cell death of the contacted epidermal host cells in *L. belinensis* and its F_1 progeny with *L. odoratus* as well as in backcross progenies and confirmed the monogenic resistance with the presence of a dominant resistance gene (Poulter et al. 2003). Partial resistance in powdery mildew has been reported earlier in pea and its crop wild relatives (CWR) (Fondevilla et al. 2007) and suggested the presence of one single recessive gene *er1* (Fondevilla and Rubiales 2012). However, this *er1* gene can also offer durable resistance and frequently used in pea breeding activities (Fondevilla et al. 2007). Powdery mildew fungus needs the MLO protein for successfully colonize a plant. In the loss of function of MLO (Mildew Locus O) gene, *PdMLO1* is responsible for the resistance associated with pre-haustorial plant resistance governed by *er1* gene (Hamphry et al. 2011). This *MLO1* (*PsMLO1*) from pea was well studied. According to a recent report by Santos et al. (2021b), *MLO1* genes have been isolated from *L. sativus* and *L. cicera* germplasm and characterized by an array of resistance reaction against powdery mildew. Like all dicot *MLO*, grass pea *MLO1* also positioned into clade V associated with powdery mildew susceptibility. Genotype by sequencing (GBS) approach was explored for developing high-density linkage map utilizing RIL population of resistant and susceptible cultivars of grass pea. Additionally, comparative mapping between *L. sativus* and *L. cicera* and *P. sativum*, *Lens culinaris*, and *Medicago truncatula* reference genomes revealed the conservation of the *MLO1* locus position thus validated the evolutionary significance of this locus concerning legume disease resistance breeding programs.

10.3.2 Downy Mildew (*Peronospora lathyri-palustris*)

Downy mildew incited by *Peronospora lathyri-palustris* is one of the deadliest diseases of South Asia (Campbell 1997) and reported in both the species of *L. sativus* and *L. cicera* (Asthana and Dixit 1998; Campbell 1997). Some grass pea accessions with high to moderate resistance (RLS-1, RLS-2, JRS-115, JRL-43, and JRL-16) have been reported in India (Lal et al. 1986). Some of the landraces and CWRs have been reported with having lower disease score, which may be owing to tolerance or escape mechanisms (Campbell 1997).

10.3.3 *Ascochyta* Blight (*Mycosphaerella pinodes*)

Grass pea are most oftenly affected by *Ascochyta* blight (*Mycosphaerella pinodes*), which reduces production potential of this crop, and no true resistance has been reported till date (Skiba et al. 2004a). However, previous studies noted wide range of genetic variation and detected several genetic stocks against this disease among the accessions of *L. sativus* (Weimer 1947; Pang et al. 2000; Gurung et al. 2002; Skiba et al. 2004a), *L. ochrus,* and *L. clymenum* (Gurung et al. 2002). The genetics behind

Ascochyta blight resistance in grass pea was well-studied (Skiba et al. 2004b) and confirmed the presence of two genes interacting in complementary manner toward showing resistance response. Later, Skiba et al. (2004a) constructed a linkage map using 92 backcross populations generated from a cross between "ATC80878" which is a resistant cultivar with "ATC80407," a susceptible cultivar. They have reported two major QTLs (QTL1 and QTL2) positioned on linkage group 1 and 2, respectively, that can explain 12% and 9% of the phenotypic variation among the backcross progenies utilizing a total of 64 various markers. Further for comprehensive detection of the precise resistance gene against *M. pinodes*, cDNA library from disease-infected stem and leaf tissue of grass pea was prepared for selection of defense-related EST-SSRs (expressed sequence tag-derived simple sequence repeat markers) (Skiba et al. 2005). The major bottleneck regarding resistance-breeding program in grass pea is lack of genomic resources and transcriptomic data for annotation of defense-related candidate genes and to get molecular insights about the pathways regulating host pathogen interaction. NGS-based transcriptome profiling was carried out in grass pea deploying control and inoculated leaf sample of resistant cultivar in grass pea incited with *A. lathyri*. The study detected that upregulation of ethylene pathway as well as lignin and cellulose biosynthesis pathway toward changing the cell wall chemistry was the main orchestral concerning resistance mechanism in grass pea-*Aschochyta* interaction (Almeida et al. 2015).

10.3.4 Fusarium Wilt (*Fusarium oxysporum*)

Fusarium wilt disease incited by *Fusarium oxysporum f.* sp. *pisi*, a soil-borne hemibiotrophic, is another biotic hindrance toward successful cultivation of grass pea. Wide range of variation was detected, and a substantial level of resistance was reported among Slovakian grass pea germplasm (Benkova and Zakova 2001). Recently, the race 2 was considered as a major strain of grass pea (Sampaio et al. 2021a). In another study, 172 diverse worldwide collections of grass pea were evaluated under artificial epiphytotic situation exploring confocal laser scanning microscopy (CLSM) focusing race 2-specific green fluorescent protein (GFP). It was observed that accessions from the USA, Europe, Northern Africa, Asia, and unknown origin with large seed size have less disease infestation. Physical barrier like restricted taproot penetration in resistant cultivar along with some biochemical stimuli, like the presence of pisatin and other metabolites as antinutritional component, is the key player toward resistant mechanism against *Fusarium* wilt disease in grass pea. Two accessions (ACC273 and SITNICA) with high resistance were detected by Sampaio et al. (2021b). Genome-wide association study (GWAS) was carried out utilizing 161 grass pea accessions to unveil the genomic regions associated with fusarium wilt resistance in grass pea (Sampaio et al. 2021a). The genes were mostly located on chromosomes 1, 6, and 7 and putatively regulated amino acid metabolism toward resistance response.

10.3.5 Rusts (*Uromyces pisi, U. cicieris-arietini,* and *U. viciae-sativae*)

Rusts (*Uromyces pisi*, *U. cicieris-arietini*, and *U. viciae-sativae*) are significant diseases of grass pea in Northwestern Ethiopia (Campbell 1997). Due to close phylogenetic relation, the resistance gene detected in grass pea against rust pathogens is also crucial for developing rust-resistant cultivars in other legumes like field pea, chick pea, and alfalfa. However, BG-15744 and BG-23505, two grass pea lines, have been discovered to be only moderately rust-resistant source from cultivated gene pool (Vaz Patto and Rubiales 2009, 2014). Wide range of resistance with partial to immune was detected among various genotypes of *L. sativus* and *L. cicera* (Vaz Patto et al. 2009). More specifically, both in the field and in the growth chamber, accessions of *L. sativus* were generally more resistant to rust than accessions of *L. cicera* (Vaz Patto and Rubiales 2009). At various stages of infection by rust, from spore deposition to haustoria development, the underlying mechanism of plant defense varied among the resistant and susceptible cultivars. Hypersensitivity with complete resistance was detected only in *L. cicera* accessions. Additionally, variable disease reactions ranging from absolute resistance with complete absence of symptoms to well-formed colonies with lack of hypersensitive responses were observed only in some *L. cicera* accessions (Vaz Patto et al. 2009).

Very meager information is available regarding insect resistance in grass pea (Parihar et al. 2022). Comprehensive research is the prior need to standardize the method of screening regarding insect resistance owing to the mobile nature of the insect as compared to fungal pathogen. Natural variation was not substantial, and no sources of resistance were reported among the primary gene pool of grass pea concerning aphid, pod borer, and bruchid. Thrips are one of the serious issues in grass pea, and the damage due to thrips is mostly associated with tiny buds and flowers owing to flower dropping and yield reduction. Interspecific variation was detected regarding thrips infestation, and *L. aphaca* is tolerant to thrips in comparison to *L. sativus* (Pandey et al. 1996). Several Indian accessions (RLK-1, RLK-281, RLK-617, RPL-26, RLK-273-1, RLK-273-3, JRL-6, and JRL-41) were detected with having good resistance against thrips (Asthana 1996; Pandey et al. 1996). Grass pea is severely parasitized by cyst nematode (Greco and Di Vito 1994; Thompson et al. 2000) and root knot nematode (Rumbaugh and Griffin 1992). However, resistant sources are available within *L. sativus* accessions (Campbell 1997), and the identified accession (IFLA 347) can be used for transferring the trait (Di Vito et al. 2001). However, resistance source for root knot nematode was not available in cultivated gene pool, but the valuable sources of resistance present in various CWRs (*L. latifolius*, *L. sylvestris*, and *L. hirsutus*) reported by Rumbaugh and Griffin (1992).

10.3.6 Broomrape (*Orobanche crenata*)

Broomrape (*Orobanche crenata*) is a holoparasitic weed that is a serious menace for grass pea cultivation in Mediterranean area and other growing countries including

Syria, Morocco (Kumar et al. 2021). A wide range of variation was observed regarding broomrape resistance among cultivated and CWRs of *Lathyrus*, and this issue is becoming a major priority in grass pea breeding program (Fernández-Aparicio et al. 2009). It is detected that the date of planting and environmental factors strongly influence broomrape resistance in grass pea. The early or late maturing cultivar can escape the infestation, whereas accession with high biomass is more vulnerable to infection (Fernández-Aparicio et al. 2012). Sources of broomrape resistance have been reported in the species like *L. ochrus*, *L. clymenum*, and *L. choranthus* (Linke et al. 1993; Sillero et al. 2005). Slow development of the established tubercles and low induction of parasite germination are the main resistance mechanisms associated with *Lathyrus* species. Most of the accessions of the species, viz., *L. hierosolymitanus*, *L. tingitanus*, *L. annuus*, *L. aphaca*, *L. cicera*, *L. gorgoni*, *L. inconspicuus*, *L. sativus*, and *L. szowitsii*, were found susceptible (Abdallah et al. 2021). For *O. foetida*, only three species (*L. annuus*, *L. tingitanus*, and *L. pseudocicera*) showed resistance among 13 species of grass pea evaluated (Abdallah et al. 2021). Only two accessions of *L. sativus*, viz., IG-64782 and IG-65197, exhibited durable resistance against *O. crenata*, thus can be utilized as useful donor for developing broomrape-resistant grass pea cultivar.

10.4 Interspecific Hybridization

CWRs are valuable treasure trover of super alleles conferring resistance against many biotic and abiotic stresses (Banerjee et al. 2022). The success rate of utilizing interspecific hybridization for introgression of biotic stress resistance gene is less comprehended owing to lack of systematic characterization, evaluation, and documentation. ICARDA is maintaining 45 CWRs of *Lathyrus* genus from diverse countries (Kumar et al. 2021) where most of these species are from secondary gene pool. *Lathyrus* species from secondary gene pool like *L. cicera* is considered as good resistance source against powdery mildew (Vaz Patto et al. 2006a, 2007), downy mildew (Asthana and Dixit 1998; Campbell 1997), and rust (Vaz Patto et al. 2009). *L. ochrus* and *L. clymenum* are considered as durable resistant source against *Ascochyta* blight (Gurung et al. 2002). *L. aphaca* is tolerant to thrips in comparison to *L. sativus* (Pandey et al. 1996). Several *Lathyrus* species, including *L. ochrus*, *L. clymenum*, and *L. cicera*, have been identified as conferring broomrape resistance that is not present in cultivated germplasm (Linke et al. 1993; Sillero et al. 2005; Fernández-Aparicio et al. 2012; Abdallah et al. 2021). The first interspecific hybridization in grass pea was with sweet pea (*L. odoratus*) performed by Campbell (1997). Since then, interspecific crosses were attempted by many researchers but mostly with futile venture, and only 16 successful crosses were reported (Campbell 1997). Successful interspecific crosses were reported only with *L. amphicarpos* and *L. cicera* but with very low fertility (Yunus and Jackson 1991). The success rate could be increased by opting various tissue culture techniques like embryo rescue technique or using bridge species (Campbell 1997). Earlier reports confirmed that alien gene transfer could be possible from readily crossable species like *L. cicera* and

L. amphicarpos into the cultivated background (McCutchan et al. 1999; Durieu and Ochatt 2000; Ochatt et al. 2002, 2004). Protoplast culture and somatic hybridization between grass pea and pea have been standardized with large number of regenerated plantlets (McCutchan et al. 1999; Ochatt et al. 2002, 2004). These methodologies can open up new vista for transferring *Ascochyta blight* resistance from grass pea into pea and broomrape resistance from *L. ochrus* into grass pea.

10.5 Mutation Breeding: A Valuable Tool for Resistance Breeding

Grass pea breeding program is mainly restricted owing to its narrow genetic base because of self-pollination and interspecific incompatibility behavior (Mahapatra et al. 2020). Mutagenesis is having potential for inducing spectrum of variability in grass pea toward development of resistant cultivars against plant pathogen. When it comes to producing chlorophyll mutations in grass pea, chemical mutagens like EMS (ethyl methane sulphonate) and NMU (N-nitroso-N-methyl urea) are more effective than radiation (Nerkar 1976). However, genotypic variation was observed considering the efficiency of physical mutagen like gamma radiation (Prasad and Das 1980). Similarly, the order concerning the effectiveness and efficiency of various mutagens is NMU, EMS, and gamma rays in grass pea (Singh and Chaturvedi 1989). There is no successful example of generating resistant cultivar through mutagenesis in grass pea. Wide spectrum of variation regarding morphological features (Nerkar 1976; Rybiński 2003) and abiotic stress tolerance (Banerjee et al. 2022) has been reported. Recently, the new concept like TILLING (Targeting Induced Local Lesions in Genomes) and Eco-TILLING is gaining popularity for those species, which lack genomic resources and recalcitrant in nature with low transformation efficiency. Grass pea with all these features is a good candidate for exploring the concept of TILLING and Eco-TILLING for identification of mutation in targeted genes. In this context, it is noteworthy to mention that the John Innes Centre, UK, is maintaining EMS-mutagenized populations of two grass pea cultivars that are the potential source for future utilization toward detecting valuable mutant targeting the resistant candidate genes in grass pea.

10.6 Future Breeding Strategies Concerning Biotic Stress Resistance in Grass pea

The availability of genomic tools and resources is leading to a new revolution in many crops. However, scanty information on genomic resources and nonavailability of marker repertoire in grass pea limits the genomics-enabled improvement against various biotic stresses. Chowdhury and Slinkard (1999) have mentioned the first detailed genetic maps of grass pea. In the last few years, scientist groups including national and international centers have created some genomic resources. The analysis of NGS data by means of bioinformatics tools allows discovering new genes and regulatory sequences and their positions and makes available large collections of molecular markers (Almeida et al. 2014; Chapman 2015; Hao et al. 2017; Xu et al. 2018; Rathi et al. 2019). Re-sequencing of genomes is very useful for the genome-wide discovery of markers amenable for high-throughput genotyping platforms, like SSRs and SNPs, or the construction of high-density linkage maps. All these tools and resources are necessary for studying the genetic diversity, which is important for germplasm management, enhancement, and use. In addition, these tools will facilitate in the identification of markers linked to genes and QTLs, using an array of techniques like bulked segregant analysis (BSA), fine genetic mapping, or association mapping (Xu et al. 2018; Rathi et al. 2019). These new markers have potential use in marker-assisted selection, including marker-assisted backcrossing (MAB), or new strategies, like genomic selection. A large RNA sequencing (RNAseq) dataset from African and European grass pea accessions has recently been used to develop over 87 validated polymorphic EST-SSR markers and 42 KASP markers that were successfully tested on a global collection of 43 different accessions (Hao et al. 2017). QTLs for stem resistance to *Ascochyta* blight in grass pea at the seedling stage have been detected (Skiba et al. 2004a), while the *L. cicera* transcriptome has been analyzed in response to rust (*Uromyces pisi*) infection (Almeida et al. 2014). Some genes have been identified, e.g., *PsMLO1*, conferring susceptibility to powdery mildew. Overexpressing MIXTA-like genes (MYB transcription factors) to increase the resistance to insect pests could achieve enhanced trichome density. Since the 1990s, MAB has received increasing attention and has been extensively used in different crop species, and integration of MAB into conventional breeding programs represents an optimistic strategy for future grass pea improvement program.

Since grass pea still lacks genomic resources and is not amenable to efficient transformation, it is difficult to mutate by transgenesis. The approaches like TILLING (McCallum et al. 2000) and Eco-TILLING (Comai et al. 2004) make possible to screen mutant and germplasm collections for allelic variants in target genes. Efforts are being made to develop EMS-mutagenized populations in one or two grass pea cultivars (Sarkar et al. 2019). In contrast to the random nature of the mutations in a TILLING population, gene editing technologies offer targeted approaches to induce changes in the gene(s) of interest. One of the current drawbacks to these technologies is that they depend on achieving transgenesis to deliver the gene editing system to the target cells (Das et al. 2021). Proteomic methods can be used to measure subtle changes in protein expression levels in response to selective breeding and for biotic stress tolerance studies among different germplasm or cultivars (Chattopadhyay et al. 2011; Rathi et al. 2015).

In vitro plant breeding methods including embryo rescue, somatic embryogenesis, in vitro pollination, flowering, and fertilization as well as protoplast and somatic hybridization need to be utilized for novel variations in grass pea improvement program (Tripathy et al. 2015; Barpete et al. 2020a, 2020b). A special focus has been given to exploitation of somaclonal variation in production of plants with better yield attributes as well as the ability to better cope with biotic stresses (Barpete et al. 2014, 2022). RNAi is an ancient evolutionary mechanism adopted by plants as a defense strategy against foreign invading genes (Younis et al. 2014; Parmar et al. 2017; Kaur et al. 2021). This potential tool presents new horizon in plant breeding by introducing small noncoding RNA sequences with the ability to switch off gene expression in a sequence-specific manner. RNAi applications in grass pea may acquire new traits imparting resistance against nematodes and other insect pests, which are difficult to obtain through traditional breeding and present the potential of combining this technology with conventional breeding to biotic stresses.

The development of transgenic against biotic stresses provides continuing opportunities for crop improvement. Though in vitro regeneration and transformation in grass pea are difficult, few reports are available for development of efficient protocols using standard *Agrobacterium*-mediated transformation techniques, but those are not repeatable in other experiments (Barik et al. 2005). In a transgenic grass pea, the expression of an oxalate-degrading enzyme, oxalate decarboxylase (FvOXDC) of *Flammulina velutipes*, could reduce the ODAP levels and improve tolerance of the fungal pathogen *Sclerotinia sclerotiorum* (Goldsmith et al. 2022). In several other cases, unintended and unexpected effects of random gene insertion and interaction between foreign genes and host genes were reported in expression studies in transgenics. Therefore, research focus should now shift to gene targeting to achieve site-specific mutagenesis to avoid potential risks associated with insertions at random locations (ectopic) in the genome. Though development of transgenics has potential to overcome biotic stresses, these also constrain their usage from a regulatory procedure in many countries. Acceptance of genetically modified organisms (GMOs) requires consumer awareness of safety issues with respect to the environment and human health.

10.7 Tissue Culture Approaches

Grass pea is recalcitrant to tissue culture-based regeneration. Many researchers have tried to regenerate *L. sativus* (Mehta and Santha 2007) but with limited success. This has seriously limited the application of genetic transformation and gene editing in grass pea (Barpete et al. 2010, 2016). The first studies related to tissue culture were carried out in the 1970s. Callus culture is the most frequent mode of regeneration of species in the genus *Lathyrus* using several explants (Mukhopadhyaya and Bhojwani 1978; Gharyal and Maheshwari 1983) and induced shoots from callus cultures. Initially, Roy et al. (1991, 1992) regenerated shoots on leaf and root explant of cv. P-24 of *L. sativus*. The regenerated shoots were rooted in ½ MS medium supplemented with 0.5 μM IBA. In another study, Roy et al. (1993) reported development of <0.1% ODAP in some clones obtained from internode explants. Debnath et al. (2001) reported induction of multiple buds from callus induced on stems and leaf explants of beach pea. However, Ochatt et al. (2002) regenerated shoot buds from hypocotyls of three grass pea genotypes. They noted that plant regeneration competence was genotype-dependent. Bazr-Afkan et al. (2019) recorded callus-derived shoot regeneration using internode explant in *L. odoratus*. Contrarily, Barik et al. (2005) have also regenerated shoots from epicotyl explants of grass pea cv. IC-120487 on MS medium contained BAP + NAA. However,

Sahin-Demirbag et al. (2008) have also obtained shoots from immature embryo explants of *L. cicera*. They obtained shoots on MS medium containing 0.45 mg/L TDZ and 0.4 mg/L ascorbic acid and achieved 16.25 shoots per explant in their experiment. In another experiment, Sahin-Demirbag et al. (2008) induced shoots on cotyledon node explants of in vitro grown seedlings of Turkish dwarf chickling. They obtained shoots on MS medium containing 0.2 mg/L TDZ, 300 mg/L casein hydrolysate, and 0.2 mg/L TDZ. Similarly, Kendir et al. (2009) noted shoot regeneration on immature zygotic embryos of *L. sativus* using 0.45 mg/L TDZ and ascorbic acid. The regenerated shoots were rooted on MS medium +0.90 mg/L NAA, and induced plantlets were acclimatized in greenhouse. Zambre et al. (2002) induced shoot regeneration from both axillary and vegetative tissues on apical and axillary bud explants of *L. sativus* using Gamborg's (B5) basal media containing thidiazuron, IAA, NAA, and coconut water. Shoot elongation is achieved on MS medium enriched coconut water, BA and IAA. Barik et al. (2004) regenerated multiple shoots on cotyledonary nodes of *L. sativus* derived from 1-week-old seedlings using MS medium fortified with BA. The induced shoots were rooted on ½- × MS medium having IAA. The induced plantlets were acclimatized in soil. However, Barpete et al. (2020a) reported selection of somaclonal variants with low β-ODAP and high protein contents in 19 *L. sativus* lines. Barpete et al. (2020b) have also reported in vitro approaches to shorten generation cycles and rapid breeding for low β-ODAP contents of *L. sativus*.

10.8 Genetic Transformation

First studies about genetic transformation were carried out by Barna and Mehta (1995), and transformed *L. sativus* via particle bombardment and transformation frequency was noted 8–13%. The next work was performed after 10 years, and Barik et al. (2005) successfully transformed grass pea using epicotyl explants through *Agrobacterium tumefaciens* strains, LBA4404 and EHA105 carrying binary plasmid p35SGUSINT with npt-II, and β-glucuronidase (gus) gene. The higher transformation ability of putative transformed explants showed 36% expression on the GUS assay that is confirmed by Southern hybridization. Later on, germline transformation was confirmed by progeny analysis and T1 seedlings segregated in a 3:1 ratio. Gronlund et al. (2008) used virus-induced gene silencing (VIGS) in *Lathyrus* species by GUS tagged PEBV. However, they fail to successfully transform plant regeneration. *L. japonicus* accessions displayed GUS staining on either inoculated or uninoculated leaves. They noted a bleaching phenotype suggesting a downregulation of PDS expression. Parsa et al. (2021) have reported overexpression of zinc finger-GpZF gene through *Agrobacterium*-mediated transformation and confirmed that it promotes drought tolerance in *Lathyrus*. In connection with, the use of recombinant DNA technology and plant cell and tissue culture technology could aid in efforts toward this end. Works of Yadav and Mehta (1995) could be cited for identification, purification, and raising of monoclonal antibodies against oxalyl-CoA (coenzyme A) synthetase, which is the main enzyme in the ODAP biosynthesis. They have suggested possibility of introducing ODAP-degrading gene (antisense gene of oxalyl-CoA synthetase), into grass pea by *Agrobacterium-mediated* transformation.

10.9 Future Outlooks

The methodologies and genetic principles of breeding all legume crops including grass pea are very identical. However, the grass pea breeders are facing numerous hurdles, and there is need to rectify these problems in the future studies. Grass pea is a neglected, self-pollinated crop but often cross-pollinated. There are more than 180 species in the genus *Lathyrus* (Allkin et al. 1986; Aci et al. 2020; Ramya et al. 2021). Many species in the genus *Lathyrus* have annual, biennial, and perennial behavior, and many among them are incompatible for breeding purposes but each species having some agronomical important gene. Therefore, there is need to collect all species in genus *Lathyrus* and conserve them in a more than one genebank, pollen gene banks, and live gene banks. At the moment, no single collection of all species in genus *Lathyrus* is available in a single gene bank at one place. Establishing all forms of gene banks will help in improving breeding studies of grass pea in the future.

There is a need to find the extent of limitations in interspecific cross pollination and gene recombination behavior and seed settings with focus on breeding cultivars with reduced or zero β-ODAP cultivars with tolerance to abiotic and biotic stress to meet the future food and nutrient demand among humans, cattles, and poultry without fear of neurolathyrism. The use of other complimentary and supplementary techniques like mutation breeding with physical and chemical mutagens will be desired. These conservations will help in making them available for breeding programs of grass pea for disease tolerance varieties and conserve CWRs. Comparison of differences among the species in an artificially created environment in detail is desired and will facilitate in agronomical, morphological, molecular, and genomic studies. Most grass pea breeders work on multiple forage crops together in a section of the institute. There is need to establish separate institutes or working groups on breeding of grass pea in the future in equation with institutes for major crops like chickpea, wheat, potato, cotton, millet, rice, maize, etc. This will help the collection of group of scientists at a single place for carrying out the breeding activities. There is need to optimize appropriate conditions to emasculate and hybridize grass pea interspecific and intervarietal crosses under field and greenhouse conditions for suitable agronomical with biotic stress tolerance cultivars.

Basic information on breeding behavior, diseases, methods of breeding, and evaluation of species has been developed for a very few *Lathyrus* species. Most of the researchers are working individually independent of one another. There is a need to make extended networks for cooperation in extending and precising the future studies.

Grass pea varieties and species should be extensively evaluated (for morphological and molecular studies using QTLs, RAPD, SSR, AFLP, RFLP, and others) and selected for their performance in crossability, disease tolerance in breeding nurseries in both well-spaced rows, and narrow dense thickly populated rows along with checking their performance as single and intercrop before selecting them to explore their potentials. There is a need to advance this work further. To meet the future challenge of protein deficiency world over and effect on global livestock industry, it is hoped that grass pea would be transformed into a safe crop in the near future and there is possibility of its safe uses for animal and human consumption.

References

Abdallah F, Kumar S, Amri A, Mentag R, Kehel Z, Mejri RK, Amri M (2021) Wild Lathyrus species as a great source of resistance for introgression into cultivated grasspea (*Lathyrus sativus* L.) against broomrape weeds (*Orobanche crenata* Forsk. And Orobanche foetida Poir.). Crop Sci 61:263–276

Aci MM, Lupini A, Badagliacca G, Mauceri A, Lo-Presti E, Preiti G (2020) Genetic diversity among *Lathyrus* ssp. based on agronomic traits and molecular markers. Agronomy 10:1182

Adebola T (2019) Examining plant variety protection in Nigeria: realities, obligations and prospects. J World Intellect Prop 22:36–58

Allkin R, Goyder DJ, Bisby FA, White RJ (1986) Names and synonyms of species and subspecies in the Vicieae; Issue 3. Vicieae database project. Southampton University, Southampton

Almeida NF, Krezdorn N, Rotter B, Winter P, Rubiales D, Vaz Patto MC (2015) *Lathyrus sativus* transcriptome resistance response to *Ascochyta lathyri* investigated by deepSuperSAGE analysis. Front Plant Sci 6:178

Almeida NF, Leitão ST, Krezdorn N, Rotter B, Winter P, Rubiales D, Patto MCV (2014) Allelic diversity in the transcriptomes of contrasting rust-infected genotypes of *Lathyrus sativus*, a lasting resource for smart breeding. BMC Plant Biol 14(1):376

Asthana AN (1996) Grasspea cultivation in problem areas: present approaches. In: Lathyrus genetic resources in Asia: proceedings of a regional workshop, 27–29 December 1995, Indira Gandhi agricultural university, Raipur, India. IPGRI Office for South Asia, New Delhi, p 143

Asthana AN, Dixit GP (1998) Utilization of genetic resources in Lathyrus. In: Mathur PN, Rao VR, Arora RK (eds) Lathyrus genetic resources network. Pro IPGRI-ICARDAICAR Regional Working Group Meeting, New Delhi, pp 64–70

Banerjee J, Das A, Parihar AK, Sharma R, Pramanik K, Barpete S (2022) Genomic designing towards development of abiotic stress tolerant Grasspea for food and nutritional security. In: Genomic Designing for Abiotic Stress Resistant Pulse Crops. Springer, Cham, pp 345–381

Barik DP, Mohapatra U, Chand PK (2005) Transgenic grass pea (Lathyrus sativus L.): factors influencing agrobacterium-mediated transformation and regeneration. Plant Cell Rep 24(9): 523–531

Barik DP, Naik SK, Mohapatra U, Chand PK (2004) High-frequency plant regeneration by in vitro shoot proliferation in cotyledonary node explants of grasspea (*Lathyrus sativus* L.). In Vitro Cell Dev Biol-Plant 40:467–470

Barna KS, Mehta SL (1995) Genetic transformation and somatic embryogenesis in Lathyrus sativus. J Plant Biochem 4(2):67–71

Barpete S, Gupta P, Khawar KM, Kumar S (2021) Effect of cooking methods on protein content and neurotoxin (β-ODAP) concentration in grass pea (Lathyrus sativus L.) grains. CyTA J Food 19(1):448–456. https://doi.org/10.1080/19476337.2021.1915879

Barpete S, Gupta P, Khawar KM, Özcan S, Kumar S (2020a) In vitro approaches for shortening generation cycles and faster breeding of low B-N-Oxalyl-L-A, B-diaminopropionic acid content of grasspea (*Lathyrus sativus* L.). Fresenius Environ Bull 29(04A):2698–2706

Barpete S, Gupta P, Sengupta DS, Kumar J, Bhowmik A, Kumar S (2022) Neurotoxin (β-N-Oxalyl-L-α, β-diamino propionic acid) content in different plant parts of grass pea (Lathyrus sativus L.) spanning seedling to maturity stage: does it increase over time? Molecules 27(12):3683. https://doi.org/10.3390/molecules27123683

Barpete S, Gupta P, Singh M, Kumar S (2020b) Culture selected somaclonal variants showing low-ODAP and high protein content in nineteen grass pea (Lathyrus sativus L.) genotypes. Plant Cell Tissue Organ Cult 142:625–634. https://doi.org/10.1007/s11240-020-01889-0

Barpete S, Özcan SF, Khawar KM, Özcan S (2016) Effect of plant growth regulators and physical factors on in vitro high frequency regeneration of grass pea. The J Anim Plant Sci 26(4): 1087–1093

Barpete S, Parmar D, Sharma NC (2010) Callus culture and plantlet formation using axillary shoot explant of grass pea (Lathyrus sativus L.). In: Pandey BN, Singh SP, Singh R (eds) Sustainable management and conservation of biodiversity. Narendra Publishing House, New Delhi, pp 55–68. ISBN: 9789380428017

Barpete S, Sharma NC, Kumar S (2014) Assessment of somaclonal variation and stability in vitro regenerated grass pea plants using SDS-PAGE. Legum Res 37(4):345–352. https://doi.org/10.5958/0976-0571.2014.00642.0

Baxter LL, Anderson WF, Gates RN, Rios EF, Hancock DW (2022) Moving warm-season forage bermudagrass (Cynodon spp.) into temperate regions of North America. Grass Forage Sci 77: 141–150

Bazr-Afkan M, Daneshvar MH, Salmi MS (2019) Effects of explant type, culture media and growth regulators on callus induction and plant regeneration of sweet pea (*Lathyrus odoratus* L.). Iranian J Rangelands Forest Plant Breed Gene Res 27:98–106

Benkova M, Zakova M (2001) Evaluation of selected traits in grass pea (Lathyrus sativus L.) genetic resources. Lath Lathyrism Newsl 12:27–30

Campbell CG (1997) Grass pea. Lathyrus sativus L. promoting the conservation and use of underutilized and neglected crops, vol 18. Institute of Plant Genetics and Crop Plant Research, Gatersleben/international plant genetic resources institute, Rome. https://www.bioversityinternational.org/e-library/publications/detail/grass-pea-lathyrus-sativus-l/

Campbell CG, Mehra RB, Agrawal SK, Chen YZ, Moneim AM, Khawaja HIT et al (1994) Current status and future strategy in breeding grasspea (*Lathyrus sativus*). In: Expanding the production and use of cool season food legumes. Springer, Dordrecht, pp 617–630

Chapman MA (2015) Transcriptome sequencing and marker development for four underutilized legumes. Appl Plant Sci 3(2):1400111

Chattopadhyay A, Subba P, Pandey A, Bhushan D, Kumar R, Datta A, Chakraborty N (2011) Analysis of the grasspea proteome and identification of stress-responsive proteins upon exposure to high salinity, low temperature, and abscisic acid treatment. Photochemistry 72(10): 1293–1307

Chiurugwi T, Kemp S, Powell W, Hickey LT (2019) Speed breeding orphan crops. Theor Appl Gene 132:607–616

Chowdhury MA, Slinkard AE (1999) Linkage of random amplified polymorphic DNA, isozyme and morphological markers in grasspea (Lathyrus sativus). Agric Sci 133(4):389–395

Cocks P, Siddique K, Hanbury C (2000) Lathyrus: a new legume. A report for the rural Indus-tries Research and Development Corporation, Jan 2000, RIRDC publication no 99/150, RIRDC project no UWA-21A. Rural Industries Research & Development Corporation (RIRDC), now Agri Futures Australia, Wagga

Comai L, Young K, Till BJ, Reynolds SH, Greene EA et al (2004) Efficient discovery of DNA polymorphisms in natural populations by Ecotilling. Plant J 37:778–786

Das A, Parihar AK, Barpete S, Kumar S, Gupta S (2021) Current perspectives on reducing the β-ODAP content and improving potential agronomic traits in grasspea (*Lathyrus sativus* L.). Fornt Plant Sci 12:703275

de Souza GF, Ferreira MC, Munhoz CBR (2022) Decrease in species richness and diversity, and shrub encroachment in Cerrado grasslands: a 20 years study. Appl Veg Sci 25(3):e12668

Debnath SC, McKenzie DB, McRae KB (2001) Callus induction and shoot regeneration from stem, rachis and leaf explants in beach pea (*Lathyrus japonicus* wild). J Plant Biochem Biotechnol 10(1):57–60

Di Vito M, Greco N, Malhotra RS, Singh KB, Saxena MC, Catalano F (2001) Reproduction of eight populations of *Heterodera ciceri* on selected plant species. Nematol Mediterr 29:79–90

Dixit GP, Parihar AK, Bohra A, Singh NP (2016) Achievements and prospects of grass pea (*Lathyrus sativus* L.) improvement for sustainable food production. Crop J 4:407–416. https://doi.org/10.1016/j.cj.2016.06.008

Duguma B, Janssens GP (2021) Assessment of livestock feed resources and coping strategies with dry season feed scarcity in mixed crop–livestock farming systems around the gilgel gibe catchment, Southwest Ethiopia. Sustainability 13(19):10713

Durieu P, Ochatt SJ (2000) Efficient intergeneric fusion of pea (*Pisum sativum* L.) and grasspea (*Lathyrus sativus* L.) protoplasts. J Exp Bot 51:1237–1242

Eastwood RJ, Tambam BB, Aboagye LM, Akparov ZI, Aladele et al (2022) Adapting agriculture to climate change: a synopsis of coordinated National Crop Wild Relative Seed Collecting Programs across five continents. Plan Theory 11(14):1840

Enneking D (2011) The nutritive value of grasspea (*Lathyrus sativus*) and allied species, their toxicity to animals and the role of malnutrition in neurolathyrism. Food Chem Toxicol 49:694–709

Fernández-Aparicio MF, Flores F, Rubiales D (2009) Field response of *Lathyrus cicera* germplasm to crenate broomrape (*Orobanche crenata*). Field Crop Res 113:321–327

Fernández-Aparicio MF, Flores F, Rubiales D (2012) Escape and true resistance to crenate broomrape (*Orobanche crenata* Forsk.) in grasspea (*Lathyrus sativus* L.) germplasm. Field Crop Res 125:92–97

Fondevilla S, Carver TLW, Moreno MT, Rubiales D (2007) Identification and characterisation of sources of resistance to *Erysiphe pisi* Syd. In Pisum spp. Plant Breed 126:113–119

Fondevilla S, Rubiales D (2012) Powdery mildew control in pea. a review. Agron Sustain Dev 32: 401–409. https://doi.org/10.1007/s13593-011-0033-1

Gautam PL, Singh IP, Karihaloo JL (1997) Need for a crop network on Lathyrus genetic resources for conservation and use. In: Mathur PN, Ramanatha Rao V, Arora RK (eds) Lathyrus Genetic Resources Network. Proc IPGRI- ICARDA-ICAR Regional Working Group Meeting., 8—10 December 1997, New Delhi, India, pp 15–21

Gharyal PK, Maheshwari SC (1983) Genetic and physiological influences on differentiation in tissue cultures of a legume, *Lathyrus sativus*. Theor Appl Genet 66:123–126

Girma D, Korbu L (2012) Genetic improvement of grasspea (*Lathyrus sativus*) in Ethiopia: an unfulfilled promise. Plant Breed 131(2):231–236

Glawe DA (2008) The powdery mildews: a review of the world's most familiar (yet poorly known) plant pathogens. Annu Rev Phytopathol 46:27–51. https://doi.org/10.1146/annurev.phyto.46.081

Goldsmith M, Barad S, Peleg Y, Albeck S, Dym O et al (2022) The identification and characterization of an oxalyl-CoA synthetase from grasspea (*Lathyrus sativus* L.). RSC Chem Biol 3:320–333

Gonçalves L, Rubiales D, Bronze MR, Vaz Patto MC (2022) Grasspea (*Lathyrus sativus* L.)-a sustainable and resilient answer to climate challenges. Agronomy 12:1324

Gongke Z, Yingzhen K, Kairong C, Zhixiao L, Yafu W (2001) Hydroxyl radical scavenging activity of β-N-oxalyl-L-α, β-diaminopropionic acid. Photochemistry 58:759–762

Granati E, Bisignano V, Chiaretti D, Crin P, Polignano GB (2003) Characterization of Italian and exotic Lathyrus germplasm for quality traits. Genet Resour Crop Evol 50:273–280

Greco N, Di Vito M (1994) Nematodes of food legumes in the Mediterranean Basin Eppo. Bulletin 24(2):393–398

Gronlund M, Constantin G, Piednoir E, Kovacev J, Johansen IE, Lund OS (2008) Virus-induced gene silencing in *Medicago truncatula* and *Lathyrus odorata*. Virus Res 135(2):345–349. https://doi.org/10.1016/j.virusres.2008.04.005

Gurung AM, Pang ECK, Taylor PWJ (2002) Examination of Pisum and Lathyrus species as sources of ascochyta blight resistance for field pea (*Pisum sativum*). Aust Plant Pathol 31:41–45

Hammer K, Laghetti G, Direnzo P, Castelli A, Mikić A (2019) Resources and opportunities for re-establishing Lathyrus cicera L. as a multipurpose cultivated plant. Genet Resour Crop Evol 66:523–544

Hamphry M, Reinstadler A, Ivanov S, Bisseling T, Panstruga R (2011) Durable broad spectrum powdery mildew resistance in pea er1 plants is conferred by natural loss-of-function mutations in PsMLO1. Mol Plant Pathol 12:866–878

Hao X, Yang T, Liu R, Hu J, Yao Y et al (2017) An RNA sequencing transcriptome analysis of grass pea (*Lathyrus sativus* L.) and development of SSR and KASP markers. Front Plant Sci 8: 1873. https://doi.org/10.3389/fpls.2017.01873

Hillocks RJ, Maruthi MN (2012) Grass pea (*Lathyrus sativus*): is there a case for further crop improvement? Euphytica 186(3):647–654

Kahriz PP, Kahrz MP, Barpete S, Khawar KM (2022) Brief introduction of Lathyrus cultivation in Turkey. Acta bot Boreal Occident Sin 0517–0526

Kaur R, Choudhury A, Chauhan S, Ghosh A, Tiwari R, Rajam MV (2021) RNA interference and crop protection against biotic stresses. Physiol Mol Biol Plants 27(10):2357–2377

Kendir H, Sahin-Demirbag N, Aasim M, Khawar KM (2009) In vitro plant regeneration from Turkish Narbon vetch (*Vicia narbonensis* L. var. narbonensis L.). Afr J Biotechnol 8(4): 614–618

Kenicer GJ, Kajita T, Pennington RT, Murata J (2005) Systematics and biogeography of *Lathyrus* (Leguminosae) based on internal transcribed spacer and cpDNA sequence data. Am J Bot 92: 1199–1209

Kislev ME (1989) Origins of the cultivation of *Lathyrus sativus* and *L. cicera* (Fabaceae). Econ Bot 43:262–270

Kock MA (2022) Plant breeding and intellectual property: a controversial topic. In: Intellectual property protection for plant related innovation. Springer, Cham, pp 3–35

Kopittke PM, Menzies NW, Wang P, McKenna BA, Lombi E (2019) Soil and the intensification of agriculture for global food security. Environ Int 132:105078

Kumar S, Bejiga G, Ahmed S, Nakkoul H, Sarker A (2011) Genetic improvement of grasspea for low neurotoxin (β-ODAP) content. Food Chem Toxicol 49:589–600

Kumar S, Gupta P, Barpete S, Choukri H, Maalouf F, Sarkar A (2021) Grass pea. In: Pratap A, Gupta S (eds) The beans and the peas. Woodhead Publishing, Elsevier, pp 273–287

Kumar S, Gupta P, Barpete S, Sarker A, Amri A, Mathur PN, Baum M (2013) Grass pea. In: Singh M, Upadhyaya HD, Bisht IS (eds) Genetic and genomic resources for grain legume improvement. Elsevier, pp 269–293. https://doi.org/10.1016/B978-0-12-397935-3.00011-6

Kuo YH, Bau HM, Rozan P et al (2000) Reduction efficiency of the neurotoxin beta-ODAP in low toxin varieties of Lathyrus sativus seeds by solid state fermentation with aspergillus oryzae and Rhizopus microsporus var chinensis. J Sci Food Agric 80:2209–2215

Lal MS, Agrawal I, Chitale MW (1986) Genetic improvement of chickling vetch in (Lathyrus sativus L.) Madhya Pradesh, India. In: Kaul AK, Combes D (eds) Lathyrus and Lathyrism. Third World Medical Research Foundation, New York, USA, pp 146–160

Lambein F, Haque R, Khan JK, Kebede N, Kuo YH (1994) From soil to brain: zinc deficiency increases the neurotoxicity of *Lathyrus sativus* and may affect the susceptibility for the motor neurone disease neurolathyrism. Toxicon 32:461–466

Lambein F, Kuo YH (2009) Lathyrism. Grain Legume 54:8–9

Lambein F, Travella S, Kuo YH, Van Montagu M, Heijde M (2019) Grass pea (*Lathyrus sativus* L.): orphan crop, nutraceutical or just plain food? Planta 250:821–838

Lan G, Chen P, Sun Q, Fang S, (2013) Methods for treating haemorrhagic conditions. Patent US8362081B2

Lanfranco B, Fernández E, Ferraro B, de Lima JMS (2022) Sustainable intensification of agriculture and economy in the pampas grasslands under climate change. In: Handbook of behavioral economics and climate change. Edward Elgar Publishing, pp 193–223

Linke KH, Abd El-Moneim AM, Saxena MC (1993) Variation in resistance of some forage legumes species to *Orobanche crenata* Forsk. Field Crop Res 32:277–285

Llorens J, Soler-Martín C, Saldaña-Ruíz S, Cutillas B, Ambrosio S, Boadas-Vaello P (2011) A new unifying hypothesis for lathyrism, konzo and tropical ataxic neuropathy: nitriles are the causative agents. Food Chem Toxicol 49:563–570

Mahapatra NS, Das A, Bhattacharyya P, Bhattacharya S, Pal S, Barpete S (2020) Studies on genetic variability, divergence and association of characters in grasspea. J Crop Weed 16(1):155–161

McCallum CM, Comai L, Greene EA, Henikoff S (2000) Targeted screening for induced mutations. Nat Biotechnol 18(4):455–457

McCutchan JS, Larkin PJ, Stoutjesdijk PA, Morgan ER, Taylor PWJ (1999) Establishment of shoot and suspension cultures for protoplast isolation in *Lathyrus sativus* L. J Breed Genet 31:43–50

Mehta SL, Santha M (2007) Somaclonal variation and genetic transformation in *Lathyrus sativus*. In: Kirti PB (ed) Handbook of new technologies for genetic improvement of legumes. CRC Press, Boca Raton, pp 177–184

Mitra P, Mobarak SH, Debnath R, Barik A (2019) The role of Lathyrus sativus flower surface wax in short-range attraction and stimulant for nymph laying by an adult viviparous aphid. Bull Entomol Res 110(2):231–241. https://doi.org/10.1017/S0007485319000531

Mukhopadhyaya A, Bhojwani SS (1978) Shoot-bud differentiation in tissue cultures of leguminous plants. Zeitschrift fuer Pflanzen Physiologie 88:263–268

Narsinghani VG, Kumar SM (1979) The field reaction to powdery and downy mildew in Lathyrus genetic stock. J Mycol Plant Pathol 9:252–253

Nerkar YS (1976) Mutation studies in *Lathyrus sativus*. Ind J Gen Plant Breed 36:223–229

Ochatt SJ, Benabdelmouna A, Marget P, Aubert G, Moussy F, Pontécaille C, Jacas L (2004) Overcoming hybridization barriers between pea and some of its wild relatives. Euphytica 137(3):353–359

Ochatt SJ, Muneaux E, Machado C, Jacas L, Pontécaille C (2002) The hyperhydricity of in vitro regenerants of grasspea (*Lathyrus sativus* L.) is linked with an abnormal DNA content. J Plant Physiol 159(9):1021–1028

Palit R, DeKeyser ES (2022) Impacts and drivers of smooth brome (*Bromus inermis* Leyss.) Invasion in Native Ecosystems. Plan Theory 11(10):1340

Pandey RL, Chitale MW, Sharma RN, Rastogi N (1996) Status of Lathyrus research in India. In: Lathyrus genetic resources in Asia: proceedings of a regional workshop, 27–29 December 1995, Indira Gandhi agricultural university, Raipur, India. IPGRI Office for South Asia, New Delhi, India, p 45

Pandey RL, Sharma RN, Chitale MW (1997) Status of Lathyrus genetic resources in India. In: Mathur PN, Rao VR, Arora RK (eds) Lathyrus genetic resources network. Proceedings of a IPGRI-ICARDA-ICAR regional working group meeting, New Delhi, India, pp 7–14

Pang ECK, Croser JS, Imberger KT, McCutchan JS, Taylor PWJ (2000) Tissue culture and protoplast fusion of cool-season pulses: pea (*Pisum sativum* L.) and chickpea (*Cicer arietinum* L.). In: Linking research and marketing opportunities for pulses in the 21st century. Springer, Dordrecht, pp 429–436

Parihar AK, Barpete S, Das A, Lamichaney A, Gupta S (2022) Lathyrus breeding. In: Fundamentals of field crop breeding. Springer, Singapore, pp 1323–1389

Parmar N, Singh KH, Sharma D, Singh L, Kumar P, Nanjundan J, Thakur AK (2017) Genetic engineering strategies for biotic and abiotic stress tolerance and quality enhancement in horticultural crops: a comprehensive review. 3-biotech 7(4):1–35

Parsa M, Kashanchi M, Zeinali A, Pourfakhraei E (2021) Overexpression of Zinc finger (GpZF) promotes drought tolerance in grass pea (*Lathyrus sativus*). bioRxiv; 2021–2010. DOI: https://doi.org/10.1101/2021.10.19.464954

Poulter R, Harvey L, Burritt DJ (2003) Qualitative resistance to powdery mildew in hybrid sweet peas. Euphytica 133:349–358

Prasad AB, Das AK (1980) Morphological variants in khesari. Indian J Genet Plant Breed 40:72–175

Ramya KR, Tripathi K, Pandey A, Barpete S, Gore PG, Raina AP et al (2021) Rediscovering the potential of multifaceted orphan legume Grasspea-a sustainable resource with high nutritional values. Front Nutr 8:1327. https://doi.org/10.3389/fnut.2021.826208

Rathi D, Chakraborty S, Chakraborty N (2015) Proteomics of an orphan legume, grasspea: current status and future strategy. Plant Tiss Cult Biotechnol 25(1):117–141

Rathi D, Gayali S, Pareek A, Chakraborty S, Chakraborty N (2019) Transcriptome profiling illustrates expression signatures of dehydration tolerance in developing grasspea seedlings. Planta 250(3):839–855

Robertson JR, Abd El-Moneim AM (1996) Lathyrus germplasm collection, conservation and utilization for crop improvement at ICARDA. In: Arora RK, Mathur PN, Riley KW, Adham Y (eds) Proc. Regional Workshop Lathyrus Genetic Resources in Asia. Indira Gandhi Agricultural University, Raipur. IPGRI, Office for South Asia, New Delhi, pp 97–111

Roy PK, All K, Gupta A, Barat GK, Mehta SL (1993) β-N-Oxalyl-L-α, β-diaminopropionic acid in somaclones derived from internode explants of *Lathyrus sativus*. J Plant Biochem Biotechnol 2(1):9–13

Roy PK, Barat GK, Mehta SL (1992) In vitro plant regeneration from callus derived from root explants of *Lathyrus sativus*. Plant Cell Tissue Organ Cult 29(2):135–138

Roy PK, Singh B, Mehta SL, Barat GK, Gupta N (1991) Plant regeneration from leaf disc of *Lathyrus sativus*. Indian J Exp Biol 29(4):327–330

Rumbaugh MD, Griffin GD (1992) Resistance of Lathyrus species and accessions to the northern root-knot nematode, *Meloidogyne hapla*. J Nematol 24(4S):729

Rybiński W (2003) Mutagenesis as a tool for improvement of traits in grasspea (*Lathyrus sativus* L.). Lathyrus Lathyrism Newsl 3:27–31

Sahin-Demirbag N, Kendir H, Khawar KM, Ciftci CY (2008) In vitro regeneration of Turkish dwarf chickling (*Lathyrus cicera* L) using immature zygotic embryo explant. Afr. Biotechnology 7(12):2030. https://doi.org/10.5897/AJB2008.000-5052

Sahu LC, Amin A (2022) The issue of Biopiracy & Protection of traditional knowledge. Int J Mgmt Human 5:2310

Sampaio AM, Alves ML, Pereira P, Valiollahi E, Santos C, Šatović Z et al (2021a) Grasspea natural variation reveals oligogenic resistance to *fusarium oxysporum* f. sp. pisi. The plant. Genome 14(3):e20154

Sampaio AM, Vitale S, Turrà D, Di Pietro A, Rubiales D et al (2021b) A diversity of resistance sources to *fusarium oxysporum* f. sp. pisi found within grasspea germplasm. Plant Soil 463(1): 19–38

Santos C, Polanco C, Rubiales D, Vaz Patto MC (2021b) The MLO1 powdery mildew susceptibility gene in Lathyrus species: the power of high-density linkage maps in comparative mapping and synteny analysis. Plant Genome 14(2):e20090

Santos R, Mansidão A, Mota M, Raymundo A, Prista C (2021a) Development and physicochemical characterization of a new grasspea (*Lathyrus sativus* L.) miso. J Sci Food Agric 101(6): 2227–2234

Sarkar A, Emmrich PM, Sarker A, Zong X, Martin C, Wang TL (2019) Grass pea: remodeling an ancient insurance crop for climate resilience. In: Kole C (ed) Genomic designing of climate-smart pulse crops. Springer, Cham, pp 425–469

Sarpaki A, Jones G (1990) Ancient and modern cultivation of *Lathyrus clymenum* L. in the Greek islands. Annu Br Sch Athens 85:363–368

Sastri A (2008) Varieties developed and releases by Indira Gandhi Krishi Vishwavidyalaya, Raipur, India

Sellami MH, Pulvento C, Amarowicz R, Lavini A (2022) Field phenotyping and quality traits of grasspea genotypes in South Italy. J Sci Food Agric 102(12):4988–4999

SenGupta D, Barpete S, Kumar J, Kumar S (2021) Breeding for better grain quality in Lathyrus. In: Gupta DS, Gupta S, Kumar J (eds) Breeding for enhanced nutrition and bio-active compounds in food legumes, pp 131–156. https://doi.org/10.1007/978-3-030-59215-8

Sharma R, Paramanik K, Banerjee J, Das A, Bhan K (2022) Genomic designing for biotic stress resistance in Grasspea. In: Genomic designing for biotic stress resistant pulse crops, pp 415–441

Sharma S, Sharma R, Govindaraj M et al (2021) Harnessing wild relatives of pearl millet for germplasm enhancement: challenges and opportunities. Crop Sci 61(1):177–200

Sher H, Aldosari A, Ali A, de Boer HJ (2015) Indigenous knowledge of folk medicines among tribal minorities in Khyber Pakhtunkhwa, North-Western Pakistan. J Ethnopharmacol 166:157–167

Sillero JC, Cubero JI, Fernández-Aparicio M, Rubiales D (2005) Search for resistance to crenate broomrape (*Orobanche crenata*) in Lathyrus. Lathyrus Lathyrism Newsl 4:7–9

Singh M, Chaturvedi SN (1989) Mutation induction in *Lathyrus sativus* L. edited from Krishna, T.G. and Mitra, R. (nuclear Agric. Div., BARC, Trombay, Bombay 400085, India). Phytochemistry 26:897–902

Skiba B, Ford R, Pang EC (2005) Construction of a cDNA library of *Lathyrus sativus* inoculated with Mycosphaerella pinodes and the expression of potential defence-related expressed sequence tags (ESTs). Physiol Mol Plant Pathol 66(1–2):55–67

Skiba B, Ford R, Pang ECK (2004a) Construction of a linkage map based on a *Lathyrus sativus* backcross population and preliminary investigation of QTLs associated with resistance to ascochyta blight. Theoretical Appl Gene 109(8):1726–1735

Skiba B, Ford R, Pang ECK (2004b) Genetics of resistance to *Mycosphaerella pinodes* in *Lathyrus sativus*. Aust J Agri Res 55(9):953–960

Skiba B, Gurung AM, Pang EC (2007) Genome mapping and molecular breeding in Lathyrus. In: Pulses, sugar and tuber crops. Springer, Berlin, Heidelberg, Pp.123–132

Thompson JP, Greco N, Eastwood R, Sharma SB, Scurrah M (2000) Integrated control of nematodes of cool season food legumes. In: Linking research and marketing opportunities for pulses in the 21st century. Springer, Dordrecht, pp 491–506

Tittonell P, Piñeiro G, Garibaldi LA, Dogliotti S, Olff H, Jobbagy EG (2020) Agroecology in large scale farming-a research agenda. Front Sustain Food Syst 4:584605

Tripathi K, Gore PG, Ramya KR, Sarker A (2022) Grasspea an inherent abiotic stress-tolerant legume: current status and future scope under changing environment. In: Developing climate resilient grain and forage legumes. Springer, Singapore, pp 125–139

Tripathy SK, Ranjan R, Dash S, Bharti R, Lenka D, Sethy YD, Mishra DR, Mohapatra BR, Pal S (2015) Genetic analysis of BOAA content in grasspea (*Lathyrus sativus* L.). Legum Res 38(4): 465–468

Tsialtas IS, Theologidou GS, Bilias F, Irakli M, Lazaridou A (2020) Ex situ evaluation of seed quality and bruchid resistance in Greek accessions of red pea (*Lathyrus cicera* L.). Genet Resour Crop Evol 67:985–997. https://doi.org/10.1007/s10722-020-00896-6

Vaz Patto MC, Fernández-Aparicio M, Moral A, Rubiales D (2006a) Characterization of resistance to powdery mildew (*Erysiphe pisi*) in a germplasm collection of *Lathyrus sativus*. Plant Breed 125:308–310

Vaz Patto MC, Fernández-Aparicio M, Moral A, Rubiales D (2007) Resistance reaction to powdery mildew (*Erysiphe pisi*) in a germplasm collection of *Lathyrus cicera* from Iberian origin. Genetic Res Crop Evol 54:1517–1521

Vaz Patto MC, Fernández-Aparicio M, Moral A, Rubiales D (2009) Pre and posthaustorial resistance to rusts in *Lathyrus cicera*. Euphytica 165:27–34

Vaz Patto MC, Rubiales D (2009) Identification and characterization of partial resistance to rust in a germplasm collection of *Lathyrus sativus*. Plant Breed 128:495–500

Vaz Patto MC, Skiba B, Pang ECK, Ochatt SJ, Lambein F, Rubiales D (2006b) Lathyrus improvement for resistance against biotic and abiotic stresses: from classical breeding to marker assisted selection. Euphytica 147:133–147

Vaz Patto MV, Rubiales D (2014) Resistance to rust and powdery mildew in Lathyrus crops. Czech J Genet Plant Breed 50(2):116–122

Verma A (2021) Plant breeders' rights v farmers' rights: a contradistinction between the Indian approach and international protection. J Intellectual Property Law and Practice 16(9):958–967

Weimer JL (1947) *Ascochyta pinodella* and mycosphaerella. J Agric Res 75(5):181
White CL, Hanbury CD, Young P et al (2002) The nutritional value of *Lathyrus cicera* and *Lupinus angustifolius* grain for sheep. Anim Feed Sci Technol 99:45–64
Xu Q, Liu F, Qu R, Gillman JD, Bi C, Hu X, Krishnan HB (2018) Transcriptomic profiling of *Lathyrus sativus* L. metabolism of β-ODAP, a neuroexcitatory amino acid associated with neurodegenerative lower limb paralysis. Plant Mol Biol Report 36(5):832–843
Yadav VK, Mehta SL (1995) *Lathyrus sativus*: a future pulse crop free of neurotoxin. Curr Sci 66: 288–292
Yan ZY, Spencer PS, Li ZX, Liang YM, Wang YF, Wang CY, Li FM (2006) *Lathyrus sativus* (grasspea) and its neurotoxin ODAP. Phytochemistry 67:107–121. https://doi.org/10.1016/j.phytochem.2005.10.022
Younis A, Siddique MI, Kim CK, Lim KB (2014) RNA interference (RNAi) induced gene silencing: a promising approach of hi-tech plant breeding. Int J Biol Sci 10:1150
Yunus AG, Jackson MT (1991) The gene pools of the grasspea (*Lathyrus sativus* L.). Plant Breed 106(4):319–328
Zambre M, Chowdhury B, Kuo YH, Van-Montagu M, Angenon G, Lambein F (2002) Prolific regeneration of fertile plants from green nodular callus induced from meristematic tissues in *Lathyrus sativus* L.(grasspea). Plant Sci 163(6):1107–1112
Zhang J, Xing GM, Yan ZY, Li ZX (2003) β-N-oxalyl-L-α, β-diaminopropionic acid protects the activity of glycolate oxidase in *Lathyrus sativus* seedlings under high light. Russ J Plant Physiol 50(5):618–622
Zhao G (2012) Fourth military medical university. Not ginseng factor in preparing medicine for treating neurodegenerative diseases drugs application of Dencichine in preparation of medicament for treating neurodegenerative diseases, China Patent. CN 102579418A

Understanding Fungal Diseases and Their Mitigation in Lentils 11

Asish Kumar Padhy, Baljinder Singh, and Sabhyata Bhatia

Abstract

Lentil, an important grain legume, is gaining global attention due to its nutritional attributes and minimal water requirement during growth. It is a significant contributor in alleviating micronutrient deficiencies especially in developing nations. The global demand of lentil is increasing, which can be seen as an uptrend in the lentil production during last five decades from 0.85 to 5.73 Mt. Although India is the leading producer and consumer of lentils, the productivity gap is still 30–105% with an average of 42%. Incidence of biotic stresses (especially diseases) is becoming the limiting factor in gaining the optimal productivity and accounts for major yield losses in lentil. Several fungal diseases such as rust, *Stemphylium* blight, *Fusarium* wilt, anthracnose, white mold, gray mold, mildews, root rots, etc. are among the major hurdles in obtaining the maximum output from the crop. Management of fungal diseases in crop plants relies on the use of chemical fungicides. However, development of fungicide resistance in many fungi has become an ambush toward sustainable food production. Therefore, exploring the genetic resistance mechanisms along with development of resistant varieties is the best alternative to tackle these problems. Recently, efforts have been made toward disease identification, exploring the genetics and pathway of infection along with the development of disease mitigation strategies in lentil. Wild resistance sources have been identified for several fungal diseases, and a number of varieties have been successfully released which can withstand the impact of the diseases. An understanding of the diseases in lentils, their symptoms, epidemiology, current mitigation practices, and the

A. K. Padhy · B. Singh · S. Bhatia (✉)
National Institute of Plant Genome Research, New Delhi, India
e-mail: baljinder@nipgr.ac.in; sabhyatabhatia@nipgr.ac.in

U. C. Jha et al. (eds.), *Diseases in Legume Crops*,
https://doi.org/10.1007/978-981-99-3358-7_11

research progress will help in determining the future research focus which is summarized in this chapter.

Keywords

Lentil · Fungal diseases · Epidemiology · Mitigation

11.1 Introduction

Lentil, a diploid ($2n = 14$) self-pollinating annual crop which is also known as red dahl, masur, massar, etc. is gaining attention due to its nutritive values. The cultivated lentil, *Lens culinaris* ssp. Culinaris, originated from the wild progenitor *Lens culinaris* ssp. orientalis (Boiss.) (Ladizinsky 1979). Canada contributes maximum, i.e., four megatons of lentil (25%), to the world lentil pool followed by India, while consumption of lentils is highest in India. Due to the cold tolerance, shorter duration, atmospheric nitrogen fixing (30–40 kg/ha available for next crop) capacity, and increasing demand, lentil became a good substitute for chickpea (Zou et al. 2011). India had a production of 1.51 million tons of lentil from an area of 1.49 million hectares with a productivity of 1008 kg/ha in 2017–2018. Madhya Pradesh is the leading lentil producing state (30.94%) followed by Uttar Pradesh (28.72%), Bihar (15.24%), and West Bengal (5.81%), while Rajasthan has the highest productivity (1408 kg/ha) from the variety CG325 in India (MoA&*FW*, GOI, 2017–18). Still, 30–105% yield gap has been reported in India with an average of 42% in different production zones (Ali and Gupta 2012)

Lentil is an excellent source of protein having a protein content of 24–26% and is rich in lysine and leucine. Lentil is rich in low fat carbohydrates, oleic, linoleic, palmitic acid, vitamin B, calcium, potassium, etc. and also an outstanding source of macro and micronutrients (Duenas et al. 2006). According to the studies, lentil consumption in regular diet gives immense health benefits and help in protection against the chronic diseases like cardiovascular diseases, coronary heart diseases, cancer, type II diabetes mellitus, and aging (Amarowicz and Pegg 2008).

Stresses can be majorly categorized into two kinds, namely, biotic and abiotic stress. Abiotic stress includes the stresses due to environmental factors such as temperature, water, heavy metals, nutrients, etc., whereas biotic stresses involve the diseases and insect pests and weeds. Crop lentil suffers from major diseases such as rust, anthracnose, gray mold, *Fusarium* wilt, powdery mildew, root rot, *Stemphylium* blight, white mold, *Ascochyta* blight, etc. (Table 11.1) in India, that limit the lentil production; however, it is comparatively free from any kind of serious infestation of insect pests except pod borer (*Etiella zinckenella*) and aphids (*Aphis craccivora*). These diseases are discussed with a focus on their causal organism/s, symptoms, epidemiology, and research efforts made toward their mitigation.

Table 11.1 Table of lentil diseases and causal organism with references (adapted from book "The Lentil: Botany, Production and Uses" by Erskine 2009)

Disease name	Pathogen	References
Alternaria blight	*Alternaria alternata* (fries)	Chen et al. (2009); Kaiser (1992)
Aphanomyces root rot	*Aphanomyces euteiches* C. Drechsler	Lamari and Bernier (1985)
Black root rot	*Fusarium solani* (Mart.) Sacc.	Ahmed and Shahab (2017)
Black streak root rot	*Thielaviopsis basicola* (Berk. and Broome) Ferraris	Bowden et al. (1985)
Cercospora leaf spot	*Cercospora lensii*	Sharma et al. (1978)
Collar rot	*Sclerotium rolfsii* Sacc.	Beniwal et al. (1993); Khare (1981)
Cylindrosporium leaf spot and stem canker	*Cylindrosporium* spp.	Bellar and Kebabeh (1983)
Downy mildew	*Peronospora lentis*	Mittal (1997)
Dry root rot	*Macrophomina phaseolina* (Tassi) Goidanich	Kaiser (1992); Vishunavat and Shukla (1979)
Helminthosporium leaf spot	*Helminthosporium* sp.	Karahan and Katırcıoğlu (1993)
Leaf yellowing	*Cladosporium herbarum* (Pers.) link	Kaiser (1992); Karahan and Katırcıoğlu (1993)
Phoma leaf spot	*Phoma medicaginis* Malbr. and Roum.	Kaiser (1992)
Pythium root and seedling rot	*Pythium ultimum* Trow, *Pythium* spp.	Paulitz et al. (2002)
Wet root rot	*Rhizoctonia solani* Kühn,	Kaiser (1992); Karahan and Katırcıoğlu (1993)

11.2 Rust

11.2.1 Introduction

Rust is a foliar disease which is caused by *Uromyces viciae-fabae* (Erskine et al. 1994). It is a macrocyclic autoecious fungus that exhibits all five spore forms of common rust fungi and is formed on a single host. Rust infection cycle starts in the spring season with diploid teliospores to form four haploid basidiospores of two mating types. Basidiospores after germination produce pycniospores and receptive hyphae. Spermatization and dikaryotization produce aeciospores inside aecia at the abaxial side of the leaf. Germination of aeciospores produces uredia in which urediospores are formed which are brown in color and develop on both sides of the leaflets, pods, and stem (Negussie and Pretorius 2012). Uredia finally differentiate into firm, raised, black-colored telia. Up to 60–69% loss in yield have been reported in India by Singh (1986), while in Ethiopia, a complete crop failure (Beniwal et al. 1993), i.e., up to 100% yield loss, has been reported (Negussie et al. 2005).

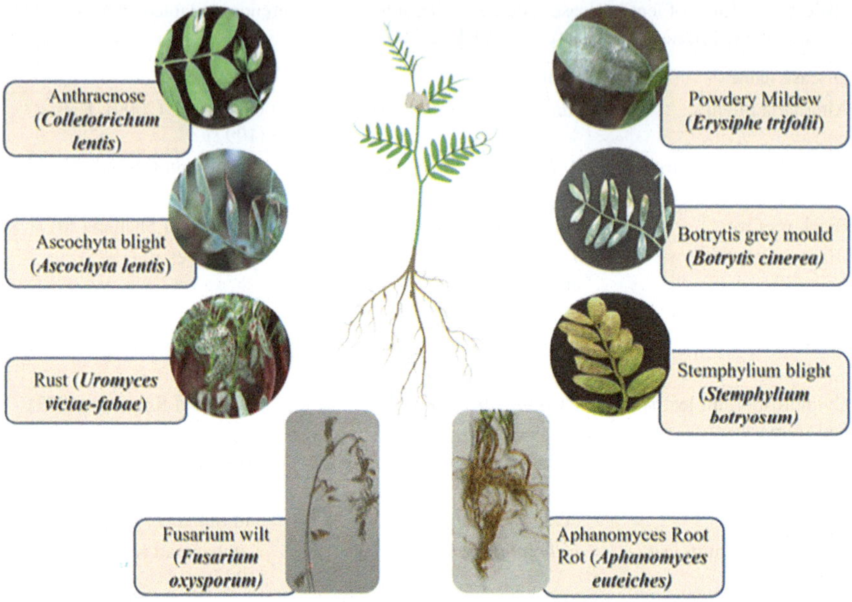

Fig. 11.1 Diagrammatic representation of major fungal diseases of lentil and their causal organisms

11.2.2 Disease Symptoms

The disease initiates with the development of yellow to white pycnidia and aecia (cup like structure) below the leaflet and pod surfaces, independently or in little groups in a circular pattern as described by Agarwal et al. (1993). Afterward, brown uredo pustules show up on both the surfaces of pods, stem, and leaflets. Pustules are round to oval with about 1mm diameter. They can unite into bigger pustules (Fig. 11.1). Late in the season, dark brown- to black-colored teliospores are formed which are elongated and generally present on the stem (Bayaa et al. 1997). The plant has a dim earthy colored to blackish appearance, obvious in influenced regions field or throughout the field if completely tainted (Beniwal et al. 1993). Under extreme contamination conditions, leaves are shed off and wrinkle, or no seed is found in the pods (Dikshit et al. 2016).

11.2.3 Epidemiology

The infection as a rule begins with low patches in the meadow and emanates to outskirts (Bayaa et al. 1997). For recurrence, lentil seed might be debased with rust-contaminated leaf, stem, and pericarps, which in many years can fill in as the essential inoculum for disease (Agarwal et al. 1993; Khare 1981). In weedy hosts, rust can likewise be proceeded from which it can contaminate the field lentil crop with the help of teliospores in the breeze. The disease is favored by high RH, shady, or rainy climate with a temperature range of 20–22 °C (Bashi and Rotem 1975). During the early blooming time frame/podding, the disease peaks up. Aeciospores sprout at 17–22 °C and taint different plants that help in formation of secondary aeciospores at 17–22 °C or urediospores at 25 °C. The uredosori grow later in the season and are immediately trailed by teliospores (Beniwal et al. 1993). The aecia and uredia present in the lentil litter bite the dust after harvesting; however, the teliospores endure high temperatures and permit the survival of disease inoculum throughout the year. Uredospores can be significant means for survival at low-temperature conditions (Bayaa et al. 1997). The uredo mycelium is profoundly impervious to warmth and daylight and is probably going to be significant for the fungus to survive in hot and dry conditions. The teliospores sprout without a break at 17–22 °C and cause new flare-ups of the infection every season. Altogether, 70 enrolled hosts of *U. viciae-fabae* are found to affect legumes including lentils. Chickpeas, *Lathyrus* spp., and *Vicia* spp. are among them.

11.2.4 Mitigation Strategies

As of now, suggested cultural control strategies for controlling *U. viciae-fabae* include summer volunteer plants control; distancing old host stubble from seasonal crops; and annihilation of old stubble (Prasada and Verma 1948). Previously, Agrosan (phenylmercury acetate)-treated seeds found to be used to control the lentil rust in India (Prasada and Verma 1948). Vigil (diclobutrazol) forestalled the presence of *U. viciae fabae* as long as 70 days after inoculation and applied as a fungicide for seed treatment (Singh 1985). Analyses exploring the utilization of foliar fungicides to control rust by Agarwal et al. (1977) identified that Hexaferb (ferric dimethyldithiocarbamate) and Dithane M-45 both give the best mitigation over *U. viciae-fabae* as tested in the test plots in Jabalpur, India. Likewise, Dithane M-45 additionally expanded bundle yield by 82% and grain weight by 24% contrasted with untreated control because of extreme disease when left untreated. The use of resistant varieties is one of the best methods for controlling rust (Bayaa et al. 1997).

Singh et al. (1995) have shown the variability in the pathogen and its pathogenicity. Effective rust-resistant donors such as ILL6002, ILL4605, ILL5604, ILL358, and ILL6209 have been identified in the hotspots, i.e., India, Pakistan, and Ethiopia (Bejiga et al. 1995; Singh and Sandhu 1988). The most commonly used donor for rust resistance in South Asia was Precoz. As described by Sinha and Yadav (1988), the resistance to rust in lentils is controlled by a single dominant gene. Studies reported that F, S, K, Zn, Cu, and Fe predominantly present on the leaf surfaces with excess N, Mn, amino acids, and sugars in the resistant cultivars than the susceptible ones; no difference has been reported on structural basis between resistant and susceptible cultivars (Reddy and Khare 1984). Chahota et al. (2002) studied the inheritance of rust resistance in two crosses of lentil and concluded that two duplicate dominant genes control the resistance. Various markers and genes were identified by the scientists which are linked with rust resistance such as SRAP marker F7XEM4a (Saha et al. 2010a, b), SSR marker Gllc 527 at 5.9 cM distance (Dikshit et al. 2016), genes urf1, urf2, urf3 (Kumar et al. 2001), etc. Gupta et al. (2012) constructed a linkage map for two parents of the lentil population for the rust resistance trait using 199 markers (SSR, ISSR, and RAPD). Among the rust-resistant varieties developed around the world, Barimasur 4 in Bangladesh (Sarker et al. 1999a, b), Pant lentil 4 in NE India (Singh et al. 1994), Assano and Teshale in Ethiopia (Fikre et al. 2007), NIAB masor 2006 in Pakistan (Sadiq et al. 2008), and Calpun-INIA in Chile (Peñaloza et al. 2007) are important ones. Asghar et al. (2018) selected 466 lentil genotypes for resistance to potential fungal diseases and identified LPP 11002 as highly resistant to rust. Kumar et al. (2015) developed seven lines of elite lentils (LL1217, LL1218, LL1219, LL1239, LL1240, LL1241, and LL1243) that were highly resistant to rust.

11.3 *Stemphylium* Blight

11.3.1 Introduction

Stemphylium botryosum is the causal organism of *Stemphylium* blight of lentil. It is among the major diseases of lentil which has been reported in Northeast India (Sinha and Singh 1991) and caused about 42% loss in the yield in Bangladesh (Bakr 1991). According to Sinha and Singh (1993), about 83% disease intensity and more than 90% yield loss have been reported in Bihar, India. Recently, the disease has also been reported in Canada and Nepal. The conidia of *Stemphylium botryosum* has 1–3 transverse and 1–3 longitudinal septa with occasional constrictions at the septation point and are ovoid to subdoliiform with pale brown to brown in color. Asci contains about eight ascospores and are cylindrical to club shaped. Ascospores are seven septate and ellipsoid to club-shaped and yellowish brown in color (Booth and Pirozynski 1967).

11.3.2 Disease Symptoms

The disease symptom begins with light brown to tan-colored spots on the leaflets with pin heads. The small spots then merge and cover the entire leaflet in 2–3 days under the favorable conditions. Lighter and darker angular areas are spread over the entire leaflet which appears light cream color in the infected tissue (Fig. 11.1) (Morrall 2003). Leaving the terminal leaflets, all leaflets shed off rapidly, and the plant turns dull yellow gradually with the spread of disease (Bakr 1991). The stem then dries, slowly turns ashy white, and bends down leaving the green pods on the plant with occasional white mycelium visible on the stem.

11.3.3 Epidemiology

Resting mycelia or the air-borne ascospores serve as the primary inoculum which comes from the infected plant debris or the seed or the weedy host. Although the seed-borne inoculum and its significance are not clearly understood (Mwakutuya 2006), according to Nasir and Bretag (1997), the pathogen has been often isolated from the lentil seeds in Australia. The inoculum is also supposed to be spread through the sea (Booth and Pirozynski 1967) from its weedy hosts such as alfalfa (Gilchrist 1990). A temperature range of 8–22 °C with canopy RH of 94% is favorable for the spread of infection (Bakr 1991), while according to Sinha and Singh (1993), relative humidity of 85–90% and about 18 °C average temperature in the morning are ideal for disease development. According to Mwakutuya (2006), temperature above 25 °C coupled with 48 h of leaf wetness optimizes the symptom development of *S. botryosum.* The large host range of the pathogen includes lentil (Bakr 1991), alfalfa, clover (Smith 1940), and tomato (Bashi and Rotem 1975). The disease has also been reported in lupin (Tate 1970), spinach (Koike et al. 2001), apple, gladiolus, and onion (Booth and Pirozynski 1967).

11.3.4 Mitigation Strategies

The major inoculum source of this disease is the infected debris and the stubble. Hence, destruction of crop residues and crop rotation are the major cultural methods to control the disease and to decrease the inoculum load. Delayed sowing although reduces the overall yield, but it was found to control the infection of *Stemphylium botryosum* positively in Bangladesh; but again, it invites the heavy infection of rust (Bakr 1991). Foliar fungicides such as Royal 50 WP were found to be very effective in controlling the damage due to this disease when applied thrice in weekly interval starting from the incidence of disease (Bakr 1991). *L. ervoides*, *L. tomentosus*, *L. nigricans*, *L. odemensis*, *L. lamottei*, and *L. culinaris* ssp. orientalis were found to be the resistance sources of *Stemphylium* blight (Podder et al. 2013). The resistance is also found to be associated with the morphological traits such as thicker cuticle, epidermal cells, and cortical cells with increase in epidermal hair density (Chowdhury et al. 1997). Barimasur 3 and Barimasur 4 were released in Bangladesh as resistant varieties to this disease (Sarker et al. 1999a, b).

Saha et al. (2010a, b) have done wonderful work for identifying the markers related to disease resistance and constructed a linkage map using 139 markers (RAPDs, SSRs, and SRAPs). He detected a QTL (QLG480–81) and reported a tightly linked SRAP marker (ME4XR16c) that is associated with resistance to *S. botryosum.* In 2017, Bhadauria et al. (2017) used 2180 high-quality SNPs to construct a linkage map and identified three QTLs (qSB-2.1, qSB-2.2, and qSB-3) for resistance to *Stemphylium* blight. “Barimasur-3” was released as resistant variety for both rust and *Stemphylium* blight diseases (Sarker et al. 1999a, b). Barimasur 4 and Binamasur 7 were released in Bangladesh as resistant varieties to this disease (Roy et al. 2018). A complex (Mihov and Stoyanova 1998) and quantitative control of resistance to this disease and its inheritance were studied by Kumar (2007) using a RIL population derived from cross between Barimasur 4 and CDC Milestone.

11.4 *Fusarium* Wilt

11.4.1 Introduction

Except Australia, *Fusarium* wilt is present and causes losses to lentil almost in every country and continent especially in rainfed areas (Beniwal et al. 1993). According to Chaudhary and Kaur (2002), this disease can cause severe loss, i.e., up to complete crop failure in certain areas under the favorable conditions. Due to difficulty in identification of species in *Fusarium*, many causal organisms were reported for this disease and which describe many wilting and dying symptoms (Khare 1981). As far as the vascular wilt of lentil is concerned, the causal organism is *Fusarium oxysporum* f.sp. lentis. In in vitro culture conditions, the mycelia of pathogen are branched, septate and hyaline. The pathogen produces microconidia, macroconidia, and chlamydospores (Khare 1980). Losses due to this disease vary depending upon the crop stage and its extent. Generally 5–12% loss and 72% loss in Syria (Bayaa et al. 1986) have been reported, while a complete crop failure has been reported in India at seedling stage (Khare 1981). 50–78% loss in India has been reported at pod filling stage (Agarwal et al. 1993), while 8.8% loss for 10% wilted plants has been reported in Syria (Erskine and Bayaa 1996).

11.4.2 Disease Symptoms

Disease symptoms for vascular wilt include stunting of the plants along with curling and shrinking of leaves at the lower parts of infected plant. Leaves at top also show symptoms like water deficiency. In progression, the symptoms come to the stem, and it becomes yellow and the plant dies (Fig. 11.1). Brown-colored discoloration and reduced growth are symptomatic to the root. Tap root tips are injured, and above it, the secondary roots start to flourish. In the lower stem, vascular streak is not always evident. Although the disease has been reported to cause damage at reproductive or pod filling stage, in India, this disease has also been reported in the seedling stage. At seedling stage, the symptoms mostly resemble to that of root rot and damping off which also includes sudden drooping of the plant and sometimes seed rot. External fungal growth near the root zone indicates the presence of other saprophytic fungus such as fungus of collar rot (Khare 1980).

11.4.3 Epidemiology

F. oxysporum f.sp. lentis is a soil-borne pathogen (Chaudhary and Kaur 2002) with a limited host range which is only lentil (Khare 1980). The disease proliferates and spreads at average temperature of 22–25 °C with a warm and dry condition (Erskine and Bayaa 1996). Chlamydospore is the resting structure of the pathogen, and it can survive in the soil for several years without a suitable host. De et al. (2001) found a synergistic collaboration between *F. oxysporum* f.sp. lentis and *Meloidogyne javanica* in both susceptible and the resistant cultivars which results in reduction in nodulation, root, and shoot length. Also the presence of the nematode *Meloidogyne* increases the wilt incidence (De et al. 2001).

11.4.4 Mitigation Strategies

Crop rotation and destruction of infected plants are the important cultural control methods. The released lentil cultivars, OPL 58, DPL 61, and DPL62 which are resistant or moderately resistant to the wilt, should be used (Chaudhary and Kaur 2002). Early maturing varieties or early sowing help in escaping the wilt by escaping the favorable conditions to some extent. Seed treatment also helps in mitigating the disease. According to the study by El-Hassan and Gowen (2006), talc and glucose formulation of biocontrol agent *Bacillus subtilis* is very effective in controlling the disease there by increasing the root length.

Kumar et al. (2010) identified 325 resistant accessions of lentil out of 20000 lines which were screened in a wilt sick plot. High level of resistance was also observed in wild species, *Lens culinaris* ssp. orientalis (ILWL 113) and *L. culinaris* ssp. ervoides (ILWL 138) (Table 11.2). According to Bayaa et al. (1995), utilization of resistance sources such as ILL5883, ILL5588, ILL4400, and ILL590 has been used widely in various countries as a resistance source to *Fusarium* wilt to develop a number of improved varieties. Lack of stable resistance, effective screening techniques, and appearance of other diseases such as collar rot, dry root rot, etc. are the main hurdle in the progress of development of an effective resistance source against the *Fusarium*

Table 11.2 Table of lentil diseases and their identified resistance sources with references (adapted from paper by Coyne and McGee (2013) doi: 1016/b978-0-12-397935-3.00007-4)

Disease	Resistance source	Reference
Anthracnose	*L. nigricans*, *L. ervoides*, *L. lamottei*, *L. odemensis*, *L. orientalis*, *L. tomentosus*	Tullu et al. (2006a, b), Tullu et al. (2011), Fiala et al. (2009), Vail and Vandenberg (2011) and Vail et al. (2012)
Ascochyta blight	*L. orientalis*, *L. nigricans*, *L. odemensis*, *L. ervoides*,	Bayaa et al. (1994), Nguyen et al. (2001) and Tullu et al. (2006a, b, 2011)
Rust	*L. orientalis*, *L. nigricans*, *L. odemensis*, *L. ervoides*	Gupta and Sharma (2006)
Powdery mildew	*L. orientalis*, *L. nigricans*,	Gupta and Sharma (2006)
Fusarium wilt	*L. orientalis*, *L. ervoides*	Bayaa et al. (1995), Gupta and Sharma (2006) and Mohammadi et al. (2012)
Stemphylium blight	*L. orientalis*, *L. nigricans*, *L. odemensis*, *L. ervoides*, *L. tomentosus*, *L. lamottei*	Podder et al. (2013)

wilt. Some prominent wilt resistant varieties are such Talya 2, Hala, and Rachayya in Lebanon; Idleb 2, 3, 4, and Ebla 1 in Syria; IPA 98 in Iraq; Pant L406, Pant L4, Priya, Seri, JL3, Noori, and VL507 in India; Assano, Alemtina, and Teshale in Ethiopia; and Firat 87 and Syran 96 in Turkey.

Chaudhary (2008) identified 43 cultural and morphological groups by studying on 333 isolates from various states of India. On the basis of disease reaction against 7 different lentil differentials, these 43 isolates were clustered in 3 clusters. Similar study by Datta et al. (2009) identified two subpopulations on molecular analysis of 24 isolates from NE-Indo-Gangetic plains with the help of 12 SSR marker pairs and 40 RAPD markers. Kamboj et al. (1990) reported five genes to confer resistance against *Fusarium* wilt which are independent of each other. Two duplicate genes in PL 234 and two complementary genes in PL 286 and JL 446 and a single dominant gene have been reported by ICARDA scientists to control wilt resistance (Abbas 1995).

Monogenic inheritance pattern of wilt resistance gene *Fw* has been recorded by Eujayl et al. (1998) in the variety ILL 5588. He also identified a RAPD marker OPK-15900 (10.8 cM) linked with *Fw* gene which was further linked with RAPD marker OP-C04650 in repulsion phase and OPB-17800 and OPD-15500 in coupling phase. By converting them into SCAR markers, these RAPD markers can be made more useful to be used in marker-assisted breeding of lentil. Hamwieh et al. (2005) further studied about *Fw* gene and identified two linked markers, i.e., SSR59-2B (at 8.0 cM) and AFLP marker p17m30710 (3.5 cM). However, in India, mapping of wilt-resistant genes is yet to be done, and works are in progress by making populations crossing Precoz and Sehore 74-3 (susceptible) with PL 2 and IPL406 (resistant). If the genes will be mapped without any segregation distortion, it will be very useful in the Indian context (Solanki et al. 2010). According to Choudhary et al. (2013), work is going on in ICARDA for mapping the race-specific genes.

11.5 *Ascochyta* Blight

11.5.1 Introduction

Ascochyta blight, which is one of the most devastating biotic production constraints of lentil, is caused by *Ascochyta lentis*. It can attack the crop at any growth stage and to any plant part above the ground under advantageous conditions. The disease has been reported to cause loss in seed quality and yield up to 70% in Canada, 30–50% in the USA, and 50% in Australia (Gossen and Morrall 1983; Kaiser 1992; Brouwer et al. 1995). The pathogen has two stages in its life cycle, i.e., asexual stage represented by pycnidia found in lesions of infected plants which release cylindrical septate conidia and the sexual phage (*Divdvmella lentis*) which was found to be heterothallic in nature (Kaiser et al. 1997) and produce ascospores. Kaiser and Hellier (1993) in Idaho, USA, for the first time, reported the sexual form of the pathogen in the wintered lentil straw in 1992.

11.5.2 Disease Symptoms

The disease symptom starts with irregularly shaped lesions on pods, stem, leaves, and petioles. The color of the lesions is tan in leaves, stem, and petioles while dark brown on the seed and pods. On heavy infection, the lesions merge to girdle the stem, and the part above it break and die up by wilting. On older lesions, pycnidia can be found at the center, while seeds become wrinkled and bleached with whitish mycelium (Kaiser and Hannan 1986). Symptoms first appear on the leaves and spread up on the plant parts near the soil near the flowering stage. Later the leaves dry up and shed off.

11.5.3 Epidemiology

The disease is seed borne (Kaiser and Hannan 1986) and spread by wind (long distances) or rain splashes (15 cM) (Pedersen et al. 1994). Experiment by Kaiser et al. (1997) showed that the pathogen is able to survive on infected seed as long as the seed remain viable, while some scientists proposed that *A. lentis* loses its viability after 29 weeks on pods and 21 weeks on seed when buried in soil. The pathogen survives by its sexual/asexual form through the infected seeds, leaflets, and plant debris. Roundhill et al. (1995) studied the early infection process of pathogen by inoculating conidia suspension on the detached leaves. He observed conidia germination, appresoria within 6 and 10 h of inoculation, respectively. Then he found the penetration peg piercing the cuticle mostly near the link of two epidermal cells. A disrupted plasmalemma and broken nucleus and cytoplasm were observed after 40 and 52 h of infection. By day 9, the pathogen enters the mesophyll cells after covering the entire epidermal layer, and the macroscopic symptoms become evident.

11.5.4 Mitigation Strategies

Among the cultural practices, crop rotation, the use of disease-free seed, early sowing to avoid moist weather, and destroying the diseased debris are very useful to control the disease (Nene 1988). Dry heat treatment (70 °C for 24 h), hot water treatment (55 °C for 25 min), and sun drying of lentil seed have been found to be very effective in controlling the seed-borne infection, while hot water treatment drastically decreases the seed germination (Ahmed and Beniwal 1991; Beniwal 1989). Among the fungicides used for seed treatment, metalaxyl, thiram (Bretag 1989), benomyl, and thiobendazole (Kaiser and Hannan 1987) are found to be very effective to control the disease, while their foliar sprays were not found to be much effective. At 3 g or more AI/kg of seed thiobendazole, if applied, showed phytotoxic effects. At early bloom or pod setting stage, the foliar fungicides were found to be very effective even on a single application. Captafol, folpet, chlorothalonil, and metiram are some of the foliar fungicides recommended to control *A. lentis* (Beauchamp et al. 1986).

Study by Ahmed and Morrall (1996) reported that over the time, virulence of the pathogen increases possibly due to genetic recombination and/or natural selection for virulent pathotypes during host pathogen interaction. Vandenberg et al. (2006) found out natural diversity of aggressiveness in a pathogen population without any discrete pathotype and host specificity. Nasir and Bretag (1997) for the first time divided *A. lentis* isolates into six pathotypes. This helped in finding resistance sources in lentil germplasm for *A. lentis* where success has been reported in wild *L. orientalis*, *L. nigricans*, *L. odemensis*, and *L. ervoides* (Table 11.2) (Bayaa et al. 1994). Resistance sources such as Indianhead, ILL358, ILL857, ILL5562, ILL5588, ILL5684, ILL5883, and ILL6024 have been used as donors, and prominent varieties such as Pant L406 in India; Masoor 93 in Pakistan; Laird, CDC Redwing, CDC Milestone and CDC Matador in Canada; and Nugget, Nipper, and Cassab in Australia have been developed. Markers AbR1, SCAR W19, and RB18 developed by Ford et al. (1999) are one of the fore most attempts to start MAS in lentil for *Ascochyta* blight resistance. Similarly, UBC227 (ral2), OPD 10 (Chowdhury et al. 2001), and QTLs C-TT/M-AC (QTL1 and 2) and M20 (QTL 3) (Rubeena et al. 2003) were also found out for *Ascochyta* blight resistance. Tullu et al. 2006a) developed four markers for identifying QTL for seedling resistance to *A. lentis* (ITAP marker DK 225, SSR AC097a, ISSR UBC890a, and RAPD V20a) and four for resistance to pod blight (SSR ILMs 25, ISSRs UBC 855a, UBC 807a, and UBC 830b). Two more markers for seedling resistance (P06a and V14a) were also identified by Tullu et al. (2006a).

11.6 Anthracnose

11.6.1 Introduction

Colletotrichum truncatum f.sp. lentis is the causal organism of anthracnose in all legumes including lentil (Kaiser et al. 1998). It was isolated by Andrus and Moore from the infected lentil stem in 1992. During the maturing season, warm, humid, and wet climatic condition favors the disease development with an optimum temperature of 20–24 °C, and spread is mediated by wind and water splashes. Dense canopy, disease debris, and infested seed are the important means of recurrence of disease from one season to another. The pathogen produces microsclerotia which are the resting structure and able to survive up to 4 years in soil which germinate to infect the suitable host on onset of favorable conditions.

11.6.2 Disease Symptoms

As the disease starts, light brown to tan-colored lesions appear on the stem in the lower part. The lesions have black dots, and as the disease advances, it covers larger area and spread upward. On the leaves, symptoms become evident near to flowering on 8–12 node stage, and they fall off prematurely. On the pods, lesions of light to dark brown colored with dark margin are produced with discolored seeds within the pod. In stem, it shows girdling symptom with evident wilting and dying and dead plant blacken on severe infection conditions (Buchwaldt et al. 2018).

11.6.3 Epidemiology

The pathogen is polycyclic, and conidia are the repeating structures which are spread through the wind and rain splashes (Buchwaldt et al. 2018). By biphasic intracellular hemibiotrophic infection strategy, the pathogen colonizes the crop. The asexual spore, conidia under favorable condition, germinates on the host tissue to form a round melanized appressorium from which infection peg emerges to rupture the host tissue by forming an infection vesicle. Unlike other biotrophic and hemibiotrophic fungi, this pathogen has weak plant-pathogen interphase (Bhadauria et al. 2011). The extent of biotrophic phase depends on the temperature and humidity conditions and is asymptomatic. The necrotrophic phase is symptomatic and characterized by secondary hypha developed from primary hypha of biotrophic phase. The water-soaked lesions which had started with infection cycle turn into necrotic lesions (Bhadauria et al. 2015). The acervuli on the lesions produce orange to salmon colored, single-celled conidia, and the whole process takes a total of 7 days. The conidia serve as the secondary inoculum, and pathogen thus completes multiple life cycles in a single cropping season (Chongo et al. 2002).

11.6.4 Mitigation Strategies

Using disease-free seeds, removing diseased plants, avoiding the harvesting in wet conditions, and removing the slightly diseased pods help in controlling the disease effectively. 50% chlorothalonil at 0.8–1.6 L mixed with 90–640 L of water/acre is very useful to control the disease, and it is to be sprayed before the onset of disease. A second spray after a fortnight is recommended if the disease appears (https://cropgenebank.sgrp.cgiar.org/).

There are two races of *Colletotrichum truncatum* f.sp. lentis, i.e., Ct1 and Ct0, to which wild species *L. nigricans*, *L. lamottei*, *L. odemensis*, *L. orientalis*, and *L. tomentosus* are resistant to one or the other race, while *L. ervoides* is found to be resistant to both the races (Table 11.2) (Tullu et al. 2006b). Three dominant (CtR3, CtR4, and CtR5) and two recessive genes (ctr1 and ctr2) were found so far which control resistance to anthracnose in lentil (Buchwaldt et al. 2013). A *Colletotrichum truncatum*-resistant locus LCt-2 was also identified by Tullu et al. (2003) for which three AFLP markers (EMCTTACA350 and EMCTTAGG375 in coupling; EMCTAAAG175 in repulsion) and two RAPD markers (OPE061250 in repulsion and UBC-704700 in coupling) were also found out. Some resistant varieties such as Indianhead, PI320937, and PI345629 are released which are showing resistant to race Ct1.

11.7 Botrytis Gray Mold

11.7.1 Introduction

Botrytis cinerea Ex Fr. and *Botrytis fabae* Sard are reported to cause Botrytis gray mold disease (Erskine and Bayaa 1996; Davidson and Krysinska-Kaczmarek 2007). *Botrytis cinerea* is ubiquitous and nonspecific with a host range of over 200 plant species (Ellis and Waller 1974), while *Botrytis fabae* is restricted to Fabaceae family mostly on lentil, faba bean, and common vetch. The disease has been reported in many countries such as Australia, Canada, Nepal, India, Pakistan, New Zealand, and Colombia. 50% yield loss to complete crop failure has been reported in severe disease conditions (Erskine and Bayaa 1996; Elad et al. 2004).

11.7.2 Disease Symptoms

The disease starts with dark green discrete lesions on the lower plant parts which turn grayish brown to cream-colored large patches with age infecting the whole leaflet. On severe infection, leaflets are shed off, and a furry layer of conidiophores comes to the stem and girdle it infecting the whole plant. Before onset of flowering and pod initiation, the plants die, and it can be seen as patches in the field (Erskine and Bayaa 1996). In dry conditions, conidia are released, and secondary spread is caused which further infects the flowers causing the flower death by forming gray mold. Infected pods too show gray-colored moldy growth and rot. The seed filling is very poor or shriveled, and discolored seeds are formed (Elad et al. 2004; Erskine and Bayaa 1996). When sowing is done with the infected seeds, gray mycelial growth on the hypocotyl of the seedling becomes evident causing seedling blight (Morrall 1997).

11.7.3 Epidemiology

Both the pathogens have similar white, cottony growth pattern which turns gray with age. Hyaline, ovoid-spherical, single-celled, thin-walled conidia are formed in clusters in the conidiophores. The size of conidia of *B. cinerea* is smaller than that of *B. fabae*, while the size of sclerotia is exactly opposite (Ellis and Waller 1974). The size and shape vary according to the environmental conditions. The favorable conditions are dense canopy, humid with temperature of 15–25 °C in which the disease can be an epidemic in a short span of time with the help of wind-borne conidia (Elad et al. 2004; Davidson and Krysinska-Kaczmarek 2007). Lentil is susceptible to the pathogen at almost every growth stage, but flowering and pod setting stage are economically more important. Although the mycelium can survive saprophytically in the diseased plant debris, the main resting structure is sclerotia which can survive a long time if buried in soil (Erskine and Bayaa 1996).

11.7.4 Mitigation Strategies

Among the cultural measures, avoiding dense canopy, delayed sowing, reduced seed rate while sowing, increased row spacing, weed control, optimum nitrogen use, crop rotation (3 years), destroying the disease debris, and the use of disease-free seed are most important (Erskine and Bayaa 1996; Elad et al. 2004). Among the fungicides, carboxin, chlorothalonnil, thiram, benomyl, or thiabendazole can be used as seed treatments to minimize seedling blight and the seed-borne infection. Carbendazim, mancozeb, vinclozolin, tridemorph, etc. can be used as foliar fungicides. However, care must be taken to prevent fungicide resistance as the pathogen evolves rapidly to synthetic or systemic fungicide (Erskine and Bayaa 1996). Breeding program is going on to develop varieties with better resistance to Botrytis gray mold.

Although the resistance mechanism is poorly understood in case of Botrytis gray mold, some resistant germplasm has been identified. A variety "Nipper" released in Australia shows resistance to both gray mold and *Ascochyta* blight (Materne et al. 2002). According to Kuchuran et al. (2003), CDC Redcap, CDC Milestone, and CDC Robin have been reported to show consistently lower disease incidence in artificial epidemic conditions. In Pakistan, are accessions, ILL6024, ILL6004, and ILL6016 (Erskine et al. 1994) and in Nepal are LG 171, LG198, Aarial, LN0038, and Simrik (Karki 1993), whereas in Canada, variety Indianhead (ILL418) and Matador (Materne et al. 2002) were found to show resistance to Botrytis gray mold.

11.8 Powdery Mildew

11.8.1 Introduction

The disease powdery mildew has several causal organisms such as *Erysiphe pisi* DC and *Erysiphe polygoni* DC which are ectoparasites and *Leveillula taurica* (Lev.) Arnaud, an endoparasite. According to Attanayake et al. (2009), it is also caused by *Erysiphe trifolii*. The disease has been reported in India, Ethiopia, Tanzania, Russia, Cyprus, Sudan, Syria, etc. India and Sudan are the hotspots for this disease. Due to dry weather conditions, Rajasthan became the most affected state in India due to powdery mildew. Although in greenhouse the disease can come at any growth stage, in the field, the disease mostly comes at flowering. In Trans-Himalayan regions of India (Lahaul Spiti and Sangla) and in Syria, the disease also comes in off-season nurseries of lentil (Attanayake et al. 2009).

11.8.2 Disease Symptoms

Symptoms are evident in the older leaves on the upper surface, where affine powdery, white growth mycelium and conidia initiate as small spots. The disease spreads rapidly and covers the entire leaflets, stem, and pods. On severe infection, the leaves shed off becoming dry and curled. The infected pod bears small and shriveled seed causing reduction in seed quality and yield (Beniwal et al. 1993).

11.8.3 Epidemiology

The asexual (anamorph) stage is responsible for disease proliferation and spread. The conidia are formed in clusters, and infection is spread by the means of wing and water splashes. Moderately high temperature and RH favor the disease development. The teleomorph stage of the pathogen is found in India and Sudan (Chitale 1981).

11.8.4 Mitigation Strategies

Cultural methods should be followed as described earlier in other diseases. Some fungicides, benomyl, tridemorph, karathane, calixin, were found to be effective when sprayed at recommended dose. As foliar sprays, some insecticides such as phoxim, triazophos, and quinalphos were also found to be effective in controlling the disease (Beniwal et al. 1993). Many resistant lentil genotypes were also developed by Tikoo et al. (2005) against powdery mildew.

Some wild accessions tested at CSKHPKV, Palampur, were found to be effective against the disease. Those are *L. orientalis* (ILWL 230, ILWL 476), *L. odemensis* (ILWL 39, ILWL 203 and IG 136788), and *L. tomentosus* (ILWL 480, ILWL 198) (Singh et al. 2020). Mildew resistance locus O (MLO) family genes were analyzed and mined by

Polanco et al. (2018), and two genes, namely, LcMLO1 and LcMLO3, were characterized for possible powdery mildew resistance in lentil wild germplasm. Not much has been carried out in powdery mildew research in lentil, but many institutions in Australia, Canada, and India have ongoing research projects to find out the genetics of resistance to this disease.

11.9 Sclerotinia White Mold

11.9.1 Introduction

Sclerotinia sclerotiorum (Lib.) De bary is the causal organism of white mold disease also known as stems rot of lentil. The pathogen is ubiquitous and has more than 400 host species from 75 families (Boland and Hall 1994). The disease is favored by high plant density, excessive growth, and high precipitation especially during flowering and pod setting (Akem et al. 2006).

11.9.2 Disease Symptoms

Bleached lesions are formed in stem, leaves, pods, and pedicels in the beginning of disease. Sometimes, these are covered by cottony white mycelium with occasional dark spots which are the resting structure, the sclerotia. If ascospores are germinated, then senescent flower petals are perfect sites for invasion. Infection on stem can result in wilting of the plants (Bolton et al. 2006).

11.9.3 Epidemiology

Sclerotia are black and round to elongate with about 1cm in length. Sclerotia can germinate myceliogenically and can rapidly grow over the host tissue. No spores are formed on microconidiophores. Carpogenic germination of sclerotia for which specific soil temperature and moisture are required (Morrall 1997) forms light to dark brown apothecia. Asci can be found on the upper layer of apothecia with eight elliptical to ovoid, unicellular, and hyaline ascospores (Bolton et al. 2006; Clarkson et al. 2003). Both the ascospores and sclerotia are able to start the primary infection. At maturity, the plants become more susceptible to this disease. When infected plant touches the healthy plants, cell wall-degrading enzymes and oxalic acid formed by the pathogen help in invasion of pathogen to healthy tissue and develop yellowish and bleached lesions. Sclerotia formed return to soil during harvesting and threshing and provide the inoculum for next season (Bolton et al. 2006).

11.9.4 Mitigation Strategies

Cultural methods have limited efficacy due to wide host range of pathogen, and the sclerotia once formed has very long survival period. Early fungicide application has been found to be effective in controlling flower and petal infection where as it is not always economical. Delayed fungicide application is often not recommended as it cannot penetrate lower plant canopy due to the dense foliage (Bolton et al. 2006).

11.10 Next-Generation Strategies for Resistance Breeding in Lentil

Global food security is the primary goal of agriculture to ensure enough agricultural production for continuously growing population. However, yield of majority of crop plants is still limited due to various biotic stresses. In terms of productivity, lentil lags significantly behind other legumes (Tiwari et al. 2022). Conventional breeding strategies for disease resistance in plants have been thoroughly utilized, and hence there is a need to use next-generation breeding strategies to develop crop plants for better resistance and yield. Next-generation breeding techniques have been successfully employed in major crop plants; however, for less studies, crop plants, such as lentil, conventional breeding methods are still being used (Kole et al. 2015). Classical breeding strategy involves the process of hybridization and continuous screening to develop an elite variety (Kumar et al. 2021). However, this strategy is of limited use due to genetic drag, hybridization bottlenecks, and its laborious nature. To overcome the shortcomings of classical methods, integration of next-generation sequencing data, high-throughput phenomics, and the use of artificial intelligence are necessary to develop improved cultivars. Plant genomics is a crucial component of next-generation breeding techniques and helps in identification and selection of superior alleles for plant breeding through genetic engineering and marker-assisted selection (MAS). Several efforts have been made in the recent past to develop genomic resources in lentil including development of SNP and SSR markers, transcriptomes, QTL mapping, etc. (Kumar et al. 2018; Singh et al. 2019; Ma et al. 2020; Gela et al. 2021; Singh et al. 2021). Genotyping data generated through NGS platforms along with high-throughput phenotyping data could accelerate precise identification of genomic regions controlling agronomical traits (Tiwari et al. 2022). So far, classical phenotyping methods are being utilized for screening of agronomically important traits; these methods are labor-intensive, less precise, and time-consuming. Recent development of advanced imaging-based phenotyping methods has been instrumental in generating accurate phenotyping data of complex quantitative traits. A HTP method for screening salt toxicity was developed and applied for screening 276 lentil accessions which demonstrated improved accuracy of image-based phenotyping method (Dikshit et al. 2020). Similarly, another study utilized digital RGB, hyperspectral, and multispectral imaging for identification of *Aphanomyces* root rot-resistant genotypes in lentil (Marzougui et al. 2019). Although several high-throughput phenotyping techniques have been developed, their use for lentil phenotyping studies is still very limited. Therefore, there is a need to develop precise multisensor phenotyping platforms for lentil. Data generated through these platforms could be integrated with NGS-based genotyping data for identification of QTL regions and markers through QTL mapping and GWAS studies. Several QTLs and markers have been identified in lentil with the advent of NGS-based sequencing platforms. Next step should be the conversion of these markers into breeder friendly PCR based markers and their utilization for future lentil breeding programs through marker-assisted selection. Further, regulatory role and mechanism of action for genes identified in QTL mapping studies could be investigated through transcriptomic studies. Also, several other strategies such as CRISPR-Cas-based gene editing, transgenics, speed breeding, mutation breeding, etc. could also be implemented in lentil breeding programs for developing improved cultivars.

11.11 Conclusion and Future Perspective

Considering the severity and occurrence of various diseases, the fungal diseases account for major losses in lentil production. After analysis of the mitigation practices used to control fungal diseases, it can be concluded that several countries use fungicides to control the diseases. Most of the times, this becomes uneconomical for the producer, while on the other hand, it may bring about increased levels of resistance among the pathogens leading to more virulent and fungicide resistant strains. Phytotoxicity and environmental deterioration are also becoming major problems, although the crop requires minimal fungicide application as compared to other cultivated cereals and legumes.

In the Indian context, occurrence of rust, *Fusarium* wilt, *Ascochyta* blight, and powdery mildew is widespread, while in terms of disease severity and history of crop damage, lentil is most devastated due to *Fusarium* wilt, *Stemphylium* blight, rust, and Botrytis gray mold. In case of lentil rust, the robust and reproducible STS and SNP markers need to be developed, whereas artificial inoculation procedure in greenhouse and in the offseason needs to be standardized for the disease screening purpose. In case of *Fusarium* wilt, standard race identification procedures need to be developed. Identification and resistance genes, their genetics and mechanism of resistance, need to be studied thoroughly. As this disease has no effect of fungicide application, resistance breeding needs to be strengthened.

Ascochyta blight is the most researched disease, and advances have been made on both conventional and molecular aspects including gene, marker, and QTL identification for resistance breeding programs. More focus is required on diseases like *Stemphylium* blight, powdery mildew, anthracnose, Botrytis gray mold, and *Sclerotinia* white mold to identify the markers and map the QTL and genic regions responsible for disease resistance. Resistance breeding must be focused on development of area-specific multiple disease-resistant varieties. Large-scale screening of germplasm for possible resistance to most diseases has helped in identifying known sources of resistance, and there is a need to utilize these in lentil improvement programs. Moreover major efforts focusing on minor diseases like wet and dry root rot, black root rot, bacterial spots, and viral diseases are desirable in order to counter possible future outbreaks.

References

Abbas A (1995) Variation in some cultural and physiological characters and host/pathogen interaction of *Fusarium oxysporum* f. sp. lentis, and inheritance of resistance to lentil wilt in Syria. Faculty Agric. Univ, Aleppo

Agarwal SC, Singh K, Lal SS (1993) Plant protection of lentil in India. In: Lentil in South Asia. International Center for Agricultural Research in the Dry Area, Aleppo, pp 147–165

Agrawal SC, Khare MN, Agrawal PS (1977) Control of lentil rust by use of fungicides [India]. Indian Phytopathol (India) 29(1):90–91

Ahmed S, Beniwal SPS (1991) Ascochyta blight of lentil and its control in Ethiopia. Int J Pest Manag 37(4):368–373

Ahmed S, Morrall RA (1996) Field reactions of lentil lines and cultivars to isolates of *Ascochyta fabae* f. sp. lentis. Can J Plant Pathol 18(4):362–369

Ahmed D, Shahab S (2017) Effect of different inoculum levels of *Fusarium solani* (Mart.) Sacc on plant growth, biochemical and nutrient parameters of lentil (Lens culinaris Medik.). Int J Agric Environ Biotech 10(2):239–243

Akem C, Bellar M, Bayaa B (2006) Comparative growth and pathogenicity of geographical isolates of *Sclerotinia sclerotiorum* on lentil genotypes. Plant Pathol J 5(1):67–71

Ali M, Gupta S (2012) Carrying capacity of Indian agriculture: pulse crops. Curr Sci 102:874–881

Amarowicz R, Pegg RB (2008) Legumes as a source of natural antioxidants. Eur J Lipid Sci Technol 110(10):865–878

Asghar MJ, Akhtar KP, Abbas G, Rizwan M, Iqbal M, Idrees M, Ali S, Aslam M, Aziz-ur-Rehman MR, Saleem U, Khan ME (2018) Identification of multiple sources of resistance in lentil against some potential fungal diseases. Pakistan J Agric Sci 55(4)

Attanayake RN, Glawe DA, Dugan FM, Chen W (2009) *Erysiphe trifolii* causing powdery mildew of lentil (*Lens culinaris*). Plant disease 93(8):797–803

Bakr MA (1991) Plant protection of lentil in Bangladesh. In: Proceedings of the seminar on lentil in South Asia, pp 11–15

Bashi E, Rotem J (1975) Host and biotic factors affecting sporulation of *Stemphylium botryosum* f. sp. lycopersici on tomatoes and of alternaria porri f. sp. solani on potatoes. Phytoparasitica 3(1):27–38

Bayaa B, Erskine W, Khoury L (1986) Survey of wilt damage on lentils in northwest Syria. Arab J Plant Protect

Bayaa B, Erskine W, Hamdi A (1994) Response of wild lentil to *Ascochyta fabae* f. sp. lentis from Syria. Genetic Res Crop Evolution 41(2):61–65

Bayaa B, Erskine W, Hamdi A (1995) Evaluation of a wild lentil collection for resistance to vascular wilt. Genetic Resources and Crop Evolution 42(3):231–235

Bayaa B, Erskine W, Singh M (1997) Screening lentil for resistance to Fusarium wilt: methodology and sources of resistance. Euphytica 98(1-2):69–74

Beauchamp CJ, Morrall RA, Slinkard AE (1986) The potential for control of ascochyta blight of lentil with foliar-applied fungicides. Canadian journal of plant pathology 8(3):354–359

Bejiga G, Tullu A, Tsegaye S, Ahmed S, Anbessa Y (1995) Evaluation of lentil genotypes (*Lens culinaris* L.) for resistance to rust (Ethiopia), Petria (Italy)

Bellar M, Kebabeh S (1983) A list of diseases, injuries and parasitic weeds of lentils in Syria (survey 1979-1980). Lens 10(1):30–31

Beniwal SPS (1989) Effect of sun drying of lentil seeds on the control of seed-borne *Ascochyta lentis* (Lens culinaris). Lentil Experimental News Service.

Beniwal SPS, Bayaa B, Weigand S, Makkouk KH, Saxena MC (1993) Field guide to lentil diseases and insect pests. International Center for Agricultural Research in the Dry Areas (ICARDA)

Bhadauria V, Banniza S, Vandenberg A, Selvaraj G, Wei Y (2011) Cataloging proteins putatively secreted during the biotrophy-necrotrophy transition of the anthracnose pathogen *Colletotrichum truncatum*. Plant Signal Behav 6(10):327

Bhadauria V, MacLachlan R, Pozniak C, Banniza S (2015) Candidate effectors contribute to race differentiation and virulence of the lentil anthracnose pathogen *Colletotrichum lentis*. BMC genomics 16(1):628

Bhadauria V, Ramsay L, Bett KE, Banniza S (2017) QTL mapping reveals genetic determinants of fungal disease resistance in the wild lentil species *Lens ervoides*. Sci Rep 7(1):1–9

Boland GJ, Hall R (1994) Index of plant hosts of *Sclerotinia sclerotiorum*. Can J Plant Pathol 16(2): 93–108

Bolton MD, Thomma BP, Nelson BD (2006) *Sclerotinia sclerotiorum* (Lib.) de Bary: biology and molecular traits of a cosmopolitan pathogen. Mol Plant Pathol 7(1):1–16

Booth C, Pirozynski KA (1967) Pleospora herharuni. C.M.I. Descriptions of Pathogenic Fungi and Bacteria, No. 150

Bowden RL, Wiese MV, Crock JE, Auld DL (1985) Root rot of chickpeas and lentils caused by *Thielaviopsis basicola*. Plant disease 69(12):1089–1091

Bretag TW (1989) Evaluation of fungicides for the control of ascochyta blight in lentils. Ann Appl Biol 114:44–45

Brouwer JB, Bretag TW, Materne MA (1995) Coordinated improvement program for Australian lentils. In: 2nd European conference on grain legumes, Copenhagen, Denmark, p 25

Buchwaldt L, Shaikh R, Adam J, Tullu A, Slinkard AE (2013) Recessive and dominant genes confer resistance to *Colletotrichum truncatum* in cultivated lentil. Can J Plant Pathol 35(2): 222–231

Buchwaldt L, Dzananovic E, Durkin J (2018) Lentil anthracnose: epidemiology, fungicide decision support system, resistance and pathogen races. Can J Plant Pathol 40(2):189–198

Chahota RK, Gupta VP, Sharma SK (2002) Inheritance of rust resistance in lentil (*Lens culinaris* Medik.). Indian J Genet Plant Breed 62(3):226–227

Chaudhary RG (2008) Pathogenic variability in *Fusarium oxysporum* f. sp. lentis. Pulses Newsl 19(4):3

Chaudhary RG, Kaur A (2002) Wilt disease as a cause of shift from lentil cultivation in Sangod Tehsil of Kota (Rajasthan). Indian J Pulse Res 15(2):193–194

Chen W, Basandrai AK, Basandrai D, Banniza S, Bayaa B, Buchwaldt L, Davidson J, Larsen R, Rubiales D, Taylor PW (2009) 17 Diseases and their Management. In: The Lentil, p 262

Chitale K. (1981) Perfect stage of *Erysiphe polygoni* on lentil from India.

Chongo G, Gossen BD, Bernier CC (2002) Infection by *Colletotrichum truncatum* in resistant and susceptible lentil genotypes. Canadian Journal of Plant Pathology 24(1):81–85

Choudhary AK, Kumar S, Patil BS, Sharma M, Kemal S, Ontagodi TP, Datta S, Patil P, Chaturvedi SK, Sultana R, Hegde VS (2013) Narrowing yield gaps through genetic improvement for Fusarium wilt resistance in three pulse crops of the semi-arid tropics. SABRAO J Breed Genet 45(03):341–370

Chowdhury AM, Ahmed A, Zaman M (1997) Studies on the defence structural factors of some susceptible and resistant varieties of lentil plants. J Mycopathological Research 35:35–40

Chowdhury MA, Andrahennadi CP, Slinkard AE, Vandenberg A (2001) RAPD and SCAR markers for resistance to ascochyta blight in lentil. Euphytica 118(3):331–337

Clarkson JP, Staveley J, Phelps K, Young CS, Whipps JM (2003) Ascospore release and survival in *Sclerotinia sclerotiorum*. Mycological Res 107(2):213–222

Coyne C, McGee R (2013) Lentil. In: Genetic and genomic resources of grain legume improvement, pp 157–180

Datta S, Rai R, Dhar V, Chaudhary RG, Gurha SN (2009) RAPD based diagnosis and diversity analysis of Fusarium wilt pathogen of pulse crops. J Food Legum 22(2):77–81

Davidson JA, Krysinska-Kaczmarek M (2007) Effects of inoculum concentration, temperature, plant age and interrupted wetness on infection of lentil (*Lens culinaris*) by *Botrytis* spp. conidia. Australas Plant Pathol 36(4):389–396

De RK, Ali SS, Dwivedi RP (2001) Effect of interaction between *Fusarium oxysporum* f. sp. lentis and *Meloidogyne javanica* on lentil. Indian J Pulse Res 14(1):71–73

Dikshit HK, Sharma TR et al (2016) Tagging and mapping of SSR marker for rust resistance gene in lentil (*Lens culinaris* Medikus subsp. culinaris). Indian J Exp Biol 54(6):394–399

Dikshit HK, Singh A, Singh D, Aski M, Jain N, Hegde VS, Basandrai AK, Dissanayake R, Kahrood HV, Dimech AM, Noy DM, Rosewarne GM, Smith KF, Cogan NO, Kaur S (2020) Development and application of image-based high-throughput phenotyping methodology for salt tolerance in lentils. Agronomy 10(12):1992

Duenas M, Hernandez T, Estrella I (2006) Assessment of in vitro antioxidant capacity of the seed coat and the cotyledon of legumes in relation to their phenolic contents. Food Chemistry 98(1): 95–103

Elad Y, Williamson B, Tudzynski P, Delen N (eds) (2004) Botrytis: biology, pathology and control. Springer Science & Business Media

El-Hassan SA, Gowen SR (2006) Formulation and delivery of the bacterial antagonist *Bacillus subtilis* for management of lentil vascular wilt caused by *Fusarium oxysporum* f. sp. lentis. J Phytopathol 154(3):148–155

Ellis MB, Waller JM (1974) Sclerotinia fuckeliana (conidial state: Botrytis cinerea). In: CMI descriptions of pathogenic fungi and bacteria

Erskine W (2009) The lentil: botany, production and uses. CABI

Erskine W, Bayaa B (1996) Yield loss, incidence and inoculum density associated with vascular wilt of lentil. Phytopathologia Mediterranea 35:24–32

Erskine W, Tufail M, Russell A, Tyagi MC, Rahman MM, Saxena MC (1994) Current status and future strategy in breeding lentil for resistance to biotic and abiotic stresses. In: Muehlbauer FJ, Kaiser WJ (eds) Expanding the Production and Use of Cool Season Food Legumes

Eujayl I, Erskine W, Bayaa B, Baum M, Pehu E (1998) Fusarium vascular wilt in lentil: inheritance and identification of DNA markers for resistance. Plant breeding 117(5):497–499

Fiala JV, Tullu A, Banniza S, Séguin-Swartz G, Vandenberg A (2009) Interspecies transfer of resistance to anthracnose in lentil (*Lens culinaris* Medic.). Crop Science 49(3):825–830

Fikre A, Sarker A, Ahmed S, Ali K, Erskine W (2007) Registration of 'Teshale' Lentil. J Plant Reg 1(1):45

Ford R, Pang ECK, Taylor PWJ (1999) Genetics of resistance to ascochyta blight (*Ascochyta lentis*) of lentil and the identification of closely linked RAPD markers. Theoretical and Applied Genetics 98(1):93–98

Gela TS, Koh CS, Caron CT, Chen LA, Vandenberg A, Bett KE (2021) QTL mapping of lentil anthracnose (Colletotrichum lentis) resistance from Lens ervoides accession IG 72815 in an interspecific RIL population. Euphytica 217(4):1–11

Gilchrist DG (1990) Stemphylium leaf spot. In: Compendium of Alfalfa diseases, 2nd edn, pp 17–20

Gossen BD, Morrall RAA (1983) Effect of ascochyta blight on seed yield and quality of lentils. Can J Plant Pathol 5(3):168–173

Gupta D, Sharma SK (2006) Evaluation of wild Lens taxa for agro-morphological traits, fungal diseases and moisture stress in North Western Indian Hills. Genet Resour Crop Evol 53(6): 1233–1241

Gupta M, Verma B, Kumar N, Chahota RK, Rathour R, Sharma SK, Bhatia S, Sharma TR (2012) Construction of intersubspecific molecular genetic map of lentil based on ISSR, RAPD and SSR markers. J Genet 91(3):279–287

Hamwieh A, Udupa SM, Choumane W, Sarker A, Dreyer F, Jung C, Baum M (2005) A genetic linkage map of Lens sp. based on microsatellite and AFLP markers and the localization of fusarium vascular wilt resistance. Theor Appl Genet 110(4):669–677

Kaiser WJ (1992) Fungi associated with the seeds of commercial lentils from the US Pacific Northwest. Plant Dis 76(6):605–610

Kaiser WJ, Hannan RM (1986) Incidence of seed borne *Ascochyta lentis* in lentil germ plasm. Phytopathol 76(3):355–360

Kaiser WJ, Hannan RM (1987) Seed-treatment fungicides for control of seedborne Ascochyta lentis on lentil. Plant Dis 71(1):58–62

Kaiser WJ, Hellier BC (1993) *Didymella* sp., the teleomorph of *Ascochyta lentis* on lentil straw. Phytopathol 83:692

Kaiser WJ, Wang BC, Rogers JD (1997) *Ascochyta fabae* and *A. lentis*: host specificity, teleomorphs (*Didymella*), hybrid analysis, and taxonomic status. Plant Dis 81(7):809–816

Kaiser WJ, Mihov M, Muehlbauer FJ, Hannan RM (1998) First report of anthracnose of lentil incited by *Colletotrichum truncatum* in Bulgaria. Plant Dis 82(1):128–128

Kamboj RK, Pandey MP, Chaube HS (1990) Inheritance of resistance to Fusarium wilt in Indian lentil germplasm (*Lens culinaris* Medik.). Euphytica 50(2):113–117

Karahan A, Katırcıoğlu YZ (1993) Occurrence and distribution of fungal diseases on lentil in Ankara Province. J Turk Phytopathol 22(1):27–33

Karki PB (1993) In: Erskine W, Saxena MC (eds) Plant protection of lentil in Nepal. Lentil in South Asia, pp 187–193

Khare MN (1980) Wilt of lentil. Jawaharlal Nehru Krishi Vishwa Vidyalaya, Jabalpur, Madhya Pradesh

Khare MN (1981) Diseases of lentils

Koike ST, Henderson DM, Butler EE (2001) Leaf spot disease of spinach in California caused by *Stemphylium botryosum*. Plant Dis 85(2):126–130

Kole C, Muthamilarasan M, Henry R, Edwards D, Sharma R, Abberton M, Batley J, Bentley A, Blakeney M, Bryant J, Cai H (2015) Application of genomics-assisted breeding for generation of climate resilient crops: progress and prospects. Front Plant Sci 6:563

Kuchuran M, Banniza S, Vandenberg B (2003) Evaluation of lentil varieties for resistance to botrytis grey mould. In: 'Meeting the challenge: opportunities and pressures within the pulse industry. Proceedings of Pulse Field Days, Saskatoon, pp 6–7

Kumar P (2007) Genetics of resistance to Stemphylium leaf blight of lentil (*Lens culinaris*) in the Barimasur-4× CDC Milestone. (Doctoral dissertation, Thesis of Masters of Science, Department of Plant Sciences, University of Saskatchewan, Saskatoon).

Kumar R, Mishra SK, Sharma B (2001) Genetics of rust resistance in lentil (Lens culinaris Medik.). Ratio 151(122):29

Kumar S, Kumar J, Singh S, Ahmed S, Chaudhary RG, Sarker A (2010) Vascular wilt disease of lentil: a review. J Lentil Res 4:1–14

Kumar A, Bal RS, Gill RK, Singh S (2015) Incorporation of resistance against rust (*Uromyces viciae-fabae*) in lentil. J Food Legum 28(2):13–14

Kumar J, Gupta S, Gupta DS, Singh NP (2018) Identification of QTLs for agronomic traits using association mapping in lentil. Euphytica 214(4):1–15

Kumar J, Sen Gupta D, Baum M, Varshney RK, Kumar S (2021) Genomics-assisted lentil breeding: Current status and future strategies. Legum Sci 3(3):e71

Ladizinsky G (1979) The origin of lentil and its wild gene pool. Euphytica 28(1):179–187

Lamari L, Bernier CC (1985) Etiology of seedling blight and root rot of faba bean (*Vicia faba*) in Manitoba. Can J Plant Pathol 7(2):139–145

Ma Y, Marzougui A, Coyne CJ, Sankaran S, Main D, Porter LD, Mugabe D, Smitchger JA, Zhang C, Amin MN, Rasheed N (2020) Dissecting the genetic architecture of Aphanomyces root rot resistance in lentil by QTL mapping and genome-wide association study. Int J Mol Sci 21(6):2129

Marzougui A, Ma Y, Zhang C, McGee RJ, Coyne CJ, Main D, Sankaran S (2019) Advanced imaging for quantitative evaluation of Aphanomyces root rot resistance in lentil. Front Plant Sci 10:383

Materne MA , McMurray L , Nitschke S , Regan K , Heuke L , Dean G , Carpenter D (2002) The future of Australian lentil production. In: 'Proceedings of Lentil Focus 2002'. Horsham, Vic. (Ed. JB Brouwer) pp. 14–18

Mihov M, Stoyanova M (1998) Results from improving investigation in lentil breeding in Bulgaria. "Rasteniev" dni Nauki 35(1):18–21

Mittal RK (1997) Effect of sowing dates and disease development in lentil as sole and mixed crop with wheat. J Mycol Plant Pathol 27(2):203–209

Mohammadi N, Puralibaba H, Goltapeh EM, Ahari AB, Sardrood BP (2012) Advanced lentil lines screened for resistance to *Fusarium oxysporum* f. sp. lentis under greenhouse and field conditions. Phytoparasitica 40(1):69–76

Morrall RAA (1997) Evolution of lentil diseases over 25 years in western Canada. Can J Plant Pathol 19(2):197–207

Morrall RAA (2003) Diseases of Lentil. In: Bailey KL, Gossen BD, Gugel RK, Morrall RAA (eds) Diseases of Field Crops in Canada, 3rd edn. The Canadian Phytopathological Society, pp 191–200

Mwakutuya E (2006) Epidemiology of Stemphylium Blight on Lentil (*'Lens Culinaris'*) in Saskatchewan. (Doctoral dissertation. University of Saskatchewan)

Nasir M, Bretag TW (1997) Pathogenic variability in Australian isolates of *Ascochyta lentis*. Australas. Plant Pathol. 26:217–220

Negussie T, Pretorius ZA (2012) Lentil rust: present status and future prospects. Crop Prot 32:119–128

Negussie T, Pretorius ZA, Bender CM (2005) Components of rust resistance in lentil. Euphytica 142(1-2):55–64

Nene YL (1988) Multiple-disease resistance in grain legumes. Ann Rev Phytopathol 26(1):203–217

Nguyen TT, Taylor PWJ, Brouwer JB, Pang ECK, Ford R (2001) A novel source of resistance in lentil (*Lens culinaris* ssp. culinaris) to ascochyta blight caused by *Ascochyta lentis*. Aus Plant Pathol 30(3):211–215

Paulitz TC, Huang HC, Gracia-Garza JA (2002) *Pythium* spp., damping-off, root and crown rot (Pythiaceae). In: Mason PG, Huber JT (eds) Biological Control Programmes in Canada 1981–2000. CABI Publishers, Wallingford, UK, pp 478–483

Pedersen EA, Morrall RAA, McCartney HA, Fitt BD (1994) Dispersal of conidia of *Ascochyta fabae* f. sp. lentis from infected lentil plants by simulated wind and rain. Plant Pathol 43(1): 50–55

Peñaloza E, Tay JU, France AI (2007) Calpun-INIA, a large seed size and rust-resistant lentil (Lens culinaris Medik.) cultivar. Agricultura Técnica 67(1):68–71

Podder R, Banniza S, Vandenberg A (2013) Screening of wild and cultivated lentil germplasm for resistance to Stemphylium blight. Plant Genetic Resour 11(1):26–35

Polanco C, de Miera LES, Bett K, de la Vega MP (2018) A genome-wide identification and comparative analysis of the lentil MLO genes. PloS One 13(3):e0194945

Prasada R, Verma UN (1948) Studies on Lentil rust, *Uromyces fabae* (Pers.) de Bary, in India. Indian Phytopathol 1(2):142–146

Reddy RR, Khare MN (1984) Further studies on factors influencing the mechanism of resistance in lentil *(Lens culinaris* M.) to rust (*Uromyces fabae* (Pers.) De Bary). Lentil Experimental News Service

Roundhill SJ, Fineran BA, Cole ALJ, Ingerfeld M (1995) Structural aspects of ascochyta blight of lentil. Can J Bot 73(3):485–497

Roy S, Ullah MA, Ghosh SR, Sarker PC, Islam R (2018) Binamasur-7: a released high yielding lentil variety in Bangladesh. Bangladesh J 34:45–50

Rubeena TP, Ford R, Taylor PWJ (2003) Molecular mapping the lentil (*Lens culinaris)* genome. Theor Appl Genet 107:910–916

Sadiq MS, Haidar S, Haq MA, Abbas G (2008) A high yielding and disease resistant mutant of lentil developed through seed irradiation of an exotic germplasm. Can J Pure Appl Sci 2(2):411

Saha GC, Sarker A, Chen W, Vandemark GJ, Muehlbauer FJ (2010a) Identification of markers associated with genes for rust resistance in *Lens culinaris* Medik. Euphytica 175(2):261–265

Saha GC, Sarker A, Chen W, Vandemark GJ, Muehlbauer FJ (2010b) Inheritance and linkage map positions of genes conferring resistance to stemphylium blight in lentil. Crop Sci 50(5): 1831–1839

Sarker A, Erskine W, Hassan MS, Afzal MA, Murshed ANMM (1999a) Registration of Barimasur-4' Lentil. Crop Sci 39(3):876–876

Sarker A, Kumar J, Rahman MM, Hassan MS, Zaman W, Afzal MA, Murshed ANMM (1999b) Registration of Barimasur-3' Lentil. Crop Sci 39(5):1536–1536

Sharma MD, Mishra RP, Jain AC (1978) *Cercospora lensii* sp. Nov. A new species of *Cercospora* on *lentil.* Curr Sci 47:774–775

Singh K (1985) Effects of seed treatment on lentil rust (*Uromyces fabae*) development. Lens Newsl

Singh K, Sandhu TS (1988) Screening of cultivars of lentil for resistance to rust (*Lens culinaris*). Lentil Experimental News Service

Singh IS, Singh JP, Singh AK, Chauhan MP (1994) Pant Lentil 4: a high yielding, rust, wilt-and blight-resistant variety for the North-Western Plains of India. Lens Newsl 21(1):8–9

Singh K, Rawal HS, Singh S, Gumber RK (1995) Physiological specialization in Uromyces fabae causing rust of lentil. Lens Newsl 22:46–47

Singh A, Dikshit HK, Mishra GP, Aski M, Kumar S (2019) Association mapping for grain diameter and weight in lentil using SSR markers. Plant Gene 20:100204

Singh M, Kumar S, Basandrai AK, Basandrai D, Malhotra N, Saxena DR, Gupta D, Sarker A, Singh K (2020) Evaluation and identification of wild lentil accessions for enhancing genetic gains of cultivated varieties. PloS One 15(3):e0229554

Singh D, Singh CK, Taunk J, Sharma S, Gaikwad K, Singh V, Sanwal SK, Singh D, Sharma PC, Pal M (2021) Transcriptome skimming of lentil (Lens culinaris Medikus) cultivars with contrast reaction to salt stress. Funct Integr Genomics 21(1):139–156

Sinha JN, Singh AP (1991) Two new records. Indian Phytopathol 44(3):421

Sinha JN, Singh AP (1993) Effect of environment on the development and spread of Stemphylium blight of lentil. Indian Phytopathol 46(3):252–253

Sinha RP, Yadav BP (1988) Inheritance of resistance to rust in lentil. Lens Newsl 16(1):41

Smith OF (1940) Stemphylium leaf spot of Red Clover and Alfalfa. J Agric Res 12:831–846

Solanki RK, Singh S, Kumar J (2010) Molecular marker assisted testing of hybridity of F1 plants in lentil. J Food Legum 23(1):21–24

Tate KG (1970) A foliage disease of blue lupin caused by *Stemphylium botryosum* Wallr. New Zealand J Agric Res 13(3):710–716

Tikoo JL, Sharma B, Mishra SK, Dikshit HK (2005) Lentil (*Lens culinaris* L.) in India: present status and future perspectives. Indian J Agric Sci 75:539–562

Tiwari M, Singh B, Min D, Jagadish SK (2022) Omics path to increasing productivity in less-studied crops under changing climate—lentil a case study. Front Plant Sci 13:1239

Tullu A, Buchwaldt L, Warkentin T, Taran B, Vandenberg A (2003) Genetics of resistance to anthracnose and identification of AFLP and RAPD markers linked to the resistance gene in PI 320937 germplasm of lentil (*Lens culinaris* Medikus). Theor Appl Genet 106(3):428–434

Tullu A, Buchwaldt L, Lulsdorf M, Banniza S, Barlow B, Slinkard AE, Sarker A, Tar'an B, Warkentin T, Vandenberg A (2006a) Sources of resistance to anthracnose (*Colletotrichum truncatum*) in wild *Lens* species. Genet Resour Crop Evol 53(1):111–119

Tullu A, Tar'an B, Breitkreutz C, Banniza S, Warkentin TD, Vandenberg A, Buchwaldt L (2006b) A quantitative-trait locus for resistance to ascochyta blight (*Ascochyta lentis*) maps close to a gene for resistance to anthracnose (*Colletotrichum truncatum*) in lentil. Can J Plant Pathol 28(4): 588–595

Tullu A, Diederichsen A, Suvorova G, Vandenberg A (2011) Genetic and genomic resources of lentil: status, use and prospects. Plant Genet Res 9(1):19–29

Vail S, Vandenberg A (2011) Genetic control of interspecific-derived and juvenile resistance in lentil to *Colletotrichum truncatum*. Crop Sci 51(4):1481–1490

Vail S, Strelioff JV, Tullu A, Vandenberg A (2012) Field evaluation of resistance to *Colletotrichum truncatum* in *Lens culinaris*, *Lens ervoides*, and *Lens ervoides* x *Lens culinaris* derivatives. Field Crop Res 126:145–151

Vandenberg A, Banniza S, Warkentin TD, Ife S, Barlow B, McHale S, Brolley B, Gan Y, McDonald C, Bandara M, Dueck S (2006) CDC Redberry lentil. Can J Plant Sci 86(2):497–498

Vishunavat K, Shukla P (1979) Fungi associated with lentil seed. Ind. Phytopath 32(2):279–280

Zou Y, Chang SK, Gu Y, Qian SY (2011) Antioxidant activity and phenolic compositions of lentil (*Lens culinaris* var. Morton) extract and its fractions. J Agric Food Chem 59(6):2268–2276

Biotic Stresses in Multipurpose Legume: Rice Bean

12

Susmita Oraon, Sabyasachi Mukhopadhyay, Krishnendu Kundu, Jasmeen Khandakar, Anshu Kumar, Soumik Mukherjee, Saptarshi Mondal, Uday Chand Jha, and Kousik Atta

Abstract

Rice bean [*Vigna umbellata* (Thunb.) Ohwi and Ohashi] is a multipurpose underutilized legume crop having profuse pod-bearing ability, wider adaptability and high resilience to biotic and abiotic stresses. Besides its nutritionally rich high seed yield even under limited management inputs, it could also be used as a fodder crop. Biotic stress which is one of the major limiting factors in crop cultivation includes damage to the vegetative and reproductive parts of a crop caused by other living organisms such as bacteria, viruses, fungi, harmful insects and weeds. Plants are adapted with various defence mechanisms to combat biotic stresses. Pattern-triggered immunity (PTI) and effector-triggered immunity (ETI) are the two molecular defence mechanisms recruited by plants to evade pathogenic attacks. In this chapter, we discussed the negative impacts of various diseases on rice bean and how prevalent genetic variability across the various gene pools could be harnessed to develop disease-resistant rice bean cultivars.

S. Oraon · S. Mukhopadhyay · K. Kundu · J. Khandakar · A. Kumar · S. Mukherjee
Department of Plant Pathology, Bidhan Chandra Krishi Viswavidyalaya, Mohanpur, Nadia, West Bengal, India

S. Mondal (✉)
Department of Crop and Soil Sciences, University of Georgia, Griffin, GA, USA
e-mail: Saptarshi.Mondal@uga.edu

U. C. Jha
Indian Institute of Pulses Research, Kanpur, India

K. Atta (✉)
Department of Plant Pathology, Bidhan Chandra Krishi Viswavidyalaya, Mohanpur, Nadia, West Bengal, India

Mineral Nutrition Lab, Division of Plant Physiology, ICAR-Indian Agricultural Research Institute, Pusa, New Delhi, India

U. C. Jha et al. (eds.), *Diseases in Legume Crops*,
https://doi.org/10.1007/978-981-99-3358-7_12

Integration of molecular advances along with conventional breeding approaches for identifying various QTLs associated with disease resistance through bi-parental QTL mapping and genome-wide association mapping and their transfers are discussed in this chapter. Finally, prospects of various emerging breeding tools including genomic selection, speed breeding and genome editing tools are briefed.

Keywords

Biotic stress · GWAS · Speed breeding · Rice bean · QTLs · ROS · Oxidative stress

12.1 Introduction

People who live in underdeveloped nations often include pulses and grain legumes in their meals (Katoch 2013). Grain legumes are known as "poor man's meat" since they not only bring variety to the diet but also serve as inexpensive sources of nutritional protein and other nutrients (Modgil et al. 2012; Thongram et al. 2016). Mung bean and black gram are the two most popular dry beans in India (Goyal et al. 2014; Goyal and Siddiqui 2014), while their production is often low since they are vulnerable to biotic and abiotic threats and grow on deficient soils, in both established and emerging nations (Sultana et al. 2014). The current over-dependence on a few main legume crops calls for unsustainable agronomic, ecological, nutritional and economic problems that can be reduced by expanding the cultivation of underutilized minor crops and using improved cultivars and modern production technologies (Schmidt et al. 2010). Crop diversification helps in incorporating spatial and temporal variety into uniform cropping systems and will assist increase tolerance to biotic and abiotic stressors.

Rice beans are annual legume crop with yellowish coloured blossoms that are utilized as a green manure and feed in addition to being used as a dried pulse (Atta et al. 2021; Atta et al. 2022a). It has historically been grown with maize, sorghum and cowpea in India, Bangladesh, Nepal and China. It is believed that cross-fertilization created rice beans in the wild species *Vigna umbellata var. gracilis*, which is naturally grown in southern northern Vietnam, China, Burma, Laos, Thailand, and India (Tomooka et al. 1991; Atta et al. 2022b). It has lately undergone further domestication in south- and north-east Asia due to its nutritional advantages.

Some additional names for rice bean are red bean, mambi bean, climbing mountain bean, ohwi, oriental bean, *Vigna calcarata* (Roxb.), *Dolichos umbellatus* Thunb, Kurz and ohashi. The rice bean is well adapted to sub-humid climates with a precipitation of 1000–1500 mm, along with moisture distribution, rainfall pattern, temperature, relative humidity, soil properties, pests and diseases, etc. (Rachie and Roberts 1974). It serves as a complete nutrition package with a good amount of crude protein, 59–93% digestible protein, all essential amino acids (especially methionine, tryptophan, lysine, tyrosine and valine), vitamins and minerals, as well as an excessively high proportion of healthy, unsaturated fatty acids (Kaur and Kapoor 1990; Mohan and Janardhanan 1994; Katoch 2013; Atta et al. 2022c).

Lack of scientific study and consumer knowledge has kept consumption to a few small communities in a few underdeveloped nations, such India and Nepal, and affected its global commercialization despite having immense potential to alleviate economic and nutritional inadequacies (Buergelt et al. 2009)

12.2 Diseases of Rice Bean

Farmers believe that rice bean is the least susceptible grain legume to disease. However, the most common rice bean diseases are common rust, powdery mildew, *Rhizoctonia* blight, bacterial blight, anthracnose and *Cercospora* leaf spot.

12.3 Various Effects and Management of the Diseases in Rice Bean

12.3.1 Rust

12.3.1.1 *Uromyces appendiculatus* (Pers.) Unger

Uromyces spp. affects a diverse and heterogeneous variety of hosts, including forest trees, ornamental plants, vegetables, grain crops and primitive ferns. Rust pustules can be seen on rice bean leaves, but they are less common on stems and pods. Rust pustules start off small, white and slightly elevated. Pustules change colour to a reddish brown later on and may be seen clearly under the leaves (Fig. 12.1). Premature leaf drop, which directly impacts yield, is the main issue if rust damages the leaves during the vegetative growth cycle. Rust has less of an effect on yield if it develops during the maturing period. To control rust, the crop must be destroyed soon after harvest. Rust can continue to grow and act as a significant source of

Fig. 12.1 *Uredinia* of rust (*Uromyces appendiculatus*) on the upper surface of a bean leaf. Source: https://www.ipmimages.org/browse/detail.cfm?imgnum=5361405#collapseseven

inoculum if this is not done. Crop rotation prevents the accumulation of inoculum of pathogen (Mersha and Hau 2008; Katoch 2020).

The most crucial strategy for rust control at the moment is the routine use of protectant fungicides. It might be advantageous to apply Maneb (Indofil M 45) spray at rates of 3 g/L of water before or as soon as rust pustules emerge (Van den Berg et al. 2013).

12.3.2 Powdery Mildew

12.3.2.1 *Oidiopsis taurica* (Lev.) Salmon

On leaves, stems and pods, the fungus assaults and coats them in a powdery layer. Plant components that are severely harmed become twisted and withered. Before defoliation, the afflicted leaves turn yellow, brown coloration and ultimately dry up (Fig. 12.2). Pods do not form, and those that do are deformed, sparse and few in number (Saharan et al. 2019).

It is frequently seen in regions with high humidity and temperature between 20 and 35 °C. It is imperative that the following crop not be planted on the same affected field. Crops that are planted a little bit later can assist to prevent the spread of disease. To lessen the sickness, spraying of triadimefon (Bayleton) 25% EC at 0.03% and carbendazim 0.5 g/L of water is effective for controlling this disease (Li et al. 2020).

12.3.3 *Rhizoctonia* Blight

12.3.3.1 *Rhizoctonia solani*

On the lower leaves, little, uneven, water-soaked, pale greenish patches with a wet appearance first appear and then spread upward to the apex, indicating the presence of the disease (Fig. 12.3). When there is a lot of humidity, the illness spreads quickly

Fig. 12.2 Powdery mildew on beans. Source: David Trinklein (2019)

Fig. 12.3 Symptoms of blight on the foliage of beans. Source: https://www.forestryimages.org/browse/detail.cfm?imgnum=5358993#collapseseven

and affects more of the leaf blade and stem. The disease is currently in a very noticeable and devastating phase. The leaflets and pods dry up, shrivel, and turn brown. In damp weather, damaged plant parts decompose quickly. The seeds do not form in the case of early infection and do not form well in the case of late infection in the pods. The plants may occasionally be killed before blossoming (Senapati et al. 2022).

For controlling this disease, it is preferable to combine cultural treatment with chemical application. It is advantageous to gather sick plants and burn them, sow late, plant sparingly, maintain crop rotation, and ensure good drainage. Bavistin is applied to seeds at a dosage of 1 g/kg of seed which could be an important approach to minimize the seed infection of this disease (Panth et al. 2020).

12.3.4 Bacterial Blight

12.3.4.1 *Pseudomonas* spp.

The principal source of primary inoculum in a new location comes from the pathogen, which is seed-borne. Small, flat, irregularly shaped, water-soaked patches that are first surrounded by a greenish-yellow zone emerge on leaves (Fig. 12.4). The infection causes the vein lets and veins to become brown and necrotic as it moves along them, distorting the leaflet in the process (Sun et al. 2017).

Planting seeds free of disease is the major defence against bacterial leaf blight. Changing the timing of sowing to prevent planting when unfavourable weather for the pathogen prevails and crop rotation are efficient ways to reduce the illness. Streptomycin seed treatment could aid in lowering the pathogen in seed (Yuliar et al. 2015).

Fig. 12.4 Symptoms of common bacterial blight. Source: https://www.ipmimages.org/browse/detail.cfm?imgnum=5363006#collapseseven

Fig. 12.5 Anthracnose of beans. Source: https://www.greenlife.co.ke/bean-anthracnose/

12.3.5 Anthracnose

12.3.5.1 *Colletotrichum truncatum*

On the underside of the leaves, symptoms take the form of dark red to black lesions (Fig. 12.5). On the pods, petioles, and stems, they appear as depressed lesions bordered by a raised brown-black border. Older lesions typically show the fungus's tiny, black fruiting bodies. Leaf shedding, flower and pod abortion are further signs. Infected seeds have a discoloured appearance with brown, grey or dark flecking (Falconí et al. 2013).

For management procedures, it is advised that affected plants be removed and destroyed. Application of a copper-based fungicide to infected area of field with care without harming earthworms and bacteria is an important management practice for controlling this disease (Falconí et al. 2013).

12.3.6 *Cercospora* Leaf Spot

12.3.6.1 *Cercospora* spp.

The *Cercospora* fungus is a member of the Order Moniliales' Deuteromycetes class of imperfect fungi. Several legumes, including rice bean, are affected by the disease, which results in leaf spots and defoliation. Small dry spots (which could be circular or amorphous) that initially develop on leaves eventually become larger and cause the plants to lose their leaves (Fig. 12.6).

The most crucial element in managing the *Cercospora* disease is choosing seeds from healthy plants. Similar advises against using diseased fields the next season. Spraying Maneb (IndofilM 45) at 3 g/L of water before or right away as dry spots start to form may be helpful (Bakhshi et al. 2018).

12.4 Genetic Resources Rice Bean Tolerance to Various Diseases and Pests

12.4.1 Source of Different Varieties of Rice Bean

Variety	Speciality of the variety	References
RBL-1	• Highly resistant to YMV and stored grain insect pests	Khadka and Acharya (2009); Sehrawat et al. (2016); Raiger et al. (2009)
RBL-6	• Resistant to YMV • Immune to stored grain insect pest	Sujayanand et al. (2021); Sehrawat et al. (2016); Raiger et al. (2009); https://www.leafconagro.com/rice-bean-red-bean-varieties/
PRR-2		Khadka and Acharya (2009); Raiger et al. (2009)

(continued)

Fig. 12.6 *Cercospora* leaf spot. Source: https://guide.utcrops.com/wp-content/uploads/2017/04/Cercospora-Leaf-Spot Foliar.jpg

Variety	Speciality of the variety	References
	• Tolerant to *Ascochyta* and resistant to yellow mosaic disease	
Resplant3	• Resistant to mung bean yellow mosaic virus	Pandiyan et al. (2020)
VM2011 and VM2164	• Resistant to bruchid infestation	Fernandez and Talekar (1990)
RBL35, RBL50,	• Resistant to mung bean yellow mosaic virus	Sehrawat et al. 2016
Dharwad local, IC-524075 and IC-341983	• Highly resistant bruchid	Sanjeev and Vishwas (2018)
DC 15	• Moderately resistant bruchid	Vishwas and Deshpande (2018)
Megha Rumbaija 1 (RCRB 1-6)	• Resistant MYMV	Raiger et al. (2009)
Bidhan rice bean 2 (KRB 4)	• Resistant MYMV	https://www.leafconagro.com/rice-bean-red-bean-varieties/

12.5 Disease Resistance of Rice Bean

The yellow mosaic virus (YMV) disease caused by a single-stranded DNA virus, i.e. begomo virus having a bipartite genome transmitted by white fly (*Bemisia tabaci*) is a momentous disease of all the legume plants causing approximately 80–100% yield loss (Nene 1973; Gilmer et al. 1974; and Williams 1977; and Khattak et al. 2000). The disease appears as small irregular yellow-coloured specs along the veins of the leaves that gradually enlarge, and the leaves become chlorotic (Qazi et al. 2007). If the infection takes place at reproductive stage, then the pods remain empty or become yellow coloured filled with infected seeds (Sehrawat and Yadav 2014). Several experiments have exhibited the significant resistance of rice bean against YMV infection (Kashiwaba et al. 2003; Sudha et al. 2015). *V. umbellate*, the wild relatives of green gram, bears several desirable genes related to yield, as well as resistance to YMV and bruchid pest. Numerous strategies to transfer the *Vigna umbellata* resistance gene to other *Vigna* species that are vulnerable have proven highly successful. Several methods have been fairly successful in transferring the *Vigna umbellata* resistance gene to other *Vigna* species that are vulnerable to mosaic disease (Chaisan et al. 2013; Pandiyan et al. 2010; Bhanu et al. 2017; Mathivathana et al. 2019). The population mapping that involves the cross between mung bean and rice bean was raised and used to study the YMV resistance (Bhanu et al. 2017). Micro-satellite markers, or simple sequence repeats (SSRs), are frequently utilized in the formation of related marker traits in the segregating population, the validation of hybridity, phylogeny and gene mapping.

Due to difficulties like linkage drag and barrier crossing, rice bean is still not used too much, but good sources of resistance in mung bean against YMV have not been found. Thus, in order to manage mung bean yellow mosaic virus (MYMV) disease severity, it is being preferable to transfer the MYMV resistance gene from rice bean to mung bean. Several researchers faced difficulty in transferring the resistance genes from rice bean (*V. umbellate*) to mung bean (*V. radiata*), but Bharathi et al. in the year 2006 and Pandiyan et al. in 2010 succeeded in producing fertile hybrids between particular rice bean and mung bean genotypes. The mapping population derived from the cross between rice bean and mung bean was utilized to explore the MYMV resistance (Bhanu et al. 2017). Using the single seed descent (SSD) approach, a recombinant inbred line (RIL) population of 108 lines was successively produced from a mung bean and rice bean hybrid. To generate entire homozygous lines that would serve as a stable genetic resource for future research, nine selfings have been carried out. The polygenic control shown in mung bean was compatible with the distribution of MYMV resistance (Alam et al. 2014).

Bhanu et al. in the year 2017 conducted an experiment in which they crossed rice bean genotype RBL 1 with two mung bean genotypes, viz., K 851 and TM 96-2. Early in the day, the mung bean stigma remains to be quite responsive. Therefore, emasculation took place between 4:00 and 6:00 pm, and pollination took place the following morning between 6:00 and 8:00 am. During the Kharif season of 2016–2017, the pure seeds of two varieties of mung bean, K 851 and TM 96-2, and one variety of rice bean, RBL 1, as well as crossed seeds (F0) of two interspecific hybrids, K851× RBL 1 and TM 96-2 × RBL 1, were grown. Mung bean and rice bean genotypes were screened, and rice bean genotype RBL 1 was found extremely resistant to MYMV, whereas mung bean genotypes, viz., K 851 and TM 96-2, were highly susceptible and moderately resistant to MYMV, respectively. According to the findings of the current study, rice bean genotypes show significant levels of resistance to yellow mosaic disease (YMD) and can be used as donor parents in interspecific hybridization programmes to create resistant varieties with high yield potential for other *Vigna* species that are more vulnerable.

In a different study, conducted in 2013, Sudha et al. used two MYMV susceptible mung bean lines, VBN (Gg) 2 and VRM (Gg) 1, as well as a MYMV-resistant mung bean line, KMG 189, and a rice bean line, TNAU RED. All the lines were native to Tamil Nadu, India. In the experiment the F1, F2 and F3 generation plants that were created by crossing resistant and susceptible parents were used. From 2006 to 2010, all the plants were sown during the Kharif and Zaid or summer seasons of the year, with an appropriate row to row and plant to plant distance of 50 cm and 10 cm, respectively. When 80% plants showed MYMV incidence, then scoring was started following the rating scale revealed by Singh et al. (1988). The mung bean plants were categorized into five categories based on the MYMY score: highly susceptible (HS), moderately susceptible (MS), susceptible (S), moderately resistant (MR) and resistant (R). The susceptible group was composed of plants identified as susceptible (S), moderately susceptible (MS) and highly susceptible (HS), while the resistant group consisted of plants classified as resistant (R) and moderately resistant (MR). Breeding for resistant cultivars is a popular and effective method for managing the

MYMV diseases and reducing the spread of the virus. The ability to breed MYMV-resistant varieties will be greatly aided by the mung bean breeders' understanding the inheritance of genes for resistance and the role of these genes in the development of susceptibility or resistance. Sensitive reactions (S), or symptoms that could be seen on both leaves and pods, were present in the two susceptible parents, VBN (Gg) 2, VRM (Gg) 1 of the crossing and in F1 plants. Resistance was assigned to the resistant parents TNAU RED and KMG 189 since no symptoms were seen in them until maturity. According to the rating scale, five reactions— moderately susceptible (MS), susceptible (S), highly susceptible (HS), resistant (R) and moderately resistant (MR) were recorded in the two F_2 and F_3 generations. In the F_2 population, the chi-square tests for the two crossings showed good fitness to 3 (susceptible): 1 (resistance) and showed the dominance of susceptibility over the resistance, confirming a monogenic inheritance known as MYMV. However, the 3:1 segregation in F_2 generation is completely different from the present investigation. Nevertheless, the F_3 offspring of two crossings showed a segregation trend of 1 (none segregating susceptible); 2 (segregating); and 1 (non-segregating resistant). Collectively, the findings demonstrated that MYMV disease resistance was regulated by a single recessive gene. Numerous researchers have documented similar outcomes of the mung bean's single recessive gene inheritance (Reddy and Singh 1995). However, reports have suggested that one dominant gene, two recessive genes and complementing recessive genes were also involved. These inconsistent results may be caused by the variations in the host genotype, viral strains and due to the interactions between virus and host. Another crucial aspect that contributes to the variations in inheritance is the relationship between meteorological conditions and vector activities. But the current research will be helpful for creating DNA markers connected to the MYMV resistance gene. Additionally, we indicate that future research will be required to enhance MYMV resistance in mung bean viruses using rice bean crop.

12.6 Disease Resistance in Rice Bean Through Agro-inoculation Technique

The screening of MYMV resistance among different pulse crops under field condition is still not satisfactory due to several hindrances such as host characteristics, environmental exposures and viral load. Thus, a modern technique named agro-inoculation has been reported by several researchers like Biswas and Varma (2001) and Sudha et al. (2013a, 2013b) for efficient identification of MYMV-resistant bean germplasms. Agro-inoculation is an extra special strategy in which Ti (tumour-inducing) plasmid of *Agrobacterium tumefaciens* is used for inoculation of infectious viral clone into a healthy plant to express the symptoms of MYMV through encapsidation and replication (Madhumitha et al. 2019). The efficacy of agro-inoculation technique in expression of MYMV resistance or susceptibility was studied by Madhumitha et al. (2019) against 15 mung bean (*Vigna radiata*) germplasms, 6 urd bean (*Vigna mungo*) germplasms and one rice bean (*Vigna *umbellata*) germplasm. Agro-inoculation of *Agrobacterium tumefaciens* carrying infectious viral clone VA 239 (KA30 DNA A + KA27 DNA B) was done on 2 days old germinated seedlings of selected crop germplasms following the protocol of Jacob et al. (2003). It was observed that after 15–17 days of agro-inoculation, out of 15 mung bean germplasms 12 were highly susceptible to MYMV showing the characteristics yellow mosaic symptoms on the leaves, and the rest 3 were found susceptible with mild yellow mosaic symptoms. The rice bean variety named TNAU yellow was again reported resistant to yellow mosaic virus throughout the replications, and no symptoms were developed till maturity. In case of urd bean, five germplasms were found susceptible, whereas rest one named CO 5 showed highly susceptibility with characteristics puckering and stunting symptoms. The same symptoms and susceptibility were also reported by Sudha et al. (2013a, b). Thereafter, the resistant rice bean germplasm and one susceptible germplasm from each of mung bean and urd bean were further studied for PCR confirmation. All the susceptible germplasms showed the anticipated amplicon size, i.e. 703 bp., whereas the resistant TNAU yellow cultivar and healthy plants from the control plots asserted no amplification. It clearly indicated the multiplication of viral genome inside the susceptible germplasms, whereas in the resistant plants, there may be certain mechanism that prevented the multiplication of viral particle leading to the absence of viral genome. The development of resistance in rice beans may be the result of a tissue-specific host tolerance mechanism that is working inside the plant. The studies of Wyatt and Brown (1996) and Maheswari (2008) have revealed that there is a gene named CP gene which is highly conserved in the members of the family *Geminiviridae* and has also shown rich homology to MYMV. The ICTV has also accepted this gene sequence as a covetable marker of MYMV in the absence of full-length genomic sequence (Rybicki et al. 2000). The presence of this CP gene sequence in agro-inoculated plants showing mosaic symptoms is also reported in the experiments of Madhumitha et al. (2019), Sudha et al. (2013a, b) and Usharani et al. (2005). The absence of CP gene in rice bean cultivar strongly supports its MYMV resistance, which can be utilized in disease-resistant breeding programme as a donor parent.

12.7 Genomic Resources Including QTLs for Disease Resistance in Rice Bean

Rice bean (*Vigna umbellate*), a diploid crop with 2n chromosome no 22, is used as a Kharif legume. It is resistant to mung bean yellow mosaic virus or MYMV (Kashiwaba et al. 2003; Sudha et al. 2015). But, most leguminous plants, particularly mung bean, are seriously afflicted by the yellow mosaic virus, a single-stranded DNA carrying, bipartite, white-fly transmitted Begomovirus (Nene 1973; Khattak et al. 2000). However, rice bean and mung bean (*Vigna radiata*) share a genetic ancestor (Mathivathana et al. 2019). As a result, it has a great chance of being used, along with mung bean, as a MYMV resistant gene source for breeding purposes. Previously, interspecific crosses between mung bean and rice bean showed evidence of a reproductive isolation barrier (Chen et al. 1983). However, Bharathi et al. (2006), Pandiyan et al. (2010), Chaisan et al. (2013), Bhanu et al. (2017) and Mathivathana et al. (2019) all showed consecutive interspecific hybridization between mung bean and rice bean. Study of Sehrawat and Yadav (2014) has demonstrated that rice bean and *Vigna mungo* and *Vigna sublobata* have considerable cross-compatibility. Modern sophisticated biotechnology offers a variety of molecular markers. Utilizing molecular markers, modern methods like marker-assisted breeding (MAB) speed up traditional breeding methods (Ashraf and Foolad 2013). SSRs (simple sequence repeats) or micro-satellite markers are frequently utilized in formation of related marker traits in a segregating population, the validation of hybridity, phylogeny and gene mapping. This also aids in choosing certain resistance genes to include in target species (Michelmore et al. 1991). Singh et al. (2013) from Punjab Agricultural University (PAU), India, created mung bean yellow mosaic India virus or MYMIV-resistant line Mash114. MYMIV resistance from rice bean was transmitted to Mash114 by multiple backcrossing and selection (introgression). Thus, black gram variety Mash114 has shown resistant to the mung bean yellow mosaic India virus (MYMIV) for the past 9 years (Dhaliwal et al. 2022). From RILs of cross KUG253 X Mash114, they also discovered a large-effect QTL (qMYMIV6.1.1) on chromosome 6 (Dhaliwal et al. 2022). Further analysis by them revealed that this area was an inter-specific introgression from rice bean. KASP markers, which were created from strong candidate genes implicated in viral resistance, were used in linkage mapping to identify the 500 kb genomic area on the genetic map associated with MYMIV. BAK1/BRI1-associated receptor kinase genes and Serine threonine kinase, UBE2D2, are the source of the three KASP indicators that are intimately linked to MYMIV. Therefore, Dhaliwal et al. proposed that KASPs may be utilized to transfer introgressed segments into appropriate *Vigna* species backgrounds with the use of markers. In another experiment, Sehrawat et al. (2016) used the MYMV-susceptible PS 1 urd bean line and 4 MYMV-resistant rice bean genotypes, including RBL1, RBL35, RBL6 and RBL50 as males in a hand emasculation and pollination experiment in the field. Additionally, they used SSRs to validate the MYMV resistance F1 urdbean progeny. In order to find quantitative trait loci or QTL linked to mung bean yellow mosaic virus or MYMV resistance, Mathivathana et al. (2019) utilized mung bean (*Vigna radiata*) and rice bean (*Vigna umbellata*) to create an inter-specific recombinant inbred line or RIL population. Initially, they prepared accurate genetic linkage map using the genotyping-by-sequencing (GBS) technique. The genetic map and 2 years' worth of phenotyping data (2015 and 2016) were used by them in the QTL analysis, which revealed five QTLs (qMYMV4_1, qMYMV4_1, qMYMV5_1, qMYMV10_1 and qMYMV6_1) with PVEs ranging from 10.11 to 20.04%. One of them, known as qMYMV4_1, was significant and consistently found in the identical marker interval in both the years. They suggested that possible potential genes for regulating MYMV resistance can be found in this QTL area for future references.

12.8 Rice Bean Genome Sequence Assembly

Recent advances in next-generation genome sequencing technologies have greatly assisted in completing genome sequence of various crop plants. Hence, these genome sequences have open up new avenues for identifying the causal genetic variants/genomic regions conferring disease resistance in various crops. Recently, Guan et al. (2022) constructed the genome assembly of FF25 landrace of rice bean using three sequencing technologies: PacBio single molecule real-time (SMRT) long-read sequencing, Illumina short-read sequencing and chromosome conformation capture sequencing data (Hi-C). The genome assembly was predicted to be 475.64 Mb with 26,736 protein coding genes. The authors also sequenced a total of 440 landraces of rice bean. Thus, the genomic resources developed in rice bean could greatly assist in uncovering the genomic regions or causal genetic variants conferring various disease and pest resistance.

12.9 Speed Breeding for Developing Disease-Resistant Cultivar

Speed breeding is a revolutionary breeding approach for accelerating the creation of new enhanced generations by adjusting different environmental variables in the growth chamber to decrease typical lagging caused by the agricultural breeding cycle (Begna 2022). In traditional breeding approach, only one to two generations per year can be achieved, whereas speed breeding can result three to nine generations per year (Wanga et al. 2021; Ghosh et al. 2018; Ochatt et al. 2002). Rice bean is a photosensitive, indeterminate crop (Pattanayak et al. 2019). Thus, multiple crop cycle of such crop in a year is very challenging. Speed breeding method can be very effective technique for fast rice bean crop improvement programme. To reduce the breeding cycle and make better use of resources, many selection approaches, such as single seed descent (SSD), single pod descent (SPD), single plant selection (SPS), clonal selection and marker-assisted selection (MAS), can be combined into speed breeding (Samineni et al. 2019; Watson et al. 2018; Hickey et al. 2017). Samineni et al. (2019) observed that modification of photoperiod and immature seed germination and SPD selection method in chickpea seeds results seven generation per year. In another study on Soybean, Jähne et al. (2020) achieved five of generation in a year after regulating photoperiod, temperature and immature seed germination. However, different environmental variables that required regulation for successful speed breeding operation include photoperiod regime, temperature, soil moisture, plant density, carbon dioxide level, different nutrient, hormone levels, etc. (Wanga et al. 2021). Thus, future optimization of speed breeding protocol in rice bean could enable in increasing genetic gain.

Rice bean naturally bears many important pest resistance genes like bruchid resistance, *Cercospora* leaf spot, bacterial leaf spot and yellow mosaic virus (Pandiyan et al. 2008; Kashiwaba et al. 2003; Arora et al. 1980). Moreover, rice bean crop is also tolerant to several abiotic stressors like acid soil and aluminium toxicity (Yang et al. 2006; Fan et al. 2014). VuMATE1 and VuMATE2 genes are mainly responsible for aluminium-induced citrate transporter and have been already cloned from this crop (Liu et al. 2018). Arora et al. (1980) reported rice bean as relatively disease-free crop. However, many disease and insect, namely, rust (*Uromyces* spp.), *Rhizoctonia* blight (*Rhizoctonia solani*), pod borer (*Helicoverpa armigera*) and soybean hairy caterpillar (*Spilarctia casignata*), blister beetle, etc. are observed by Pattanayak et al. (2019). Moreover, other challenges that hindrance wide acceptability of rice bean crop cultivation include abrupt crop duration which makes problem for next crop, low harvest index of 25.8–27.3% and seed coat types (Andersen 2012; Pattanayak et al. 2019). Therefore, speed breeding can be the possible way out for rice bean crop improvement programme for eliminating those problems as early as possible.

12.10 Conclusion

There is an urgent need to upgrade and exploit the nutritional potential of novel crops to ensure food and nutritional security. Biotic stresses are of huge concern leading to crop failure and reduced productivity. Rice bean is an underrated multipurpose leguminous crop with high nutritional quality that has the potential to meet the need for increasing pulse production supplanting conventionally grown pluses. The diseases mentioned in this chapter are prime contributors towards the low productivity of rice bean. Cultural practices and chemical managements are not enough to continue the steady production of pulses unless farmers are provided with disease-resistant varieties. However, several species of rice bean possess excellent resistance against YMV which are great resources to transfer resistance to other *Vigna* species. Identification of resistance genes in different rice bean species along with closely flanked molecular markers will expedite the breeding efforts facilitating marker assisted selections. Additionally, further understanding of underlying genetics and gene cloning will broaden the scope for genome editing and the development of transgenic or cisgenics in the near future.

References

Alam AM, Somta P, Srinives P (2014) Identification and confirmation of quantitative trait loci controlling resistance to mungbean yellow mosaic disease in mungbean [*Vigna radiata* (L.) Wilczek]. Mol Breed 34(3):1497–1506

Andersen P (2012) Challenges for under-utilized crops illustrated by rice bean (*Vigna umbellata*) in India and Nepal. Int J Agric Sustain 10:164–174

Arora RK, Chandel PS, Joshi BS, Pant KC (1980) Rice bean: tribal pulse of eastern India. Econ Bot 34:260–263

Ashraf M, Foolad MR (2013) Crop breeding for salt tolerance in the era of molecular markers and marker-assisted selection. Plant Breed 132:10–20

Atta K et al (2022c) A review on stress physiology and breeding potential of an underutilized, multipurpose legume: Rice bean (*Vigna umbellata*). In: Jha UC, Nayyar H, Agrawal SK, Siddique KHM (eds) Developing climate resilient grain and forage legumes. Springer, Singapore. https://doi.org/10.1007/978-981-16-9848-4_11

Atta K, Pal A, Karmakar S, Dutta D, Jana K, Pal AK (2022b) Assessment of physiological and biochemical traits of rice bean [*Vigna umbellata* (Thunb.) Ohwi and Ohashi] seedling in response to equimolar concentration of copper and lead stress. Legum Res. https://doi.org/10.18805/LR-4820. https://arccjournals.com/journal/legume-research-an-international-journal/LR-4820

Atta K, Pal AK, Jana K (2021) Effects of salinity, drought and heavy metal stress during seed germination stage in rice bean [*Vigna umbellata* (Thunb.) Ohwi and Ohashi]. Plant Physiol 26(1):109–115. https://doi.org/10.1007/s40502-020-00542-4

Atta K, Sen J, Chettri P, Pal AK (2022a) Antioxidant responses of rice bean [*Vigna umbellata* (Thunb.) Ohwi and Ohashi] seedling under iso-osmotic potential of salinity and drought stress. Legum Res 45(4):429–434. https://doi.org/10.18805/LR-4551. https://arccjournals.com/journal/legume-research-an-international-journal/LR-4551

Bakhshi M, Arzanlou M, Babai-ahari A, Groenewald JZ, Crous PW (2018) Novel primers improve species delimitation in *Cercospora*. IMA Fungus 9:299. https://doi.org/10.5598/imafungus.2018.09.02.06

Begna T (2022) Speed breeding to accelerate crop improvement. Int J Agric Sc Food Technol 8(2): 178–186. https://doi.org/10.17352/2455-815X.000161

Bhanu AN, Kumar P, Singh MN, Srivastava K, Hemantaranjan A (2017) Assessment of genetic purity of inter-specific F1 hybrids involving *Vigna radiate* and *Vigna umbellata*. J Exp Biol Agric Sci 5:636–643

Bharathi A, Vijay Selvaraj KS, Veerabadhiran P, Subba Lakshmi B (2006) Crossability barriers in mungbean (*Vigna radiata* L. Wilczek): with its wild relatives. Indian J Crop Sci 1:120–124

Biswas K, Varma A (2001) Agro-inoculation: a method of screening germplasm resistance to Mungbean yellow mosaic geminivirus. Indian Phytopathol 54:240–245

Buergelt D, Von Oppen M, Yadavendra JP (2009) Quality parameters in relation to consumer's preferences in ricebean. In: Presentation at the international conference on India. Grain legumes: quality improvement, value addition and trade held at IIPR, Kanpur

Chaisan T, Somta P, Srinives P, Chanprame S, Kaveeta R, Dumrongkittikule S (2013) Development of tetraploid plants from an interspecific hybrid between mungbean (*Vigna radiata*) and ricebean (*Vigna umbellata*). J Crop Sci Biotechnol 116(45):51

Chen NE, Baker LR, Honma S (1983) Interspecific cross ability among four species of *Vigna* food legumes. Euphytica 32:925–937

Dhaliwal SK, Gill RK, Sharma A, Kaur A, Bhatia D, Kaur S (2022) A large-effect QTL introgressed from ricebean imparts resistance to mung bean yellow mosaic India virus in blackgram (*Vigna mungo* (L.) Hepper). Theor Appl Genet 135(12):4495–4506

Falconí CE, Visser RGF, van Heusden AW (2013) Phenotypic, molecular, and pathological characterization of *Colletotrichum acutatum* associated with andean lupine and tamarillo in the ecuadorianandes. Plant Dis 97(6):819–827. https://doi.org/10.1094/pdis-02-12-0175-re

Fan W, Lou HQ, Gong YL, Liu MY, Wang ZQ, Yang JL, Zheng SJ (2014) Identification of early al-responsive genes in rice bean (*Vigna umbellata*) roots provides new clues to molecular mechanisms of Al toxicity and tolerance. Plant Cell Environ 37:1586–1597

Fernandez GCJ, Talekar NS (1990) Genetics and breeding for bruchid resistance in Asiatic Vigna species. *Bruchids and legumes: economics, ecology and coevolution*. Springer, Dordrecht, pp 209–217

Ghosh S, Watson A, Gonzalez-Navarro OE, Ramirez-Gonzalez RH, Yanes L, Mendoza-Suárez M, Simmonds J, Wells R, Rayner T, Green P, Hafeez A, Hayta S, Melton RE, Steed A, Sarkar A, Carter J, Perkins L, Lord J, Tester M, Hickey LT (2018) Speed breeding in growth chambers and glasshouses for crop breeding and model plant research. Nat Protoc 13(12):2944–2963. https://doi.org/10.1038/s41596-018-0072-z

Gilmer RM, Whitney WK, Williams RJ (1974) Epidemiology and control of cowpea mosaic in Western Nigeria. In: Pro. 1st IITA grain legume improvement workshop. International Institute of Tropical Agriculture, Ibadan, Nigeria, p 325

Goyal A, Siddiqui S (2014) Effects of ultraviolet irradiation, pulsed electric field, hot water dip and ethanol vapours treatment on keeping and sensory quality of mung bean (*Vigna radiata* L. Wilczek) sprouts. J Food Sci Technol 51(10):2664–2670

Goyal A, Siddiqui S, Upadhyay N, Soni J (2014) Effects of ultraviolet irradiation, pulsed electric field, hot water and ethanol vapours treatment on functional properties of mung bean sprouts. J Food Sci Technol 51(4):708–714

Guan J, Zhang J, Gong D et al (2022) Genomic analyses of rice bean landraces reveal adaptation and yield related loci to accelerate breeding. Nat Commun 13:5707

Hickey LT, Germán SE, Pereyra SA, Diaz JE, Ziems LA, Fowler RA, Platz GJ, Franckowiak JD, Dieters MJ (2017) Speed breeding for multiple disease resistance in barley. Euphytica 213:1–14

Jacob SS, Vanitharani R, Karthikeyan AS, Chinchore Y, Thillaichidambaram P, Veluthambi K (2003) Mungbean yellow mosaic virus-vi agro infection by co delivery of DNA A and DNA B from one agrobacterium strain. Plant Dis 87:247–251

Jähne F, Hahn V, Würschum T, Leiser WL (2020) Speed breeding short-day crops by LED-controlled light schemes. Theor Appl Genet 133:2335–2342. https://doi.org/10.1007/s00122-020-03601-4

Kashiwaba K, Tomooka N, Kaga A, Han OK, Vaughan DA (2003) Characterization of resistance to three bruchid species (*Callosobruchus* spp., Coleoptera, Bruchidae) in cultivated rice bean (*Vigna umbellata*). J Econ Entomol l96:207–213. https://doi.org/10.1093/jee/96.1.207

Katoch R (2013) Nutritional potential of rice bean (*Vigna umbellata*): an underutilized legume. J Food Sci 78(1):8–16

Katoch R (2020) Incriminating factors in Rice bean. In: Rice bean. Springer, Singapore. https://doi.org/10.1007/978-981-15-5293-9_9

Kaur D, Kapoor AC (1990) Some anti-nutritional factors in rice bean (*Vigna umbellata*): effects of domestic processing and cooking methods. Food Chem 37(3):171–179

Khadka K, Acharya, B.D. (2009) Cultivation practices of rice bean. Research and Development (LI-BIRD), local initiatives for biodiversity, Pokhara, Nepal

Khattak GSS, Haq MA, Ashraf M, Elahi T (2000) Genetics of mungbean yellow mosaic virus (MYMV) in mungbean (*Vigna radiata (L.)*Wilczek). J Genet Breed 54:237–243

Li J, Zhou X, Zhang C, Zhao Y, Zhu Y, Zhang J, Xiao X (2020) The effects of Carbendazim on acute toxicity, development, and reproduction in *Caenorhabditis elegans*. J Food Qual 2020:1–6. https://doi.org/10.1155/2020/8853537

Liu MY, Lou HQ, Chen WW et al (2018) Two citrate transporters co-ordinately regulate citrate secretion from rice bean root tip under aluminium stress. Plant Cell Environ 41:809–822. https://doi.org/10.1111/pce.13150

Madhumitha B, Eraivan Arutkani Aiyanathan K, Kanimoli Mathivathana M, Sudha M (2019) Field and Agroinoculation screening for resistance against MYMV in Mungbean backcross populations. Int J Curr Microbiol App Sci 8(04):2154–2159

Maheswari R (2008) DNA finger printing of Mungbean yellow mosaic virus (MYMV) affecting pulses. (M.Sc Thesis),. Tamil Nadu Agricultural University

Mathivathana MK, Murukarthick J, Karthikeyan A, Jang W, Dhasarathan M, Jagadeeshselvam N et al (2019) Detection of QTLs associated with mungbean yellow mosaic virus (MYMV) resistance using the interspecific cross of *Vigna radiata Vigna umbellata*. J Appl Genet 60: 255–268. https://doi.org/10.1007/s13353-019-00506-x

Mersha Z, Hau B (2008) Effects of bean rust (*Uromyces appendiculatus*) epidemics on host dynamics of common bean *(Phaseolus vulgaris)*. Plant Pathol 57(4):674–686. https://doi.org/10.1111/j.1365-3059.2008.01827.x

Michelmore RW, Paranand I, Kessele RV (1991) Identification of markers linked to disease resistance genes by bulk segregant analysis: a rapid method to detect markers in specific genome using segregant population. Proc Nat Acad Sci 88:9828–9832

Modgil R, Mankotia K, Gupta R, Malhotra SR, Tanwar B (2012) Carbohydrates, minerals, antinutritional constituents and in vivo protein quality of domestically processed cow peas (*Vigna unguiculata*). Legum Res 35(3):214–219

Mohan VR, Janardhanan K (1994) Chemical composition and nutritional evaluation of raw seeds of six rice bean varieties. J Indian Bot Soc 73(3–4):259–263

Nene YL (1973) Viral diseases of some warm weather pulse crops in India. Plant Dis Rep 57:463–467

Ochatt SJ, Sangwan RS, Marget P, Ndong YA, Rancillac M, Perney P (2002) New approaches towards the shortening of generation cycles for faster breeding of protein legumes. Plant Breed 121(5):436–440. https://doi.org/10.1046/j.1439-0523.2002.746803.x

Pandiyan M, Karthikeyan A, Senthil N, Sudha M, Sivakumar C et al (2020) Development of *Mungbean yellow mosaic virus* resistant genotypes in Mungbean through interspecific crosses of wild *Vigna* species. J Plant Sci Curr Res 4:011

Pandiyan M, Ramamoorthi N, Ganesh SK, Jebraj S, Pagarajan P, Balasubramanian P (2008) Broadening the genetic base and introgression of MYMY resistance and yield improvement through unexplored genes from wild relatives in mungbean. Plant Mut Rep 2:33–38

Pandiyan M, Senthil N, Ramamoorthi N, Muthiah AR, Tomooka N, Duncan V, Jayaraj T (2010) Interspecific hybridization of Vigna radiata 13 wild *Vigna* species for developing MYMV donar. Electron J Plant Breed 1:600–610

Panth M, Hassler SC, Baysal-Gurel F (2020) Methods for Management of Soil borne diseases in crop production. Agriculture 10:16. https://doi.org/10.3390/agriculture10010016

Pattanayak A, Roy S, Sood S, Iangrai B, Banerjee A, Gupta S, Joshi DC (2019) Rice bean: a lesser known pulse with well-recognized potential. Planta 250:873. https://doi.org/10.1007/s00425-019-03196-1

Qazi J, Ilyas M, Mansoor S, Briddon B (2007) Legume yellow mosaic viruses: genetically isolated begomo viruses. Mol. Plant Patho, 18:343–348

Rachie KO, Roberts LM (1974) Grain legumes of the lowland tropics. Adv Agron 26:1–132

Raiger HL et al. 2009. Improved varieties and cultivation practices of Rice bean. https://krishi.icar.gov.in/jspui/handle/123456789/45049

Reddy KR, Singh DP (1995) Inheritance of resistance to Mungbean yellow mosaic virus. Madras Agric J 88(2):199–201

Rybicki EP, Briddon RW, Brown JK, Fauquet CM, Maxwell DP, Harrison BD, Markham PG, Bisaro DM, Robinson D, Stanley J (2000) Family Geminiviridae. In: van Regenmortel MHV, Fauquet CM, Bishop DHL et al (eds) Virus taxonomy: classification and nomenclature of viruses. Academic Press, New York, N.Y, pp 285–297

Saharan GS, Mehta NK, Meena PD (2019) The disease: powdery mildew. In: Powdery mildew disease of crucifers: biology, ecology and disease management. Springer, Singapore. https://doi.org/10.1007/978-981-13-9853-7_2

Samineni S, Sen M, Sajja SB, Gaur PM (2019) Rapid generation advance (RGA) in chickpea to produce up to seven generations per year and enable speed breeding. Crop J 8(1):164–169. https://doi.org/10.1016/j.cj.2019.08.003

Sanjeev K. Deshpande and Vishwas, A.R. (2018) Stability Analysis for Yield and Yield Attributing Characters among Advanced Stabilized Promising Lines in Cowpea [Vigna unguiculata (L.) Walp.]. IntJCurrMicrobiolAppSci 7(10): 887–894. https://doi.org/10.20546/ijcmas.2018.710.098

Schmidt M, Lam NT, Hoanh MT, Padulosi S, Quy GL (2010) Promoting neglected and underutilised tuberous plant species in Vietnam. Looking east looking west: organic and quality food Marketing in Asia and Europe. Wageningen academic publishers, Wageningen, Netherlands. ISBN: 9789086860951, pp 183–193

Sehrawat N, Yadav M (2014) Screening and cross-compatibility of various Vigna species for yellow mosaic virus resistance. J Innov Biol 1(1):031–034

Sehrawat N, Yadav M, Bhat KV, Sairam RK, Jaiwal PK (2016) Introgression of mungbean yellow mosaic virus resistance in *Vigna mungo* (L.) Hepper and purity testing of F1 hybrids using SSRs. Turk J Agric For 40:95–100

Senapati M, Tiwari A, Sharma N, Chandra P, Bashyal BM, Ellur RK, Bhowmick PK, Bollinedi H, Vinod KK, Singh AK, Krishnan SG (2022) *Rhizoctonia solani* Kühn pathophysiology: status and prospects of sheath blight disease management in rice. Front Plant Sci 13:881116. https://doi.org/10.3389/fpls.2022.881116

Singh G, Kapoor S, Singh K (1988) Multiple disease resistance in mungbean with special emphasis on mungbean yellow mosaic virus. In: Proceedings of the second international symposium on Mungbean vegetable Research and Development Centre, pp 290–296

Singh I, Sandhu JS, Gupta SK, Singh S (2013) Introgression of productivity and other desirable traits from rice bean (*Vigna umbellata*) into black gram (*Vigna mungo*). Plant Breed 132:401–406. https://doi.org/10.1111/pbr.12068

Sudha M, Karthikeyan A, Anusuya P, Ganesh NM, Pandiyan M, Senthil N, Angappan K (2013a) Inheritance of resistance to mungbean yellow mosaic virus (MYMV) in inter and intra specific crosses of mungbean (*Vigna radiata*). Am J Plant Sci 4(10):1924

Sudha M, Karthikeyan A, Nagarajan P, Raveendran M, Senthil N, Pandiyan M, Angappan K, Ramalingam J, Bharathi M, Rabindran R, Veluthambi K, Balasubramanian P (2013b) Screening of mungbean (*Vigna radiata*) germplasm for resistance to Mungbean yellow mosaic virus using agroinoculation. Can J Plant Pathol 35(3):424–430

Sudha M, Karthikeyan A, Shobhana VG, Nagarajan P, Raveendran M, Senthil N (2015) Search for Vigna species conferring resistance to Mungbean yellow mosaic virus in mungbean. Plant Genet Res 13:162–167

Sujayanand GK, Chandra A, Jagadeeswaran R, Dubey S, Sheelamary S (2021) Rice bean: potential vine legume for achieving nutritional self-sufficiency in India. Vigyan Varta 2(12):10–14

Sultana R, Choudhary AK, Pal AK, Saxena KB, Prasad BD, Singh R (2014) Abiotic stresses in major pulses: current status and strategies. In: Approaches to plant stress and their management. Springer, pp 173–190

Sun S, Zhi Y, Zhu Z, Jin J, Duan C, Wu X, Xiaoming W (2017) An emerging disease caused by pseudomonas syringae pv. Phaseolicola threatens mung bean production in China. Plant Dis 101(1):95–102. https://doi.org/10.1094/pdis-04-16-0448-re

Thongram S, Tanwar B, Chauhan A, Kumar V (2016) Physicochemical and organoleptic properties of cookies incorporated with legume flours. Cogent Food and Agric 2(1):1172389

Tomooka N, Lairungreang C, Nakeeraks P, Egawa Y, Thavarasook C (1991) Mung bean and the genetic resources, the subgenus Ceratotropis. Tropical Agric Res Center, Tsukuba, Japan

Trinklein D (2019) University of Missouri. Plant Sci Technol 573:882–9631

Usharani KS, Surendranath B, Haq QMR, Malathi VG (2005) Infectivity analysis of a soybean isolate of Mungbean yellow mosaic India virus by agro-inoculation. J Gen Plant Pathol 71:230–237

Van den Berg F, van den Bosch F, Paveley ND (2013) Optimal fungicide application timings for disease control are also an effective anti-resistance strategy: a case study for Zymoseptoria tritici (Mycosphaerella graminicola) on wheat. Phytopathology 103(12):1209–1219. https://doi.org/10.1094/PHYTO-03-13-0061-R

Wanga MA, Shimelis H, MashiloJ LMD (2021) Opportunities and challenges of speed breeding: A review. Plant Breed 140(2):185–194

Watson A, Ghosh S, Williams MJ, Cuddy WS, Simmonds J, Rey MD, Hatta AM, M., Hinchliffe, A., Steed, A., Reynolds, D., Adamski, N. M., Breakspear, A., Korolev, A., Rayner, T., Dixon, L. E., Riaz, A., Martin, W., Ryan, M., Edwards, D., Batley, J., . . . Hickey, L. T. (2018) Speed breeding is a powerful tool to accelerate crop research and breeding. Nature plants 4(1):23–29. https://doi.org/10.1038/s41477-017-0083-8

Williams RJ (1977) Identification of multiple disease resistance in cowpea. Trop Agric 54(1):53

Wyatt SD, Brown JK (1996) Detection of subgroup III Gemini virus isolates in leaf extracts by degenerate primers and polymerase chain reaction. Phytopathology 86:1288–1291

Yang JL, Zhang L, Li YY, You JF, Wu P, Zheng SJ (2006) Citrate transporters play a critical role in aluminium-stimulated citrate efflux in rice bean (*Vigna umbellata*) roots. Ann Bot 97:579–584

Yuliar, Nion YA, Toyota K (2015) Recent trends in control methods for bacterial wilt diseases caused by *Ralstonia solanacearum*. Microbes Environ 30(1):1–11. https://doi.org/10.1264/jsme2.ME14144

Genomic Approaches for Resistance Against Fungal Diseases in Soybean

13

Rintu Jha, Menka Tiwari, Bandana Devi, Uday Chand Jha, Shailesh Tripathi, and Prashant Singh

Abstract

Soybean (*Glycine max* (L.) Merr.), an essential leguminous crop, is plagued by several fungal diseases, which is a major worry for soybean farmers worldwide. Significant progress has been made in recent decades in the identification of pathogen-caused diseases, the sources of resistance, and the determination of genomic loci granting resistance to various diseases on linkage maps of soybean. To maintain the sustainability and expansion of soybean production globally, the application of genomics to disease-resistant soybean cultivars is a common goal. Marker-assisted selection and genomic selection have been shown to be effective methods for quickly integrating vertical resistance or horizontal resistance into improved soybean varieties. Vertical resistance is defined as R genes and major effect QTLs, whereas horizontal resistance is a combination of major and minor effect genes or QTLs. In this chapter, we have focused on some important fungal

Menka Tiwari and Bandana Sharma contributed equally with all other contributors.

R. Jha
Department of Botany, Institute of Science, Banaras Hindu University, Varanasi, Uttar Pradesh, India

Division of Genetics, ICAR-Indian Agricultural Research Institute, New Delhi, India

M. Tiwari · B. Devi · P. Singh (✉)
Department of Botany, Institute of Science, Banaras Hindu University, Varanasi, Uttar Pradesh, India
e-mail: p.singh@bhu.ac.in

U. C. Jha
Crop Improvement Division, ICAR-Indian Institute of Pulses Research, Kanpur, Uttar Pradesh, India

S. Tripathi
Division of Genetics, ICAR-Indian Agricultural Research Institute, New Delhi, India

U. C. Jha et al. (eds.), *Diseases in Legume Crops*,
https://doi.org/10.1007/978-981-99-3358-7_13

diseases of soybean, and genomic approaches like breeding, identification of QTLs, transcriptomics for differentially expressed genes (DEGs), metabolomics, and proteomics that confer resistance to fungal diseases in all major soybean production regions of the world are provided. We also emphasized the use of modern genomic tools by providing a thorough summary of significant resistance genes and QTLs for soybean improvement. The condensed genetic knowledge also illuminates the future directions for translational genomics research and expedited soybean breeding. The primary goals of soybean crop improvement are centred on the discovery of sources of resistance to various biotic as well as abiotic stresses and the use of these sources for additional hybridization and transgenic processes to generate new cultivars for stress management.

Keywords

Soybean · Fungi · Resistance · QTLs · Genomics

13.1 Introduction

Soybean (*Glycine max*) is an important legume crop recognized for its high seed protein and oil content (Chander et al. 2019). Having diverse climate adaptability and high protein content, it is cultivated in most part of the globe. A variety of food products and industrial food items are made from soybeans; in addition, it is also utilized as animal feed (Ratnaparkhe et al. 2022). In India, soya bean (*Glycine max* (L) Merrill) has been the most cultivated oil seed crop in terms of both production and area since 2005 (Gawai and Mangnalikar 2018). Soybean seeds are high in protein, oil, vitamins, and minerals, and they are an excellent source of vegetable oil and nutritious plant protein (Patil et al. 2018). Soybean accounted for 42% of total oil seed production in India and 25% of edible oil production. In India, soya bean is primarily produced in Madhya Pradesh, Maharashtra, Rajasthan, Karnataka, Telangana, Chhattisgarh, Nagaland, and Gujarat during a kharif season crop (Gawai and Mangnalikar 2018). According to Tripathi et al. (2022), about ten of the total number of fungal pathogens are consistently present in different regions of the world. Six of which are harmful particularly in India, namely, *Sclerotium rolfsii*, *Macrophomina phaseolina*, *Colletotrichum truncatum*, *Phakopsora pachyrhizi*, *Cercospora sojina*, and *Cercospora kikuchii*. Most of the diseased plants are treated with various chemicals to protect the crops and left their residual effects to the environment. It is better to find some resistant genotypes rather than using hazardous chemicals. Although identification of disease-resistant cultivar is difficult task, the modern molecular breeding tools could increase the efficiency to develop disease-resistant cultivars by transferring resistant gene to the genotype of our interest, developing mapping population, identification genomic regions/QTLs, etc. The resistance nature in soybean was found to be monogenic or polygenic (Tripathi et al. 2022). The present study offers a glimpse into the genomic strategies used to identify the genes/markers linked to the targeted genes in soybeans that are resistant to fungal diseases.

13.2 Soybean Rust (SBR), Its Causal Organism, Important Symptoms, and Economic Loss

In the southern hemisphere, primarily in Asia (Taiwan, Thailand, Japan, and India), Africa, and South America, a potentially fatal foliar disease caused by two meticulously associated obligate fungal species, *Phakopsora pachyrhizi* Sydows and *P. meibomiae* (Arthur), is posing a serious threat to soybean cultivation (Langenbach et al. 2016). The specific ability of *P. pachyrhizi* to infect a wide variety of crop species, a total of 95 plants from 42 genera of the family *Papillionaceae*, presents significant management issues for soybean rust disease (Bromfield 1984). This disease has varying impact on soybean output as it may cause up to 80% yield loss in the zones favourable for growth and proliferation of the causative organism (Hartman et al. 2005).

13.3 *Rhizoctonia* Root Rot, Its Causal Organism, Important Symptoms, and Economic Loss

Rhizoctonia root rot is a soil-borne fungal disease caused by *Rhizoctonia solani* Küuhn. It causes up to 60–70% yield losses in India, 30–60% yield losses in Brazil, and 30–45% yield losses in the USA (Ciampi et al. 2008).

13.4 Brown Stem Rot (BSR), Its Causal Organism, Important Symptoms, and Economic Loss

The soil-borne fungus *Cadophora gregata* is the primary cause of BSR, a serious disease of soybeans (Harrington and McNew 2003). The fungus prevents water and nutrients from moving through the stem of soybean plants, which is essential for their normal growth and development. The majority of BSR illness cases are only detectable after complete pod formation (McCabe et al. 2018). Nutrient deficiency is the most common diagnosis for this illness. Recently, McCabe and Graham (2020) presented a diagnostic strategy based on genes and their network for quick and precise identification to combat misidentification of BSR. The management of this condition may benefit from this strategy. BSR has been cited as the cause of a 38% yield reduction in soybean harvests (McCabe et al. 2018).

13.5 Powdery Mildew (PMD), Its Causal Organism, Important Symptoms, and Economic Loss

The fungus *Microsphaera diffusa* (Paxton and Rogers (1974) causes powdery mildew. The main signs of this illness are infected soybean leaves that have a white, powdery coating. The rate of photosynthesis is decreased by more than 50% as a result of this coating (Dunleavy 1978). In addition, approximately 35% yield reduction occurs along with deteriorated soybean seed quality (Phillips 1984). The powdery patches are first visible on the leaves, but after a few days, they quickly cover the entire leaf and defoliate (Silva 2004).

13.6 *Fusarium* Wilt (FW), Its Causal Organism, Important Symptoms, and Economic Loss (FW, Also Known as Sudden Death Syndrome, SDS)

For the first time, wilted soybean plants were diagnosed in May 2014 in commercial fields at Osijek (Slavonia County) and are caused by the fungal pathogen *Fusarium oxysporum* Schlecht. emend. Snyder and Hansen (Duvnjak et al. 2016). The symptoms of wilting in soybean plants were interveinal chlorosis of leaves, mortality of shoots, and external and internal browning at the base of stems but no symptoms in roots. Due to SDS, yield reductions of up to 5–15% have been seen in the USA (Luo et al. 2001). Due to the disease's frequently environment-sensitive, unpredictable, and irregular disease appearance as well as its time-consuming and expensive treatment, sudden death syndrome resistance is difficult to control in the field (Gibson et al. 1994). Resistance to SDS is partial, and partial disease resistance has advantages over total resistance in terms of consistency and yield compatibility (Yuan et al. 2002).

13.7 Downy Mildew, Its Causal Organism, Important Symptoms, and Economic Loss

Soybean downy mildew (SDM) is one of the major fungal diseases caused by *Peronospora manshurica* (Dong et al. 2018). The onset of symptoms is greatly influenced by the environment and is favoured by high humidity and temperatures of 20–22 °C (Phillips 1999). According to Taguchi-Shiobara et al. (2019), 33 different downy mildew races have been identified so far in the USA. In epidemic years, the average yield loss ranged from 6 to 15% (Dong et al. 2018).

13.8 Anthracnose, Its Causal Organism, Important Symptoms, and Economic Loss

The common soybean disease anthracnose is brought on by the fungus *Colletotrichum truncatum* (Schw.) Andrus & W.D. Moore (Sinclair and Backman 1989). The anthracnose disease caused a yield loss of 16–25% in India (Boufleur et al. 2021). Although several other species are also recognized as anthracnose causal agents, *C. truncatum* has been thought to be the primary cause of the anthracnose disease in soybeans.

13.9 Soybean White Mould (SWM), Its Causal Organism, Important Symptoms, and Economic Loss

One of the most devastating fungal diseases is soybean white mould (SWM) caused by the fungus *Sclerotinia sclerotiorum* (Lib) de Barry which can be found in southern Canada and the Upper Midwest of the USA (Kandel et al. 2018). According to Koenning and Wrather (2010), SWM causes significant yield losses and ranked fourth from the top 28 soybean producing US states. *Sclerotinia sclerotiorum* overwinters in resting structures known as sclerotia in the soil and debris (Yang et al. 1998). However, ascospores that initially touch down on fragile plant parts, such flower petals, are what caused infections of soybean in field situations. They become colonized by ascospores, which subsequently move downhill to infect and girdle the main stem, causing the plant to eventually perish. In addition, necrotic leaves, bleached lesions on stems and pods, white fluffy mycelial growth, and the appearance of black sclerotia on the leaves, stems, and pods are the common symptoms of infected plants (Chen and Wang 2005).

13.10 *Phomopsis* Seed Decay, Its Causal Organism, Important Symptoms, and Economic Loss

Phomopsis seed decay (PSD) of soybean is the primary cause of poor seed quality and causes a significant yield loss in most soybean-growing countries (Sinclair 1993). PSD is more likely to occur in environments that are hot and humid, and it typically gets worse when early maturing cultivars are planted early in the season. Significant symptoms include shrivelled, elongated, or cracked look and a chalky texture, but seed infection is typically asymptomatic.

13.11 *Cercospora* Leaf Blight (CLB)/Purple Seed Stain, Its Causal Organism, Important Symptoms, and Economic Loss

Cercospora leaf blight (CLB)/purple seed stain is a foliar fungal disease of soybean caused by *Cercospora kikuchii* (Albu et al. 2016). Reddish patches on leaves are one of the symptoms. Additionally, these hues intensify and cause soybeans to flower too early. *Cercospora kikuchii* also reduces the marketability, processing potentials, germination, and vigour of seed (Kashiwa and Suzuki 2021). In soybeans, this fungus is the source of causing both *Cercospora* leaf blight (CLB) and/or purple seed stain (PSS) disease. In contrast to CLB, which affects leaves and petioles, PSS affects seed pods and seeds. A distinctive abrasion with a dark purple colour is one of these signs. The pathogen's synthesis of cercosporin led to the development of this lesion (Callahan et al. 1999). Because it degrades the quality of the seed, purple seed stain is a major barrier to its profitable marketability (Li et al. 2019). Various study groups in India have observed yield loss due to purple seed discolouration at different percentages, including 15–30% (Gupta et al. 1999) and 36–80% (Gupta et al. 2014). It is a disease that Americans find undesirable due to economic yield losses (Doupnik 1993).

13.12 Charcoal Rot, Its Causal Organism, Important Symptoms, and Economic Loss

Charcoal rot is caused by soil-borne fungus *Macrophomina phaseolina* and also causes significant yield reduction in soybean (Tripathi et al. 2022). This disease was first time reported in 1949 in the USA, and it was assumed that the presence of two toxins, *phaseolina* and *botryodiplodin*, are responsible for the infection caused by *M. phaseolina* in crops (Ramezani et al. 2007). *M. phaseolina* can infect the vascular system by growing and multiplying under favourable environmental circumstances in plants. It obstructs the movement of water and nutrients toward the leaves in the second step, which results in disease symptoms and further premature leaf death (Gupta and Chauhan 2005). Microsclerotia return to the soil after the crop is harvested and remain there for at least 2 years (Reis et al. 2014). Soybean crops have only exhibited little resistance to *M. phaseolina* (Pawlowski et al. 2015). Due to polygenic inheritance, it is challenging to breed soybean cultivars resistant to charcoal rot (Coser et al. 2017).

13.13 *Phytophthora* Rot and Stem Rot, Its Causal Organism, Important Symptoms, and Economic Loss

Phytophthora sojae is a soil-borne pathogen that causes *Phytophthora* root and stem rot diseases. The soybean crop is affected throughout the years by this disease. It is more devastating in flooded areas (Bernard et al. 1957). *Phytophthora root rot* often results in a yield loss of 35–40%, but under extreme circumstances, it can even result in a loss of 100% of the crop. The most effective strategy for controlling this disease is the creation of resistant cultivars.

In disease management strategies, it is better to find resistance genes or screening resistant cultivars rather than going for chemical application. Advances in plant breeding techniques, application of molecular markers, identification, and expression analysis of target genes linked to disease resistance have opened multiple ways for the modification of the targeted genomic regions of desired genotypes or cultivars (Fig. 13.1). Here we have attempted to explain integrated genomics for several fungal disease management and identification of some resistant lines/ genotypes cultivars (Tables 13.1 and 13.2).

13.14 Integrated Genomic Approaches for Developing Resistance Against Fungal Disease in Soybean

13.14.1 Screening and Identification of Soybean Genotype/Germplasm Resistant to Fungal Disease

Screening of different genetic materials like pre-breeding lines, germplasms, accessions, etc. has tremendous importance in search of sources for disease resistance in both fields as well as in laboratory condition.

There have been numerous research organization attempts to screen soybean germplasm for the presence of fungal disease resistance in soybean. Recently, Nataraj et al. (2020) evaluated 225 soybean genotypes and identified five genotypes as highly resistant, and they are EC 538828, EC 34372, EC 457254, AKSS 67, and Karune. In addition to this, the genetics of anthracnose resistance in three F2 populations descended from the resistant parents EC 34372 × JS 95-60, EC 457254 × JS 95-60, and AKSS 67 JS 95-60 which showed that the resistance in all three resistant parents was controlled by two key genes interacting in a complimentary manner. Similar study by Sajeesh et al. (2014) identified DSb 12 as an anthracnose-resistant genotype.

13.14.2 Identification of QTL(S)/Genomic Loci Conferring Resistance to Fungal Disease in Soybean

Biparental mapping populations are made up of a group of individuals resulting from inter- or intraspecific crossing between two parents. Such recombinant lines are mostly used to provide pre-breeding sources for use in crop improvement, and they constitute a potent technique for analysing the genetic underpinnings of complex traits in crops (Tripodi 2021). Recently, Chanchu et al. (2022) reported a single QTL, *qSBR18.1*, for SBR resistance by evaluating a recombinant inbred line (RIL) population comprising of 108 lines developed from a cross between a susceptible cultivar Sukhothai 2 (SKT2) and CM5.

For BSR, the BSR resistance genes in soybean have been mapped by a number of researchers using marker-assisted breeding. The *Rbs3* gene was initially mapped by Lewers et al. (1999) using 320 recombinant inbred lines (RIL) developed from a cross between BSR 101 and PI 437.654. The same study was also verified by Klos et al. (2000) using SSR markers. Later study, SSR markers were used by Bachman et al. (2001) to map the *Rbs1* and *Rbs2* genes on chromosome 16 in soybean. In addition, Perez et al. (2010) have identified some novel sources of BSR resistance. In

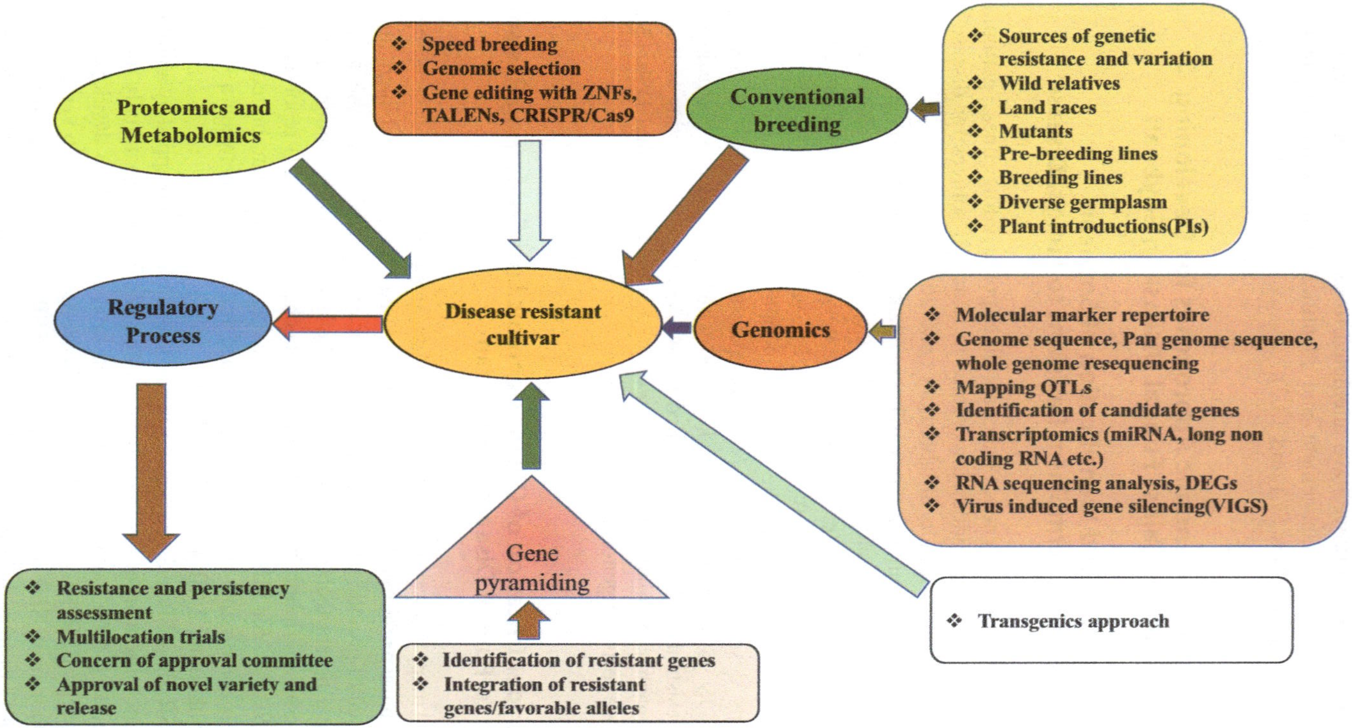

Fig. 13.1 Different genomic approaches for disease improvement in soybean

Table 13.1 List of identified genomic regions/quantitative trait loci (QTLs) conferring resistance to various fungal diseases in soybean

Fungal diseases	Causal organism	Material used	Linkage groups/ chromosome number	Tightly linked markers/ flanking markers	Molecular marker used	Position of QTLs on chromosome (cM)/bp	Locus/ QTLs name	Isolates used	% Phenotypic variation	References
Rust	*Phakopsora pachyrhizi*	F2:3 (173)	Chr.03	Sat_275 and Sat_280	SSR	40.81–43.45 cM	*Rpp5*	*BRSMS Bacuri*		Garcia et al. (2008)
		F2 (86)	Chr. 06	Sat_263 and Sat_238	SSR	118.67–117.45 cM	*Rpp3*	*Japanese T1–2*	70%	Hossain et al. (2015)
		F6 (240)	Chr. 08	Satt409 and Satt429	SSR	145.57–162.02 cM	*QTL Asian soybean rust 2–1*	Local isolate (Georgia, USA)	10%	Harris et al. (2015)
		F6:7 (250)	Chr. 06	Sat_312 and BARC-203517–05,442	SSR	27,940,542–36,131,665 bp	Rpp3	Field tests (FL, US)	8.40%	Vuong et al. (2015)
		F2 (106)	Chr. 18	Sat_064 and AF162283	SSR	108.69–87.94 cM		*Isolate E1-4–12*	65–67%	Yamanaka et al. (2016)
		F5:6 (184)	Chr. 18	GSM0374 and GSM0427	SSR	5,998,461–6,160,481 bp		*Isolate GA12*		King et al. (2016)
		F2:3 (90/100)	Chr. 19	GSM0546 and GSM0463	SSR	39,462,291–39,616,643 bp	Rpp7	*AU79-1 (Australia), CO04-2 (Armenia, Columbia), GA12-1 (Georgia, USA),*		Childs et al. (2018)
		RIL(108)	Chr. 18(G)	T001855631m—sc21_3420	SSR	123 cM	qSBR18.1	–	37.55	Chanchu et al. (2022)
Rhizoctonia root rot	*Rhizoctonia solani*	PI 442031	A2	Satt177	SSR	–	–	–	–	Ishiwata and Furuya (2020)
			C2 M	Satt281	SSR	–	–	–	–	Ishiwata and Furuya (2020)
			M	Satt245	SSR	–	–	–	–	Ishiwata and Furuya (2020)
Brown stem rot	*Cadophora gregata*	F2:3 (73)	Chr. 16	Satt215	SSR	28,944,536–28,944,665 bp	Rbs1	Green house assay	28%	Bachman et al. (2001)

(continued)

Table 13.1 (continued)

Fungal diseases	Causal organism	Material used	Linkage groups/ chromosome number	Tightly linked markers/ flanking markers	Molecular marker used	Position of QTLs on chromosome (cM)/bp	Locus/ QTLs name	Isolates used	% Phenotypic variation	References
		F2:3 (73)	Chr. 16	Satt431	SSR	36,221,174–36,221,397 bp		Green house assay	74%	Bachman et al. (2001)
		F2:3 (77)	Chr. 16	Satt244	SSR	33,818,897–33,819,094 bp	Rbs2	Green house assay	67%	Bachman et al. (2001)
		F2:3 (77)	Chr. 16	Satt431	SSR	36,221,174–36,221,397 bp		Green house assay	46%	Bachman et al. (2001)
		F6:7 (320)	Chr. 16	K375	SSR	67.3–69.3 cM bp	Rbs3	Green house assay	62%	Lewers et al. (1999)
		F6:7 (320)	Chr. 16	B122	SSR	53.8–55.8 cM bp		Green house assay	45%	Lewers et al. (1999)
Powdery mildew	*Microsphaera diffusa*	BRS135 (cultivar)	Chr. 16	GMES6959 Satt_393	SSR	4.6 cM–7.1 cM	*PMD*	–	–	Gordon et al. (2007)
		PI 567301B	Chr. 16	BARCSOYSSR_16_1291	SSR	3.3 cM	*PMD*	–	–	Jun et al. (2012)
		PI 243540	Chr. 16	Sat_224 BARC-021875-04228	SSR	1.3 cM–9.6 cM	*Rmd*	–	–	Kang and Mian (2010)
		ZH24		Gm16_428	SSR	–	*ASR*	–	–	Zhou et al. (2022)
Fusarium wilt/SDS	*Fusarium oxysporum / fusarium virguliforme*	F6:13 (50)	Chr. 2	ss107920774–ss107912689	SSR	30.0–36.0 cM	qRfv02-01	*Isolate Mont1*	5.20%	Abdel Majid et al. (2012)
		Advanced breeding lines (300)	Chr. 2	ss244884978	SSR	49,773,810 bp	qRfv02-01	*Field test (MI, US)*	6.40%	Wen et al. (2014)
		F7 derived RIL (200)	Chr. 2	BARC-041581–08,046—BARC-046084–10,230	SSR	93.341–02.59 cM	qRfv02-01	*Isolates Clinton 1B, Scott F2II 1a and Scott B2*	8.40%	Swaminathan et al. (2016)
		F5:11 (100)	Chr. 3	OF041600	SSR	–	qRfv03-01	*Field test (IL, US)*	10%	Chang et al. (1996), Chang et al. (2018)

		F6:13 (50)	Chr. 3	ss107912585–ss107920575	SSR	38.3–42.6 cM	qRfv03-02	*Isolate Mont1*	9.90%	Chang et al. (2018)
		F4 derived (129)	Chr. 4	ss245526764–ss245561373	SSR	48.56–83.86 cM	qRfv04-01	Field test (MI, US)	3.7–5.3%	Tan et al. (2018)
		F4 derived (153)	Chr. 4	ss245560843–ss245567348	SSR	47.29 Mb–48.08 Mb	qRfv04-01	Field test (MI, US)	6.36%	Tan et al. (2019)
Downy mildew	*Peronospora manshurica*	F6 and F7 (112)	Chr. 2	WGSP02_0160–WGSP02_0170	SSR	50 Mb	QRpm2-1	Field test (Japan)	8–10%	Taguchi-Shiobara et al. (2019)
		F6 and F7 (155), F5 and F6 (190), F6 and F7 (112)	Chr. 3	WGSP03_0040–WGSP03_0070	SSR	–	QRpm3-1	Field test (Japan)	18–72%	Taguchi-Shiobara et al. (2019)
		F6 and F7 (155)	Chr. 4	WGSP04_0120–WGSP04_0140	SSR	–	QRpm4-1	Field test (Japan)	4%	Taguchi-Shiobara et al. (2019)
		F6 and F7 (155)	Chr. 6	WGSP06_0200–WGSP06_0210	SSR	–	QRpm6-1	Field test (Japan)	8%	Taguchi-Shiobara et al. (2019)
		F5 and F6 (189), F9 and F10 (231), (F5 and F6 (190)	Chr. 7	WGSP07_0060–WGSP07_0070	SSR	5 Mb	QRpm7-1	Field test (Japan)	6–91%	Taguchi-Shiobara et al. (2019)
		F6 and F7 (155)	Chr. 8	WGSP08_0110–WGSP08_0130	SSR	20 Mb	QRpm8-1	Field test (Japan)	13–24%	Taguchi-Shiobara et al. (2019)
		F9 and F10(231)	Chr. 11	WGSP11_0100–WGSP11_0120	SSR		QRpm11-1	Field test (Japan)	4%	Taguchi-Shiobara et al. (2019)
		F6 and F7 (112)	Chr. 12	WGSP12_0120–WGSP12_0130	SSR	35 Mb	QRpm12-1	Field test (Japan)	6–8%	Taguchi-Shiobara et al. (2019)

(continued)

Table 13.1 (continued)

Fungal diseases	Causal organism	Material used	Linkage groups/ chromosome number	Tightly linked markers/ flanking markers	Molecular marker used	Position of QTLs on chromosome (cM)/bp	Locus/ QTLs name	Isolates used	% Phenotypic variation	References
		F5 and F6 (189)	Chr. 13	WGSP13_0080–WGSP13_0120	SSR	–	QRpm13-1	Field test (Japan)	3%	Taguchi-Shiobara et al. (2019)
		F5 and F6 (189)	Chr. 14	WGSP14_0050–WGSP14_0060	SSR	–	QRpm14-1	Field test (Japan)	4%	Taguchi-Shiobara et al. (2019)
		F5 and F6 (190)	Chr. 15	WGSP15_0130–WGSP15_0140	SSR	–	QRpm15-1	Field test (Japan)	3%	Taguchi-Shiobara et al. (2019)
		F5 and F6 (190)	Chr. 16	WGSP16_0090–WGSP16_0100	SSR	–	QRpm16-1	Field test (Japan)	3%	Taguchi-Shiobara et al. (2019)
		F5 and F6 (189)	Chr. 18	WGSP18_0150–WGSP18_0160	SSR	50–60 Mb	QRpm18-1	Field test (Japan)	11–16%	Taguchi-Shiobara et al. (2019)
		F6 and F7 (155)	Chr. 19	WGSP19_0150–WGSP19_0170	SSR	–	QRpm19-1	Field test (Japan)	7%	Taguchi-Shiobara et al. (2019)
		F5 and F6 (189)	Chr. 20	WGSP20_0100–WGSP20_0130	SSR	–	Rpm20-1	Field test (Japan)	4%	Taguchi-Shiobara et al. (2019)
		F6 and F7 (112)	Chr. 20	WGSP20_0090–WGSP20_0100	SSR	–	Rpm20-2	Field test (Japan)	5%	Taguchi-Shiobara et al. (2019)
Cercospora leaf blight/ purple seed stain	*Cercospora kikuchii*	F2 (148)	Chr. 18	Sat_308 and Satt594	SSR	6.6 cM and 11.6 cM	Rpss1	Field test	–	Jackson et al. (2008)
		PI 80837	Chr. 18	Satt115 Satt340	SSR	44.1 cM–46.8 cM	*Rpss*1	–	–	Albu et al. (2016)
Charcoal rot	*Macrophomina phaseolina*	Maturity group I-V (130)	Chr. 2	Satt644	SSR	38,221,027 bp	–	Isolate S8	12%	Ghorbanipour et al. (2019)
		USDA PI lines (459)	Chr. 4	ss715588228	SSR	4,307,731 bp	–	Isolate from Iowa soybean field	–	Coser et al. (2017)

	F2:3 (140)	Chr. 5	–	SSR	25,338,390 bp	–	Isolate Conway	–	da Silva et al. 2019
	USDA PI lines (459)	Chr. 6	ss715593307	SSR	14,918,492 bp	–	Isolate from Iowa soybean field	–	Coser et al. (2017)
	F2:3 (140)	Chr. 8	–	SSR	7,511,708 bp	–	Isolate Conway	–	da Silva et al. 2019
	USDA PI lines (459)	Chr. 9	ss715604575	SSR	45,369,206 bp	–	Isolate from Iowa soybean field	–	Coser et al. (2017)
	Maturity group I-V (130)	Chr. 11	Satt359	SSR	32,411,307 bp	–	Isolate S8	–	Ghorbanipour et al. (2019)
	USDA PI lines (459)	Chr. 12	ss715613120	SSR	492,020 bp	–	Isolate from Iowa soybean field	–	Coser et al. (2017)
	USDA PI lines (459)	Chr. 14	ss715618004	SSR	219,725 bp	–	Isolate from Iowa soybean field	–	Coser et al. (2017)
	F2:3 (140)	Chr. 15	Gm15_01842053 and Gm15_03051337	SSR	1,842,060 bp	–	Isolate Conway	29.40%	da Silva et al. 2019
	F2:3 (140)	Chr. 16	Gm16_28961127 and Gm16_30493887	SSR	29,328,591–30,862,012 bp	–	Isolate Conway	25.40%	da Silva et al. (2019)
	USDA PI lines (459)	Chr. 18	ss715631726	SSR	51,751,797 bp	–	Isolate from Iowa soybean field	–	Coser et al. (2017)
	Maturity group I-V (130)	Chr. 19	Sat_124	SSR	50,728,020 bp	–	Isolate S8	11%	Ghorbanipour et al. (2019)
	USDA PI lines (459)	Chr. 20	ss715638424	SSR	43,471,723 bp	–	Isolate from Iowa soybean field	–	Coser et al. (2017)

(continued)

Table 13.1 (continued)

Fungal diseases	Causal organism	Material used	Linkage groups/ chromosome number	Tightly linked markers/ flanking markers	Molecular marker used	Position of QTLs on chromosome (cM)/bp	Locus/ QTLs name	Isolates used	% Phenotypic variation	References
Phytopthora rot		L88-8470	Chr. 3	Satt159	SSR	1.2 cM	*Rps*1a	–	–	Gordon et al. (2007)
		L76-1988	Chr. 16	Sat_393	SSR	0.5 cM	*Rps2*	–	–	Lewers et al. (1999)
		L83-570	Chr. 13	Sat_317	SSR	0.1 cM	*Rps3*	–	–	Bernard et al. (1957)
		L852352	Chr. 16	Sat_004	SSR	4.3 cM	*Rps4*	–	–	Klos et al. (2000)
		L85-3059	Chr. 16				*Rps5*	–	–	Bachman et al. (2001)
		L89-1581	Chr. 16	Sct-187 Sat_372	SSR	0.3 cM–0.4 cM	*Rps6*	–	–	Mueller et al. (1978)
		L93-3258	Chr. 3	Satt152	SSR	6.7 cM	*Rps7*	–	–	Rincker et al. (2016)
		PI 399073	Chr. 13	Satt154	SSR	4.0 cM	*Rps8*	–	–	Paxton and Rogers (1974)

SSR simple sequence repeat, *cM* centi Morgan *bp* base pair

Table 13.2 List of some identified different fungal disease-resistant genotypes/lines/cultivars of soybean developed through various genomic approaches

Name of diseases	Genotypes/lines/cultivars/ resistant genes	Approaches	References
Soybean rust	SRE-Z-11A, SRE-Z-11B, SRE-Z-15A	Breeding	Langenbach et al. (2016)
	PI 441001	Breeding	Bromfield (1984)
	USP 97-08135	Breeding	Hartman et al. (2005)
	PI 416764, PI 462312, KS 1034	Breeding	McLean and Byth (1980)
	TGx 1993 4FN, TGx 1995 5FN, PI 594538A	Breeding	Cheng and Chan (1968)
	PI 594723, PI 594538A, PI 587880A, PI 230970, PI 459025A	Breeding	Hidayat and Somaatmadja (1977)
	PI 200492	Breeding	McLean and Byth (1980)
	PI 230970	Breeding	Cheng and Chan (1968)
	PI 462312	Breeding	Bromfield and Hartwig (1980)
	PI 459025B	Breeding	Hartwig (1986)
	PI 200456	Breeding	Wilcox et al. (1975)
	PI 567102B	Breeding	Li et al. (2012)
	PI 605823	Breeding	Alloatti et al. (2015)
	PI 594538A	Breeding	Calvo et al. (2008)
	70 differentially expressed proteins	Proteomics	Zhang et al. (2014)
Rhizoctonia root rot	AGS-129, G00056	Breeding	Kofsky et al. (2021)
	PI 442031	Breeding	Ishiwata and Furuya (2020)
Brown stem rot	PI 84946-2, PI 437833,PI 437970, L84-5873, and PI 86150	Breeding	Rincker et al. (2016)
Powdery mildew	BRS135 (cultivar)	Breeding	Gordon et al. (2007)
	PI 567301B	Breeding	Jun et al. (2012)
	PI 243540	Breeding	Kang and Mian (2010)
	ZH24	Breeding	Zhou et al. (2022)
	Djakl	Breeding	Dunn and Gaynor (2020)
Downy mildew	52 differentially expressed genes	Transcriptomics	Zhu et al. (2018)
Cercospora leaf blight/ purple seed stain	PI 417361, PI 504488, PI 88490, PI 346308, PI 416779, PI 417567, PI 381659, PI 417567, PI 407749	Breeding	Rahman et al. (2018)
	PI 80837	Breeding	Alloatti et al. (2015)

(continued)

Table 13.2 (continued)

Name of diseases	Genotypes/lines/cultivars/ resistant genes	Approaches	References
Phomopsis seed decay	PI 82,264		Walters and Caviness (1973)
	PI 181,550		Athow (1987)
	Delmar		Crittenden and Cole (1967)
	PI 200,501 and Arksoy		Ross (1986)
	PI 80,837, PI 417,479, and PI 360,841		Brown et al. (1987)
Anthracnose	EC 538828,EC 34372, EC 457254, AKSS 67, and Karune	Breeding	Nataraj et al. (2020)
	DSb 12	Breeding	Sajeesh et al. (2014)
Charcoal rot	JS 20-98, JS 20-34, MAUS 162	Breeding	Zhang et al. (2014)
	1219 DEGs	Transcriptomics	Deshmukh and Tiwari (2021)
Phytophthora rot	L88-8470	Breeding	Athow and Laviolette (1982)
	L76-1988	Breeding	Lewers et al. (1999)
	L83-570	Breeding	Bernard et al. (1957)
	L85-2352	Breeding	Klos et al. (2000)
	L85-3059	Breeding	Bachman et al. (2001)
	L89-1581	Breeding	Mueller et al. (1978)
	L93-3258	Breeding	Rincker et al. (2016)
	PI 399073	Breeding	Paxton and Rogers (1974)
	Zaoshu18	Breeding	Moellers et al. (2017)
	E00003	Breeding	Boudhrioua et al. (2020)
	46 differentially expressed proteins	Proteomics	Zhang et al. (2011)
	90 differentially accumulated metabolites	Metabolomics	Gordon et al. (2007)

further study, the genes conferring resistance to BSR were mapped on chromosome 16. These results led to the conclusion that soybean BSR resistance is caused by just one gene (McCabe and Graham 2020). Using mapping populations developed by crossing the resistant sources "Bell," PI 84946-2, PI 437833, PI 437970, L84-5873, and PI 86150 with either the susceptible cultivar Colfax or Century 84, three BSR resistant genes, Rbs1, Rbs2, and Rbs3, have been discovered and located on chromosome 16 (Rincker et al. 2016).

For powdery mildew (PMD), according to study, three alleles were present at the *Rmd* locus on the inheritance of host plant resistance to PMD, and they are *Rmd*, *Rmd-c*, and *rmd* (Lohnes and Bernard 1992). In the soybean cultivar PI 243540,

Kang and Mian (2010) found that a single dominant gene contributes to PMD resistance at all stages of soybean plant development. They discovered the gene *Rmd* PI 243540 from the cultivar PI 243540 to be situated between the SSR marker *Sat 224* and SNP marker *BARC-021875-04228* over the course of their investigation. The PMD resistance gene *Rmd* was linked to both markers at distances of 9.6 and 1.3 cM, respectively. The use of genetic markers for molecular characterization and diversity analysis among soybean genotypes for powdery mildew resistance has undergone a number of attempts. SSR analysis was utilized by DeMore et al. (2009) to find PMD resistance gene-linked markers in an F_2 population, derived from a cross between MGBR95-20937 IAC-Foscarin 31 and MGBR-46 EMBRAPA 48. In their investigation, two SSRs *Sat 366* and *Sat 393* were discovered and situated 9.41 cM and 12.45 cM away from PMD resistance genes, respectively. More recently, Zhou et al. (2022) examined adult plant resistance (APR) to PMD in soybean using recombinant inbred lines (RILs) populations created from crossing Zhonghuang 24 (ZH24) and Huaxia 3 (HX3). The outcomes showed that a single dominant locus controlled PMD resistance.

In case of FW, quantitative trait loci have significant role of controlling *Fusarium* wilt resistance. Studies reported four genes in a cluster with two duos in close proximity or two genes in a cluster with each gene exhibiting pleiotropy are responsible for triggering resistance to *Fusarium* wilt (Stephens et al. 1993). Another study under greenhouse conditions reported single dominant gene *Rfs1* may be responsible for controlling SDS resistance (Hnetkovsky et al. 1996). Similarly, a small number of significant QTLs govern some levels of resistance (Triwitayakorn et al. 2005). Consequently, a number of QTLs may also function as a qualitative locus (Anderson 2012). In same study, the candidate genes *QRfs1* and *QRfs2* are identified for two loci, and both offered resistance against root infection and leaf scorch, respectively (Anderson 2012). According to Fronza et al. (2004), QTLs on linkage group G conferred nine LGs (A2, C2, D2, F, G, I, J, L, and N) for resistance to root infection (*Rfs1*). According to Soybase (2010) report, more than 56 records of QTLs for *Fusarium* wilt in soybean have registered. Similarly, Fronza et al. (2004) reported multiple trait loci for resistance on chromosome number 18 (linkage group G), using four populations of almost isogenic lines and nine DNA markers. In linkage group G, it was hypothesized that three to four genes, namely, QRfs-, *QRfs1*-, *QRfs2*-, and *QRfs3*-rich islands, transfer resistance (Anderson et al. 2014). Similar study reported QTLs, namely, *BARC-Satt163*, *BARC-Satt080*, and *BARC-Satt307* for resistance that were identified on linkage groups G, N, and C2, respectively (Zou et al. 2005). Recombinant inbred lines with presentations that were environmentally stable and contained all three QTLs for resistance were considerably more resistant than other recombinant inbred lines. With a significant impact on the QTL *Rfs1*, the SSR marker *Satt183* has been found to provide resistance to SDS on molecular linkage group J (Sanitchon et al. 2004). The SSR marker *Satt183* has been found to provide SDS resistance (56% variance) on linkage group J. The SSR marker *Satt183* found to be most significant robust marker associated with QTL for *Rfs1* (Sanitchon et al. 2004).

For DM, 31 quantitative trait loci (QTL) were identified using five populations of RILs, derived from ('Natto-shoryu' × 'Tachinagaha' (NT), 'Nattoshoryu' × 'Suzumaru' (NS), 'Satonohohoemi' × 'Fukuibuki' (SF), 'Kinusayaka' × 'COL/Akita/2009/TARC/1' (KC), and 'YR-82' × 'Harosoy' (YH) grown across location and years (Taguchi-Shiobara et al. 2019).

For SWM, the use of molecular markers in conjunction with field studies has opened up new possibilities because they are independent of environmental factors. Recently Kandel et al. (2018) reported ten significant QTLs by single marker analysis that could be used as source of resistance to develop SWM-resistant cultivars. Another study by Moellers et al. (2017) reported 58 SNP-based loci had main effects, and some others had epistatic effects that were related to SWM resistance.

For PSD, employing progenies derived from the cross between resistant cultivar 'Taekwangkong' and the susceptible cultivar 'SS2-2' yielded two QTLs for PSD resistance under greenhouse condition (Sun et al. 2013).

For purple seed stain disease (PSS), the only partially resistant sources for PSS that have been reported are PI 80, 837, and SJ2 (Roy and Abney 1976; Ploper et al. 1992). According to Jackson et al. (2006), a single dominant gene *Rpss1*, on linkage group G, was shown to be responsible for resistance to *C. kikuchii* in the cultivar PI80837. In this study, the potential resistant gene was located between the flanking markers *Sat 308* and *Satt594* away from resistant genomic loci of 6.6 cM and 11.6 cM, respectively, on linkage group G. The use of such molecular markers in PSS resistance study will aid the advantages in marker-assisted breeding and selection (Jackson et al. 2008). Similarly, two SSR molecular markers *Satt115* and *Satt340* that are associated with resistance of purple seed stain have been identified in an association mapping study by evaluating two population derived from the cross of PI 80,837 (resistant) with AP 350 and MO/PSD-0259 (Alloatti et al. 2015).

For charcoal rot, in a recent study, a total of 140 $F_{2:3}$ lines derived from the cross PI 567562A (resistant) PI 567437 (susceptible) were genotyped, and QTL mapping analysis revealed one QTL on chromosome 15 and two QTLs on chromosome 16 for resistance to *M. phaseolina* (da Silva et al. 2019) (Table 13.1).

For *Phytophthora* root and stem rot of soybean, the mapping of molecular markers conferring resistance to the disease on different linkage groups has advanced since the introduction of the soybean linkage map. Several studies reported different genes responsible for resistance against of *P. sojae*. The resistant genes *Rps1*, *Rps2*, *Rps3*, *Rps4*, *Rps5*, *Rps6*, *Rps7*, and *Rps8* have been mapped on linkage groups N, J, F, G, G, G, N, and F, respectively, from different studies (Cregan et al. 1999; Sugimoto et al. 2007; Bernard and Cremeens 1981; Demirbas et al. 2001; Buzzell and Anderson 1981; Sugimoto et al. 2011). In addition, one RFLP marker, *pT-5*, found to be associated with the *Rps5* gene (Athow and Laviolette 1982). However, in addition to these resistant genes, soybean also has several partial resistance-related genes (Akem 1996). Some more genomic loci/QTLs have been reported by several researchers and documented which are elaborated in Table 13.1.

13.15 Genome-Wide Association Studies (GWAS) for Identification of Potential Candidate Gene(s) Associated with Resistance to Fungal Disease

With the help of molecular markers, GWAS have been successfully used to understand the genetic architecture of panels of germplasm lines and to pinpoint regions of the soybean genome linked to various disease resistance and also useful for marker-assisted selection in breeding programme. In an association study, 256 germplasm accessions from various countries were examined with several years across the location for their responses to soybean rust (SBR) along with susceptible controls and plant introductions (PIs) with *Rpp* genes at known loci (Walker et al. 2022). According to GWAS analysis, 31,114 SNPs were found, and 8 significant SNPs in 8 genomic areas on 7 chromosomes were found. Eight genomic areas, including previously unreported parts of chromosomes 1, 4, 6, 9, 13, and 15, as well as the *Rpp3* and *Rpp6* locus, were found to be related with SBR resistance on 7 chromosomes (Walker et al. 2022). Linkage map analysis with SSR markers revealed significant marker association to rust resistance in the linkage group (LG) C2 in cultivar FT 2 (Cregan et al. 1999). Another study reported a resistance gene situated in between the flanking marker *Satt134* and *Satt460* on LG-C2 and has been mapped in the cultivar Hyuuga (Monteros et al. 2007). Similarly, *Rpp3* was also located at the same location as reported by Hyten et al. (2007). On LG-G, between flanking markers *Sct 187* and *Sat 064*, 1 cM interval has been identified as the location of the *Rpp1* locus (Hyten et al. 2007). The *Rpp4* locus was located on chromosome 18 in linkage group G by 1.9 cM distance (Silva et al. 2008) and 2.8 cM (Garcia et al. 2008) from SSR marker *Satt288*, respectively. Meyer et al. (2009) reported *Rpp4C4* (PI 459025B) was highly expressed in the resistant genotype, while the expression of the other intrusive genes was essentially undetectable. According to the results of reverse transcription polymerase chain reaction sequencing, *Rpp4C4* considered to be the single candidate gene for *Rpp4C4*-mediated rust resistance. Molecular marker was used to increase the resistance against SBR of Vietnamese elite soybean cultivar (Khanh et al. 2013). In the same study, the *Rpp5* gene of SBR resistance was successfully incorporated into a popular Vietnamese soybean variety HL203 by using molecular markers in a backcross breeding technique. The *Rpp5* locus was discovered and to be located in the N linkage group between the flanking markers *Sat 275* and *Sat 280*. Further, based on the molecular information, Maphosa et al. (2012) asserted that the three resistance genes *Rpp2*, *Rpp3*, and *Rpp4* were effectively pyramided in pairwise combinations in the soybean F_2 generation.

For *Rhizoctonia* root rot, the development of resistant genotype was aided by marker-assisted selection in combination with phenotypic selection in later generations. According to an association study, the identified SSR markers, *Satt177* on linkage group A2, *Satt281* on linkage group C2, and *Satt245* on linkage group M, found to be associated with the resistance to *Rhizoctonia* root rot (Tomar et al. 2011). Utilizing these three SSR markers for further screening revealed the allelic variation for resistance (Sserunkuma 2016). In this study, five alleles were amplified by each of the three markers. These markers amplified uncommon alleles and were found to be highly polymorphic.

For FW, an association mapping strategy by using 282 soybean lines along with 1536 SNP markers was used by Bao et al. (2015) to locate the loci that differ in SDS resistance and were employed, and two new loci were identified on chromosomes 3 and 18. The findings of these studies have accelerated the value of association mapping in locating significant loci in soybean.

For SWM, genome-wide association study revealed a novel QTL on chromosome 1 which is associated with SWM resistance (Boudhrioua et al. 2020).

For charcoal rot, a set of 459 different plant introductions from the USDA soybean germplasm core collection were screened in the field and greenhouse, and GWAS revealed some putative candidate genes led to new source of resistance (Coser et al. 2017). Similarly, in an association mapping study using 130 different soybean varieties and lines, *Sat_252*, *Satt359*, *Satt190*, *Sat_169*, *Sat_416*, and *Sat460* markers were identified that are associated with the charcoal rot disease (Ghorbanipour et al. 2019).

13.15.1 Virus-Induced Gene Silencing

The molecular identification of resistance in plants uses a virus-induced gene silencing approach and can be used as an alternative transgenic approach for disease resistance. Using this method, Meyer et al. (2009) discovered the *P. pachyrhizi*-resistant accession PI459025B in soybean. Additionally, Pedley et al. (2018) employed this method to characterize *Rpp1* in a recent study. According to this study, *Rpp1* was situated on chromosome 18 between the flanking markers *Sct 187* and *Sat 064*. According to results, the *Rpp1* gene was found to be distinct among other *Rpp* genes as it provides an immune response to isolates of avirulent *P. pachyrhizi* and is known to produce ULP1-NBSLRR protein which is essential for the immunological response.

13.15.2 Gene Pyramiding

There are reports on the use of gene pyramiding in soybean to create resistance to soybean rust. Combining *Rpp2*, *Rpp4*, and *Rpp5* in one soybean genotype demonstrated greater resistance to SBR (Lemos et al. 2011). Similarly, *Rpp2*, *Rpp3*, and *Rpp4* were combined with cumulative resistance using the gene pyramiding strategy (Pedley et al. 2018). The gene pyramiding strategy to promote disease resistance in the soybean crop is clearly reflected in these results (Chander et al. 2019). Recently, it has been discovered that using marker-assisted selection in conjunction with line breeding can help create soybean cultivars that have ASR resistance genes. It contributed to the introduction of two new soybean varieties in Paraguay, namely, JFNC 1 and JFNC 2. Three all-stage resistance (ASR) genes, *Rpp2*, *Rpp4*, and *Rpp5*, were present in both cultivars (Kato et al. 2022).

13.15.3 Transcriptomics

RNA-seq analysis identified 52 differentially expressed genes (DEGs) DEGs, demonstrating soybean downy mildew (SDM) defence-responsive genes (Dong et al. 2018). These discoveries have opened the door for additional functional evaluation of potential candidate genes, which can then be exploited to create superior soybean cultivars with improved SDM resistance.

The differential expression of WRKY transcription factors (TFs) in SDM-high resistant (HR) and SDM-high susceptible (HS) genotypes was examined in order to provide new insights regarding the defence mechanism of soybean response to Pm infection. In addition, a total of 16 WRKY TFs were discovered to be specific in response to fungal inoculation, and 22 WRKY TFs were shown to be differentially expressed in HR and HS genotypes. The yeast one-hybrid (Y1H) experiment was used to test the capacity of the GmWRKY31 to bind the cis-acting W-box element in the promoter region of the GmSAGT1 gene, whose higher transcriptional expression was associated with increased SDM resistance (Dong et al. 2018).

13.16 Conclusion and Future Perspectives

Throughout this content, it has been discussed how fungal diseases affect soybean production globally and how much yield is lost as a result. Diseases have been consistently documented to cause significant yearly output losses in the millions of dollars in the literature for decades (Savary et al. 2019; Bandara et al. 2020). The most efficient and long-lasting method for managing disease in soybeans worldwide is genetic resistance, which serves as a crucial tenet supporting the global soybean value chain and food security. Since the discovery and use of molecular markers are intimately related to resistance genes, public and private soybean breeding programmes have consistently introduced vertical resistance through MAS. Despite the fact that our evaluation identified hundreds of key genomic areas that confer resistance to numerous fungal diseases, there are still other aspects of genetic resistance that need to be clarified and actively researched.

The development of high-density molecular markers based on next-generation sequencing (NGS) was made possible by advances in genomics. These markers quickly advanced and were affordable for use in both public and private breeding programmes (Song et al. 2013, 2020). The soybean genome has many novel regions that are significantly associated with resistance to various pathogens, according to genome-wide studies. Traits that were previously thought to be qualitative in nature have somewhat changed into quantitative traits, with major and minor alleles having small effects contributing to the observed phenotypes.

However, a successful genetic transformation mechanism is necessary for the generation of CRISPR/Cas9 transformants, though. Unfortunately, soybeans are a difficult commodity for plant transformation technology, and the majority of GE research are still in the early stages of development. Although a few studies have successfully demonstrated the introduction of ribonucleoprotein complex (Cas12a-

RNP) in soybean protoplast (Kim et al. 2017), significant efforts may be required to incorporate these tools into soybean.

References

Abdel Majid KM, Ramos L, Leandro L, Mbofung G, Hyten DL, Kantartzi SK, Gier RL IV, Njiti VN, Meksem K (2012) The 'PI 438489B' by 'Hamilton' SNP-based genetic linkage map of soybean [Glycine max (L.) Merr.] identified quantitative trait loci that underlie seedling SDS resistance. J Plant Genome Sci 1:18–30

Akem CN (1996) Management of soybean diseases. In: IITA research guide 40, training program, 3rd edn. International Institute of Tropical Agriculture (IITA), Ibadan, Nigeria

Albu S, Schneider RW, Price PP, Doyle VP (2016) Cercospora cf. flagellaris and Cercospora cf. sigesbeckiae are associated with Cercospora leaf blight and purple seed stain on soybean in North America. Phytopathology 106:1376–1385

Alloatti J, Chen P, Zeng A, Li S, Rupe J, Florez-Palacios L, Orazaly M (2015) Inheritance of and molecular markers for purple seed stain resistance in soybean. Euphytica 206:701–709

Anderson J (2012) Evaluation of soybean recombinant inbred lines for yield potential and resistance to sudden death syndrome. Southern Illinois University Carbondale, Carbondale, IL, USA, p Paper 837

Anderson J, Akond M, Kassem MA, Meksem K, Kantartzi SK (2014) Quantitative trait loci underlying resistance to sudden death syndrome (SDS) in MD96–5722 by 'Spencer' recombinant inbred line population of soybean. 3 Biotech 5:203–210

Athow KL, Laviolette FA (1982) Rps6, a major gene for resistance to Phytophthora megasperma f. sp. glycinea in soybean. Phytopathology 72:1564–1567

Athow KL (1987) Fungal diseases. In: Wilcox JR (ed) Soybeans: improvement, production, and uses, 2nd edn. American Society of Agronomy, Crop Science Society of America, Soil Science Society of America, Madison, Wisconsin, pp 687–727

Bachman MS, Tamulonis JP, Nickell CD, Bent AF (2001) Molecular markers linked to brown stem rot resistancegenes,Rbs1and Rbs2, in soybean. Crop Sci 41:527–535

Bao Y, Kurle JE, Anderson G, Young ND (2015) Association mapping and genomic prediction for resistance to sudden death syndrome in early maturing soybean germplasm. Mol Breed 35:128

Bandara AY, Weerasooriya DK, Bradley CA et al (2020) Dissecting the economic impact of soybean diseases in the United States over two decades. PLoS One 15:1–28

Bernard RL, Smith PE, Kaufmann MJ, Schmitthenner AF (1957) Inheritance of resistance to *Phytophthora* root and stem rot in soybean. Agron J 49:391

Bernard RL, Cremeens CR (1981) An allele at the rps1 locus from the variety 'Kingwa. Soybean Genet Newsl 8:40–42

Boudhrioua C, Bastien M, Torkamaneh D, Belzile F (2020) Genome wide association mapping of Sclerotinias sclerotiorum resistance in soybean using whole-genome resequencing data. BMC Plant Biol 20:195

Boufleur TR, CiampiGuillardi M, Tikami Í, Rogério F, Thon MT, Sukno SA, MassolaJúnior NS, Baroncelli R (2021) Soybean anthracnose caused by Colletotrichum species: current status and future prospects. Mol Plant Pathol 22:393–409

Bromfield KR, Hartwig EE (1980) Resistance to soybean rust and mode of inheritance. Crop Sci 20: 254–255

Bromfield KR (1984) Soybean rust, monograph no. 11. American Phytopathological society, St. Paul, MN, USA

Brown EA, Minor HC, Calvert OH (1987) A soybean genotype resistant to Phomopsis seed decay. Crop Sci 27:895–898

Buzzell RI, Anderson TR (1981) Another major gene for resistance to Phytophthora megasperma var. sojae in soybean. Soybean Genet Newsl 18:30–33

Callahan TM, Rose MS, Meade MJ, Ehrenshaft M, Upchurch RGCFP (1999) The putative cercosporin transporter of *Cercospora kikuchii*, is required for wild type cercosporin production, resistance, and virulence on soybean. Mol Plant-Microbe Interact 12:901–910

Calvo ES, de Souza Kiihl RA, Garcia A, Harada A, Hiromoto DM (2008) Two major recessive soybean genes conferring soybean rust resistance. Crop Sci 48:1350–1354

Chanchu T, Yimram T, Chankaew S, Kaga A, Somta P (2022) Mapping QTLs controlling soybean rust disease resistance in Chiang Mai 5, an induced mutant cultivar. Genes (Basel) 14(1):19. https://doi.org/10.3390/genes14010019

Chander S, Ortega-Beltran A, Bandyopadhyay R, Sheoran P, Ige GO, Vasconcelos MW, Garcia-Oliveira AL (2019) Prospects for durable resistance against an old soybean enemy: a four-decade journey from *Rpp1* (*resistance* to *Phakopsora pachyrhizi*) to *Rpp7*. Agronomy 9:348

Chang SJC, Doubler TW, Kilo V et al (1996) Two additional loci underlying durable field resistance to soybean sudden death syndrome (SDS). Crop Sci 36:1684–1688

Chang HX, Roth MG, Wang D et al (2018) Integration of sudden death syndrome resistance loci in the soybean genome. Theor Appl Genet 131:757–773

Cheng YW, Chan KL (1968) The breeding of rust resistant soybean Tainung 3. J Taiwan Agric Res 17:30–34

Chen Y, Wang D (2005) Two convenient methods to evaluate soybean for resistance to *Sclerotinia sclerotiorum*. Plant Dis 89:1268–1272

Childs SP, Buck JW, Li Z (2018) Breeding soybeans with resistance to soybean rust (Phakopsora pachyrhizi). Plant Breed 137:250–261

Ciampi MB, Meyer MC, Costa MJ, Zala NM, McDonald BA, Ceresini PC (2008) Genetic structure of populations of Rhizoctonia solani anastomosis group-1 IA from soybean in Brazil. Phytopathology 98:932–941

Coser SM, Reddy RVC, Zhang J, Mueller DS, Mengistu A, Wise KA, Allen TW, Singh A, Singh AK (2017) Genetic architecture of charcoal rot (Macrophomina phaseolina) resistance in soybean revealed using a diverse panel. Front Plant Sci 8:1626

Cregan PB, Jarvik T, Bush AL et al (1999) An integrated genetic linkage map of the soybean genome. Crop Sci 39:1464–1490

Crittenden HW, Cole RH (1967) Registration of Delmar and Bethel soybeans. Crop Sci 7:279–280

da Silva MP, Klepadlo M, Gbur EE, Pereira A, Mason RE, Rupe JC, Bluhm BH, Wood L, Mozzoni LA, Chen P (2019) QTL mapping of charcoal rot resistance in PI 567562A soybean accession. Crop Sci 59:474–479

Demirbas A, Rector B, Lohnes D et al (2001) Simple sequence repeat markers linked to the soybean Rps genes for Phytophthora resistance. Crop Sci 41:1220–1227

DeMore P, UnêdaTrevisoli S, Mauro A, Morceli T, Muniz F, Costa M, Sarti D, Mancini M (2009) Validation of microsatellite markers for assisted selection of soybean genotypes resistant to powdery mildew. Crop Breed Appl Biotechnol 9:45–51

Deshmukh R, Tiwari S (2021) Analysis of soybean transcriptome uncovers differential genomic regulations in response to charcoal rot in resistant and susceptible genotypes. Physiol Mol Plant Pathol 115:101672

Dong H, Shi S, Zhang C, Zhu S, Li M, Tan J, Yu Y, Lin L, Jia S, Wang X et al (2018) Transcriptomic analysis of genes in soybean in response to Peronospora manshurica infection. BMC Genomics 19:366

Doupnik B Jr (1993) Soybean production and disease loss estimates for north Central United States from 1989 to 1991. Plant Dis 77:1170–1171

Dunleavy JM (1978) Soybean seed yield losses caused by powdery mildew1. Crop Sci 18:337–339

Dunn MW, Gaynor LG (2020) Impact and control of powdery mildew on irrigated soybean varieties grown in Southeast Australia. Agronomy 10:514

Duvnjak T, Sudaric A, Matosa Kocar M, Cosic J, Vrandecic K (2016) First Report of Soybean Fusarium Wilt Caused by Fusarium oxysporum in Croatia. APS Publications, p 249. https://doi.org/10.1094/PDIS-05-16-0665-PDN

Fronza V, Vello NA, Camargo LEA (2004) Genetic analysis of soybean resistance to fusarium solani f.sp.glycines. Genet Mol Biol 27:400–408

Garcia A, Calvo ES, de Souza Kiihl RA, Harada A, Hiromoto DM, Vieira LG (2008) Molecular mapping of soybean rust (*Phakopsora pachyrhizi*) resistance genes: discovery of a novel locus and alleles. Theor Appl Genet 117:545–553

Gawai DU, Mangnalikar SS (2018) Effect of temperature and pH on growth of Alternaria alternata, leaf spot pathogen of soyabean. Biosci Discov 9(1):162–165

Ghorbanipour A, Rabiei B, Rahmanpour S, Khodaparast SA (2019) Association analysis of charcoal rot disease resistance in soybean. Plant Pathol 35:189–199

Gibson PT, Shenaut MA, Njiti VN, Suttner RJ, Myers O (1994) Jr. soybean varietal response to sudden death syndrome. In: Wilkinson D (ed) Proceedings of the twenty-fourth soybean seed research conference, Chicago, IL, USA. American seed trade association, Washington, DC, USA, pp 436–446

Gordon SG, Kowitwanich K, Pipatpongpinyo W et al (2007) Molecular marker analysis of soybean plant introductions with resistance to Phytophthora sojae. Phytopathology 97:113–118

Gupta GK, Ansari M, Karmarkar P, Husain S, Ramteke R (1999) Resurrection of soybean rust *(Phakopsora pachyrhizi)*in India. In: Kauffman HE (ed) World soybean research conference VI, vol 617. University of Illinois, Chicago, IL, USA

Gupta GK, Chauhan GS (2005) Symptoms, identification and Management of Soybean Diseases; GaoTechnical bulletin 10. National Research Centre for Soybean, Indore, India

Gupta KL, Girish KL, Sharma S, Dhyani S, Vineet K, Rajkumar R (2014) Effect of purple seed stain disease on physical and biochemical traits of soybean. Soybean Res 12:77–84

Harris DK, Abdel-Haleem H, Buck JW et al (2015) Soybean quantitative trait loci conditioning soybean rust-induced canopy damage. Crop Sci 55:2589–2597

Hartman GL, Miles MR, Frederick RD (2005) Historical viewpoint and soybean resistance to soybean rust. In: Proceedings of the 2005 Illinois crop protection conference, Urbana, IL, USA; pp. 16–20

Harrington TC, McNew DL (2003) Phylogenetic analysis places the Phialophora-like anamorph genus Cadophora in the Helotiales. Mycotaxon 87:141–152

Hartwig EE (1986) Identification of a fourth major gene conferring resistance to soybean rust. Crop Sci 26:1135–1136

Hidayat OO, Somaatmadja S (1977) Screening of soybean breeding lines for resistance to soybean rust (Phakopsora pachyrhizi Sydow). Soybean Rust Newsl 1:9–22

Hnetkovsky N, Chang SC, Gibson TWDPT, Lightfoot DA (1996) Genetic mapping of loci underlying field resistance to sudden- death syndrome. Crop Sci 36:392–400

Hossain MM, Akamatsu H, Morishita M et al (2015) Molecular mapping of Asian soybean rust resistance in soybean landraces PI 594767A, PI 587905 and PI 416764. Plant Pathol 64:147–156

Hyten DL, Hartman GL, Nelson RL, Frederick RD, Concibido VC, Cregan PB (2007) Map location of the *Rpp1* locus that confers resistance to soybean rust in soybean. Crop Sci 47:837–838

Ishiwata YI, Furuya J (2020) Evaluating the contribution of soybean rust- resistant cultivars to soybean production and the soybean market in Brazil: a supply and demand model analysis. Sustainability 12:1422

Jackson S, Rokhsar D, Stacey G, Shoemaker RC, Schmutz J, Grimwood J (2006) Toward a reference sequence of the soybean genome: a multiagency effort. Crop Sci 46:55–61

Jackson EW, Feng C, Fenn P, Chen P (2008) Genetic mapping of resistance to purple seed stain in PI 80837 soybean. J Hered 99:319–322

Jun TH, Mian MAR, Kang ST, Michel AP (2012) Genetic mapping of the powdery mildew resistance gene in soybean PI567301B. Theor Appl Genet 125:1159–1168

Kandel R, Chen CY, Grau CR, Dorrance AE, Liu JQ, Wang Y, Wang D (2018) Soybean resistance to white Mold: evaluation of soybean germplasm under different conditions and validation of QTL. Front Plant Sci 9:505

Kang ST, Mian MR (2010) Genetic map of the powdery mildew resistance gene in soybean PI243540. Genome 53:400–405

Kashiwa T, Suzuki T (2021) High-quality genome assembly of the soybean fungal pathogen *Cercospora kikuchii*. G3 Genes 11:jkab277

Kato M, Morel A, Yamanaka N (2022) Resistance to Asian soybean rust and yield of new soybean cultivars, JFNC 1 and JFNC 2, harboring three resistance genes. Trop Plant Pathol 47:1–9

Khanh TD, Anh TQ, Buu BC, Xuan TD (2013) Applying molecular breeding to improve soybean rust resistance in Vietnamese elite soybean. Am J Plant Sci 4:27573

Kim H, Kim S-T, Ryu J et al (2017) CRISPR/Cpf1-mediated DNA- free plant genome editing. Nat Commun 8:14406

King ZR, Harris DK, Pedley KF et al (2016) A novel Phakopsora pachyrhizi resistance allele (Rpp) contributed by PI 567068A. Theor Appl Genet 129:517–534

Klos KLE, Paz MM, FredrickMarek L, Cregan PB, Shoemaker RC (2000) Molecular markers useful for detecting resistance to brown stem rot in soybean. Crop Sci 40:1445–1452

Koenning SR, Wrather JA (2010) Suppression of soybean yield potential in the continental United States by plant diseases from 2006 to 2009. Plant Health Prog 11:5

Kofsky J, Zhang H, Song BH (2021) Novel resistance strategies to soybean cyst nematode (SCN) in wild soybean. Sci Rep 11:7967

Langenbach C, Schultheiss H, Rosendahl M, Tresch N, Conrath U, Goellner K (2016) Interspecies gene transfer provides soybean resistance to a fungal pathogen. Plant Biotechnol J 14:699–708

Lemos NG, Braccini ADLE, Abdelnoor RV, de Oliveira MCN, Suenaga K, Yamanaka N (2011) Characterization of genes Rpp2, Rpp4, and Rpp5 for resistance to soybean rust. Euphytica 182: 53–64

Lewers K, Crane E, Bronson C, Schupp J, Keim P, Shoemaker R (1999) Detection of linked QTL for soybean brown stem rot resistance in 'BSR 101^0 as expressed in a growth chamber environment. Mol Breed 5:33–42

Li S, Smith JR, Ray JD, Frederick RD (2012) Identification of a new soybean rust resistance gene in PI 567102B. Theor Appl Genet 125:133–142

Li S, Sciumbato G, Boykin D, Shannon G, Chen P (2019) Evaluation of soybean genotypes for reaction to natural field infection by Cercospora species causing purple seed stain. PLoS One 14: e0222673

Lohnes DG, Bernard RL (1992) Inheritance of resistance to powdery mildew in soybeans. Plant Dis 76:964–965

Luo Y, Chong SK, Myers O (2001) Spatio-temporal analysis of soybean root colonization by fusarium solani f. sp. Glycines in Fields Plant Dis 85:303–310

Maphosa M, Talwana H, Tukamuhabwa P (2012) Enhancing soybean rust resistance through Rpp2, Rpp3 andRpp4 pairwise gene pyramiding. Afr J Agric Res 7:4271–4277

McCabe CE, Cianzio SR, O'Rourke JA, Graham MA (2018) Leveraging RNA Seq to characterize resistance to Brown stem rot and the Rbs3 locus in soybean. Mol Plant-Microbe Interact 31: 1083–1094

McCabe CE, Graham MA (2020) New tools for characterizing early brown stem rot disease resistance signaling in soybean. Plant Genome 13:e20037

McLean RJ, Byth DE (1980) Inheritance of resistance to rust (Phakopsora pachyrhizi) in soybean. Aust J Agric Res 31:951–956

Meyer JDF, Silva CGD, Yang C, Pedley FK, Zhang C, van de Mortel M, Hill JH, Shoemaker RC, Abdelnoor RV, Whitham SA et al (2009) Identification and analyses of candidate genes for *Rpp4*-mediated resistance to Asian soybean rust in soybean (*Glycine max* (L.) Merr.). Plant Physiol 150:295–307

Moellers TC, Singh A, Zhang J, Brungardt J, Kabbage M, Mueller DS, Grau CR, Ranjan A, Smith DL, Chowda-Reddy RV et al (2017) Main and epistatic loci studies in soybean for Sclerotinia sclerotiorum resistance reveal multiple modes of resistance in multi-environments. Sci Rep 7: 3554

Monteros MJ, Missaoui AM, Phillips DV et al (2007) Mapping and confirmation of the "Hyuuga" red-brown lesion resistance gene for Asian soybean rust. Crop Sci 47:829–836
Mueller EH, Athow KL, Laviolette FA (1978) Inheritance of resistance to four physiologic races of Phytophthora mega- sperma var sojae. Phytopathology 68:1318–1322
Nataraj V, Maranna S, Kumawat G, Gupta S, Rajput LS, Kumar S, Sharma AN, Bhatia VS (2020) Genetic inheritance and identification of germplasm sources for anthracnose resistance in soybean [Glycine max (L.) Merr.]. Genet Resour Crop Evol 67:1449–1456
Paxton JD, Rogers DP (1974) Powdery mildew of soybean. Mycologia 66:894–896
Patil G, Vuong TD, Kale S, Valliyodan B, Deshmukh R, Zhu C et al (2018) Dissecting genomic hotspots underlying seed protein, oil, and sucrose content in an interspecific mapping population of soybean using high-density linkage mapping. Plant Biotechnol J 16(11):1939–1953
Pawlowski ML, Hill CB, Hartman GL (2015) Resistance to charcoal rot identified in ancestral soybean germplasm. Crop Sci 55:1230–1235
Pedley KF, Pandey AK, Ruck A, Lincoln LM, Whitham SA, Graham MA (2018) Rpp1 encodes a ULP1-NBS-LRR protein that controls immunity to Phakopsora pachyrhizi in soybean. Mol Plant-Microbe Interact 32:120–133
Perez PT, Diers BW, Lundeen P, Tabor GM, Cianzio SR (2010) Genetic analysis of new sources of soybean resistance to brown stem rot. Crop Sci 50:2431–2439
Phillips DV (1984) Stability of Microsphaera diffusa and the effect of powdery mildew on yield of soybean. Plant Dis 68:953–956
Phillips DV (1999) Downy mildew. In: Hartman GL, Sinclair JB, Rupe JC (eds) Compendium of soybean diseases, 4th edn. APS Press, St. Paul, pp 230–257
Ploper LD, Abney TS, Roy KW (1992) Influence of soybean genotype on rate of seed maturation and its impact on seed borne fungi. Plant Dis 76:287–292
Rahman MT, Bhuiyan MKA, Karim MA, Rubayet MT (2018) Screening of soybean resistance genotypes against fusarium oxysporum, Macrophomina phaseolina, Rhizoctonia solani AND Sclerotium rolfsii. Res Agric Vet Sci 2:139–156
Ramezani M, Shier WT, Abbas HK, Tonos JL, Baird RE, Sciumbato GL (2007) Soybean charcoal rot disease fungus Macrophomina phaseolina in Mississippi produces the phytotoxin (−) botryodiplodin but no detectable phaseolinone. J Nat Prod 70:128–129
Ratnaparkhe MB, Satpute GK, Kumawat G, Chandra S, Kamble VG, Kavishwar R, Singh V, Singh J, Singh AK, Ramesh SV et al (2022) Genomic designing for abiotic stress tolerant soybean. In: Kole C (ed) Genomic designing for abiotic stress resistant oilseed crops. Springer, Cham, Switzerland
Reis EM, Boaretto C, Danelli ALD (2014) Macrophomina phaseolina: density and longevity of microsclerotia in soybean root tissues and free on the soil, and competitive saprophytic ability. Summa Phytopathol 40:128–133
Rincker K, Hartman GL, Diers BW (2016) Fine mapping of resistance genes from five Brown stem rot resistance sources in soybean. Plant Genome 9(1):plantgenome2015-08
Ross JP (1986) Registration of four soybean germplasm lines resistant to BPMV. Crop Sci 6:210
Roy KW, Abney TS (1976) Purple seed strain of soybeans. Phytopathology 66:1045–1049
Sajeesh PK, Rao MSL, Jahagirdar S (2014) Screening of soybean (Glycine max(L.) Merill) genotypes against purple seed stain and anthracnose diseases. Environ Ecol 32:1092–1095
Sanitchon J, Vanavichit A, Theerayut SC (2004) Identification of simple sequence repeat markers linked to sudden death syndrome resistance in soybean. Sci Asia 30:205–209
Savary S, Willocquet L, Pethybridge SJ et al (2019) The global burden of pathogens and pests on major food crops. Nat Ecol Evol 3:430–439
Silva OC (2004) Danose Controledo Complexode Doençasfoliaresda Soja. TheUniversity of Passo Fundo, Passo Fundo, Brazil, Master'Thesis
Silva DCG, Yamanaka N, Brogin RL, Arias CAA, Nepomuceno AL, Di Mauro AO, Pereira SS, Nogueira LM, Passianotto ALL, Abdelnoor RV (2008) Molecular mapping of two loci that confer resistance to Asian rust in soybean. Theor Appl Genet 117:57–63

Sinclair JB, Backman PA (1989) Compendium of Soybean Diseases, 3rd edn. American Phytopathological Society, St.Paul,MN,USA

Sinclair JB (1993) Control of seed borne pathogens and diseases of soybean seeds and seedlings. Pestic Sci 37:15–19

Song Q, Hyten DL, Jia G et al (2013) Development and evaluation of SoySNP50K, a high-density genotyping array for soybean. PLoS One 8(1):e54985

Song Q, Yan L, Quigley C et al (2020) Soybean BARCSoySNP6K: an assay for soybean genetics and breeding research. Plant J 104:800–811

Sserunkuma H (2016) Identification of Qtls for resistance against Rhizoctonia Solani and Phomaglycinicola in soybeans (*Glycine max* L. Merr). Ph.D. Thesis. University of Nebraska, Lincoln, NE, USA

Stephens PA, Nickell CD, Vodkin LO (1993) Pink flower color associated with increased protein and seed size in soybean. Crop Sci 33:1135–1137

Sugimoto T, Yoshida S, Watanabe K et al (2007) Identification of SSR markers linked to the Phytophthora resistance gene Rps1-d in soybean. Plant Breed 127:154–159

Sugimoto T, Yoshida S, Kaga A et al (2011) Genetic analysis and identification of DNA markers linked to a novel Phytophthora sojae resistance gene in the Japanese soybean cultivar Waseshiroge. Euphytica 182:133

Sun S, Kim MY, Van K et al (2013) QTLs for resistance to Phomopsis seed decay are associated with days to maturity in soybean (Glycine max). Theor Appl Genet 126:2029–2038

Swaminathan S, Abeysekara NS, Liu M et al (2016) Quantitative trait loci underlying host responses of soybean to fusarium virguli- forme toxins that cause foliar sudden death syndrome. Theor Appl Genet 129:495–506

Taguchi-Shiobara F, Fujii K, Sayama T, Hirata K, Kato S, Kikuchi A, Ishimoto M (2019) Mapping versatile QTL for soybean downy mildew resistance. Theor Appl Genet 132(4):959–968

Tan R, Serven B, Collins PJ et al (2018) QTL mapping and epistatic interaction analysis of field resistance to sudden death syndrome (fusarium virguliforme) in soybean. Theor Appl Genet 131:1729–1740

Tan R, Collins PJ, Wang J et al (2019) Different loci associated with root and foliar resistance to sudden death syndrome (fusarium virguliforme) in soybean. Theor Appl Genet 132:501–513

Tomar J, Saini N, Goyal BS, Tripathi N, Shrivastava AN, Verma RK, Tiwari S (2011) Assessment of genetic diversity among rhizoctonia root rot resistant soybean genotypes. J Food Legumes 24: 267–272

Tripathi N, Tripathi MK, Tiwari S, Payasi DK (2022) Molecular breeding to overcome biotic stresses in soybean: update. Plan Theory 11:1967. https://doi.org/10.3390/plants11151967

Tripodi P (2021) Methods of development of Biparental mapping populations in horticultural crops. In: Tripodi P (ed) Crop breeding. Methods in molecular biology, vol 2264. Humana Press, New York, NY. https://doi.org/10.1007/978-1-0716-1201-9_1

Triwitayakorn K, Njiti VN, Iqbal MJ, Yaegashi S, Town C, Lightfoot DA (2005) Genomic analysis of a region encompassing trait loci for resistance to soybean sudden death syndrome (SDS) *QRfs1* and *QRfs2*: genes that underlie soybean resistance to sudden death syndrome. Genome 48:125–138

Vuong TD, Sonah H, Meinhardt CG et al (2015) Genetic architecture of cyst nematode resistance revealed by genome-wide association study in soybean. BMC Genomics 16:593–605

Walker DR, McDonald SC, Harris DK et al (2022) Genomic regions associated with resistance to soybean rust (Phakopsora pachyrhizi) under field conditions in soybean germplasm accessions from Japan, Indonesia and Vietnam. Theor Appl Genet 135:3073–3086

Walters HJ, Caviness CE (1973) Breeding for improved soybean seed quality. Arkansas Farm Res 23(5):5

Wen Z, Tan R, Yuan J et al (2014) Genome-wide association mapping of quantitative resistance to sudden death syndrome in soybean. BMC Genomics 15:809

Wilcox JR, LaViolette FA, Martin RJ (1975) Heritability of purple seed stain resistance in Soybeans1. Crop Sci 15:525–526

Yamanaka N, Morishita M, Mori T et al (2016) The locus for resistance to Asian soybean rust in PI 587855. Plant Breed 135:621–626

Yang XB, Workneh F, Lundeen P (1998) First report of sclerotium production by *Sclerotinia sclerotiorum* in soil on infected soybean seeds. Plant Dis 8:264. https://doi.org/10.1094/PDIS.1998.82.2.264B

Yuan J, Njiti VN, Meksem K, Iqbal MJ, Triwitayakorn K, Kassem MA, Davis GT, Schmidt ME, Lightfoot DA (2002) Quantitative trait loci in two soybean recombinant inbred line populations segregating for yield and disease resistance. Crop Sci 42:271–277

Zhang Y, Zhao J, Xiang Y, Bian X, Zuo Q, Shen Q, Gai J, Xing H (2011) Proteomics study of changes in soybean lines resistant and sensitive to Phytophthora sojae. Proteome Sci 9:52

Zhang ZN, Hao JJ, Yuan JZ, Song QJ, Hyten DL, Cregan PB, Zhang GR, Gu CH, Ming L, Wang DC (2014) Phytophthora root rot resistance in soybean E00003. Crop Sci 54:492–499

Zhou Q, Jiang B, Cheng Y, Ma Q, Xia Q, Jiang Z, Cai Z, Nian H (2022) Fine mapping of an adult-plant resistance gene to powdery mildew in soybean cultivar Zhonghuang 24. Crop J 10:1103

Zhu L, Zhou Y, Li X, Zhao J, Guo N, Xing H (2018) Metabolomics analysis of soybean hypocotyls in response to *Phytophthora sojae* infection. Front Plant Sci 9:1530

Zou J, Rodriguez-Zas S, Aldea M, Li M, Zhu J, Gonzalez DO, Vodkin LO, DeLucia E, Clough SJ (2005) Expression profiling soybean response to *pseudomonas syringae* reveals new defense-related genes and rapid HR-specific downregulation of photosynthesis. Mol Plant-Microbe Interact 18:1161–1174

14 Disease Resistance and Seed Production in Two Common New England Grain Legumes

Eric J. B. von Wettberg, Emmanuel Brefo, Jasmine Hart, Bailey Kretzler, and Gianna Sassi

Abstract

New England and the broader Northeastern region of North America have very limited grain legume production. Common beans and peas are the two leading legumes for direct human consumption. Both suffer from a range of diseases, particularly soil-borne pathogens.

14.1 Introduction

Seeds, along with soil, are the foundation of crop production systems. In traditional agricultural systems, seeds for planting are obtained by seed saving or through informal networks of family and community (Hodgkin et al. 2007; Stromberg et al. 2010). The rise of industrialized seed production systems and the formal seed sector over the last few centuries have had immense impacts on crop genetic diversity (Hodgkin et al. 2007; Khoury et al. 2022, Chen et al. in review). In recent decades, the seed industry has undergone extensive consolidation and concentration (e.g., Bonny 2017) leading to a homogenization and erosion of crop genetic diversity (Khoury et al. 2022). This erosion of genetic diversity considerably decreases the prevalence of disease resistance genes within crop species, increasing their vulnerability to disease outbreaks. This has enormous consequences for the resilience and sustainability of agricultural production systems.

E. J. B. von Wettberg (✉) · J. Hart · B. Kretzler
Department of Plant and Soil Science, The University of Vermont, Burlington, VT, USA

The Gund Institute for the Environment, University of Vermont, Burlington, VT, USA
e-mail: Eric.Bishop-Von-Wettberg@uvm.edu

E. Brefo · G. Sassi
Department of Plant and Soil Science, The University of Vermont, Burlington, VT, USA

U. C. Jha et al. (eds.), *Diseases in Legume Crops*,
https://doi.org/10.1007/978-981-99-3358-7_14

Industrial seed systems have increased disease susceptibility because they breed and produce seeds in environments far removed from where they are ultimately grown, restricting the adaptation of commercial crop varieties to local environments. To be commercially successful, many private and public sector breeders select varieties to be adapted to a broad range of climates, soils, and palates but also diseases. While the locations where international seed companies choose to produce seeds are proprietary information, they may choose to produce seeds in certain regions due to production costs, jurisdictional issues, as well as disease and pest pressures. Informally, it appears that much of seed production for global seed companies is in drier climates, such as Mediterranean climate regions of Argentina, Chile, New Zealand, South Africa, and the US Inland Pacific Northwest. Many of these climates are slightly more arid than those in the centers of domestication of these crops. However, diseases are particularly problematic for legumes, particularly in wet and cool environments like northeastern North America that are conducive to fungal and oomycota growth.

In the Northeastern United States and eastern Canada, legume production predates European settlement. Beans are indigenous crops, grown by groups like the Haudenosaunee (Iroquois) and Wabanaki confederation since ~1300 AD (Hart et al. 2002; Hart 2008). Although a later introduction than maize and squash (Hart et al. 2002), beans have been and are grown as part of the Three Sisters or Seven Sisters complex (e.g., Hart 2008; Wiseman 2018). With the onset of the Columbian exchange, beans, maize, and squash were quickly adapted by European settlers in northeastern North America, while peas and faba beans along with other Eurasian crops were introduced to the Northeast. The increased movement of people was accompanied by increased movement of disease and insect pests and by a shift toward increased monocultural production of genetically uniform varieties—conditions that lead to more potential for disease epidemics. While beans and peas remain culturally significant in Northeastern North America, featured in regional dishes like Boston baked beans and Maine "bean-hole suppahs" (e.g., Wiseman 2018), by the twentieth century, most production of these crops has shifted to other regions. In 2005, fewer than 750 acres of pulses were planted commercially in New England (Siligato 2007; Siligato and Koehler 2003; Padder et al. 2017). Currently New England, the region of the six most northeastern states in the USA, produces less than 15% of its food and imports nearly all its plant-based protein foods, including pulses and oil-seed legumes (Griffin et al. 2015; Peters et al. 2022). It is thought that high cost of production on smaller farms than in flatter regions, combined with limited adaptation of pulses to the region, contributes to low self-sufficiency in legume production in New England (Peters et al. 2022). Further, with the exception of small groups of seed savers, almost no legume seed for planting is produced in the region. Disease pressure is likely a major contributor to this low production. Here we review the prevalence of diseases on pulses with a focus on New England. We also expand out to the broader northeastern North America to encompass neighboring regions with flatter and less forested landscapes and more intensive production of beans, peas, and other legumes.

14.2 Prevalence of Diseases of Legumes in Northeastern North America

We performed a literature review to summarize diseases of pulses (primarily common beans and garden peas) in New England and more broadly in Northeastern North America. We began with Google scholar with the search terms "Bean, disease, New England" and "Pea disease New England." For any hits, we examined related articles. Due to a small number of articles, we expanded search terms to include "northeast." We used Google scholar to increase our coverage of white papers and other literature that might not show up in Web of Science.

White mold (Sclerotinia), rust, and anthracnose are the most important fungal and oomycete diseases of mature bean plants, while a range of "damping off" pathogens like *Pythium* can be severe challenges for bean seedlings (e.g., Siligato 2007; Padder et al. 2017). On peas, the most common diseases are damping off, seed decay (fungal), root rots, stem canker, and powdery mildew (Siligato and Koehler 2003). In both crops, all of these diseases are common in other regions, where similarly accommodating climatic and agronomic conditions exist that allow the diseases to thrive. Furthermore, since nearly all seeds of these crops are produced outside of the region, most producers are restricted to cultivars that may lack suitable resistance to these diseases. In general, we found a substantial lack of research on these diseases in northeastern North America, consistent with their limited economic importance in the region. We describe superficially each of these diseases, with links to relevant research, almost entirely from other regions. We have not striven to review these diseases thoroughly, as each has already been more carefully reviewed by researchers in regions where they are more prevalent (see below).

White mold, caused by *Sclerotinia sclerotiorum*, is common to beans across any wet habitat. It is made more problematic in that the pathogen infects greater than 360 species (Abawi and Hunter 1979). There is population differentiation across strains (Kamvar et al. 2017). As a necrotrophic pathogen, it tends to start infections in decaying tissue and then infect living tissue (Hegedus and Rimmer 2005). Bean rust is caused by *Uromyces appendiculatus* (Souza et al. 2008; Liebenberg and Pretorius 2010), a global disease, with variation in symptoms, but the potential for devastating outbreaks. It also affects other grain legumes, from cowpea to faba bean. Anthracnose is caused by *Colletotrichum lindemuthianum* in beans. It is most prevalent in cool and humid environments like northeastern North America.

Powdery mildew in peas is caused by *Erysiphe pisi* (and sometimes other *Erysiphe*) and ruins the quality of green peas. There is genetic variation in susceptibility to powdery mildew, with at least three loci known to confer resistance (Fondevilla and Rubiales 2012). Only one of these loci, *er-1*, has been widely used (Devi et al. 2022). Chemical control is possible although comes with risks. Growing peas in a polyculture, with a larger crop like barley or potentially rye, does reduce disease incidence and spread (Villegas-Fernández et al. 2021). Pea stem canker can refer to many things. In some instances, it may be *Phomopsis* (Telomorphe *Diaporthe*) (e.g., Ondřej et al. 2006). It can also look like *Ascochyta*, or *Mycosphaerella pinodes* and *Phoma pinodella*, making identification difficult even for trained experts.

Root rots of grain legumes can be caused by a range of pathogens, including *Rhizoctonia solani*, *Phythium*, and *Fusarium*. Diagnostically they can be hard to distinguish, as infection with one can allow the others to thrive on the same dying plant. *Pythium* spp. is exceptionally important with a potential yield loss of 100% (Singh and Schwartz 2010). Pythium root rot (PRR) can impact the plant all stages of growth but is particularly catastrophic at the seedling stage and is known generally as "damping off" or when the seedling dies before emergence or shortly after. PRR causes stem lesions, girdled stem at the soil surface, and can cause seed or preemergence rot (Laemmlen 2002). *Fusarium* causes leaf curling and wilting of leaves, along with yellowing or reddening of stems (e.g., Markell et al. 2022). Despite the cold winters and cool conditions of spring and fall, summers in New England can be sufficient hot and dry for *Macrophomina*, another fungal pathogen (e.g., Kaur et al. 2012; Pandey and Basandrai 2021).

Breeding approaches for disease resistance in beans (Beaver and Osorno 2009; Singh and Schwartz 2010; Meziadi et al. 2016) and peas (e.g., Rubiales et al. 2009; Rubiales et al. 2015; Jha et al. 2021; Wohor et al. 2022) are well reviewed. Even with improving molecular markers, progress toward disease-resistant germplasm remains slow, with relatively few resistance genes having been identified for many diseases. With limited public sector support breeding, particularly of grain legumes, the resources to maintain resistance breeding programs only exist in a few public locations globally.

For example, although two major genes have been identified for quantitative inheritance of PRR resistance in the *P. vulgaris* Mesoamerican gene pool, introducing these genes has been unsuccessful for the large-seeded Andean varieties while maintaining other agronomically important characteristics (Dramadri et al. 2020). Dramadri et al. (2020) screened the Andean *P. vulgaris* gene pool and discovered several other genes that confer some level of PRR resistance. However, for stable and enhanced stress tolerance, the "stacking" or "pyramiding" of several genes is an important tool for ensuring inheritance for future generations (Shehryar et al. 2021).

14.3 Epigeal-Hypegeal Differentiation as a Source of Disease Resistance

We hypothesize that fundamental aspects of plant seed emergence strategy may have an underappreciated role in disease incidence and resistance. Two major strategies for seedling emergence are epigeal and hypogeal emergence. The *Fabaceae* is unique in that there are variable germination styles within the genera, *Phaseolus* and *Vigna* (Ibrahim and Coyne 1975; Tomooka 2002). Hypogeal germination is when a plant's cotyledons or embryonic first leaves within the seed stay belowground during germination, and epigeal germination is when the cotyledons emerge above the soil (Gates 1951; Burridge et al. 2020). Hypogeal germination has been attributed to increased seedling fitness when faced with environmental pressures like soil compaction, cold tolerance, and overall stress response (Beckham and Rutger 1972; Laskar et al. 2019; Pujol et al. 2005). We explore this difference further here as a potential source of resistance to root rot.

Though there have been resistance genes identified in *P. vulgaris* for PRR, *P. coccineus* has commonly been used as a genetic resource for resistance (R) genes and other stress tolerance traits in *P. vulgaris* (Rodino et al. 2007). *P. coccineus*, known as scarlet runner bean in English, is a crop in Mexico and other regions but is only grown ornamentally or at a very small scale in New England. However, it has substantial pathogen resistance, perhaps due to its origin in humid regions of southern Mexico or its outcrossing nature, allowing for more gene flow between wild and domesticated varieties, unlike *P. vulgaris* (Spataro et al. 2011). Interestingly, *P. vulgaris* exhibits an epigeal germination style, while *P. coccineus* exhibits a hypogeal germination style and can express an intermediate phenotype when crossed (Ibrahim and Coyne 1975).

We hypothesize that some aspects of the hypogeal germination of *P. coccineus* are associated with aspects of disease resistance that may be distinct from those traditionally looked at by pathologists. To have disease incidence, we need the disease triangle: appropriate environmental conditions, susceptible host, and a virulent pathogen. Traits such as growth habit are able to influence the microclimate around the plant, encouraging or mitigating disease severity and incidence. In addition to molecular R genes, these morphological R genes are particularly promising when pyramided together for more complete and stable resistance. These morphological traits also have the potential to increase yield and decrease labor inputs (i.e., staking and trellising), in addition to their benefit to overall plant health. Plant pathogens use several strategies to invade healthy plant tissues despite the natural protective mechanisms that are observed in plants. Fungi invade crops either penetrating directly into the epidermal leaf cell or forming hyphae between and over the epidermal cells. On the other hand, bacteria and virus invade plant cells through specialized structures such as stomata. Wounds also expose plant tissues, making them vulnerable to pathogen infection. Plants are often subjected to mechanical and wind damages in the field. Other vector organisms such as protozoa, fungi, and insects are carriers of many viral and bacterial diseases (Nazarov et al. 2020). It is worth investigating whether hypogeal taxa harbor traits that increase their resistance to colonization by oomycete pathogens that cause root rot and damping off. The genera *Phaseolus* and *Vigna* offer unique opportunities to compare these strategies, as they include interfertile species that vary in these two germination and emergence strategies.

The epidermis of plants has waxy cuticle and trichomes that function as a barrier defense from pathogenic invasion. Trichomes are outgrowths from epidermis that are visible with microscope. Plant cell wall has a distinct property that performs structural and protective functions. The cell wall contains cellulose, microfibrils, hemicellulose, pectin, soluble protein, and lignin (reinforced agent). Plant cells synthesize antimicrobial compounds as immune response to fight disease infections. The antimicrobial compound inhibits the activities of pathogen hydrolytic enzymes involved in damaging cell walls. The different kinds of plant antimicrobial compounds are categorized into two groups, namely, phytoanticipins and phytoalexins. The former is pre-synthesized, and the latter only synthesized in response to pathogen attack (Nazarov et al. 2020). There is an absence of research examining whether these structures differ among epigeal and hypogeal taxa.

In general, plants use two main types of innate immunity systems to fight pathogens. The primary immune system uses transmembrane recognition receptors to recognize the presence of pathogen molecule patterns and then trigger a defensive response (de Wit 2007). With the secondary immune system which happens intracellular, the R genes recognize effector molecules through proteins and then trigger a very strong defensive response to stop pathogen growth (de Wit 2007). In *Phaseolus*, most prior work that has used *P. coccineus* as a resistance source has searched for genes of this type. We believe this remains a research need but that it may be positively informed by understanding the habitat differences and germination strategy differences.

14.4 The Disease Triangle in New England: Detection Challenges in a Reforested Landscape

The ecological history of the New England region has received considerable attention from ecological historians (e.g., Cronon 2011). Although many aspects of indigenous settlement and land use remain debated, it is clear that European settlement after 1620 led to a period of extensive land use change, with clearing of forested areas and extensive plowing of glacial soils. Following the opening of flatter and more fertile land farther west to colonists of European ancestry, much of New England was not farmed intensively after the mid-1800s, allowing extensive forest regrowth. To this date, agriculture in general is more limited than in other regions and has little production of pulses. Consequently, long-lived soil pathogens that can persist 10 or more years may be absent in many places. With the absence of a pathogen, a susceptible cultivar can be grown in a region like New England that is conducive to disease until the pathogen is reintroduced. We suspect this has been the case with small-scale trials of other grain legumes, such as chickpea, which have been successful in recent years despite low resistance to diseases that thrive in moist climates like *Ascochyta* blight (e.g., Gan et al. 2006).

An alternative perspective on disease incidence on grain legumes in New England is that the leading production area for dry beans is in the potato producing region (Aroostok County) of Maine. Potatoes are particularly susceptible to a range of soil-borne pathogens, and beans grown in rotation with potatoes may suffer particularly severe pathogen incidence. We are not aware of studies to date examining the rotational impacts of potatoes in bean diseases in Maine or other regions, although anecdotally it seems plausible that carryover of pathogens from potatoes limits farmer choices for developing profitable crop rotations. We believe questions of rotational impact need further investigation to develop more sustainable rotations.

As a region with very limited but expanding grain legume production, the potential for disease introduction with seeds produced in other regions increases. New England has few laboratories prepared and certified to perform ISTA testing of seed lots for disease. Efforts are needed to maintain and expand diagnostic capacity in the region. Most resources for identification have been developed in other regions (e.g., Markell et al. 2022). Although accurate and useful, they may not fit the needs of farmers or technical service provides well. Perhaps of greatest concern, the land-grant universities of New England have very few plant pathologists, with several looming retirements that may further thin the ranks of a very important profession. Training, retaining, and promoting extension-focused plant pathologists are critical.

In this chapter, we have taken the approach of highlighting the very limited information available in our region about diseases of pulses to plan ahead for a time when these crops may be more widely grown to meet regional plant protein needs. The food system shocks of the COVID pandemic have illustrated the need to increase regional self-reliance (e.g., Niles et al. 2020). New England is a region that currently imports over 85% of its food (e.g., Peters et al. 2022) and that historically has produced considerable amounts of dairy products. Increased production of grain legumes would provide greater amounts of protein on less acreage with fewer inputs. Among other limiting factors, disease control and increased resistance are needed for legumes in New England.

Acknowledgments We thank Uday Jha for the encouragement to venture into an area where we have learned many new things in crafting this review. We are supported by funds from USDA multistate hatch NE2210, Northeast SARE LNE21-428R, Northeast SARE GNE22-284, and the UVM Food Systems Research Center for research on grain legumes.

References

Abawi G and Hunter J (1979) White mold of beans in New York. New Yorks food and lif e Science Bulletin No. 77. https://ecommons.cornell.edu/bitstream/handle/1813/5083/FLS-077.pdf; sequence=1

Beaver JS, Osorno JM (2009) Achievements and limitations of contemporary common bean breeding using conventional and molecular approaches. Euphytica 168:145–175

Beckham LS, Rutger JN (1972) Emergence of hypogeal and epigeal germinating species of Phaseolus and Pisum in compacted and uncompacted soil1. HortScience 7(3):276–277

Bonny S (2017) Corporate concentration and technological change in the global seed industry. Sustainability 9(9):1632. https://pdfs.semanticscholar.org/2ef2/753fc842cec210f6f8a17c67769977e0684e.pdf

Burridge JD, Rangarajan H, Lynch JP (2020) Comparative phenomics of annual grain legume root architecture. Crop Sci 60(5):2574–2593

Cronon W (2011) Changes in the land: Indians, colonists, and the ecology of New England. Hill and Wang

de Wit PJGM (2007) How plants recognize pathogens and defend themselves. Cell Mol Life Sci 64: 2726–2732. https://doi.org/10.1007/s00018-007-7284-7

Devi J, Mishra GP, Sagar V, Kaswan V, Dubey RK, Singh PM, Sharma SK, Behera TK (2022) Gene-based resistance to Erysiphe species causing powdery mildew disease in peas (Pisum sativum L.). Gene 13(2):316

Dramadri IO, Amongi W, Kelly JD, Mugisha Mukankusi C (2020) Genome-wide association analysis of resistance to Pythium ultimum in common bean (Phaseolus vulgaris). Plant Breed 139(6):1168–1180

Fondevilla S, Rubiales D (2012) Powdery mildew control in pea. a review. Agron Sustain Dev J 32: 401–409. https://doi.org/10.1007/s13593-011-0033-1

Gan YT, Siddique KHM, MacLeod WJ, Jayakumar P (2006) Management options for minimizing the damage by ascochyta blight (Ascochyta rabiei) in chickpea (Cicer arietinum L.). Field Crop Res 97(2–3):121–134

Gates RR (1951) Epigeal germination in the Leguminosae. Bot Gaz 113(2):151–157

Griffin T, Conrad Z, Peters C, Ridberg R, Tyler EP (2015) Regional self-reliance of the northeast food system. Renew Agric Food Syst 30(4):349–363

Hart JP (2008) Evolving the three sisters: the changing histories of maize, bean, and squash in New York and the greater northeast. In: Current Northeast Paleoethnobotany II, vol 512, pp 87–99

Hart JP, Asch DL, Scarry CM, Crawford GW (2002) The age of the common bean (Phaseolus vulgaris L.) in the northern eastern woodlands of North America. Antiquity 76(292):377–383

Hegedus DD, Rimmer SR (2005) Sclerotinia sclerotiorum: when "to be or not to be" a pathogen? FEMS Microbiol Lett 251(2):177–184

Hodgkin T, Rana R, Tuxill J, Balma D, Subedi A, Mar I, Karamura D, Valdivia R, Collado L, Latournerie L, Sadiki M (2007) 4. Seed systems and crop genetic diversity in agroecosystems. In: Managing biodiversity in agricultural ecosystems. Columbia University press, pp 77–116

Ibrahim AM, Coyne DP (1975) Genetics of Stigma shape, Cotyledon position, and flower color in reciprocal crosses between Phaseolus vulgaris L. and Phaseolus coccineus (Lam.) and implications in breeding1. J Am Soc Hortic Sci 100(6):622–626

Jha AB, Gali KK, Alam Z, Lachagari VR, Warkentin TD (2021) Potential application of genomic technologies in breeding for fungal and oomycete disease resistance in pea. Agronomy 11(6): 1260

Kamvar ZN, Amaradasa BS, Jhala R, McCoy S, Steadman JR, Everhart SE (2017) Population structure and phenotypic variation of Sclerotinia sclerotiorum from dry bean (Phaseolus vulgaris) in the United States. Peer J 5:e4152

Kaur S, Dhillon GS, Brar SK, Vallad GE, Chand R, Chauhan VB (2012) Emerging phytopathogen Macrophomina phaseolina: biology, economic importance and current diagnostic trends. Crit Rev Microbiol 38(2):136–151

Khoury CK, Brush S, Costich DE, Curry HA, de Haan S, Engels JM, Guarino L, Hoban S, Mercer KL, Miller AJ, Nabhan GP (2022) Crop genetic erosion: understanding and responding to loss of crop diversity. New Phytol 233(1):84–118

Laemmlen F (2002) Damping-off diseases. UC Davis Extension Factsheet. http://anrcatalog.ucdavis.edu

Laskar RA, Khan S, Deb CR, Tomlekova N, Wani MR, Raina A, Amin R (2019) Lentil (Lens culinaris Medik.) diversity, cytogenetics and breeding. In: Advances in plant breeding strategies: legumes, vol 7, pp 319–369

Liebenberg MM, Pretorius ZA (2010) 1 common bean rust: pathology and control. Hortic Rev 37:1

Markell S, Pasche J, Porter L (2022) Pea Disease Diagnostic Series. North Dakota State Extension. https://www.ndsu.edu/agriculture/sites/default/files/2022-11/pp1790.pdf

Meziadi C, Richard MM, Derquennes A, Thareau V, Blanchet S, Gratias A, Pflieger S, Geffroy V (2016) Development of molecular markers linked to disease resistance genes in common bean based on whole genome sequence. Plant Sci 242:351–357

Nazarov PA, Baleev DN, Ivanova MI, Sokolova LM, Karakozova MV (2020) Infectious plant diseases: etiology, current status, problems and prospects in plant protection. Acta Nat 12(3): 46–59. https://doi.org/10.32607/actanaturae.11026. PMID: 33173596; PMCID: PMC7604890

Niles MT, Bertmann F, Belarmino EH, Wentworth T, Biehl E, Neff R (2020) The early food insecurity impacts of COVID-19. Nutrients 12(7):2096

Ondřej M, Odstrčilová L, Dostálová R (2006) Phomopsis pisi–a new species causing pea stem canker. Plant Prot Sci 42(3):95–98

Padder BA, Sharma PN, Awale HE, Kelly JD (2017) Colletotrichum lindemuthianum, the causal agent of bean anthracnose. J Plant Pathol:317–330

Pandey AK, Basandrai AK (2021) Will Macrophomina phaseolina spread in legumes due to climate change? A critical review of current knowledge. J Plant Dis Protect 128(1):9–18

Peters et al. (2022) New England Feeding New England Volume 2: Estimating production for 30% regional self-reliance. In Press

Pujol B, David P, McKey D (2005) Microevolution in agricultural environments: how a traditional Amerindian farming practice favours heterozygosity in cassava (Manihot esculenta Crantz, Euphorbiaceae). Ecol Lett 8(2):138–147

Rodino AP, Lema M, Pérez-Barbeito M, Santalla M, De Ron AM (2007) Assessment of runner bean (Phaseolus coccineus L.) germplasm for tolerance to low temperature during early seedling growth. Euphytica 155(1–2):63–70

Rubiales D, Fernandez-Aparicio M, Moral A, Barilli E, Sillero JC, Fondevilla S (2009) Disease resistance in pea (Pisum sativum L.) types for autumn sowings in Mediterranean environments-a review. Czech J Genet Plant Breed 45(4):135–142

Rubiales D, Fondevilla S, Chen W, Gentzbittel L, Higgins TJ, Castillejo MA, Singh KB, Rispail N (2015) Achievements and challenges in legume breeding for pest and disease resistance. Crit Rev Plant Sci 34(1–3):195–236

Shehryar K, Khan RS, Iqbal A, Hussain SA, Imdad S, Bibi A, Hamayun L, Nakamura I (2020) Transgene stacking as effective tool for enhanced disease resistance in plants. Mol Biotechnol 62:1–7

Siligato M (2007) New England bean Pest management strategic plan. New England pest management network. PMSP database. Accessed 28 Feb 2023. https://ipmdata.ipmcenters.org/documents/pmsps/NewEnglandBeanPMSP.pdf

Siligato M and Koehler G (2003) New England pea crop profile. PMSP database. Accessed 28 Feb 2023. https://citeseerx.ist.psu.edu/document?repid=rep1&type=pdf&doi=eeba78a2b218ff2c7420f0d9fd6145d5d7ec04c2

Singh SP, Schwartz HF (2010) Breeding common bean for resistance to diseases: a review. Crop Sci 50(6):2199–2223

Souza TLPO, Alzate-Marin AL, Faleiro FG, Barros EG (2008) Pathosystem common bean-Uromyces appendiculatus: host resistance, pathogen specialization, and breeding for rust resistance. Pest Technol 2(2):56–69

Spataro G, Tiranti B, Arcaleni P, Bellucci E, Attene G, Papa R, Spagnoletti Zeuli P, Negri V (2011) Genetic diversity and structure of a worldwide collection of Phaseolus coccineus L. Theor Appl Genet 122:1281–1291

Stromberg PM, Pascual U, Bellon MR (2010) Seed systems and farmers' seed choices: the case of Maize in the Peruvian Amazon. Hum Ecol 38:539–553

Tomooka N (2002) The Asian Vigna: genus Vigna subgenus ceratotropis genetic resources. Springer Science & Business Media

Villegas-Fernández ÁM, Amarna AA, Moral J, Rubiales D (2021) Crop diversification to control powdery mildew in pea. Agronomy 11(4):690

Wiseman FM (2018) Seven sisters and the heritage food Systems of the Wabanaki People and of the Chesapeake Bay region. EHLC Publishing, Thomasburg Ontario, CA

Wohor OZ, Rispail N, Ojiewo CO, Rubiales D (2022) Pea breeding for resistance to rhizospheric pathogens. Plan Theory 11(19):2664

The manufacturer's authorised representative in the EU is Springer Nature Customer Service Centre GmbH, Europaplatz 3, 69115 Heidelberg, Germany. If you have any concerns regarding our products, please contact ProductSafety@springernature.com

Printed and bound by CPI Group (UK) Ltd, Croydon, CR0 4YY
30/09/2026
02225319-0012